D0789098

Actuarial Modelling of Claim Counts

Actuarial Modelling of Claim Counts

Risk Classification, Credibility and Bonus-Malus Systems

Michel Denuit
Institut de Statistique, Université Catholique de Louvain, Belgium

Xavier Maréchal
Reacfin, Spin-off of the Université Catholique de Louvain, Belgium

Sandra Pitrebois
Secura, Belgium

Jean-François Walhin
Fortis, Belgium

John Wiley & Sons, Ltd

Telephone (+44) 1243 779777

Email (for orders and customer service enquiries): cs-books@wiley.co.uk
Visit our Home Page on www.wiley.com

Other Wiley Editorial Offices

John Wiley & Sons Inc., 111 River Street, Hoboken, NJ 07030, USA

Jossey-Bass, 989 Market Street, San Francisco, CA 94103-1741, USA

Wiley-VCH Verlag GmbH, Boschstr. 12, D-69469 Weinheim, Germany

John Wiley & Sons Australia Ltd, 42 McDougall Street, Milton, Queensland 4064, Australia

John Wiley & Sons (Asia) Pte Ltd, 2 Clementi Loop #02-01, Jin Xing Distripark, Singapore 129809

John Wiley & Sons Canada Ltd, 6045 Freemont Blvd, Mississauga, ONT, Canada, L5R 4J3

Wiley also publishes its books in a variety of electronic formats. Some content that appears in print may not be available in electronic books.

Anniversary Logo Design: Richard J. Pacifico

Library of Congress Cataloging in Publication Data

Actuarial Modelling of Claim Counts : Risk Classification, Credibility and Bonus-Malus Systems /
Michel Denuit . . . [et al.].
p. cm.
Includes bibliographical references and index.
ISBN 978-0-470-02677-9 (cloth)
1. Insurance, Automobile—Rates—Europe. 2. Automobile insurance claims—Europe.
I. Denuit, M. (Michel)
HG9970.2.A25 2007
368′.092094—dc22

2007019885

British Library Cataloguing in Publication Data

A catalogue record for this book is available from the British Library

ISBN-13 978-0-470-02677-9

Typeset in 10/12pt Times by Integra Software Services Pvt. Ltd, Pondicherry, India
Printed and bound in Great Britain by Antony Rowe Ltd, Chippenham, Wiltshire
This book is printed on acid-free paper responsibly manufactured from sustainable forestry in which at least two trees are planted for each one used for paper production.

Contents

Foreword

Belgium has a long and distinguished history in actuarial science. One of its leading centres in the area is the Institut des Sciences Actuarielles at l'Université Catholique de Louvain (UCL). Since its tender beginnings in the 1970s, the Institute has grown to critical mass and now boasts an internationally renowned faculty conducting research and education in a broad range of actuarial subjects – newish ones in the interface of insurance and finance as well as more traditional ones that used to form the core of insurance mathematics. Among the latter is risk classification and experience rating in general insurance, which is the subject matter of the present book. This is an area of applied statistics that has been fetching tools from various kits of theoretical statistics, notably empirical Bayes, regression, and (generalized) linear models. However, the complexity of the typical application, featuring unobservable risk heterogeneity, imbalanced design, and nonparametric distributions, inspired independent fundamental research under the label 'credibility theory', now a cornerstone in contemporary insurance mathematics. Quite naturally, the present book is a tribute to Florian (Etienne) De Vylder, who was a Professor at UCL and one of the greatest minds in insurance mathematics and, in particular, credibility theory. The book grew out of years of studies by a collective of researchers based at UCL and its industrial environment. The lead author, Michel Denuit, is one of the most prolific researchers in contemporary actuarial science, who has publicized widely in actuarial and statistical journals on topics in risk theory and actuarial science and related basic disciplines. Also Jean-François Walhin is a well established researcher with a long list of publications in the scientific actuarial press. Together with their (even) younger co-authors Xavier Maréchal and Sandra Pitrebois, they have formed a team that is well placed to write a comprehensive reference text on risk classification and premium rating. Their combined expertise covers all theory areas that are at the base of the topic – risk theory and insurance mathematics, but also modern statistics and scientific computation. The team's total contribution to theoretical research in the subject matter of the book is substantial, and it is merged with sound practical insights gained through commitment to applicability and also through career experience outside the purely academic walks of life.

The book will be welcomed by practitioners and researchers who need a broad introduction to the titular subject area or an update aided by modern statistical methodology for complex models and high-dimensional data. The book may also serve as a textbook at graduate level further to an introduction to basic principles explained in simple models.

The opening Chapters 1 and 2 present basic notions of risk and risk characteristics and their theoretical representation in stochastic models with fixed and random effects and more or less specified classes of distributions. Anticipating the orientation of the book, emphasis is placed on parametric models for the number of claims. This gives clarity to the exposition and also sets a suitable framework for discussion of model choice and model calibration that goes way beyond what is usually found in conventional tutorials. Poisson conditional distributions with varying exposures are merged with different mixing distributions on the individual proportional hazards, and there are extensions to generalized linear (regression) models, time trends, and spatial patterns. Statistical calibration is carried out with maximum likelihood methods but also with alternative schemes like generalized estimating functions. Ample numerical examples with authentic data gives real life to the theoretical ideas throughout. Claim counts remain a main theme, but the remainder of the book nevertheless presents a wealth of material, partly based on recent research by the authors: credibility theory, Bayes estimation with exponential loss, bonus-malus systems in a number of variations, elements of heavy-tailed distributions, bonus hunger and other 'behavioural' problems related to individual experience rating, optimal design of bonus-malus systems for aggregates of sub-portfolios, and much more. The final chapter is devoted to a carefully conducted case study of the French bonus-malus system.

I would like to thank the authors for soliciting my views on a draft version of the book and for inviting my preface to their work. Most of all I would like to thank them for undertaking the formidable task of collecting and making accessible to a wide readership an area of actuarial science that has undergone great changes over the past few decades while remaining essential to decision making in insurance.

Ragnar Norberg
London, March 2007

Preface

Motor Insurance

This book is devoted to the analysis of the number of claims filed by an insured driver over time. Property and liability motor vehicle coverage is broadly divided into first and third party coverage. First party coverage provides protection in the event the vehicle owner is responsible for the accident and protects him and his property. Third party coverage provides protection in the event the vehicle owner causes harm to another party, who recovers their cost from the policyholder. First party coverages may include first party injury benefits such as medical expenses, death payments and comprehensive coverages.

A third party liability coverage is required in most countries for a vehicle to be allowed on the public road network. The compulsory motor third party liability insurance represents a considerable share of the yearly nonlife premium collection in developed countries. This share becomes even more prominent when first party coverages are considered (such as medical benefits, uninsured or underinsured motorist coverage, and collision and other than collision insurance). Moreover, large data bases recording policyholders' characteristics as well as claim histories are maintained by insurance companies. The economic importance and the availability of detailed information explain why a large body of the nonlife actuarial literature is devoted to this line of business.

Tort System Versus No Fault System

The liability insurance provides coverage to the policyholder if, as the driver of a covered vehicle, the policyholder injures a third party's property. If the policyholder is sued with respect to negligence for such bodily injury or property damage, the insurer will provide legal defense for the policyholder. If the policyholder is found to be liable, the insurer will pay, on behalf of the policyholder, damages assessed against the policyholder.

In the tort system, the insurer indemnifies the claim only if it believes the insured was at fault in the accident or the third party sues the insured and proves that he/she was at fault in the accident. Needless to say, a large part of the premium income is consumed by legal fees, court costs and insurers' administration expenses in such a system. Because of this, several North-American jurisdictions have implemented a no-fault motor insurance system.

Even in a pure no-fault motor environment, the police still ask which driver was at fault (or the degrees to which the drivers shared the fault) because at-fault events cause the insurance premium to rise at the next policy renewal.

Insurance Ratemaking

Cost-based pricing of individual risks is a key actuarial ratemaking principle. The price charged to policyholders is an estimate of the future costs related to the insurance coverage. The pure premium approach defines the price of an insurance policy as the ratio of the estimated costs of all future claims against the coverage provided by the insurance policy while it is in effect to the risk exposure, plus expenses.

The property/casualty ratemaking is based on a claim frequency distribution and a loss distribution. The claim frequency is defined as the number of incurred claims per unit of earned exposure. The exposure is measured in car-year for motor third party liability insurance (the rate manual lists rates per car-year). The average loss severity is the average payment per incurred claim. Under mild conditions, the pure premium is then the product of the average claim frequency times the average loss severity. The loss models for motor insurance are reviewed in Chapters 1–2 (frequency part) and 5 (claim amounts).

In liability insurance, the settlement of larger claims often requires several years. Much of the data available for the recent accident years will therefore be incomplete, in the sense that the final claim cost will not be known. In this case, loss development factors can be used to obtain a final cost estimate. The average loss severity is then based on incurred loss data. In contrast to paid loss data (that are purely objective, representing the actual payments made by the company), incurred loss data include subjective reserve estimates. The actuary has to carefully analyse the large claims since they represent a considerable share of the insurer's yearly expenses. This issue will be discussed in Chapter 5, where incurred loss data will be analysed and appropriately modelled.

Risk Classification

Nowadays, it has become extremely difficult for insurance companies to maintain cross subsidies between different risk categories in a competitive market. If, for instance, females are proved to cause significantly fewer accidents than males and if a company disregarded this variable and charged an average premium to all policyholders regardless of gender, most of its female policyholders would be tempted to move to another company offering better rates to female drivers. The former company is then left with a disproportionate number of male policyholders and insufficient premium income to pay for the claims.

To avoid lapses in a competitive market, actuaries have to design a tariff structure that will fairly distribute the burden of claims among policyholders. The policies are partitioned into classes with all policyholders belonging to the same class paying the same premium. Each time a competitor uses an additional rating factor, the actuary has to refine the partition to avoid losing the best drivers with respect to this factor. This explains why so many factors are used by insurance companies: this is not required by actuarial theory, but instead by competition among insurers.

In a free market, insurance companies need to use a rating structure that matches the premiums for the risks as closely as possible, or at least as closely as the rating structures used by competitors. This entails using virtually every available classification variable correlated to the risks, since failing to do so would mean sacrificing the chance to select against competitors, and incurring the risk of suffering adverse selection by them. It is thus the competition between insurers that leads to more and more partitioned portfolios, and not actuarial science. This trend towards more risk classification often causes social disasters: bad drivers (or more precisely, drivers sharing the characteristics of bad drivers) do not find a coverage for a reasonable price, and are tempted to drive without insurance. Note also that even if a correlation exists between the rating factor and the risk covered by the insurer, there may be no causal relationship between that factor and risk. Requiring that insurance companies establish such a causal relationship to be allowed to use a rating factor is subject to debate.

Property and liability motor vehicle insurers use classification plans to create risk classes. The classification variables introduced to partition risks into cells are called *a priori* variables (as their values can be determined before the policyholder starts to drive). Premiums for motor liability coverage often vary by the territory in which the vehicle is garaged, the use of the vehicle (driving to and from work or business use) and individual characteristics (such as age, gender, occupation and marital status of the main driver of the vehicle, for instance, if not precluded by legislation or regulatory rules). If the policyholders misrepresent any of these classification variables in their declaration, they are subject to loss of coverage when they are involved in a claim. There is thus a strong incentive for accurate reporting of risk characteristics.

As explained in Chapter 2, it is convenient to achieve *a priori* classification with the help of generalized regression models. The method can be roughly summarized as follows: One risk classification cell is chosen as the base cell. It normally has the largest amount of exposure. The rate for the base cell is referred to as the base rate. Other rate cells are defined by a variety of risk classification variables, such as territory and so on. For each risk classification variable, there is a vector of differentials, with the base cell characteristics always assigned 100 %.

In this book, we make extensive use of the generalized linear models (better known under the acronym GLM) developed after NELDER & WEDDERBURN (1972). These authors discovered that regression models with a response distribution belonging to the exponential family of probability distributions shared the same characteristics. Members of this family include Normal, Binomial, Poisson, Gamma and Inverse Gaussian distributions that have been widely used by actuaries to model the number of claims, or their severities. Working in the exponential family allows the actuary to relax the very restrictive hypotheses behind the Normal linear regression model, namely:

- the response variable takes on the theoretical shape of a Normal distribution;
- the variance is constant over individuals;
- the fitted values are obtained from linear combinations of the explanatory variables (called linear predictors, or scores).

Specifically, the Normal distribution can be replaced with another member of the exponential family, heteroscedasticity can be allowed for, and fitted values can be obtained from a nonlinear transformation (called the link function) of linear predictors. Efficient algorithms are available under most statistical packages to estimate the regression parameters by maximum likelihood.

Pay-As-You-Drive System

Every kilometer travelled by a vehicle transfers risk to its insurer: the total cost of the coverage thus increases kilometer by kilometer. This is why several authors, including Butler (1993) suggested charging a cents-per-kilometer based premium; the car-kilometer should be adopted as the exposure unit instead of the car-year that is currently used. Motor insurance companies are adopting a new scheme called 'pay as you drive' (henceforth referred to as PAYD for the sake of brevity). Under a PAYD system, a driver pays for every kilometer driven at a rate varying from a premium to use the busiest roads at peak hours to a lower rate for the rural roads.

Several insurance companies (including the pioneering company Norwich Union, http://www.norwichunion.com/pay-as-you-drive/) have now started to offer a motor insurance policy under a PAYD system after successful pilot schemes involving thousands of motorists. With PAYD systems, drivers are provided with in-car Global Positioning System (GPS) devices coupled with maps, enabling the insurance company to calculate insurance premiums for each journey, depending on time of day, type of road and distance travelled. A 'black box' is installed in the car and receives signals from GPS technology to determine the vehicle's current position, speed, and time and direction driven. The black box then acts as a wireless modem to transmit these inputs through standard mobile phone networks to the insurer. The insurer sends a monthly bill to the customer based on vehicle usage, including time of day, type of road and distance travelled. Historical data then provide detailed information of how, when and where cars are actually used, and whether accidents and claims can be identified with particular factors. Moreover, the tracker detects speed infringements and more generally, the aggressiveness behind the wheel. Dangerous driving habits could lead to higher premiums for car insurance, increasing road safety. In addition to the static measures of risk, such as the driver's age, dynamic measures, such as speed, time of day, and location, are used to give the best possible overall risk assessment.

The generalization of the PAYD system is also expected to change motorists' attitudes: like petrol, insurance is bought on a pay-as-you-drive basis, and people think of their insurance costs as related to their actual use of their vehicle. Several North-American studies demonstrate that PAYD systems could reduce motoring by more than 10 %. The PAYD rating system is expected to decrease congestion and pollution (since the busier roads usually attract the higher rates).

Experience Rating

The trend towards more classification factors has lead the supervising authorities to exclude from the tariff structure certain risk factors, even though they may be significantly correlated to losses. Many states consider banning classification based on items that are beyond the control of the insured, such as gender or age. The resulting inadequacies of the *a priori* rating system can be corrected for by using the past number of claims to reevaluate future premiums. This is much in line with the concept of fairness: as it will be seen from Chapter 2, *a priori* ratemaking penalizes individuals who 'look like' bad drivers (even if they are in reality excellent drivers who will never cause any accident) whereas experience rating uses the individual claim record to adjust the amount of premium. Actuarial credibility models make a balance between the likelihood of being an unlucky good driver (who suffered a claim) and the likelihood of being a truly bad driver (who should suffer an increase in the premium paid to the insurance company for coverage). It seems fair to correct the inadequacies of the *a priori* system by using an adequate experience rating plan; such a 'crime and punishment' system may be more acceptable to policyholders than seemingly arbitrary *a priori* classifications.

Moreover, many important factors cannot be taken into account in the *a priori* risk classification. Think for instance of swiftness of reflexes, drinking habits or respect for the highway code. Consequently, tariff cells are still quite heterogeneous despite the use of many classification variables. This heterogeneity can be modelled by a random effect in a statistical model. It is reasonable to believe that the hidden characteristics are partly revealed by the number of claims reported by the policyholders. Several empirical studies have shown that, if insurers were allowed to use only one rating variable, it should be some form of merit rating: the best predictor of the number of claims incurred by a driver in the future is not age or vehicle type but past claims history. Hence the adjustment of the premium from the individual claims experience in order to restore fairness among policyholders as explained in Chapter 3. In that respect, the allowance of past claims in a rating model derives from an exogeneous explanation of serial correlation for longitudinal data. In this case, correlation is only apparent and results from the revelation of hidden features in the risk characteristics.

It is worth mentioning that serial correlation for claim numbers can also receive an endogeneous explanation. In this framework, the history of individuals modifies the risk they represent; this mechanism is termed 'true contagion', referring to epidemiology. For instance, a car accident may modify the perception of danger behind the wheel and lower the risk of reporting another claim in the future. Experience rating schemes also provide incentives to careful driving and should induce negative contagion. Nevertheless, the main interpretation for automobile insurance is exogeneous, since positive contagion (that is, policyholders who reported claims in the past being more likely to produce claims in the future than those who did not) is always observed for numbers of claims, whereas true contagion should be negative.

Bonus-Malus Systems

In many European and Asian countries, as well as in North-American states or provinces, insurers use experience rating in order to relate premium amounts to individual past claims experience in motor insurance. Such systems penalize insured drivers responsible for one or more accidents by premium surcharges (or *maluses*) and reward claim-free policyholders by awarding them discounts (or *bonuses*). Such systems are called no-claim discounts, experience rating, merit rating, or bonus-malus systems.

Discounts for claim-free driving have been awarded in the United Kingdom as early as 1910. At that time, they were intended as an inducement to renew a policy with the same company rather than as a reward for prudent driving. The first theoretical treatments of bonus-malus systems were provided in the pioneering works of GRENANDER (1957a,b). The first ASTIN colloquium held in France in 1959 was exclusively devoted to no-claim discounts in insurance, with particular reference to motor business.

There are various bonus-malus systems used around the world. A typical form of no-claim bonus in the United Kingdom is as follows:

one claim-free year 25 % discount
two claim-free years 40 % discount
three claim-free years 50 % discount
four claim-free years 60 % discount.

Drivers earn an extra year of bonus for each year they remain without claims at fault up to a maximum of four years, but lose two years bonus each time they report a claim at fault. In such a system, maximum bonus is achieved in only a few years and the majority of mature drivers have maximum bonus.

Bonus-malus systems used in Continental Europe are often more elaborate. Bonus-malus scales consist of a finite number of levels, each with its own relativity (or relative premium). The amount of premium paid by a policyholder is then the product of a base premium with the relativity corresponding to the level occupied in the scale. New policyholders have access to a specified level. After each year, the policy moves up or down according to transition rules of the bonus-malus system. If a bonus-malus system is in force, all policies in the same tariff class are partitioned according to the level they occupy in the bonus-malus scale. In this respect, the bonus-malus mechanism can be considered as a refinement of *a priori* risk evaluation splitting each risk class into a number of subcategories according to individual past claims histories.

As explained in Chapter 4, bonus-malus systems can be modelled using (conditional) Markov chains provided they possess a certain memoryless property that can be summarized as follows: the knowledge of the present level and of the number of claims of the present year suffices to determine the level to which the policy is transferred. In other words, the bonus-malus system satisfies the famous Markov property: the future (the level for year $t+1$) depends on the present (the level for year t and the number of accidents reported during year t) and not on the past (the claim history and the levels occupied during years $1, 2, \ldots, t-1$). This allows us to determine the optimal relativities in Chapter 4 using an asymptotic criterion based on the stationary distribution, and in Chapter 8 using transient distributions. Several performance measures for bonus-malus systems are reviewed in Chapters 5 and 8.

During the 20th century, most European countries imposed a uniform bonus-malus system on all the companies operating in their territory. In 1994, the European Union decreed that all its member countries must drop their mandatory bonus-malus systems, claiming that such systems reduced competition between insurers and were in contradiction to the total rating freedom implemented by the Third Directive. Since that date, Belgium, for instance, dropped its mandatory system, but all companies operating in Belgium still apply the former uniform system (with minor modifications for the policyholders occupying the lowest levels in the scale). In other European countries, however, insurers compete on the basis of bonus-malus systems. This is the case for instance in Spain and Portugal.

However, the mandatory French system is still in force. Quite surprisingly, the European Court of Justice decided in 2004 that both the French and Grand Duchy of Luxembourg mandatory bonus-malus systems were not contrary to the rating freedom imposed by the European legislation. The French law thus still imposes on the insurers operating in France a unique bonus-malus system. That bonus-malus system is not based on a scale. Instead the

French bonus-malus system uses the concept of an increase-decrease coefficient (*coefficient de réduction-majoration* in French). More precisely, the French bonus-malus system implies a malus of 25 % per claim and a bonus of 5 % per claim-free year. So each policyholder is assigned a base premium and this base premium is adapted according to the number of claims reported to the insurer, multiplying the premium by 1.25 each time an accident at fault is reported to the company, and by 0.95 per claim-free year. The French-type bonus-malus systems will be studied in Chapter 9.

Actuarial and Economic Justifications for Bonus-Malus Systems

Bonus-malus systems allow premiums to be adapted for hidden individual risk factors and to increase incentives for road safety, by taking into consideration the past claim record. This can be justified by asymmetrical information between the insurance company and the policyholders. Asymmetric information arises in insurance markets when firms have difficulties in judging the riskiness of those who purchase insurance coverage. There are mainly two aspects of this phenomenon: adverse selection and moral hazard. Adverse selection occurs when the policyholders have a better knowledge of their claim behaviour than the insurer does. Policyholders take advantage of information about their driving patterns, known to them but unknown to the insurer. In the context of compulsory motor third party liability insurance, adverse selection is not a significant problem compared to moral hazard when the insurance companies charge similar amounts of premium to all policyholders. Things are more complicated in a deregulated environment with companies using many risk classification factors. Since very heterogeneous driving behaviours are observed among policyholders sharing the same *a priori* variables, adverse selection cannot be avoided. For all the related coverages, such as comprehensive damages for instance, adverse selection always plays an important role.

Considering adverse selection in the vein of Rotschild and Stiglitz, individuals partly reveal their underlying risk through the contract they choose, a fact that has to be taken into account when setting an adequate tariff structure. In the presence of unobservable heterogeneity, riskier agents will choose a more comprehensive coverage and low risk insurance applicants have an interest in signalling their quality, by selecting high deductibles (excesses) for instance.

It is interesting to compare economists' and actuaries' approaches to experience rating. In the economic literature, discounts and penalties are introduced mainly to counteract the inefficiency which arises from moral hazard. In the actuarial literature, the main purpose is to better assess the individual risk so that everyone will pay, in the long run, a premium corresponding to his own claim frequency. Actuaries are thus more interested in adverse selection than moral hazard.

Cost of Claims

The vast majority of bonus-malus systems in force around the world penalize the number of at-fault accidents reported to the company, and not their amounts. A severe accident involving bodily injuries is penalized in the same way as a fender-bender. The reason to base motor risk classification on just claim frequencies is the long delay to access the cost of bodily injury and other severe claims. Not incorporating claim sizes in bonus-malus systems and *a priori* risk classification requires an (implicit) assumption of independence between the random variables 'number of claims' and 'cost of a claim', as well as the belief that the latter does not depend on the driver's characteristics. This means that the actuarial practice considers that the cost of an accident is, for the most part, beyond the control of a driver: a cautious driver reduces the number of accidents, but for the most part cannot control the cost of these accidents (which is largely independent of the mistake that caused it). This belief will be challenged in Chapter 5.

The penalty induced by the majority of bonus-malus systems being independent of the claim amount, policyholders have to decide whether it is profitable or not to report small claims (in order to avoid an increase in premium). Cheap claims are likely to be defrayed by the policyholders themselves, and not to be reported to the company. This phenomenon is known as the hunger for bonus and censors claim amounts and claim frequencies. In Chapter 5, a statistical model is specified, that takes into account the fact that only 'expensive' claims are reported to the insurance company. Retention limits for the policyholders are determined using the Lemaire algorithm.

In a few bonus-malus systems, however, reporting a 'severe' claim (typically, a claim with bodily injuries) entails a more severe penalty than reporting a 'minor' claim (typically, a claim with material damage only). In the system in force in Japan before 1993, claims involving bodily injuries were penalized by four levels, while claims with property damage only were penalized by only two levels. Bonus-malus systems using different types of events to update premium amount will be examined in Chapter 6.

In Chapter 7, we examine an innovative system using variable deductibles rather than premium relativities. It differs from the systems studied in preceding chapters in that it mixes elements of both a conventional bonus-malus system and a set of deductibles depending on the level occupied in the bonus-malus scale. The first system is a conventional discount system with loss of discount in the case where a claim at fault is reported. The second system also has a variable discount scale, which can increase with claim-free experience. However, there is no stepback of the discount on claim, only a stepback of the deductible.

Aims of This Book

About ten years after the seminal book 'Bonus-Malus Systems in Automobile Insurance' by Professor Jean Lemaire, we aim to offer a comprehensive treatment of the various experience rating systems applicable to automobile insurance and their relationships with risk classification.

We hope that the present book will be useful for students in actuarial science, actuaries (both practitioners and in academia) and more generally for all the persons involved in technical problems inside insurance companies or consulting firms. For the first time, systems taking into account the exogeneous information are presented in an actuarial textbook. Many numerical illustrations carried out with advanced statistical softwares allow for a deep understanding of the concepts.

The present book is the result of a close and fruitful collaboration between the Institute of Actuarial Science of the Université Catholique de Louvain, Louvain-la-Neuve, Belgium, its spin-off consulting firm Reacfin SA and the reinsurance company Secura, based in Brussels.

This collaboration brings together academic expertise and practical experience to provide efficient solutions to motor ratemaking.

Software

The numerical illustrations presented in this book use SAS® (standing for Statistical Analysis System), a powerful software package for the manipulation and statistical analysis of data. SAS® is widely used in the insurance industry and practicing actuaries should be familiar with it. Among the large range of modules that can be added to the basic system (known as SAS®/BASE), we concentrate on the SAS®/STAT module. When no built-in procedures were available, we have coded programs in the SAS®/IML environment.

The computations of bonus-malus scales are performed with the software BM-Builder developed by Reacfin. This is a computer solution running on SAS® that enables creation of a new bonus-malus scale by choosing the number of levels, the transition rules, etc. This scale, tailored to the insurer's portfolio, is financially balanced.

Here and there, comments about software available to perform the analyses detailed in this book will be provided to help the readers interested in practical implementation. Appropriate references to the websites of the providers are given for further information.

Acknowledgements

The present text originated from a series of lectures by Michel Denuit and Jean-François Walhin to Masters students in actuarial science in different universities (including UCL, Louvain-la-Neuve, Belgium; UCBL, Lyon, France; ULP, Strasbourg, France; and INSEA, Rabat, Morocco). Both Michel Denuit and Jean-François Walhin would like to thank the students, who have worked through the nonlife ratemaking courses over the past years and supplied invaluable reactions and comments.

Training sessions with insurance professionals provided practical insights in the contents of the lectures. The feedback we received from short course audiences in Bucharest, Niort, Paris, and Warsaw, helped to improve the presentation of the topic.

The authors' own research in this area has benefited at various stages from discussions or collaborations with esteemed colleagues, including Jean-Philippe Boucher, Arthur Charpentier, Christophe Crochet, Jan Dhaene, Montserrat Guillén, Philippe Lambert, Stefan Lang, José Paris, Christian Partrat, Jean Pinquet, and Richard Verrall.

We gratefully acknowledge the financial support of the *Communauté française de Belgique* under contract 'Projet d'Actions de Recherche Concertées' ARC 04/09-320, of the *Région Wallonne* under project First Spin-off 'ActuR&D # 315481', of Secura, the Belgian reinsurance company, and of the *Banque Nationale de Belgique* under grant 'Risk measures and Economic capital'.

We would like to express our deepest gratitude to Professor Ragnar Norberg for kindly accepting to preface this book, as well as for his careful reading of a previous version of this manuscript and for the numerous resulting comments. Any errors or omission, however, remain the responsibility of the authors. Professor Norberg's pioneering works are among the most influential contributions to credibility theory and bonus-malus systems. It is a real honour that, thirty years after his seminal work appeared in the Scandinavian Actuarial

Journal which integrates bonus-malus scales in the framework of Markov chains, Professor Norberg introduces the present work.

As always, it has been a pleasure to work with Wendy Hunter and Susan Barclay, Project Editors; Simon Lightfoot, Publishing Assistant; Kathryn Sharples, Commissioning Editor in Statistics and Mathematics; and Kelly Board and Sarah Kiddle, Content Editors, Engineering and Statistics at John Wiley & Sons, Ltd and Sunita Jayachandran at Integra Software Services Pvt. Ltd.

Last but not least, we apologize to our families for the time not spent with them during the preparation of this book, and we are very grateful for their understanding.

Michel Denuit
Xavier Maréchal
Sandra Pitrebois
Jean-François Walhin
Louvain-la-Neuve and Brussels, January 2007.

Notation

Here are a few words on the notation and terminology used throughout the book. For the most part, the notation used in this book conforms to what is usual in mathematical statistics as well as nonlife insurance mathematics.

The real line $(-\infty, +\infty)$ is denoted as $\mathbb{R}$. The half positive real line is $\mathbb{R}^+ = [0, +\infty)$. The set of the nonnegative integers is $\mathbb{N} = \{0, 1, 2, \ldots\}$. The real n-dimensional space is denoted as $\mathbb{R}^n$, and $\mathbb{N}^n$ is the set of all the n-tuples of nonnegative integers. A point of $\mathbb{R}^n$ is an n-dimensional vector with real coordinates. It is represented by a bold letter $\boldsymbol{x}$; the ith component of $\boldsymbol{x}$ is x_i, $i = 1, 2, \ldots, n$. All the vectors are tacitly assumed to be column vectors, that is,

$$\boldsymbol{x} = \begin{pmatrix} x_1 \\ \vdots \\ x_n \end{pmatrix}.$$

A superscript 'T' is used to indicate transposition. Hence, $\boldsymbol{x}^T$ is a row vector with components $x_1, \ldots, x_n$. The dimension of $\boldsymbol{x}$ is denoted as $\dim(\boldsymbol{x})$.

The matrices are denoted by a capital letter in boldface, for instance $\boldsymbol{M}$; $\boldsymbol{M}^T$ denotes the transposition of $\boldsymbol{M}$. The vector of ones, that is $(1, 1, \ldots, 1)^T$, will be denoted by $\boldsymbol{e}$. The identity matrix (with entries 1 on the main diagonal and 0 elsewhere) is denoted by $\boldsymbol{I}$. The determinant of the matrix $\boldsymbol{M}$ is denoted as $\det(\boldsymbol{M})$, its inverse by $\boldsymbol{M}^{-1}$.

We denote as $o(h)$ a function of h that tends to 0 faster than the identity, that is, such that

$$\lim_{h \searrow 0} \frac{o(h)}{h} = 0.$$

Intuitively, $o(h)$ is negligible when h becomes sufficiently small.

The factorial of the positive integer n is denoted as $n!$ and defined by $n! = n(n-1)\ldots 1$. By convention, $0! = 1$. The binomial coefficient $\binom{n}{k}$ denotes the number of different possible combinations of k items from n different items:

$$\binom{n}{k} = \frac{n!}{k!(n-k)!} = \binom{n}{n-k}.$$

Note that the binomial coefficient is sometimes denoted as C_n^k, especially in the French-written mathematical literature, but here we adhere to the more standard notation $\binom{n}{k}$. The Gamma function $\Gamma(\cdot)$ is defined as

$$\Gamma(x) = \int_0^{\infty} t^{x-1} \exp(-t)dt, \quad x > 0.$$

As for any positive integer n, we have $\Gamma(n) = (n-1)!$, the Gamma function can be considered as an interpolation of the factorials defined for positive integers. Integration by parts shows that $\Gamma(x+1) = x\Gamma(x)$ for any positive real x. When a and b are positive real numbers, the definition of the binomial coefficient is extended to positive integers as

$$\binom{a}{b} = \frac{\Gamma(a+1)}{\Gamma(a-b+1)\Gamma(b+1)}.$$

The incomplete Gamma function $\Gamma(\cdot, \cdot)$ is defined as

$$\Gamma(t, \xi) = \frac{1}{\Gamma(t)} \int_0^{\xi} x^{t-1} \exp(-x)dx, \quad \xi, t \geq 0.$$

A real-valued random variable is denoted by a capital letter, for instance X. The mathematical expectation operator is denoted as $\mathbb{E}[\cdot]$. For instance, $\mathbb{E}[X]$ is the expectation of the random variable X. The variance is $\mathbb{V}[X]$, given by $\mathbb{V}[X] = \mathbb{E}[X^2] - (\mathbb{E}[X])^2$. A random vector is denoted by a bold capital letter, for instance $\boldsymbol{X} = (X_1, \ldots, X_n)^T$. Matrices should not be confused with random vectors (the context will make this clear). The variance-covariance matrix $\boldsymbol{\Sigma}$ of $\boldsymbol{X}$ has covariances $\mathbb{C}[X_i, X_j] = \mathbb{E}[X_iX_j] - \mathbb{E}[X_i]\mathbb{E}[X_j]$ outside the main diagonal (that is, for $i \neq j$) and the variances $\mathbb{V}[X_i]$ along the main diagonal.

The probability distributions used in this book are summarized next:

- the Bernoulli distribution with parameter $0 < q < 1$, denoted as $\mathcal{Ber}(q)$, has probability mass function

$$p(k) = q^k(1-q)^{1-k}, \quad k \in \{0, 1\};$$

- the Binomial distribution with parameters $m \in \{1, 2, \ldots\}$ (usually called the exponent, or size) and $0 < q < 1$, denoted as $\mathcal{Bin}(m, q)$, has probability mass function

$$p(k) = \binom{m}{k} q^k(1-q)^{m-k}, \quad k \in \{0, 1, \ldots, m\};$$

- the Poisson distribution with parameter $\lambda > 0$, denoted as $\mathcal{Poi}(\lambda)$, has probability mass function

$$p(k) = \exp(-\lambda)\frac{\lambda^k}{k!}, \quad k \in \mathbb{N};$$

- the Negative Binomial distribution with parameters $a > 0$ and $\lambda > 0$, denoted as $\mathcal{NB}in(a, \lambda)$, has probability mass function

$$p(k) = \binom{a+k-1}{k} \left(\frac{a}{a+\lambda}\right)^a \left(\frac{\lambda}{a+\lambda}\right)^k, \quad k \in \mathbb{N};$$

- the Normal distribution with parameters $\mu \in \mathbb{R}$ and $\sigma^2 > 0$, denoted as $\mathcal{N}or(\mu, \sigma^2)$, has probability density function

$$f(x) = \frac{1}{\sigma\sqrt{2\pi}} \exp\left(-\frac{1}{2\sigma^2}(x-\mu)^2\right), \quad x \in \mathbb{R};$$

- the LogNormal distribution with parameters $\mu \in \mathbb{R}$ and $\sigma^2 > 0$, denoted as $\mathcal{LN}or(\mu, \sigma^2)$, has probability density function

$$f(x) = \frac{1}{x\sigma\sqrt{2\pi}} \exp\left(-\frac{1}{2\sigma^2}(\ln(x)-\mu)^2\right), \quad x \in \mathbb{R}^+;$$

- the Negative Exponential distribution with parameter $\theta > 0$, denoted as $\mathcal{E}xp(\theta)$, has probability density function

$$f(x) = \theta \exp(-\theta x), \quad x \in \mathbb{R}^+;$$

- the Gamma distribution with parameters $\alpha > 0$ and $\tau > 0$, denoted as $\mathcal{G}am(\alpha, \tau)$, has probability density function

$$f(x) = \frac{x^{\alpha-1}\tau^\alpha \exp(-x\tau)}{\Gamma(\alpha)}, \quad x \in \mathbb{R}^+;$$

- the Inverse Gaussian distribution, with parameters $\mu > 0$ and $\beta > 0$, denoted as $\mathcal{IG}au(\mu, \beta)$, has probability density function

$$f(x) = \frac{\mu}{\sqrt{2\pi\beta x^3}} \exp\left(-\frac{1}{2\beta x}(x-\mu)^2\right), \quad x \in \mathbb{R}^+;$$

- the Pareto distribution, with parameters $\alpha > 0$ and $\theta > 0$, denoted as $\mathcal{P}ar(\alpha, \theta)$, has probability density function

$$f(x) = \frac{\alpha\theta^\alpha}{(x+\theta)^{\alpha+1}}, \quad x \in \mathbb{R}^+;$$

- the Uniform distribution, with parameters $a < b \in \mathbb{R}$, denoted as $\mathcal{U}ni[a, b]$, has probability density function

$$f(x) = \frac{1}{b-a}, \quad x \in [a, b].$$

We use the symbol $\sim$ to mean 'is distributed as'. For instance, $X \sim \mathcal{LN}or(\mu, \sigma^2)$ means that X is distributed according to the LogNormal distribution with parameters μ and σ^2.

Part I

Modelling Claim Counts

1

Mixed Poisson Models for Claim Numbers

1.1 Introduction

1.1.1 Poisson Modelling for the Number of Claims

In view of the economic importance of motor third party liability insurance in industrialized countries, many attempts have been made in the actuarial literature to find a probabilistic model for the distribution of the number of claims reported by insured drivers. This chapter aims to introduce the basic probability models for count data that will be applied in motor insurance. References to alternative models are gathered in the closing section to this chapter.

The Binomial distribution is the discrete probability distribution of the number of successes in a sequence of n independent yes/no experiments, each of which yields success with probability q. Such a success/failure experiment is also called a Bernoulli experiment or Bernoulli trial. Two important distributions arise as approximations of Binomial distributions. If n is large enough and the skewness of the distribution is not too great (that is, q is not too close to 0 or 1), then the Binomial distribution is well approximated by the Normal distribution. When the number of observations n is large, and the success probability q is small, the corresponding Binomial distribution is well approximated by the Poisson distribution with mean $\lambda = nq$. The Poisson distribution is thus sometimes called the law of small numbers because it is the probability distribution of the number of occurrences of an event that happens rarely but has very many opportunities to happen. The parallel with traffic accidents is obvious.

The Poisson distribution was discovered by Siméon-Denis Poisson (1781–1840) and published in 1838 in his work entitled *Recherches sur la Probabilité des Jugements en Matières Criminelles et Matière Civile* (which could be translated as 'Research on the Probability of Judgments in Criminal and Civil Matters'). Typically, a Poisson random

Actuarial Modelling of Claim Counts: Risk Classification, Credibility and Bonus-Malus Systems M. Denuit, X. Maréchal, S. Pitrebois and J.-F. Walhin

variable is a count of the number of events that occur in a certain time interval or spatial area. For example, the number of cars passing a fixed point in a five-minute interval, or the number of claims reported to an insurance company by an insured driver in a given period. A typical characteristic associated with the Poisson distribution is certainly equidispersion: the variance of the Poisson distribution is equal to its mean.

1.1.2 Heterogeneity and Mixed Poisson Model

The Poisson distribution plays a prominent role in modelling discrete count data, mainly because of its descriptive adequacy as a model when only randomness is present and the underlying population is homogeneous. Unfortunately, this is not a realistic assumption to make in modelling many real insurance data sets. Poisson mixtures are well-known counterparts to the simple Poisson distribution for the description of inhomogeneous populations. Of special interest are populations consisting of a finite number of homogeneous sub-populations. In these cases the probability distribution of the population can be regarded as a finite mixture of Poisson distributions.

The problem of unobserved heterogeneity arises because differences in driving behaviour among individuals cannot be observed by the actuary. One of the well-known consequences of unobserved heterogeneity in count data analysis is overdispersion: the variance of the count variable is larger than the mean. Apart from its implications for the low-order moment structure of the counts, unobserved heterogeneity has important implications for the probability structure of the ensuing mixture model. The phenomena of excesses of zeros as well as heavy upper tails in most insurance data can be seen as an implication of unobserved heterogeneity (Shaked's Two Crossings Theorem will make this clear). It is customary to allow for unobserved heterogeneity by superposing a random variable (called a random effect) on the mean parameter of the Poisson distribution, yielding a mixed Poisson model. In a mixed Poisson process, the annual expected claim frequency itself becomes random.

1.1.3 Maximum Likelihood Estimation

All the models implemented in this book are parametric, in the sense that the probabilities are known functions depending on a finite number of (real-valued) parameters. The Binomial, Poisson and Normal models are examples of parametric distributions. The first step in the analysis is to select a reasonable parametric model for the observations, and then to estimate the underlying parameters. The maximum likelihood estimator is the value of the parameter (or parameter vector) that makes the observed data most likely to have occurred given the data generating process assumed to have produced the observations. All we need to derive the maximum likelihood estimator is to formulate statistical models in the form of a likelihood function as a probability of getting the data at hand. The larger the likelihood, the better the model.

Maximum likelihood estimates have several desirable asymptotic properties: consistency, efficiency, asymptotic Normality, and invariance. The advantages of maximum likelihood estimation are that it fully uses all the information about the parameters contained in the data and that it is highly flexible. Most applied maximum likelihood problems lack closed-form solutions and so rely on numerical maximization of the likelihood function. The advent of fast computers has made this a minor issue in most cases. Hypothesis testing for maximum likelihood parameter estimates is straightforward due to the asymptotic Normal distribution of maximum likelihood estimates and the Wald and likelihood ratio tests.

1.1.4 Agenda

Section 1.2 briefly reviews the basic probability concepts used throughout this chapter (and the entire book), for further reference. Notions including probability spaces, random variables and probability distributions are made precise in this introductory section.

In Section 1.3, we recall the main probabilistic tools to work with discrete distributions: probability mass function, distribution function, probability generating function, etc. Then, we review some basic counting distributions, including the Binomial and Poisson laws.

Section 1.4 is devoted to mixture models to account for unobserved heterogeneity. Mixed Poisson distributions are discussed, including Negative Binomial (or Poisson-Gamma), Poisson-Inverse Gaussian and Poisson-LogNormal models.

Section 1.5 presents the maximum likelihood estimation method. Large sample properties of the maximum likelihood estimators are discussed, and testing procedures are described. The large sample properties are particularly appealing to actuaries who usually deal with tens of thousands of observations in insurance portfolios.

Section 1.6 gives numerical illustrations on the basis of a Belgian motor third party liability insurance portfolio. The observed claim frequency distribution is fitted using the Poisson distribution and various mixed Poisson probability distributions, and the optimal model is selected on the basis of appropriate goodness-of-fit tests.

The final Section, 1.7, concludes Chapter 1 by providing suggestions for further reading and bibliographic notes about the models proposed in the actuarial literature for the annual number of claims.

1.2 Probabilistic Tools

1.2.1 Experiment and Universe

Many everyday statements for actuaries take the form 'the probability of A is p', where A is some event (such as 'the total losses exceed the threshold € 1 000 000' or 'the number of claims reported by a given policyholder is less than two') and p is a real number between zero and one. The occurrence or nonoccurrence of A depends upon the chain of circumstances under consideration. Such a particular chain is called an experiment in probability; the result of an experiment is called its outcome and the set of all outcomes (called the universe) is denoted by Ω.

The word 'experiment' is used here in a very general sense to describe virtually any process for which all possible outcomes can be specified in advance and for which the actual outcome will be one of those specified. The basic feature of an experiment is that its outcome is not definitely known by the actuary beforehand.

1.2.2 Random Events

Random events are subsets of the universe Ω associated with a given experiment. A random event is the mathematical formalization of an event described in words. It is random since we cannot predict with certainty whether it will be realized or not during the experiment. For instance, if we are interested in the number of claims incurred by a policyholder of an automobile portfolio during one year, the experiment consists in observing the driving behaviour of this individual during an annual period, and the universe Ω is simply the set $\{0, 1, 2, \ldots\}$ of the nonnegative integers. The random event $A =$ 'the policyholder reports at most one claim' is identified with the subset $\{0, 1\} \subset \Omega$.

As usual, we use $A \cup B$ and $A \cap B$ to represent the union and the intersection, respectively, of any two subsets A and B of Ω. The union of sets is defined to be the set that contains the points that belong to at least one of the sets. The intersection of sets is defined to be the set that contains the points that are common to all the sets. These set operations correspond to the 'or' and 'and' between sentences: $A \cup B$ is the event which is realized if A or B is realized and $A \cap B$ is the event realized if A and B are simultaneously realized during the experiment. We also define the difference between sets A and B, denoted as $A \setminus B$, as the set of elements in A but not in B. Finally, $\overline{A}$ is the complementary event of A, defined as $\Omega \setminus A$; it is the set of points of Ω that do not belong to A. This corresponds to the negation: $\overline{A}$ is realized if A is not realized during the experiment. In particular, $\overline{\Omega} = \emptyset$, where $\emptyset$ is the empty set.

1.2.3 Sigma-Algebra

One needs to specify a family $\mathcal{F}$ of events to which probabilities can be ascribed in a consistent manner. The family $\mathcal{F}$ has to be closed under standard operations on sets; indeed, given two events A and B in $\mathcal{F}$, we want $A \cup B$, $A \cap B$ and $\overline{A}$ to still be events (i.e. still belong to $\mathcal{F}$). Technically speaking, this will be the case if $\mathcal{F}$ is a sigma-algebra. Recall that a family $\mathcal{F}$ of subsets of the universe Ω is called a sigma-algebra if it fulfills the three following properties: (i) $\Omega \in \mathcal{F}$, (ii) $A \in \mathcal{F} \Rightarrow \overline{A} \in \mathcal{F}$, and (iii) $A_1, A_2, A_3, \ldots \in \mathcal{F} \Rightarrow \bigcup_{i \geq 1} A_i \in \mathcal{F}$.

The three properties (i)-(iii) are very natural. Indeed, (i) means that Ω itself is an event (it is the event which is always realized). Property (ii) means that if A is an event, the complement of A is also an event. Finally, property (iii) means that the event consisting in the realization of at least one of the A_is is also an event.

1.2.4 Probability Measure

Once the universe Ω has been equipped with a sigma-algebra $\mathcal{F}$ of random events, a probability measure Pr can be defined on $\mathcal{F}$. The knowledge of Pr allows us to discuss the likelihood of the occurrence of events in $\mathcal{F}$. To be specific, Pr assigns to each random event A its probability $\Pr[A]$; $\Pr[A]$ is the likelihood of realization of A. Formally, a probability measure Pr maps $\mathcal{F}$ to $[0, 1]$, with $\Pr[\Omega] = 1$, and is such that given $A_1, A_2, A_3, \ldots \in \mathcal{F}$ which are pairwise disjoint, i.e., such that $A_i \cap A_j = \emptyset$ if $i \neq j$,

$$\Pr\left[\bigcup_{i \geq 1} A_i\right] = \sum_{i \geq 1} \Pr[A_i];$$

this technical property is usually referred to as the sigma-additivity of Pr.

The properties assigned to Pr naturally follow from empirical evidence: if we were allowed to repeat an experiment a large number of times, keeping the initial conditions as equal as possible, the proportion of times that an event A occurs would behave according to the definition of Pr. Note that $\Pr[A]$ is then the mathematical idealization of the proportion of times A occurs.

1.2.5 Independent Events

Independence is a crucial concept in probability theory. It aims to formalize the intuitive notion of 'not influencing each other' for random events: we would like to give a precise meaning to the fact that the realization of an event does not decrease nor increase the probability that the other event occurs. Formally, two events A and B are said to be independent if the probability of their intersection equals the product of their respective probabilities, that is, if $\Pr[A\cap B]=\Pr[A]\Pr[B]$.

This definition is extended to more than two events as follows. The events in a family $\mathcal{A}$ of events are independent if for every finite sequence $A_1, A_2, \ldots, A_k$ of events in $\mathcal{A}$,

$$\Pr\left[\bigcap_{i=1}^{k} A_i\right]=\prod_{i=1}^{k}\Pr[A_i]. \tag{1.1}$$

The concept of independence is very important in assigning probabilities to events. For instance, if two or more events are regarded as being physically independent, in the sense that the occurrence or nonoccurrence of some of them has no influence on the occurrence or nonoccurrence of the others, then this condition is translated into mathematical terms through the assignment of probabilities satisfying Equation (1.1).

1.2.6 Conditional Probability

Independence is the exception rather than the rule. In any given experiment, it is often necessary to consider the probability of an event A when additional information about the outcome of the experiment has been obtained from the occurrence of some other event B. This corresponds to intuitive statements of the form 'if B occurs then the probability of A is p', where B can be 'March is rainy' and A 'the claim frequency in motor insurance increases by 5 %'. This is called the conditional probability of A given B, and is formally defined as follows. If $\Pr[B]>0$ then the conditional probability $\Pr[A|B]$ of A given B is defined to be

$$\Pr[A|B]=\frac{\Pr[A\cap B]}{\Pr[B]}. \tag{1.2}$$

The definition of conditional probabilities through (1.2) is in line with empirical evidence. Repeating a given experiment a large number of times, $\Pr[A|B]$ is the mathematical idealization of the proportion of times A occurs in those experiments where B did occur, hence the ratio (1.2).

It is easily seen that A and B are independent if, and only if,

$$\Pr[A|B]=\Pr[A|\overline{B}]=\Pr[A]. \tag{1.3}$$

Note that this interpretation of independence is much more intuitive than the definition given above: indeed the identity expresses the natural idea that the realization or not of B does not increase nor decrease the probability that A occurs.

1.2.7 Random Variables and Random Vectors

Often, actuaries are not interested in an experiment itself but rather in some consequences of its random outcome. For instance, they are more concerned with the amounts the insurance company will have to pay than with the particular circumstances which give rise to the claims. Such consequences, when real-valued, may be thought of as functions mapping Ω into the real line $\mathbb{R}$.

Such functions are called random variables provided they satisfy certain desirable properties, precisely stated in the following definition: A random variable X is a measurable function mapping Ω to the real numbers, i.e., $X:\Omega\to\mathbb{R}$ is such that $X^{-1}((-\infty,x])\in\mathcal{F}$ for any $x\in\mathbb{R}$, where $X^{-1}((-\infty,x])=\{\omega\in\Omega|X(\omega)\leq x\}$. In other words, the measurability condition $X^{-1}((-\infty,x])\in\mathcal{F}$ ensures that the actuary can make statements like 'X is less than or equal to x' and quantify their likelihood. Random variables are mathematical formalizations of random outcomes given by numerical values. An example of a random variable is the amount of a claim associated with the occurrence of an automobile accident.

A random vector $\boldsymbol{X}=(X_1,X_2,\ldots,X_n)^T$ is a collection of n univariate random variables, $X_1,X_2,\ldots,X_n$, say, defined on the same probability space $(\Omega,\mathcal{F},\Pr)$. Random vectors are denoted by bold capital letters.

1.2.8 Distribution Functions

In many cases, neither the universe Ω nor the function X need to be given explicitly. The practitioner has only to know the probability law governing X or, in other words, its distribution. This means that he is interested in the probabilities that X takes values in appropriate subsets of the real line (mainly intervals).

To each random variable X is associated a function F_X called the distribution function of X, describing the stochastic behaviour of X. Of course, F_X does not indicate what is the actual outcome of X, but shows how the possible values for X are distributed (hence its name). More precisely, the distribution function of the random variable X, denoted as F_X, is defined as

$$F_X(x)=\Pr[X^{-1}((-\infty,x])]\equiv\Pr[X\leq x],\quad x\in\mathbb{R}.$$

In other words, $F_X(x)$ represents the probability that the random variable X assumes a value that is less than or equal to x. If X is the total amount of claims generated by some policyholder, $F_X(x)$ is the probability that this policyholder produces a total claim amount of at most €x. The distribution function F_X corresponds to an estimated physical probability distribution or a well-chosen subjective probability distribution.

Any distribution function F has the following properties: (i) F is nondecreasing, i.e. $F(x)\leq F(y)$ if $x<y$, (ii) $\lim_{x\searrow-\infty}F(x)=0$, (iii) $\lim_{x\nearrow+\infty}F(x)=1$, (iv) F is right-continuous,

i.e. $\lim_{h\searrow 0} F(x+h) = F(x)$, and (v) $\Pr[a < X \leq b] = F(b) - F(a)$, for any $a < b$. Henceforth, we denote as $F(\cdot -)$ the left limit of F, that is,

$$F(x-) = \lim_{\xi \nearrow x} F(\xi) = \Pr[X < x].$$

Suppose that $X_1, X_2, \ldots, X_n$ are n random variables defined on the same probability space $(\Omega, \mathcal{F}, \Pr)$. Their marginal distribution functions $F_1, F_2, \ldots, F_n$ contain all the information about their associated probabilities. But how can the actuary encapsulate information about their properties relative to each other? As explained above, the key idea is to think of $X_1, X_2, \ldots, X_n$ as being components of a random vector $\boldsymbol{X} = (X_1, X_2, \ldots, X_n)^T$ taking values in $\mathbb{R}^n$ rather than being unrelated random variables each taking values in $\mathbb{R}$.

As was the case for random variables, each random vector $\boldsymbol{X}$ possesses a distribution function $F_{\boldsymbol{X}}$ that describes its stochastic behaviour. The distribution function of the random vector $\boldsymbol{X}$, denoted as $F_{\boldsymbol{X}}$, is defined as

$$\begin{aligned} F_{\boldsymbol{X}}(x_1, x_2, \ldots, x_n) &= \Pr\left[\boldsymbol{X}^{-1}((-\infty, x_1] \times (-\infty, x_2] \times \cdots \times (-\infty, x_n])\right] \\ &= \Pr[X_1 \leq x_1, X_2 \leq x_2, \ldots, X_n \leq x_n], \end{aligned}$$

$x_1, x_2, \ldots, x_n \in \mathbb{R}$. The value $F_{\boldsymbol{X}}(x_1, x_2, \ldots, x_n)$ represents the probability that simultaneously X_1 assumes a value that is less than or equal to x_1, X_2 assumes a value that is less than or equal to x_2, ..., X_n assumes a value that is less than or equal to x_n; a more compact way to express this is

$$F_{\boldsymbol{X}}(\boldsymbol{x}) = \Pr[\boldsymbol{X} \leq \boldsymbol{x}], \quad \boldsymbol{x} \in \mathbb{R}^n.$$

Even if the distribution function $F_{\boldsymbol{X}}$ does not tell us which is the actual value of $\boldsymbol{X}$, it thoroughly describes the range of possible values for $\boldsymbol{X}$ and the probabilities assigned to each of them.

1.2.9 Independence for Random Variables

A fundamental concept in probability theory is the notion of independence. Roughly speaking, the random variables $X_1, X_2, \ldots, X_n$ are mutually independent when the behaviour of one of these random variables does not influence the others. Formally, the random variables $X_1, X_2, \ldots, X_n$ are mutually independent if, and only if, all the random events built with these random variables are independent. It can be shown that the random variables $X_1, X_2, \ldots, X_n$ are independent if, and only if,

$$F_{\boldsymbol{X}}(\boldsymbol{x}) = \prod_{i=1}^{n} F_{X_i}(x_i) \text{ holds for all } \boldsymbol{x} \in \mathbb{R}^n.$$

In other words, the joint distribution function of a random vector $\boldsymbol{X}$ with independent components is thus the product of the marginal distribution functions.

1.3 Poisson Distribution

1.3.1 Counting Random Variables

A discrete random variable X assumes only a finite (or countable) number of values. The most important subclass of nonnegative discrete random variables is the integer case, where each observation (outcome) is an integer (typically, the number of claims reported to the company). More precisely, a counting random variable N is valued in $\{0, 1, 2, \ldots\}$. Its stochastic behaviour is characterized by the set of probabilities $\{p_k,\ k = 0, 1, \ldots\}$ assigned to the nonnegative integers, where $p_k = \Pr[N = k]$. The (discrete) distribution of N associates with each possible integer value $k = 0, 1, 2, \ldots$ the probability p_k that it will be the observed value. The distribution must satisfy the two conditions:

$$p_k \geq 0 \text{ for all } k \text{ and } \sum_{k=0}^{+\infty} p_k = 1,$$

i.e. the probabilities are all nonnegative real numbers lying between zero (impossibility) and unity (certainty), and their sum must be unity because it is certain that one or other of the values will be observed.

1.3.2 Probability Mass Function

In discrete distribution theory the p_ks are regarded as values of a mathematical function, i.e.

$$p_k = p(k|\boldsymbol{\xi}),\ k = 0, 1, 2, \ldots, \tag{1.4}$$

where $p(\cdot|\boldsymbol{\xi})$ is a known function depending on a set of parameters $\boldsymbol{\xi}$. The function $p(\cdot|\boldsymbol{\xi})$ defined in (1.4) is usually called the probability mass function. Different functional forms lead to different discrete distributions. This is a parametric model.

The distribution function $F_N : \mathbb{R} \to [0, 1]$ of N gives for any real threshold x, the probability for N to be smaller than or equal to x. The distribution function F_N of N is related to the probability mass function via

$$F_N(x) = \sum_{k=0}^{\lfloor x \rfloor} p_k,\ x \in \mathbb{R}^+,$$

where p_k is given by Expression (1.4) and where $\lfloor x \rfloor$ denotes the largest integer n such that $n \leq x$ (it is thus the integer part of x). Considering (1.4), F_N also depends on $\boldsymbol{\xi}$.

1.3.3 Moments

There are various useful and important quantities associated with a probability distribution. They may be used to summarize features of the distribution. The most familiar and widely used are the moments, particularly the mean

$$\mathbb{E}[N] = \sum_{k=0}^{+\infty} k p_k,$$

which is given by the sum of the products of all the possible outcomes multiplied by their probability, and the variance

$$\mathbb{V}[N] = \mathbb{E}[(N - \mathbb{E}[N])^2] = \sum_{k=0}^{+\infty} \left(k - \mathbb{E}[N]\right)^2 p_k,$$

which is given by the sum of the products of the squared differences between all the outcomes and the mean, multiplied by their probability. Expanding the squared difference in the definition of the variance, it is easily seen that the variance can be reformulated as

$$\mathbb{V}[N] = \mathbb{E}\left[N^2 - 2N\mathbb{E}[N] + (\mathbb{E}[N])^2\right] = \mathbb{E}[N^2] - (\mathbb{E}[N])^2,$$

which provides a convenient way to compute the variance as the difference between the second moment $\mathbb{E}[N^2]$ and the square $(\mathbb{E}[N])^2$ of the first moment $\mathbb{E}[N]$. The mean and the variance are commonly denoted as μ and σ^2, respectively. Considering (1.4), both $\mathbb{E}[N]$ and $\mathbb{V}[N]$ are functions of $\boldsymbol{\xi}$, that is,

$$\mathbb{E}[N] = \mu(\boldsymbol{\xi}) \text{ and } \mathbb{V}[N] = \sigma^2(\boldsymbol{\xi}).$$

The mean is used as a measure of the location of the distribution: it is an average of the possible outcomes 0, 1, ... weighted by the corresponding probabilities $p_0, p_1, \ldots$ The variance is widely used as a measure of the spread of the distribution: it is a weighted average of the squared distances between the outcomes 0, 1, ... and the expected value $\mathbb{E}[N]$. Recall that $\mathbb{E}[\cdot]$ is a linear operator. From the properties of $\mathbb{E}[\cdot]$, it is easily seen that the variance $\mathbb{V}[\cdot]$ is shift-invariant and additive for independent random variables.

The degree of asymmetry of the distribution of a random variable N is measured by its skewness, denoted as $\gamma[N]$. The skewness is the third central moment of N, normalized by its variance raised to the power 3/2 (in order to get a number without unit). Precisely, the skewness of N is given by

$$\gamma[N] = \frac{\mathbb{E}[(N - \mathbb{E}[N])^3]}{(\mathbb{V}[N])^{3/2}}.$$

For any random variable N with a symmetric distribution the skewness $\gamma[N]$ is zero. Positively skewed distributions tend to concentrate most of the probability mass on small values, but the remaining probability is stretched over a long range of larger values.

There are other related sets of constants, such as the cumulants, the factorial moments, the factorial cumulants, etc., which may be more convenient to use in some circumstances. For details about these constants, we refer the reader, e.g., to Johnson *et al.* (1992).

1.3.4 Probability Generating Function

In principle all the theoretical properties of the distribution can be derived from the probability mass function. There are, however, several other functions from which exactly the same information can be derived. This is because the functions are all one-to-one transformations of each other, so each characterizes the distribution. One particularly useful function is the probability generating function, which is defined as

$$\varphi_N(z) = \mathbb{E}[z^N] = \sum_{k=0}^{+\infty} p_k z^k, \quad 0 < z < 1. \tag{1.5}$$

When p_k is given by (1.4), $\varphi_N(\cdot)$ depends on $\boldsymbol{\xi}$.

If any function that is known to be a probability generating function is expanded as a power series in z, then the coefficient of z^k must be p_k for the corresponding distribution. An alternative way of obtaining the probabilities is by repeated differentiation of φ_N with respect to z. Specifically,

$$\varphi_N(0) = \Pr[N = 0] \text{ and } \frac{d^k}{dt^k}\varphi_N(z)\bigg|_{z=0} = k!\Pr[N = k] \text{ for } k = 1, 2, \ldots$$

1.3.5 Convolution Product

A key feature of probability generating functions is related to the computation of sums of independent discrete random variables. Considering two independent counting random variables N_1 and N_2, their sum is again a counting random variable and thus possesses a probability mass function as well as a probability generating function.

The probability mass function of $N_1 + N_2$ is obtained as follows: We obviously have that

$$\Pr[N_1 + N_2 = k] = \sum_{j=0}^{k} \Pr[N_1 = j, N_2 = k - j]$$

for any integer k. Since N_1 and N_2 are independent, their joint probability mass function factors to the product of the univariate probability mass functions. This simply comes from

$$\begin{aligned}
\Pr[N_1 = j, N_2 = k - j] &= \Pr[N_1 \le j, N_2 \le k - j] - \Pr[N_1 \le j, N_2 \le k - j - 1] \\
&\quad - \Pr[N_1 \le j - 1, N_2 \le k - j] + \Pr[N_1 \le j - 1, N_2 \le k - j - 1] \\
&= \Pr[N_1 \le j]\big(\Pr[N_2 \le k - j] - \Pr[N_2 \le k - j - 1]\big) \\
&\quad - \Pr[N_1 \le j - 1]\big(\Pr[N_2 \le k - j] - \Pr[N_2 \le k - j - 1]\big) \\
&= \Pr[N_2 = k - j]\big(\Pr[N_1 \le j] - \Pr[N_1 \le j - 1]\big) \\
&= \Pr[N_1 = j]\Pr[N_2 = k - j].
\end{aligned}$$

The probability mass function of $N_1 + N_2$ can thus be obtained from the discrete convolution formula

$$\Pr[N_1 + N_2 = k] = \sum_{j=0}^{k} \Pr[N_1 = j]\Pr[N_2 = k - j], \quad k = 0, 1, \ldots$$

For large values of k, a direct application of the discrete convolution formula can be rather time-consuming.

The probability generating function of $N_1 + N_2$ is easily obtained from

$$\varphi_{N_1+N_2}(z) = \mathbb{E}[z^{N_1+N_2}] = \mathbb{E}[z^{N_1}]\mathbb{E}[z^{N_2}] = \varphi_{N_1}(z)\varphi_{N_2}(z)$$

since the mutual independence of N_1 and N_2 ensures that

$$\begin{aligned}
\mathbb{E}[z^{N_1+N_2}] &= \sum_{k_1=0}^{+\infty}\sum_{k_2=0}^{+\infty} z^{k_1+k_2}\Pr[N_1 = k_1]\Pr[N_2 = k_2] \\
&= \sum_{k_1=0}^{+\infty} z^{k_1}\Pr[N_1 = k_1]\sum_{k_2=0}^{+\infty} z^{k_2}\Pr[N_2 = k_2] \\
&= \mathbb{E}[z^{N_1}]\mathbb{E}[z^{N_2}].
\end{aligned}$$

Summing random variables thus corresponds to a convolution product for probability mass functions and to regular products for probability generating functions. An expansion of $\varphi_{N_1}\varphi_{N_2}(\cdot)$ as a series in powers of z then gives the probability mass function of $N_1 + N_2$, usually in a much easier way than computing the convolution product of the probability mass functions of N_1 and N_2.

1.3.6 From the Binomial to the Poisson Distribution

Bernoulli Distribution

The Bernoulli distribution is an extremely simple and basic distribution. It arises from what is known as a Bernoulli trial: a single observation is taken where the outcome is dichotomous, e.g., success or failure, alive or dead, male or female, 0 or 1. The probability of success is q. The probability of failure is $1 - q$.

If N is Bernoulli distributed with success probability q, which is denoted as $N \sim \mathcal{B}er(q)$, we have

$$p(k|q) = \begin{cases} 1-q \text{ if } k=0 \\ q \text{ if } k=1 \\ 0 \text{ otherwise.} \end{cases}$$

There is thus just one parameter: the success probability q. The mean is

$$\mathbb{E}[N] = 0 \times (1-q) + 1 \times q = q \tag{1.6}$$

and the variance is

$$\mathbb{V}[N] = \mathbb{E}[N^2] - q^2 = q - q^2 = q(1-q). \tag{1.7}$$

The probability generating function is

$$\varphi_N(z) = (1-q) \times z^0 + q \times z^1 = 1 - q + qz. \tag{1.8}$$

It is easily seen that $\varphi_N(0) = p(0|q)$ and $\varphi'_N(0|q) = p(1|q)$, as it should be.

Binomial Distribution

The Binomial distribution describes the outcome of a sequence of n independent Bernoulli trials, each with the same probability q of success. The probability that success is the outcome in exactly k of the trials is

$$p(k|n,q) = \binom{n}{k} q^k (1-q)^{n-k}, \quad k = 0, 1, \ldots, n, \tag{1.9}$$

and 0 otherwise. Formula (1.9) defines the Binomial distribution. There are now two parameters: the number of trials n (also called the exponent, or size) and the success probability q. Henceforth, we write $N \sim \mathcal{B}in(n,q)$ to indicate that N is Binomially distributed, with size n and success probability q.

Moments of the Binomial Distribution

The mean of $N \sim \mathcal{B}in(n,q)$ is

$$\begin{aligned}\mathbb{E}[N] &= \sum_{k=1}^{n} \frac{n!}{(k-1)!(n-k)!} q^k (1-q)^{n-k} \\ &= nq \sum_{k=1}^{n} \Pr[M = k-1] = nq \end{aligned} \tag{1.10}$$

where $M \sim \mathcal{B}in(n-1,q)$. Furthermore, with M as defined before,

$$\begin{aligned}\mathbb{E}[N^2] &= \sum_{k=1}^{n} \frac{n!}{(k-1)!(n-k)!} k q^k (1-q)^{n-k} \\ &= nq \sum_{k=1}^{n} k \Pr[M = k-1] \\ &= n(n-1)q^2 + nq \end{aligned}$$

so that the variance is

$$\mathbb{V}[N] = \mathbb{E}[N^2] - (nq)^2 = nq(1-q). \tag{1.11}$$

We immediately observe that the Binomial distribution is underdispersed, i.e. its variance is smaller than its mean : $\mathbb{V}[N] = nq(1-q) \leq \mathbb{E}[N] = nq$.

Probability Generating Function and Closure under Convolution for the Binomial Distribution

The probability generating function of $N \sim \mathcal{B}in(n,q)$ is

$$\varphi_N(z) = \sum_{k=0}^{n} \binom{n}{k} (qz)^k (1-q)^{n-k} = (1-q+qz)^n. \tag{1.12}$$

Note that Expression (1.12) is the Bernoulli probability generating function (1.8), raised to the nth power. This was expected since the Binomial random variable N can be seen as the sum of n independent Bernoulli random variables with equal success probability q. This also explains why (1.10) is equal to n times (1.6) and why (1.11) is equal to n times (1.7) (in the latter case, since the variance is additive for independent random variables).

From (1.12), we also see that having independent random variables $N_1 \sim \mathcal{B}in(n_1, q)$ and $N_2 \sim \mathcal{B}in(n_2, q)$, the sum $N_1 + N_2$ is still Binomially distributed. This comes from the fact that the probability generating function of $N_1 + N_2$ is

$$\varphi_{N_1+N_2}(z) = \varphi_{N_1}(z)\varphi_{N_2}(z) = (1-q+qz)^{n_1+n_2}$$

so that $N_1 + N_2 \sim \mathcal{B}in(n_1 + n_2, q)$. Note that this is not the case if the success probabilities differ.

Limiting Form of the Binomial Distribution

When n becomes large, (1.9) may be approximated by a Normal distribution according to the De Moivre–Laplace theorem. The approximation is much improved when a continuity correction is applied. The Poisson distribution can be obtained as a limiting case of the Binomial distribution when n tends to infinity together with q becoming very small. Specifically, let us assume that $N_n \sim \mathcal{B}in(n, \lambda/n)$ and let n tend to $+\infty$. The probability mass at 0 then becomes

$$\Pr[N_n = 0] = \left(1 - \frac{\lambda}{n}\right)^n \to \exp(-\lambda), \text{ as } n \to +\infty.$$

To get the probability masses on the positive integers, let us compute the ratio

$$\frac{\Pr[N_n = k+1]}{\Pr[N_n = k]} = \frac{\frac{n-k}{k+1}\frac{\lambda}{n}}{1 - \frac{\lambda}{n}} \to \frac{\lambda}{k+1}, \text{ as } n \to +\infty,$$

from which we conclude

$$\lim_{n\to+\infty} \Pr[N_n = k] = \exp(-\lambda)\frac{\lambda^k}{k!}, \quad k = 0, 1, 2, \ldots$$

Poisson Distribution

The Poisson random variable takes its values in $\{0, 1, \ldots\}$ and has probability mass function

$$p(k|\lambda) = \exp(-\lambda)\frac{\lambda^k}{k!}, \; k = 0, 1, \ldots. \tag{1.13}$$

Having a counting random variable N, we denote as $N \sim \mathcal{P}oi(\lambda)$ the fact that N is Poisson distributed with parameter λ. The Poisson distribution occupies a central position in discrete distribution theory analogous to that occupied by the Normal distribution in continuous distribution theory. It also has many practical applications.

The Poisson distribution describes events that occur randomly and independently in space or time. A classic example in physics is the number of radioactive particles recorded by a Geiger counter in a fixed time interval. This property of the Poisson distribution means that it can act as a reference standard when deviations from pure randomness are suspected. Although the Poisson distribution is often called the law of small numbers, there is no need for $\lambda = nq$ to be small. It is the largeness of n and the smallness of $q = \lambda/n$ that are important. However most of the data sets analysed in the literature show a small frequency. This will be the case with motor data sets in insurance applications.

Moments of the Poisson Distribution

If $N \sim \mathcal{P}oi(\lambda)$, then its expected value is given by

$$\begin{aligned}\mathbb{E}[N] &= \sum_{k=1}^{+\infty} k \exp(-\lambda)\frac{\lambda^k}{k!} \\ &= \exp(-\lambda)\sum_{k=0}^{+\infty}\frac{\lambda^{k+1}}{k!} = \lambda.\end{aligned} \tag{1.14}$$

Moreover,

$$\begin{aligned}\mathbb{E}[N^2] &= \sum_{k=1}^{+\infty} k^2 \exp(-\lambda)\frac{\lambda^k}{k!} \\ &= \exp(-\lambda)\sum_{k=0}^{+\infty}(k+1)\frac{\lambda^{k+1}}{k!} = \lambda+\lambda^2,\end{aligned}$$

so that the variance of N is equal to

$$\mathbb{V}[N] = \mathbb{E}[N^2] - \lambda^2 = \lambda. \tag{1.15}$$

Considering Expressions (1.14) and (1.15), we see that both the mean and the variance of the Poisson distribution are equal to λ, a phenomenon termed as equidispersion.

The skewness of $N \sim \mathcal{P}oi(\lambda)$ is

$$\gamma[N] = \frac{1}{\sqrt{\lambda}}. \tag{1.16}$$

Clearly, $\gamma[N]$ decreases with λ. For small values of λ the distribution is very skewed (asymmetric) but as λ increases it becomes less skewed and is nearly symmetric by $\lambda = 15$.

Probability Generating Function and Closure Under Convolution for the Poisson Distribution

The probability generating function of the Poisson distribution has a very simple form. Coming back to the Equation (1.5) defining φ_N and replacing the p_ks with their expression (1.13) gives

$$\varphi_N(z) = \sum_{k=0}^{+\infty} \exp(-\lambda)\frac{(\lambda z)^k}{k!} = \exp\big(\lambda(z-1)\big). \tag{1.17}$$

This shows that the Poisson distribution is closed under convolution. Having independent random variables $N_1 \sim \mathcal{P}oi(\lambda_1)$ and $N_2 \sim \mathcal{P}oi(\lambda_2)$, the probability generating function of the sum $N_1 + N_2$ is

$$\varphi_{N_1+N_2}(z) = \varphi_{N_1}(z)\varphi_{N_2}(z) = \exp\big(\lambda_1(z-1)\big)\exp\big(\lambda_2(z-1)\big) = \exp\big((\lambda_1+\lambda_2)(z-1)\big)$$

so that $N_1 + N_2 \sim \mathcal{P}oi(\lambda_1 + \lambda_2)$.

The sum of two independent Poisson distributed random variables is also Poisson distributed, with parameter equal to the sum of the Poisson parameters. This property obviously extends to any number of terms, and the Poisson distribution is said to be closed under convolution (i.e. the convolution of Poisson distributions is still Poisson).

1.3.7 Poisson Process

Definition

Recall that a stochastic process is a collection of random variables $\{N(t), t \in \mathcal{T}\}$ indexed by a real-valued parameter t taking values in the index set $\mathcal{T}$. Usually, $\mathcal{T}$ represents a set of observation times. In this book, we will be interested in continuous-time stochastic processes where $\mathcal{T} = \mathbb{R}^+$.

A stochastic process $\{N(t), t \geq 0\}$ is said to be a counting process if $t \mapsto N(t)$ is right-continuous and $N(t) - N(t-)$ is 0 or 1. Intuitively speaking, $N(t)$ represents the total number of events that have occurred up to time t. Such a process enjoys the following properties: (i) $N(t) \geq 0$, (ii) $N(t)$ is integer valued, (iii) if $s < t$, then $N(s) \leq N(t)$, and (iv) for $s < t$, $N(t) - N(s)$ equals the number of events that have occurred in the interval $(s, t]$. By convention, $N(0) = 0$. The counting process $\{N(t),\ t \geq 0\}$ is said to be a Poisson process with rate $\lambda > 0$ if

(i) the process has stationary increments, that is,

$$\Pr[N(t+\Delta) - N(t) = k] = \Pr[N(s+\Delta) - N(s) = k]$$

for any integer k, instants $s \leq t$ and increment $\Delta > 0$.

(ii) the process has independent increments, that is, for any integer $k > 0$ and instants $0 \leq t_0 < t_1 < t_2 < \cdots < t_k$, the random variables $N(t_1) - N(t_0)$, $N(t_2) - N(t_1), \ldots,$ $N(t_k) - N(t_{k-1})$ are mutually independent.

(iii) and

$$\Pr[N(h) = k] = \begin{cases} 1 - \lambda h + o(h) \text{ if } k = 0 \\ \lambda h + o(h) \text{ if } k = 1 \\ o(h) \text{ if } k \geq 2 \end{cases}$$

where $o(h)$ is a function of h that tends to 0 faster than the identity, that is, $\lim_{h \searrow 0} o(h)/h = 0$. Intuitively speaking, $o(h)$ is negligible when h becomes sufficiently small.

Assumption (i) implies that the probability of causing an accident is assumed to be the same for every day during any given period (we thus neglect the fact that meteorological conditions, safety conditions and other factors could vary over time). According to assumption (ii), the occurrence of an accident at one point in time is independent of all accidents that might have occurred before: reporting one accident does not increase nor decrease the probability of causing an accident in the future. This supports the fact that traffic accidents occur randomly in time. Assumption (iii) indicates that the probability that the policyholder files two or more claims in a sufficiently small time interval is negligible when compared to the probability that he reports zero or only one claim.

Link with the Poisson Distribution

The Poisson process is intimately linked to the Poisson distribution, as precisely stated in the next result.

Property 1.1 *For any Poisson process, the number of events in any interval of length t is Poisson distributed with mean λt, that is, for all $s, t \geq 0$,*

$$\Pr[N(t+s) - N(s) = n] = \exp(-\lambda t)\frac{(\lambda t)^n}{n!}, \quad n = 0, 1, 2, \ldots$$

Proof Without loss of generality, we only have to prove that $N(t) \sim \mathcal{Poi}(\lambda t)$. For any integer k, let us denote $p_k(t) = \Pr[N(t) = k]$, $t \geq 0$. The announced result for $k = 0$ comes from

$$\begin{aligned} p_0(t+\Delta t) &= \Pr[N(t) = 0 \text{ and } N(t+\Delta t) - N(t) = 0] \\ &= \Pr[N(t) = 0] \Pr[N(t+\Delta t) - N(t) = 0] \\ &= p_0(t) p_0(\Delta t) \\ &= p_0(t)\big(1 - \lambda \Delta t + o(\Delta t)\big), \end{aligned}$$

where the joint probability factors into two terms since the increments of a Poisson process are independent random variables. This gives

$$\frac{p_0(t+\Delta t) - p_0(t)}{\Delta t} = -\lambda p_0(t) + \frac{o(\Delta t)}{\Delta t} p_0(t).$$

Taking the limit for $\Delta t \searrow 0$ yields

$$\frac{d}{dt} p_0(t) = -\lambda p_0(t).$$

This differential equation with the initial condition $p_0(0) = 1$ admits the solution

$$p_0(t) = \exp(-\lambda t), \tag{1.18}$$

which is in fact the $\mathcal{Poi}(\lambda t)$ probability mass function evaluated at the origin.

For $k \geq 1$, let us write

$$\begin{aligned} p_k(t+\Delta t) &= \Pr[N(t+\Delta t) = k] \\ &= \Pr[N(t+\Delta t) = k | N(t) = k] \Pr[N(t) = k] \\ &\quad + \Pr[N(t+\Delta t) = k | N(t) = k-1] \Pr[N(t) = k-1] \\ &\quad + \sum_{j=2}^{k} \Pr[N(t+\Delta t) = k | N(t) = k-j] \Pr[N(t) = k-j] \end{aligned}$$

$$\begin{aligned}
&= \Pr[N(t+\Delta t) - N(t) = 0]\Pr[N(t) = k] \\
&\quad + \Pr[N(t+\Delta t) - N(t) = 1]\Pr[N(t) = k-1] \\
&\quad + \sum_{j=2}^{k} \Pr[N(t+\Delta t) - N(t) = j]\Pr[N(t) = k-j].
\end{aligned}$$

Since the increments of a Poisson process are independent random variables, we can write

$$\begin{aligned}
p_k(t+\Delta t) &= p_0(\Delta t)p_k(t) + p_1(\Delta t)p_{k-1}(t) + \sum_{j=2}^{k} p_j(\Delta t)p_{k-j}(t) \\
&= (1 - \lambda\Delta t)p_k(t) + \lambda\Delta t p_{k-1}(t) + o(\Delta t).
\end{aligned}$$

This gives

$$\frac{p_k(t+\Delta t) - p_k(t)}{\Delta t} = \lambda(p_{k-1}(t) - p_k(t)) + \frac{o(\Delta t)}{\Delta t}.$$

Taking the limit for $\Delta t \searrow 0$ yields as above

$$\frac{d}{dt}p_k(t) = \lambda(p_{k-1}(t) - p_k(t)), \quad k \geq 1. \tag{1.19}$$

Multiplying by z^k each of the equation (1.19), and summing over k gives

$$\sum_{k=0}^{+\infty} \left(\frac{d}{dt}p_k(t)\right) z^k = \lambda z \sum_{k=0}^{+\infty} p_k(t)z^k - \lambda \sum_{k=0}^{+\infty} p_k(t)z^k. \tag{1.20}$$

Denoting as φ_t the probability generating function of $N(t)$, equation (1.20) becomes

$$\frac{\partial}{\partial t}\varphi_t(z) = \lambda(z-1)\varphi_t(z). \tag{1.21}$$

With the condition $\varphi_0(z) = 1$, Equation (1.21) has solution

$$\varphi_t(z) = \exp(\lambda t(z-1)),$$

where we recognize the $\mathcal{P}oi(\lambda t)$ probability generating function (1.17). This ends the proof. □

When the hypotheses behind a Poisson process are verified, the number $N(1)$ of claims hitting a policy during a period of length 1 is Poisson distributed with parameter λ. So, a counting process $\{N(t),\ t \geq 0\}$, starting from $N(0) = 0$, is a Poisson process with rate $\lambda > 0$ if

(i) The process has independent increments
(ii) The number of events in any interval of length t follows a Poisson distribution with mean λt (therefore it has stationary increments), i.e.

$$\Pr[N(t+s)-N(s)=k]=\exp(-\lambda t)\frac{(\lambda t)^k}{k!}, k=0,1,2,\ldots$$

Exposure-to-Risk

The Poisson process setting is useful when one wants to analyse policyholders that have been observed during periods of unequal lengths. Assume that the claims occur according to a Poisson process with rate λ. If the policyholder is covered by the company for a period of length d then the number N of claims reported to the company has probability mass function

$$\Pr[N=k]=\exp(-\lambda d)\frac{(\lambda d)^k}{k!}, \quad k=0,1,\ldots,$$

that is, $N\sim\mathcal{P}oi(\lambda d)$. In actuarial studies, d is referred to as the exposure-to-risk. We see that d simply multiplies the annual expected claim frequency λ in the Poisson model.

Time Between Accidents

The Poisson distribution arises for events occurring randomly and independently in time. Indeed, denote as $T_1, T_2, \ldots$ the times between two consecutive accidents. Assume further that these accidents occur according to a Poisson process with rate λ. Then, the T_ks are independent and identically distributed and

$$\Pr[T_k>t]=\Pr[T_1>t]=\Pr[N_t=0]=\exp(-\lambda t)$$

so that $T_1, T_2, \ldots$ have a common Negative Exponential distribution.

Note that in this case, the equality

$$\Pr[T_k>s+t|T_k>s]=\frac{\Pr[T_k>s+t]}{\Pr[T_k>s]}=\Pr[T_k>t]$$

holds for any s and $t\geq 0$. It is not difficult to see that this memoryless property is related to the fact that the increments of the process $\{N(t),\ t\geq 0\}$ are independent and stationary. Assuming that the claims occur according to a Poisson process is thus equivalent to assuming that the time between two consecutive claims has a Negative Exponential distribution.

Nonhomogeneous Poisson Process

A generalization of the Poisson process is obtained by letting the rate of the process vary with time. We then replace the constant rate λ by a function $t\mapsto\lambda(t)$ of time t and we define the nonhomogeneous Poisson process with rate $\lambda(\cdot)$. The Poisson process defined above (with a constant rate) is then termed as the homogeneous Poisson process. A counting process $\{N(t),\ t\geq 0\}$ starting from $N_0=0$ is said to be a nonhomogeneous Poisson process with rate $\lambda(\cdot)$, where $\lambda(t)\geq 0$ for all $t\geq 0$, if it satisfies the following conditions:

(i) The process $\{N(t),\ t \geq 0\}$ has independent increments, and
(ii)

$$\Pr[N(t+h)-N(t)=k]=\begin{cases}1-\lambda(t)h+o(h) \text{ if } k=0\\ \lambda(t)h+o(h) \text{ if } k=1\\ o(h) \text{ if } k\geq 2.\end{cases}$$

The only difference between the nonhomogeneous Poisson process and the homogeneous Poisson process is that the rate may vary with time, resulting in the loss of the stationary increment property.

For any nonhomogeneous Poisson process $\{N(t), t \geq 0\}$, the number of events in the interval $(s, t]$, $s \leq t$, is Poisson distributed with mean

$$m(s, t)=\int_s^t \lambda(u)du,$$

that is

$$\Pr[N(t)-N(s)=k]=\exp\big(-m(s, t)\big)\frac{\big(m(s, t)\big)^k}{k!}, k=0, 1, \ldots$$

In the homogeneous case, we obviously have $m(s, t)=(t-s)\lambda$.

1.4 Mixed Poisson Distributions

1.4.1 Expectations of General Random Variables

Mixed Poisson distributions involve expectations of Poisson probabilities with a random parameter. Therefore, we need to be able to compute expectations with respect to general distribution functions.

Continuous probability distributions are widely used in probability and statistics when the underlying random phenomenon is measured on a continuous scale. If the distribution function is a continuous function, the associated probability distribution is called a continuous distribution. Note that in this case,

$$\Pr[X=x]=\lim_{h\searrow 0}\Pr[x<X\leq x+h]=\lim_{h\searrow 0}F(x+h)-F(x)=0$$

for every real x. If X has a continuous probability distribution, then $\Pr[X=x]=0$ for any real x.

In this book, we often consider distribution functions possessing a derivative, $f(x)=dF(x)/dx$. The function f is called the probability density function. Then one can integrate the probability density function to recover the distribution function, that is,

$$F(x)=\int_{-\infty}^x f(y)dy.$$

One can calculate the probability of an event by integrating the probability density function; for example, if the event is an interval $[a, b]$, then

$$\Pr[a\leq X\leq b]=F(b)-F(a)=\int_a^b f(x)dx.$$

The interpretation of the probability density function is that

$$\Pr[x \leq X \leq x+h] \approx f(x)h \text{ for small } h > 0.$$

That is, the probability that a random variable, with an absolutely continuous probability distribution, takes a value in a small interval of length h is given by the probability density function times the length of the interval.

A general type of distribution function is a combination of the discrete and (absolutely) continuous cases, being continuous apart from a countable set of exception points $x_1, x_2, x_3, \ldots$ with positive probabilities of occurrence, causing jumps in the distribution function at these points. Such a distribution function F_X can be represented as

$$F_X(x) = (1-p)F_X^{(c)}(x) + pF_X^{(d)}(x), x \in \mathbb{R}, \tag{1.22}$$

for some $p \in [0, 1]$, where $F_X^{(c)}$ is a continuous distribution function and $F_X^{(d)}$ is a discrete distribution function with support $\{d_1, d_2, \ldots\}$.

Let us assume that F_X is of the form (1.22) with

$$pF_X^{(d)}(t) = \sum_{d_n \leq t} \Big(F_X(d_n) - F_X(d_n-)\Big) = \sum_{d_n \leq t} \Pr[X = d_n],$$

where $\{d_1, d_2, \ldots\}$ denotes the set of discontinuity points and

$$(1-p)F_X^{(c)}(t) = F_X(t) - pF_X^{(d)}(t) = \int_{-\infty}^{t} f_X^{(c)}(x)dx.$$

Then,

$$\begin{aligned} \mathbb{E}[X] &= \sum_{n\geq 1} d_n \Big(F_X(d_n) - F_X(d_n-)\Big) + \int_{-\infty}^{+\infty} x f_X^{(c)}(x)dx \\ &= \int_{-\infty}^{+\infty} x dF_X(x), \end{aligned} \tag{1.23}$$

where the differential of F_X, denoted as dF_X, is defined as

$$dF_X(x) = \begin{cases} F_X(d_n) - F_X(d_n-), & \text{if } x = d_n, \\ f_X^{(c)}(x)dx, & \text{otherwise.} \end{cases}$$

This unified notation allows us to avoid tedious repetitions of statements like 'the proof is given for continuous random variables; the discrete case is similar'. A very readable introduction to differentials and Riemann–Stieltjes integrals can be found in Carter & Van Brunt (2000).

1.4.2 Heterogeneity and Mixture Models

Definition

Mixture models are a discrete or continuous weighted combination of distributions aimed at representing a heterogeneous population comprised of several (two or more) distinct sub-populations. Such models are typically used when a heterogeneous population of sampling units consists of several sub-populations within each of which a relatively simpler model applies. The source of heterogeneity could be gender, age, geographical area, etc.

Discrete Mixtures

In order to define a mixture model mathematically, suppose the distribution of N can be represented by a probability mass function of the form

$$\Pr[N=k]=p(k|\boldsymbol{\psi})=q_1p_1(k|\xi_1)+\cdots+q_\nu p_\nu(k|\xi_\nu) \tag{1.24}$$

where $\boldsymbol{\psi}=(\boldsymbol{q}^T,\boldsymbol{\xi}^T)^T$, $\boldsymbol{q}^T=(q_1,\ldots,q_\nu)$, $\boldsymbol{\xi}^T=(\xi_1,\ldots,\xi_\nu)$. The model is usually referred to as a discrete (or finite) mixture model. Here ξ_j is a (vector) parameter characterizing the probability mass function $p_j(\cdot|\xi_j)$ and the q_js are mixing weights.

Example 1.1 A particular example of finite mixture is the zero-inflated distribution. It has been observed empirically that counting distributions often show excess of zeros against the Poisson distribution. In order to accommodate this feature, a combination of the original distribution $\{p_k,\ k=0,1,\ldots\}$ (be it Poisson or not) together with the degenerate distribution with all probability concentrated at the origin, gives a finite mixture with

$$\begin{aligned}\Pr[N=0]&=\omega+(1-\omega)p_0\\ \Pr[N=k]&=(1-\omega)p_k,\ k=1,2,\ldots\end{aligned}$$

A mixture of this kind is usually referred to as zero-inflated, zero-modified or as a distribution with added zeros.

Model (1.24) allows each component probability mass function to belong to a different parametric family. In most applications, a common parametric family is assumed and thus the mixture model takes the following form

$$p(k|\boldsymbol{\psi})=q_1p(k|\xi_1)+\cdots+q_\nu p(k|\xi_\nu) \tag{1.25}$$

which we assume to hold in the sequel. The mixing weight $\boldsymbol{q}$ can be regarded as a discrete probability function over $\boldsymbol{\xi}$, describing the variation in the choice of $\boldsymbol{\xi}$ across the population of interest.

This class of mixture models includes mixtures of Poisson distributions. Such a mixture is adequate to model count data (number of claims reported to an insurance company, number of accidents caused by an insured driver, etc.) where the components of the mixture are Poisson distributions with mean ξ_j. In that respect, (1.25) means that there are ν categories of policyholders, with annual expected claim frequencies $\xi_1,\xi_2,\ldots,\xi_\nu$, respectively. The proportion of the portfolio in the different categories is $q_1,q_2,\ldots,q_\nu$, respectively. Considering a given policyholder, the actuary does not know to which category he belongs, but the probability that he comes from category j is q_j. The probability mass function of the number of claims reported by this insured driver is thus a weighted average of the probability mass functions associated with the k categories.

Continuous Mixtures

Multiplying the number ν of categories in (1.25) often leads to a dramatic increase in the number of parameters (the q_js and the ξ_js). For large ν, it is therefore preferable to switch to a continuous mixture, where the sum in (1.25) is replaced with an integral with respect to some simple parametric continuous probability density function.

Specifically, if we allow ξ to be continuous with probability density function $g(\cdot)$, the finite mixture model suggested above is replaced by the probability mass function

$$p(k) = \int p(k|\xi)g(\xi)d\xi,$$

which is often referred to as a mixture distribution. When $g(\cdot)$ is modelled without parametric assumptions, the probability mass function $p(\cdot)$ is a semiparametric mixture model. Often in actuarial science, $g(\cdot)$ is taken from some parametric family, so that the resulting probability mass function is also parametric.

Mixed Poisson Model for the Number of Claims

The Poisson distribution often poorly fits observations made in a portfolio of policyholders. This is in fact due to the heterogeneity that is present in the portfolio: driving abilities vary from individual to individual. Therefore it is natural to multiply the mean frequency λ of the Poisson distribution by a positive random effect Θ. The frequency will vary within the portfolio according to the nonobservable random variable Θ. Obviously we will choose Θ such that $\mathbb{E}[\Theta] = 1$ because we want to obtain, on average, the frequency of the portfolio. Conditional on Θ, we then have

$$\Pr[N = k|\Theta = \theta] = p(k|\lambda\theta) = \exp(-\lambda\theta)\frac{(\lambda\theta)^k}{k!}, \quad k = 0, 1, \ldots, \tag{1.26}$$

where $p(\cdot|\lambda\theta)$ is the Poisson probability mass function, with mean $\lambda\theta$. The interpretation we give to this model is that not all policyholders in the portfolio have an identical frequency λ. Some of them have a higher frequency ($\lambda\theta$ with $\theta \geq 1$), others have a lower frequency ($\lambda\theta$ with $\theta \leq 1$). Thus we use a random effect to model this empirical observation.

The annual number of accidents caused by a randomly selected policyholder of the portfolio is then distributed according to a mixed Poisson law. In this case, the probability that a randomly selected policyholder reports k claims to the company is obtained by averaging the conditional probabilities (1.26) with respect to Θ. In general, Θ is not discrete nor continuous but of mixed type. The probability mass function associated with mixed Poisson models is defined as

$$\Pr[N = k] = \mathbb{E}\big[p(k|\lambda\Theta)\big] = \int_0^\infty \exp(-\lambda\theta)\frac{(\lambda\theta)^k}{k!}dF_\Theta(\theta) \tag{1.27}$$

where F_Θ denotes the distribution function of Θ, assumed to fulfill $F_\Theta(0) = 0$. The mixing distribution described by F_Θ represents the heterogeneity of the portfolio of interest; dF_Θ is often called the structure function. It is worth mentioning that the mixed Poisson model (1.27) is an accident-proneness model: it assumes that a policyholder's mean claim frequency does not change over time but allows some insured persons to have higher mean claim frequencies than others. We will say that N is mixed Poisson distributed with parameter λ and risk level Θ, denoted as $N \sim \mathcal{MP}oi(\lambda, \Theta)$ when it has probability mass function (1.27).

Remark 1.1 Note that a better notation would have been $\mathcal{MP}oi(\lambda, F_\Theta)$ instead of $\mathcal{MP}oi(\lambda, \Theta)$ since only the distribution function of Θ matters to define the associated Poisson mixture. We have nevertheless opted for $\mathcal{MP}oi(\lambda, \Theta)$ for simplicity.

Note that the condition $\mathbb{E}[\Theta] = 1$ ensures that when $N \sim \mathcal{MP}oi(\lambda, \Theta)$

$$\begin{aligned}\mathbb{E}[N] &= \int_0^\infty \sum_{k=0}^{+\infty} k \exp(-\lambda\theta) \frac{(\lambda\theta)^k}{k!} dF_\Theta(\theta) \\ &= \lambda \mathbb{E}[\Theta] = \lambda,\end{aligned}$$

or, more briefly,

$$\mathbb{E}[N] = \mathbb{E}\Big[\mathbb{E}[N|\Theta]\Big] = \mathbb{E}[\lambda\Theta] = \lambda. \tag{1.28}$$

In (1.28), $\mathbb{E}[\cdot|\Theta]$ means that we take an expected value considering Θ as a constant. We then average with respect to all the random components, except Θ. Consequently, $\mathbb{E}[\cdot|\Theta]$ is a function of Θ. Given Θ, N is Poisson distributed with mean $\lambda\Theta$ so that $\mathbb{E}[N|\Theta] = \lambda\Theta$. The mean of N is finally obtained by averaging $\mathbb{E}[N|\Theta]$ with respect to Θ. The expectation of N given in (1.28) is thus the same as the expectation of a $\mathcal{P}oi(\lambda)$ distributed random variable. Taking the heterogeneity into account by switching from the $\mathcal{P}oi(\lambda)$ to the $\mathcal{MP}oi(\lambda, \Theta)$ distribution has no effect on the expected claim number.

1.4.3 Mixed Poisson Process

The Poisson processes are suitable models for many real counting phenomena but they are insufficient in some cases because of the deterministic character of their intensity function. The doubly stochastic Poisson process (or Cox process) is a generalization of the Poisson process when the rate of occurrence is influenced by an external process such that the rate becomes a random process. So, the rate, instead of being constant (homogeneous Poisson process) or a deterministic function of time (nonhomogeneous Poisson process) becomes itself a stochastic process. The only restriction on the rate process is that it has to be nonnegative. Mixed Poisson distributions are linked to mixed Poisson processes in the same way that the Poisson distribution is associated with the standard Poisson process.

Specifically, let us assume that given $\Theta = \theta$, $\{N(t), t \geq 0\}$ is a homogeneous Poisson process with rate $\lambda\theta$. Then $\{N(t), \quad t \geq 0\}$ is a mixed Poisson process, and for any $s, t \geq 0$, the probability that k events occur during the time interval $(s, t]$ is

$$\begin{aligned}\Pr[N(t+s) - N(s) = k] &= \int_0^\infty \Pr[N(t+s) - N(s) = k|\Theta = \theta] dF_\Theta(\theta) \\ &= \int_0^\infty \exp(-\lambda\theta t) \frac{(\lambda\theta t)^k}{k!} dF_\Theta(\theta),\end{aligned}$$

that is, $N(t+s) - N(s) \sim \mathcal{MP}oi(\lambda t, \Theta)$. Note that, in contrast to the Poisson process, mixed Poisson processes have dependent increments. Hence, past number of claims reveal future number of claims in this setting (in contrast to the Poisson case).

1.4.4 Properties of Mixed Poisson Distributions

Moments and Overdispersion

If $N \sim \mathcal{MP}oi(\lambda, \Theta)$ then its second moment is

$$\mathbb{E}[N^2] = \int_0^{+\infty} (\lambda\theta + \lambda^2\theta^2) dF_\Theta(\theta) = \lambda\mathbb{E}[\Theta] + \lambda^2\mathbb{E}[\Theta^2]$$

so that

$$\begin{aligned} \mathbb{V}[N] &= \lambda\mathbb{E}[\Theta] + \lambda^2\mathbb{E}[\Theta^2] - \lambda^2\big(\mathbb{E}[\Theta]\big)^2 \\ &= \lambda + \lambda^2\mathbb{V}[\Theta]. \end{aligned} \tag{1.29}$$

It is then easily seen that the variance of N exceeds its mean, that is,

$$\mathbb{V}[N] = \lambda + \lambda^2\mathbb{V}[\Theta] \geq \lambda = \mathbb{E}[N]. \tag{1.30}$$

Therefore, unless Θ is degenerated in 1, we observe that mixed Poisson distributions are overdispersed: the variance exceeds the mean. The skewness can be expressed as

$$\gamma[N] = \frac{1}{(\mathbb{V}[N])^{3/2}} \left(3\mathbb{V}[N] - 2\mathbb{E}[N] + \frac{\gamma[\Theta]}{\sqrt{\mathbb{V}[\Theta]}} \frac{(\mathbb{V}[N] - \mathbb{E}[N])^2}{\mathbb{E}[N]} \right). \tag{1.31}$$

Shaked's Two Crossings Theorem

Recall that $\mathbb{E}[\phi(X)] \geq \phi(\mathbb{E}[X])$ for any random variable X and convex function ϕ. This inequality, known as the Jensen inequality, ensures that if $N \sim \mathcal{MP}oi(\lambda, \Theta)$ then

$$\Pr[N=0] = \int_0^{\infty} \exp(-\lambda\theta) dF_\Theta(\theta) \geq \exp\left(-\int_0^{\infty} \lambda\theta dF_\Theta(\theta) \right) = \exp(-\lambda),$$

showing that mixed Poisson distributions have an excess of zeros compared to Poisson distributions with the same mean. This is in line with empirical studies, where actuaries often observe more policyholders producing 0 claims than the number predicted by the Poisson model.

The following result that has been proved by Shaked (1980) reinforces this straightforward conclusion.

Property 1.2 *Let N be mixed Poisson distributed with mean $\mathbb{E}[N] = \lambda$. Then there exist two integers $0 \leq k_0 < k_1$ such that*

$$\Pr[N=k] \geq \exp(-\lambda)\frac{\lambda^k}{k!}, \quad k = 0, 1, \ldots, k_0,$$

$$\Pr[N=k] \leq \exp(-\lambda)\frac{\lambda^k}{k!}, \quad k = k_0+1, \ldots, k_1,$$

$$\Pr[N=k] \geq \exp(-\lambda)\frac{\lambda^k}{k!}, \quad k \geq k_1+1.$$

Shaked's Two Crossings Theorem tells us (i) that the mixed Poisson distribution has an excess of zeros compared to the Poisson distribution with the same mean and (ii) that the mixed Poisson distribution has a thicker right tail than the Poisson distribution with the same mean.

Probability Generating Function

The probability generating function of Poisson mixtures is closely related to the moment generating function of the underlying random effect. Moment generating functions are a widely used tool in many statistics texts, and also in actuarial mathematics. They serve to prove statements about convolutions of distributions, and also about limits. Recall that the moment generating function of the nonnegative random variable X, denoted as M_X, is given by

$$M_X(t) = \mathbb{E}[\exp(tX)], \quad t > 0.$$

It is interesting to mention that M_X characterizes the probability distribution of X, i.e. the information contained in F_X and M_X is equivalent.

If there exists $h > 0$ such that $M_X(t)$ exists and is finite for $0 < t < h$ then the Taylor expansion of the exponential function yields

$$M_X(t) = 1 + \sum_{n=1}^{+\infty} \frac{t^n}{n!} \mathbb{E}[X^n] \text{ for } 0 < t < h. \tag{1.32}$$

It is well known that if any moment of a distribution is infinite, the moment generating function does not exist. However, it is conceivable that there might exist distributions with moments of all orders and, yet, the moment generating function does not exist in any neighbourhood around 0. In fact, the LogNormal distribution is one such example.

Just as the probability generating function was interesting for analyzing sums of independent counting random variables, the moment generating function is a powerful tool to deal with sums of independent continuous random variables. Specifically, if X_1 and X_2 are independent random variables with respective moment generating functions $M_1(\cdot)$ and $M_2(\cdot)$, then the sum $X_1 + X_2$ has a moment generating function that is just the product $(M_1M_2)(\cdot)$ of $M_1(\cdot)$ and $M_2(\cdot)$.

Now, consider $N \sim \mathcal{MP}oi(\lambda, \Theta)$. We have

$$\begin{aligned}\varphi_N(z) &= \int_0^{+\infty} \exp(-\lambda\theta) \sum_{k=0}^{+\infty} \frac{(z\lambda\theta)^k}{k!} dF_\Theta(\theta) \\ &= \int_0^{+\infty} \exp\big(\lambda\theta(z-1)\big) dF_\Theta(\theta) \\ &= M_\Theta\big(\lambda(z-1)\big),\end{aligned}$$

or, equivalently,

$$M_\Theta(t) = \varphi_N\left(1 + \frac{t}{\lambda}\right). \tag{1.33}$$

From (1.33), we see that the knowledge of the mixed Poisson distribution $\mathcal{MP}oi(\lambda, \Theta)$ is equivalent to the knowledge of F_Θ. The mixed Poisson distributions are thus identifiable, that is, having $N_1 \sim \mathcal{MP}oi(\lambda, \Theta_1)$ and $N_2 \sim \mathcal{MP}oi(\lambda, \Theta_2)$ then N_1 and N_2 are identically distributed if, and only if, Θ_1 and Θ_2 are identically distributed.

1.4.5 Negative Binomial Distribution

Gamma Distribution

Recall that a random variable X is distributed according to the two-parameter Gamma distribution, which will henceforth be denoted as $X \sim \mathcal{G}am(\alpha, \beta)$, if its probability density function is given by

$$f(x) = \frac{x^{\alpha-1}\beta^\alpha \exp(-\beta x)}{\Gamma(\alpha)}, \quad x > 0. \tag{1.34}$$

Note that when $\alpha = 1$, the Gamma distribution reduces to the Negative Exponential one (which is denoted as $X \sim \mathcal{E}xp(\beta)$) with probability density function

$$f(x) = \beta \exp(-\beta x), \quad x > 0.$$

The distribution function F of X can be expressed in terms of the incomplete Gamma function. Specifically, if $X \sim \mathcal{G}am(\alpha, \beta)$, then $F(x) = \Gamma(\alpha, \beta x)$.

Probability Mass Function

The Negative Binomial distribution is a widely used alternative to the Poisson distribution for handling count data when the variance is appreciably greater than the mean (this condition, known as overdispersion, is frequently met in practice, as discussed above).

There are several models that lead to the Negative Binomial distribution. A classic example arises from the theory of accident proneness which was developed after GREENWOOD & YULE (1920). This theory assumes that the number of accidents suffered by an individual is Poisson distributed, but that the Poisson mean (interpreted as the individual's accident proneness) varies between individuals in the population under study. If the Poisson mean is assumed to be Gamma distributed, then the Negative Binomial is the resultant overall distribution of accidents per individual.

Specifically, completing (1.26)–(1.27) with $\Theta \sim \mathcal{G}am(a, a)$, that is, with probability density function

$$f_\Theta(\theta) = \frac{1}{\Gamma(a)} a^a \theta^{a-1} \exp(-a\theta), \quad \theta > 0, \tag{1.35}$$

yields the Negative Binomial probability mass function

$$\begin{aligned}
\Pr[N = k] &= \frac{(a+k-1)\cdots a}{k!} \left(\frac{a}{a+\lambda d}\right)^a \left(\frac{\lambda d}{a+\lambda d}\right)^k \\
&= \frac{\Gamma(a+k)}{\Gamma(a)k!} \left(\frac{a}{a+\lambda d}\right)^a \left(\frac{\lambda d}{a+\lambda d}\right)^k, \quad k = 0, 1, 2, \ldots
\end{aligned}$$

where λ is the annual expected claim number and d is the length of the observation period (the exposure-to-risk). The probability mass function can be expressed using the generalized binomial coefficient:

$$\Pr[N=k] = \frac{\Gamma(a+k)}{\Gamma(a)\Gamma(k+1)}\left(\frac{a}{a+\lambda d}\right)^{a}\left(\frac{\lambda d}{a+\lambda d}\right)^{k}$$
$$= \binom{a+k-1}{k}\left(\frac{a}{a+\lambda d}\right)^{a}\left(\frac{\lambda d}{a+\lambda d}\right)^{k}, \quad k=0,1,2,\ldots$$

Henceforth, we write $N \sim \mathcal{NB}in(a, \lambda d)$ to indicate that N obeys the Negative Binomial distribution with parameters a and λd. This model has been applied to retail purchasing, absenteeism, doctor's consultations, amongst many others.

Moments

If $X \sim \mathcal{G}am(\alpha, \beta)$, its mean is $\mathbb{E}[X] = \alpha/\beta$ and its variance is $\mathbb{V}[X] = \alpha/\beta^2$. If $N \sim \mathcal{NB}in(a, \lambda d)$ then the mean is $\mathbb{E}[N] = \lambda d$ and the variance is $\mathbb{V}[N] = \lambda d + (\lambda d)^2/a$ according to (1.29). It can be shown that $\gamma[\Theta]/\sqrt{\mathbb{V}[\Theta]} = 2$, in (1.31) for the Negative Binomial distribution.

Probability Generating Function

If $X \sim \mathcal{G}am(\alpha, \beta)$, its moment generating function is

$$M(t) = \left(1 - \frac{t}{\beta}\right)^{-\alpha} \quad \text{if } t < \beta. \tag{1.36}$$

The probability generating function of $N \sim \mathcal{NB}in(a, \lambda d)$ is

$$\varphi_N(z) = \left(\frac{a}{a - \lambda d(z-1)}\right)^{a}. \tag{1.37}$$

This result comes from (1.33) together with (1.36).

True and Apparent Contagion

Apparent contagion arises from the recognition that sampled individuals come from a heterogeneous population in which individuals have a constant but different propensity to experience accidents. A given individual may have a high (or low) propensity for accidents but occurrence of an accident does not make it more (or less) likely that another accident will occur. However, aggregation across heterogeneous individuals may generate a misleading statistical finding which suggests that occurrence of an accident increases the probability of another accident; the observed but persistent heterogeneity can be misinterpreted as due to a strong serial dependence.

True contagion refers to dependence between the occurrences of successive events. The occurrence of an event, such as an accident or illness, may change the probability of subsequent occurrences of similar events. True positive contagion implies that the occurrence of an event shortens the expected waiting time to the next occurrence of the event.

The alleged phenomenon of accident proneness can be interpreted in terms of true contagion as suggesting that an individual who has experienced an accident is more likely to experience another accident. In a longitudinal setting, actual and future outcomes are directly influenced by past values, and this causes a substantial change over time in the corresponding distribution.

Since with event count data we only observe the total number of events at the end of the period, contagion, like heterogeneity, is an unobserved, within-observation process. For research problems where both heterogeneity and contagion are plausible, the different underlying processes are not distinguishable with aggregate count data because they both lead to the same probability distribution for the counts. One can still use this distribution to derive fully efficient and consistent estimates, but this analysis will only be suggestive of the underlying process.

Poisson Limiting Form

The Negative Binomial distribution has a Poisson limiting form if $\mathbb{V}[\Theta]=\frac{1}{a}\to 0$. This result can be recovered from the sequence of the probability generating functions, noting that

$$\lim_{a\nearrow\infty}\left(\frac{a}{a-\lambda d(1-z)}\right)^{a}=\lim_{a\nearrow\infty}\left(1-\frac{\lambda d}{a}(1-z)\right)^{-a}=\exp(-\lambda d(1-z))$$

that is seen to converge to the probability generating function of the Poisson distribution with parameter λd.

Derivation as a Compound Poisson Distribution

A different type of heterogeneity occurs when there is clustering. If it is assumed that the number of clusters is Poisson distributed, but the number of individuals in a cluster is distributed according to the Logarithmic distribution, then the overall distribution is Negative Binomial. In an actuarial context, this amounts to recognizing that several vehicles can be involved in the same accident, each of the insured drivers filing a claim. Therefore, a single accident may generate several claims. If the number of claims per accident follows a Logarithmic distribution, and the number of accidents over the time interval of interest follows a Poisson distribution, then the total number of claims for the time interval can be modelled with the Negative Binomial distribution.

Let us formally establish this result. Recall that the random variable M has a Logarithmic distribution if

$$\Pr[M=k]=\frac{\theta^k}{-k\ln(1-\theta)},\qquad k=1,2,\ldots$$

where $0<\theta<1$. The probability generating function of M is given by

$$\varphi_M(z)=\frac{\ln(1-\theta z)}{\ln(1-\theta)}.$$

Now, let $M_1, M_2, \ldots$ be a sequence of independent Logarithmically distributed random variables with the same parameter θ, and let $K\sim\mathcal{P}oi(\mu)$. Define

$$N=M_1+\cdots+M_K.$$

The random variable N just defined has a compound Poisson distribution. The probability generating function of a compound distribution is given by

$$\begin{aligned}
\varphi_N(z) &= \mathbb{E}[z^{M_1+\cdots+M_K}] \\
&= \sum_{k=0}^{+\infty} \Pr[K=k]\mathbb{E}[z^{M_1+\cdots+M_k}] \\
&= \sum_{k=0}^{+\infty} \Pr[K=k]\big(\varphi_M(z)\big)^k \\
&= \varphi_K\big(\varphi_M(z)\big).
\end{aligned} \tag{1.38}$$

Note that formula (1.38) is true in general for compound distributions. Replacing φ_K and φ_M with their expressions gives the probability generating function of N

$$\begin{aligned}
\varphi_N(z) &= \exp\Big(-\mu(1-\varphi_M(z))\Big) \\
&= \left(\frac{1-\theta}{1-\theta z}\right)^{-\mu/\ln(1-\theta)}.
\end{aligned}$$

It can be checked that the probability generating function φ_N corresponds to the probability generating function (1.37) of a Negative Binomial distribution with $d=1$, $a=-\mu/\ln(1-\theta)$ and $\lambda=-\theta\mu/((1-\theta)\ln(1-\theta))$.

1.4.6 Poisson-Inverse Gaussian Distribution

There is no reason to restrict ourselves to the Gamma distribution for Θ, except perhaps mathematical convenience. In fact, any distribution with support in the half positive real line is a candidate to model the stochastic behaviour of Θ. Here, we discuss the Inverse Gaussian distribution.

Inverse Gaussian Distribution

The Inverse Gaussian distribution is an ideal candidate for modelling positive, right-skewed data. Recall that a random variable X is distributed according to the Inverse Gaussian distribution, which will be henceforth denoted as $X \sim \mathcal{IGau}(\mu, \beta)$, if its probability density function is given by

$$f(x) = \frac{\mu}{\sqrt{2\pi\beta x^3}} \exp\left(-\frac{1}{2\beta x}(x-\mu)^2\right), \quad x>0. \tag{1.39}$$

If $X \sim \mathcal{IGau}(\mu, \beta)$ then the mean is $\mathbb{E}[X]=\mu$ and the variance is $\mathbb{V}[X]=\mu\beta$. The moment generating function is given by

$$M(t) = \int_0^{+\infty} \frac{\mu}{\sqrt{2\pi\beta x^3}} \exp\left(-\frac{1}{2\beta x}(x-\mu)^2 + tx\right) dx$$

$$= \exp\left(\frac{\mu}{\beta}\right)\int_0^{+\infty}\frac{\mu}{\sqrt{2\pi\beta x^3}}\exp\left(-\frac{1}{2\beta x}\left(x^2(1-2\beta t)+\mu^2\right)\right)dx.$$

Making the change of variable $\xi = x\sqrt{1-2\beta t}$ yields

$$M(t) = \exp\left(\frac{\mu}{\beta}\right)\int_0^{+\infty}\frac{\mu}{\sqrt{2\pi\frac{\beta}{\sqrt{1-2\beta t}}\xi^3}}\exp\left(-\frac{1}{2\frac{\beta}{\sqrt{1-2\beta t}}\xi}(\xi^2+\mu^2)\right)d\xi$$

$$= \exp\left(\frac{\mu}{\beta}(1-\sqrt{1-2\beta t})\right). \tag{1.40}$$

For the last three decades, the Inverse Gaussian distribution has gained attention in describing and analyzing right-skewed data. The main appeal of Inverse Gaussian models lies in the fact that they can accommodate a variety of shapes, from highly skewed to almost Normal. Moreover, they share many elegant and convenient properties with Gaussian models. In applied probability, the Inverse Gaussian distribution arises as the distribution of the first passage time to an absorbing barrier located at a unit distance from the origin in a Wiener process.

Poisson-Inverse Gaussian Distribution
Let us now complete (1.26)–(1.27) with $\Theta \sim \mathcal{IGau}(1,\tau)$, that is,

$$f_\Theta(\theta) = \frac{1}{\sqrt{2\pi\tau\theta^3}}\exp\left(-\frac{1}{2\tau\theta}(\theta-1)^2\right), \quad \theta > 0. \tag{1.41}$$

The probability mass function is given by

$$\Pr[N=k] = \int_0^{\infty}\exp(-\lambda d\theta)\frac{(\lambda d\theta)^k}{k!}\frac{1}{\sqrt{2\pi\tau\theta^3}}\exp\left(-\frac{1}{2\tau\theta}(\theta-1)^2\right)d\theta. \tag{1.42}$$

The probability mass function can be expressed using modified Bessel functions of the second kind. Bessel functions have some useful properties that can be used to compute the Poisson-Inverse Gaussian probabilities and to find the maximum likelihood estimators, for instance.

Moments and Probability Generating Function
Considering (1.28) and (1.29), we have

$$\mathbb{E}[N] = \lambda \text{ and } \mathbb{V}[N] = \lambda + \lambda^2\tau.$$

It can be shown that $\gamma[\Theta]/\sqrt{\mathbb{V}[\Theta]} = 3$ in (1.31) for the Poisson-Inverse Gaussian distribution. Therefore the skewness of a Poisson-Inverse Gaussian distribution exceeds the skewness of the Negative Binomial distribution having the same mean and the same variance.

Setting $\mu = 1$ and $\beta = \tau$, the probability generating function of N can be obtained from (1.33) together with (1.40), which gives

$$\varphi_N(z) = \exp\left(\frac{1}{\tau}\left(1-\sqrt{1-2\tau\lambda(z-1)}\right)\right).$$

Computation of the Probability Mass Function

The probability mass at the origin is

$$\varphi_N(0) = \Pr[N=0] = \exp\left(\frac{1}{\tau}\left(1-\sqrt{1+2\tau\lambda}\right)\right).$$

Now, taking the derivatives of φ_N with respect to t, and evaluating it at 0 gives the probability mass function for positive integers. Specifically,

$$\begin{aligned}\varphi_N'(0) &= \Pr[N=1] \\ &= \frac{\lambda}{\sqrt{1-2\tau\lambda(z-1)}}\varphi_N(z)\Big|_{z=0} \\ &= \frac{\lambda}{\sqrt{1+2\tau\lambda}}\Pr[N=0]\end{aligned}$$

and

$$\begin{aligned}\varphi_N'(0) &= 2\Pr[N=2] \\ &= \frac{\lambda^2\tau}{\left(1-2\tau\lambda(z-1)\right)^{3/2}}\varphi_N(z)\Big|_{z=0} + \frac{\lambda}{\sqrt{1-2\tau\lambda(z-1)}}\varphi_N'(z)\Big|_{z=0} \\ &= \frac{\lambda^2\tau}{\left(1+2\tau\lambda\right)^{3/2}}\Pr[N=0] + \frac{\lambda}{\sqrt{1+2\tau\lambda}}\Pr[N=1] \\ &= \frac{\lambda^2\tau}{\left(1+2\tau\lambda\right)^{3/2}}\frac{\sqrt{1+2\tau\lambda}}{\lambda}\Pr[N=1] + \left(\frac{\lambda}{\sqrt{1+2\tau\lambda}}\right)^2\Pr[N=0] \\ &= \frac{\lambda\tau}{1+2\tau\lambda}\Pr[N=1] + \frac{\lambda^2}{1+2\tau\lambda}\Pr[N=0].\end{aligned}$$

In general, we have the following recursive formula

$$\begin{aligned}\Pr[N=n] =& \frac{2\lambda\tau}{1+2\lambda\tau}\left(1-\frac{3}{2n}\right)\Pr[N=n-1] \\ &+ \frac{\lambda^2}{(1+2\lambda\tau)n(n-1)}\Pr[N=n-2] \qquad (1.43)\end{aligned}$$

valid for $n = 2, 3, 4, \ldots$, which allows us to compute the probability mass function of the Poisson-Inverse Gaussian distribution. The formal proof of (1.43) is based on properties of the modified Bessel function.

1.4.7 Poisson-LogNormal Distribution

In addition to the Gamma and Inverse Gaussian distributions to model Θ, the LogNormal distribution is often used in biostatistical studies.

LogNormal Distribution

Recall that a random variable X is Normally distributed with mean μ and variance σ^2, denoted as $X \sim \mathcal{N}or(\mu, \sigma^2)$, if its distribution function is

$$F(x) = \Phi\left(\frac{x-\mu}{\sigma}\right),$$

where

$$\Phi(x) = \frac{1}{\sqrt{2\pi}} \int_{-\infty}^{x} \exp(-y^2/2)\, dy. \tag{1.44}$$

Now, a random variable X is LogNormally distributed with parameters μ and σ (notation $X \sim \mathcal{LN}or(\mu, \sigma^2)$) if $\ln X$ is Normally distributed with mean μ and variance σ^2, that is, if its probability density function is given by

$$f(x) = \frac{1}{\sqrt{2\pi} x \sigma} \exp\left(-\frac{1}{2\sigma^2}(\ln x - \mu)^2\right), \quad x > 0.$$

If $X \sim \mathcal{LN}or(\mu, \sigma^2)$, then its mean is

$$\mathbb{E}[X] = \exp\left(\mu + \frac{\sigma^2}{2}\right),$$

and its variance

$$\mathbb{V}[X] = \exp\left(2\mu + \sigma^2\right)\left(\exp(\sigma^2) - 1\right).$$

Poisson-LogNormal Distribution

Taking $\mu = -\sigma^2/2$ (to ensure that $\mathbb{E}[\Theta] = 1$), the probability density function of $\Theta \sim \mathcal{LN}or(-\sigma^2/2, \sigma^2)$ is

$$f_\Theta(\theta) = \frac{1}{\theta \sigma \sqrt{2\pi}} \exp\left(-\frac{(\ln \theta + \sigma^2/2)^2}{2\sigma^2}\right), \quad \theta > 0. \tag{1.45}$$

The probability mass function of the Poisson-LogNormal distribution is given by

$$\Pr[N = k] = \frac{1}{\sigma\sqrt{2\pi}} \frac{(\lambda d)^k}{k!} \int_0^{\infty} \exp(-\lambda d \theta) \theta^{k-1} \exp\left(-\frac{(\ln \theta + \sigma^2/2)^2}{2\sigma^2}\right) d\theta.$$

Coming back to (1.28) and (1.29), we easily see that

$$\mathbb{E}[N] = \lambda \text{ and } \mathbb{V}[N] = \lambda + \lambda^2\left(\exp(\sigma^2) - 1\right).$$

It can be shown that $\gamma[\Theta]/\sqrt{\mathbb{V}[\Theta]} = 2 + \exp(\sigma^2)$ in (1.31) for the Poisson-LogNormal distribution. Therefore the skewness of a Poisson-LogNormal distribution exceeds the skewness of the Poisson-Inverse Gaussian distribution having the same mean and the same variance.

1.5 Statistical Inference for Discrete Distributions

1.5.1 Maximum Likelihood Estimators

Maximum likelihood is a method of estimation and inference for parametric models. The maximum likelihood estimator is the value of the parameter (or parameter vector) that makes the observed data most likely to have occurred given the data generating process assumed to have produced the variable of interest.

The likelihood of a sample of observations is defined as the joint density of the data, with the parameters taken as variable and the data as fixed (multiplied by any arbitrary constant or function of the data but not of the parameters). Specifically, let $N_1, N_2, \ldots, N_n$ be a set of independent and identically distributed outcomes with probability mass function $p(\cdot|\boldsymbol{\xi})$ where $\boldsymbol{\xi}$ is a vector of parameters. The likelihood function is the probability of observing the data $N_1 = k_1, \ldots, N_n = k_n$, that is,

$$\mathcal{L}(\boldsymbol{\xi}) = \prod_{i=1}^{n} p(k_i|\boldsymbol{\xi}).$$

The key idea for estimation in likelihood problems is that the most reasonable estimate is the value of the parameter vector that would make the observed data most likely to occur. The implicit assumption is of course that the data at hand are reliable. More formally we seek a value of $\boldsymbol{\xi}$ that maximizes $\mathcal{L}(\boldsymbol{\xi})$. The maximum likelihood estimator of $\boldsymbol{\xi}$ is the random variable $\widehat{\boldsymbol{\xi}}$ for which the likelihood is maximum, that is

$$\mathcal{L}(\widehat{\boldsymbol{\xi}}) \geq \mathcal{L}(\boldsymbol{\xi}) \text{ for all } \boldsymbol{\xi}.$$

It is usually simpler mathematically to find the maximum of the logarithm of the likelihood

$$L(\boldsymbol{\xi}) = \ln \mathcal{L}(\boldsymbol{\xi}) = \sum_{i=1}^{n} \ln p(k_i|\boldsymbol{\xi})$$

rather than the likelihood itself. The function $L(\boldsymbol{\xi})$ is usually referred to as the log-likelihood. Because the logarithm is a monotonic transformation, the log-likelihood will be maximized at the same parameter value that maximizes the likelihood (although the shape of the log-likelihood is different from that of the likelihood).

When working with counting variables, it is often easier to use the observed frequencies

$$f_k = \#\{\text{observations equal to } k\}, \quad k = 0, 1, 2, \ldots \tag{1.46}$$

In other words, f_k is the number of times that the value k has been observed in the sample. Denoting the largest observation as

$$k_{\max} = \max_{i=1,\ldots,n} k_i,$$

the log-likelihood becomes

$$L(\boldsymbol{\xi}) = \sum_{k=0}^{k_{\max}} f_k \ln p(k|\boldsymbol{\xi}).$$

We may solve analytically for the maximum likelihood estimator. To maximize any regular function, we find the value of the parameters that makes the first derivatives of the function with respect to the parameters equal to zero. The first derivative of the log-likelihood is called Fisher's score, and is denoted by

$$U_j(\boldsymbol{\xi}) = \frac{\partial}{\partial \xi_j} L(\boldsymbol{\xi}), \; j = 1, \ldots, \dim(\boldsymbol{\xi}). \tag{1.47}$$

Then one can find the maximum likelihood estimator by setting the score to zero, i.e. by solving the system of equations

$$U_j(\boldsymbol{\xi}) = 0, \; j = 1, \ldots, \dim(\boldsymbol{\xi}).$$

We also check the second derivatives to ensure that this is a maximum.

Example 1.2 Assume that policyholder i has been observed during a period d_i and produced k_i claims. Assuming that the annual number of claims is Poisson distributed with mean λ (here, λ is the annual expected claim frequency, so that policyholder i is expected to produce $d_i\lambda$ claims during the observation period, under the condition of the Poisson process); the log-likelihood is

$$\begin{aligned} L(\lambda) &= \sum_{i=1}^{n} \ln\left(\exp(-\lambda d_i)\frac{(\lambda d_i)^{k_i}}{k_i!}\right) \\ &= -\lambda \sum_{i=1}^{n} d_i + \sum_{i=1}^{n} k_i\big(\ln\lambda + \ln d_i\big) - \sum_{i=1}^{n} \ln k_i! \\ &= -\lambda \sum_{i=1}^{n} d_i + \ln\lambda \sum_{i=1}^{n} k_i + \text{constant}. \end{aligned}$$

Setting the first derivative of $L(\lambda)$ with respect to λ equal to 0 gives

$$-\sum_{i=1}^{n} d_i + \frac{1}{\lambda}\sum_{i=1}^{n} k_i = 0 \Rightarrow \widehat{\lambda} = \frac{\sum_{i=1}^{n} k_i}{\sum_{i=1}^{n} d_i}.$$

The second derivative is

$$-\frac{1}{\lambda^2}\sum_{i=1}^{n} k_i < 0 \text{ for any } \lambda$$

so that $\widehat{\lambda}$ indeed corresponds to the maximum of $L(\lambda)$. The estimated annual expected claim frequency $\widehat{\lambda}$ is thus obtained as the ratio of the total number of claims to the total exposure-to-risk. It is important to note here that the total number of claims is not divided by the number of policies, because of unequal risk exposure.

Example 1.3 Assume, as above, that policyholder i has been observed during a period d_i and produced k_i claims. Assuming that the annual number of claims filed by policyholder i is Negative Binomially distributed with mean λd_i, the log-likelihood is

$$L(a, \lambda) = \sum_{i=1}^{n} \sum_{j=0}^{k_i-1} \ln(a+j) + na \ln a - \sum_{i=1}^{n}(a+k_i) \ln(a+\lambda d_i) + \ln \lambda \sum_{i=1}^{n} k_i + \text{constant}.$$

The maximum likelihood estimators for a and λ solve

$$\frac{\partial}{\partial a} L(a, \lambda) = \sum_{i=1}^{n} \sum_{j=0}^{k_i-1} \frac{1}{a+j} + n \ln a + n - \sum_{i=1}^{n} \ln(a+\lambda d_i) - \sum_{i=1}^{n} \frac{a+k_i}{a+\lambda d_i} = 0$$

$$\frac{\partial}{\partial \lambda} L(a, \lambda) = -\sum_{i=1}^{n} d_i \frac{a+k_i}{a+\lambda d_i} + \frac{1}{\lambda} \sum_{i=1}^{n} k_i = 0.$$

These equations do not possess explicit solutions, and must be solved numerically. A convenient choice is to use the Newton–Raphson algorithm (see Section 1.5.3). Initial values for the parameters are obtained by the method of moments. Specifically, the moment estimator of λ is simply

$$\hat{\lambda} = \frac{\sum_{i=1}^{n} k_i}{\sum_{i=1}^{n} d_i},$$

which is the maximum likelihood estimate of λ in the homogeneous Poisson case. For the variance, we start from $\mathbb{V}[N_i] = \mathbb{E}[N_i] + (\lambda d_i)^2 \tau$ where $\tau = \mathbb{V}[\Theta_i]$. The empirical analogue is given by

$$\hat{\tau} = \frac{\sum_{i=1}^{n} \Big((k_i - \lambda d_i)^2 - \lambda d_i \Big)}{\sum_{i=1}^{n} (\lambda d_i)^2}$$

from which we easily deduce an estimator for a in the Negative Binomial case.

1.5.2 Properties of the Maximum Likelihood Estimators

Maximum likelihood estimators enjoy a number of convenient properties that are discussed below. It is important to note that these are asymptotic properties, i.e. properties that hold only as the sample size becomes infinitely large. It is impossible to say in general at what point a sample is large enough for these properties to apply, but the majority of actuarial applications involve large data sets so that actuaries generally trust in the large sample properties of the maximum likelihood estimators.

Consistency

First, maximum likelihood estimators are consistent. There are several definitions of consistency, but an intuitive version is that as the sample size gets large the estimator is increasingly likely to fall within a small region around the true value of the parameter. This is called convergence in probability and is defined more formally as follows: A consistent estimator T_j for some parameter ξ_j computed from a sample of size n is one for which

$$\lim_{n\nearrow\infty} \Pr[|T_j - \xi_j| \geq c] = 0$$

for all positive c. This will henceforth be denoted as $T_j \to_{\text{proba}} \xi_j$ as $n \nearrow +\infty$. A consistent estimator is thus an estimator that converges to the population parameter as the sample size goes to infinity. Consistency is an asymptotic property.

Asymptotic Normality

Any estimator will vary across repeated samples. We must be able to calculate this variability in order to express our uncertainty about a parameter value and to make statistical inferences about the parameters. This variability is measured by the variance-covariance matrix of the estimators. This matrix provides the variances for each parameter on the main diagonal while the off-diagonal elements estimate the covariances between all pairs of parameters.

The asymptotic variance-covariance matrix $\boldsymbol{\Sigma}_{\widehat{\boldsymbol{\xi}}}$ for maximum likelihood estimators $\widehat{\boldsymbol{\xi}}$ is the inverse of what is called the Fisher information matrix $\mathcal{I}(\boldsymbol{\xi})$. Element ij of $\mathcal{I}(\boldsymbol{\xi})$ is given by

$$\begin{aligned} -\mathbb{E}\left[\frac{\partial^2}{\partial\xi_i\partial\xi_j}\ln L(\boldsymbol{\xi})\right] &= -n\mathbb{E}\left[\frac{\partial^2}{\partial\xi_i\partial\xi_j}\ln p(N_1|\boldsymbol{\xi})\right] \\ &= n\mathbb{E}\left[\frac{\partial}{\partial\xi_i}\ln p(N_1|\boldsymbol{\xi})\frac{\partial}{\partial\xi_j}\ln p(N_1|\boldsymbol{\xi})\right] \\ &= n\sum_{k=0}^{\infty} p(k|\boldsymbol{\xi})\frac{\partial}{\partial\xi_i}\ln p(k|\boldsymbol{\xi})\frac{\partial}{\partial\xi_j}\ln p(k|\boldsymbol{\xi}). \end{aligned}$$

Thus, $\boldsymbol{\Sigma}_{\widehat{\boldsymbol{\xi}}} = \big(\mathcal{I}(\boldsymbol{\xi})\big)^{-1}$.

An insight into why this makes sense is that the second derivatives measure the rate of change in the first derivatives, which in turn determines the value of the maximum likelihood estimate. If the first derivatives are changing rapidly near the maximum, then the peak of the likelihood is sharply defined and the maximum is easy to see. In this case, the second derivatives will be large and their inverse small, indicating a small variance of the estimated parameters. If on the other hand the second derivatives are small, then the likelihood function is relatively flat near the maximum and so the parameters are less precisely estimated. The inverse of the second derivatives will produce a large value for the variance of the estimates, indicating low precision of the estimates.

The distribution of $\widehat{\boldsymbol{\xi}}$ is usually difficult to obtain. Therefore we resort to the following asymptotic theory: Under mild regularity conditions (including that the true value of the parameter $\boldsymbol{\xi}$ must be interior to the parameter space, that the log-likelihood function must be thrice differentiable, and that the third derivatives must be bounded) that are usually fulfilled, the maximum likelihood estimator $\widehat{\boldsymbol{\xi}}$ has approximately in large samples a multivariate Normal distribution with mean equal to the true parameter and variance-covariance matrix given by the inverse of the information matrix.

Recall that having a $n \times n$ positive definite matrix $\boldsymbol{M}$ and a real vector $\boldsymbol{\mu}$, the random vector $\boldsymbol{X} = (X_1, X_2, \ldots, X_n)^T$ is said to have the multivariate Normal distribution with mean $\boldsymbol{\mu}$ and variance-covariance matrix $\boldsymbol{M}$ if its probability density function is of the form

$$f_{\boldsymbol{X}}(\boldsymbol{x}) = \frac{1}{\sqrt{(2\pi)^n \det(\boldsymbol{M})}} \exp\left(-\frac{1}{2}(\boldsymbol{x}-\boldsymbol{\mu})^T \boldsymbol{M}^{-1}(\boldsymbol{x}-\boldsymbol{\mu})\right), \quad \boldsymbol{x} \in \mathbb{R}^n. \tag{1.48}$$

Henceforth, we indicate that the random vector $\boldsymbol{X}$ has the multivariate Normal distribution with probability density function (1.48) as $\boldsymbol{X} \sim \mathcal{N}or(\boldsymbol{\mu}, \boldsymbol{M})$. A convenient characterization of the multivariate Normal distribution is as follows: $\boldsymbol{X} \sim \mathcal{N}or(\boldsymbol{\mu}, \boldsymbol{M})$ if, and only if, any random variable of the form $\sum_{i=1}^n \alpha_i X_i$ with $\boldsymbol{\alpha} \in \mathbb{R}^n$, has the univariate Normal distribution.

Coming back to the properties of the maximum likelihood estimator $\widehat{\boldsymbol{\xi}}$, we have that

$$\widehat{\boldsymbol{\xi}} \text{ is approximately } \mathcal{N}or(\boldsymbol{\xi}, \boldsymbol{\Sigma}_{\widehat{\boldsymbol{\xi}}}) \text{ distributed,} \tag{1.49}$$

that is, the distribution function of $\widehat{\boldsymbol{\xi}}$ can be approximated by integrating the Normal probability density function

$$f_{\widehat{\boldsymbol{\xi}}}(\boldsymbol{S}) = \frac{1}{\sqrt{(2\pi)^{\dim(\boldsymbol{\xi})} \det(\boldsymbol{\Sigma}_{\widehat{\boldsymbol{\xi}}})}} \exp\left(-\frac{1}{2}(\boldsymbol{S}-\boldsymbol{\xi})^T \boldsymbol{\Sigma}_{\widehat{\boldsymbol{\xi}}}^{-1}(\boldsymbol{S}-\boldsymbol{\xi})\right), \quad \boldsymbol{S} \in \mathbb{R}^{\dim(\boldsymbol{\xi})}.$$

Attribute (1.49) says that maximum likelihood estimators converge in distribution to a Normal with mean equal to the population value of the parameter and variance-covariance matrix equal to the inverse of the information matrix. This means that regardless of the distribution of the variable of interest the maximum likelihood estimator of the parameters will have a multivariate Normal distribution. Thus, a variable may be Poisson distributed, but the maximum likelihood estimate of the Poisson mean will be asymptotically Normally distributed, and likewise for any distribution. Note however that in the Poisson case, the exact distribution of the maximum likelihood estimator of the parameter derived in Example 1.2 can easily be derived from the stability of the Poisson family under convolution.

Invariance

A natural question is how the parameterization of a likelihood affects the resulting inference. Maximum likelihood has the property that any transformation of a parameter can be estimated by the same transformation of the maximum likelihood estimate of that parameter. This provides substantial flexibility in how we parameterize our models while guaranteeing that we will get the same result if we start with a different parameterization.

The invariance property can be stated formally as follows: If $\boldsymbol{\gamma} = t(\boldsymbol{\xi})$, where $t(\cdot)$ is a one-to-one transformation, then the maximum likelihood estimator of $\boldsymbol{\gamma}$ is $t(\widehat{\boldsymbol{\xi}})$. In particular, the maximum likelihood estimator of $p(k|\boldsymbol{\xi})$ is simply $p(k|\widehat{\boldsymbol{\xi}})$, that is,

$$\widehat{p(k|\boldsymbol{\xi})} = p(k|\widehat{\boldsymbol{\xi}}).$$

1.5.3 Computing the Maximum Likelihood Estimators with the Newton–Raphson Algorithm

Calculation of the maximum likelihood estimators often requires iterative procedures. Let $\boldsymbol{H}$ denote the Hessian (or matrix of second derivatives) of the log-likelihood function, with elements

$$\begin{aligned} H_{ij}(\boldsymbol{\xi}) &= \frac{\partial^2}{\partial\xi_i\partial\xi_j}L(\boldsymbol{\xi}) \\ &= \frac{\partial}{\partial\xi_i}U_j(\boldsymbol{\xi}) \\ &= -\sum_{k=0}^{k_{\max}} f_k \frac{\partial^2}{\partial\xi_i\partial\xi_j}\ln p(k|\boldsymbol{\xi}) \end{aligned} \tag{1.50}$$

for $i, j = 1, \ldots, \dim(\boldsymbol{\xi})$. For $\boldsymbol{\xi}^\star$ close enough to $\widehat{\boldsymbol{\xi}}$, a first-order Taylor expansion gives

$$0 = \boldsymbol{U}(\widehat{\boldsymbol{\xi}}) \approx \boldsymbol{U}(\boldsymbol{\xi}^\star) + \boldsymbol{H}(\boldsymbol{\xi}^\star)\Big(\widehat{\boldsymbol{\xi}} - \boldsymbol{\xi}^\star\Big)$$

yielding

$$\widehat{\boldsymbol{\xi}} \approx \boldsymbol{\xi}^\star - \boldsymbol{H}^{-1}(\boldsymbol{\xi}^\star)\boldsymbol{U}(\boldsymbol{\xi}^\star).$$

Starting from an appropriate initial value $\boldsymbol{\xi}^{(0)}$, the Newton–Raphson algorithm is based on the recurrence relation

$$\widehat{\boldsymbol{\xi}}^{(r+1)} = \widehat{\boldsymbol{\xi}}^{(r)} - \boldsymbol{H}^{-1}\left(\widehat{\boldsymbol{\xi}}^{(r)}\right)\boldsymbol{U}\left(\widehat{\boldsymbol{\xi}}^{(r)}\right). \tag{1.51}$$

This result provides the basis for an iterative approach for computing the maximum likelihood estimator known as the Newton–Raphson technique. Given a trial value, we use (1.51) to obtain an improved estimate and repeat the process until the elements of the vector of first derivatives are sufficiently close to zero.

This procedure tends to converge quickly if the log-likelihood is well-behaved in a neighbourhood of the maximum and if the starting value is reasonably close to the maximum likelihood estimator.

Remark 1.2 (Fisher Scoring) Noting that $\mathcal{J}(\boldsymbol{\xi}) = -\mathbb{E}[\boldsymbol{H}(\boldsymbol{\xi})]$, an alternative procedure is to replace minus the Hessian by its expected value, i.e. minus the Fisher information matrix. The resulting procedure takes as an improved estimate

$$\widehat{\boldsymbol{\xi}}^{(r+1)} \approx \widehat{\boldsymbol{\xi}}^{(r)} + \mathcal{J}^{-1}\left(\widehat{\boldsymbol{\xi}}^{(r)}\right)\boldsymbol{U}\left(\widehat{\boldsymbol{\xi}}^{(r)}\right)$$

and is known as Fisher Scoring.

1.5.4 Hypothesis Tests

Sample Distribution of Individual Parameters

Standard hypothesis tests about parameters in maximum likelihood models are handled quite easily, thanks to the asymptotic Normal distribution of the maximum likelihood estimator. Specifically, we use the fact that

$$\frac{\widehat{\xi}_j - \xi_j}{\sigma_{\widehat{\xi}_j}} \text{ is approximately } \mathcal{N}or(0, 1)$$

where the standard deviation $\sigma_{\widehat{\xi}_j}$ of $\widehat{\xi}_j$ is the square root of the jth diagonal element of

$$\boldsymbol{\Sigma}_{\widehat{\boldsymbol{\xi}}} = (\mathcal{I}(\boldsymbol{\xi}))^{-1}.$$

Such tests will be useful in Chapter 2 to select the relevant risk factors.

In practice, $\sigma_{\widehat{\xi}_j}$ often involves unknown parameters $\boldsymbol{\xi}$ so that it is estimated by the jth element $\widehat{\sigma}_{\widehat{\xi}_j}$ of

$$\widehat{\boldsymbol{\Sigma}}_{\widehat{\boldsymbol{\xi}}} = (\mathcal{I}(\widehat{\boldsymbol{\xi}}))^{-1}.$$

In such a case, $(\widehat{\xi}_j - \xi_j)/\widehat{\sigma}_{\widehat{\xi}_j}$ is approximately Student's distributed with $n-1$ degrees of freedom. This is the familiar z-score for a standard Normal variable developed in all introductory statistics classes. The Normality of maximum likelihood estimates means that our testing of hypotheses about the parameters is as simple as calculating the z-score and finding the associated p-value from a table or by calling a software function.

The hypothesis test is based on Student's t-distribution. However, because the maximum likelihood properties are all asymptotic we are unable to address the finite sample distribution. Asymptotically, the Student's t-distribution converges to the Normal as the degrees for freedom grow, so that using $\mathcal{N}or(0, 1)$ p-values in the maximum likelihood test is the same as the t-test as long as the number of cases is large enough. Specifically,

$$\frac{\widehat{\xi}_j - \xi_j}{\widehat{\sigma}_{\widehat{\xi}_j}} \text{ is approximately } \mathcal{N}or(0, 1) \tag{1.52}$$

if the sample size n is large enough.

In addition to the test of hypotheses about a single parameter, there are three classical tests that encompass hypotheses about sets of parameters as well as one parameter at a time: the likelihood ratio, Wald, and Lagrange multiplier tests. All are asymptotically equivalent, but they differ in the ease of implementation depending on the particular case. Here, we will present Wald, likelihood ratio and Vuong tests, as well as the Score test.

Likelihood Ratio Test

This test is based on a comparison of maximized likelihoods for nested models. Specifically, the null hypothesis H_0 corresponds to a constrained model with $\dim(\boldsymbol{\xi}) - j$ parameters, whereas the alternative H_1 corresponds to the full model with $\dim(\boldsymbol{\xi})$ parameters. Most of the time, the test is performed with $j = 1$, so that we compare the full model to a simpler one with one parameter less.

Let $\widetilde{\boldsymbol{\xi}}$ be the maximum likelihood estimator under H_0, and let $\widehat{\boldsymbol{\xi}}$ be the maximum likelihood estimator under H_1. The likelihood ratio test is based on the ratio of the likelihoods between a full and a restricted (or reduced) nested model with fewer parameters. The restricted model must be nested within (i.e., be a subset of) the full model. The likelihood ratio test statistic is

$$T = 2 \ln \frac{\mathcal{L}(\widehat{\boldsymbol{\xi}})}{\mathcal{L}(\widetilde{\boldsymbol{\xi}})} = 2\Big(L(\widehat{\boldsymbol{\xi}}) - L(\widetilde{\boldsymbol{\xi}})\Big).$$

The evidence against H_0 will be strong when T is large.

The Chi-square distribution plays a prominent role in likelihood ratio tests. Recall that the Gamma distribution with $\alpha = \nu/2$ and $\beta = 1/2$ for some positive integer ν is known as the Chi-square distribution with ν degrees of freedom (which is denoted as χ^2_ν), with associated probability density function

$$f(x) = \frac{x^{\nu/2-1} \exp(-x/2)}{\Gamma\left(\frac{\nu}{2}\right) 2^{\nu/2}}, \quad x > 0.$$

If $X \sim \chi^2_\nu$ then its mean is ν, and its variance 2ν. It is useful to recall that the χ^2_ν distribution is closely related to the Normal distribution. Specifically, the χ^2_ν arises as the distribution of the sum of ν independent squared $\mathcal{N}or(0, 1)$ random variables.

Under H_0, the test statistic T is approximatively Chi-square distributed with degrees of freedom equal to the number of parameters in the full model minus the number of parameters in the restricted model (that is, with j degrees of freedom) when the sample size n is sufficiently large (and additional mild regularity conditions are fulfilled). Note that the likelihood ratio test requires us to perform two maximum likelihood estimations, one under H_0 and another one under H_1. When the largest model H_1 is misspecified (that is, the data have not been generated by this probability model), the likelihood ratio statistic is no longer necessarily Chi-square distributed under H_0.

Unfortunately, there are cases where regularity conditions do not hold for T to be approximately χ^2_j distributed under H_0. In particular this happens when a constrained parameter is on the boundary of the parameter space, e.g., testing Poisson versus Negative Binomial. Here Poisson is a particular case of Negative Binomial when the latter has a parameter on its boundary space. In this case, the limiting distribution of the statistic T becomes a mixture of Chi-square distributions. We refer the reader to TITTERINGTON *ET AL.* (1985) for more details about these situations.

Wald Tests

The Wald test provides an alternative to the likelihood ratio test that requires the estimation of only the full model, not the restricted model. The logic of the Wald test is that if the restrictions are correct then the unrestricted parameter estimates should be close to the value hypothesized under the restricted model.

The Wald test is based on the distribution of a quadratic form of the weighted sum of squared Normal deviates, a form that is known to be Chi-square distributed. Specifically, using (1.49), we can test $H_0 : \boldsymbol{\xi} = \boldsymbol{\xi}_0$ versus $H_1 : \boldsymbol{\xi} \neq \boldsymbol{\xi}_0$ with the statistic

$$W = (\widehat{\boldsymbol{\xi}} - \boldsymbol{\xi}_0)\boldsymbol{\mathcal{I}}(\widehat{\boldsymbol{\xi}})(\widehat{\boldsymbol{\xi}} - \boldsymbol{\xi}_0)$$

which is approximately $\chi^2_{\dim(\boldsymbol{\xi})}$ distributed under H_0, in large samples. The test statistic can be interpreted as a measure of the distance between the maximum likelihood estimator $\widehat{\boldsymbol{\xi}}$ and the hypothesized value $\boldsymbol{\xi}_0$. The Wald test leads to the rejection of H_0 in favor of H_1 if $\widehat{\boldsymbol{\xi}}$ is too far from $\boldsymbol{\xi}_0$. Note that the Wald test suffers from the same problems as likelihood ratio tests when $\boldsymbol{\xi}_0$ lies on the boundary of the parametric space.

Sometimes the calculation of the expected information is difficult, and we may use the observed information instead.

Score Tests

Using the asymptotic theory, we have that $\boldsymbol{U}(\boldsymbol{\xi})$ is approximately $\mathcal{N}or(\boldsymbol{0}, \boldsymbol{\mathcal{I}}(\boldsymbol{\xi}))$ distributed. Therefore we can test $H_0: \boldsymbol{\xi} = \boldsymbol{\xi}_0$ versus $H_1: \boldsymbol{\xi} \neq \boldsymbol{\xi}_0$ with the statistic

$$Q = \boldsymbol{U}(\boldsymbol{\xi}_0)\boldsymbol{I}^{-1}(\boldsymbol{\xi}_0)\boldsymbol{U}(\boldsymbol{\xi}_0)$$

which is approximately $\chi^2_{\dim(\boldsymbol{\xi})}$ distributed under H_0, in large samples.

The advantage of the score test is that the calculation of the maximum likelihood estimator $\widehat{\boldsymbol{\xi}}$ is bypassed. Moreover, it remains applicable even if $\boldsymbol{\xi}_0$ lies on the boundary of the parametric space.

Vuong Test

The Chi-square approximation to the distribution of the likelihood ratio test statistic is valid only for testing restrictions on the parameters of a statistical model (i.e., H_0 and H_1 are nested hypotheses). With non-nested models, we cannot make use of likelihood ratio tests for model comparison. In this case, information criteria like AIC or (S)BIC are useful, as well as the Vuong test for non-nested models. Recall that the AIC (Akaike Information Criteria), is given by

$$\text{AIC} = -2L(\widehat{\boldsymbol{\xi}}) + 2\dim(\boldsymbol{\xi}),$$

and the BIC (Bayesian Information Criteria), is given by

$$\text{BIC} = -2L(\widehat{\boldsymbol{\xi}}) + \ln(n)\dim(\boldsymbol{\xi}).$$

Both criteria are equal to minus two times the maximum log-likelihood, penalized by a function of the number of observations and sample size.

VUONG (1989) proposed a likelihood ratio-based statistic for testing the null hypothesis that the competing models are equally close to the true data generating process against the alternative that one model is closer. Consider two statistical models given by the probability mass functions $p(\cdot|\boldsymbol{\xi})$ and $q(\cdot|\boldsymbol{\zeta})$ with $\dim(\boldsymbol{\xi}) = \dim(\boldsymbol{\zeta})$, and define the likelihood ratio statistic for the model $p(\cdot|\boldsymbol{\xi})$ against $q(\cdot|\boldsymbol{\zeta})$ as

$$LR(\widehat{\boldsymbol{\xi}}_n, \widehat{\boldsymbol{\zeta}}_n) = \sum_{k=0}^{k_{\max}} f_k \ln \frac{p(k|\widehat{\boldsymbol{\xi}}_n)}{q(k|\widehat{\boldsymbol{\zeta}}_n)}$$

where $\widehat{\boldsymbol{\xi}}_n$ and $\widehat{\boldsymbol{\zeta}}_n$ are the maximum likelihood estimators in each model based on the sample $\{k_1, \ldots, k_n\}$ and f_k is defined as in (1.46).

If both models are strictly non-nested (so that standard likelihood ratio tests do not apply) then under H_0

$$\frac{LR(\widehat{\boldsymbol{\xi}}_n, \widehat{\boldsymbol{\zeta}}_n)}{\widehat{\omega}_n\sqrt{n}} \text{ is approximately } \mathcal{N}or(0, 1) \text{ distributed,}$$

where

$$\widehat{\omega}_n = \frac{1}{n}\sum_{i=1}^{n}\left(\ln\frac{p(k_i|\widehat{\boldsymbol{\xi}}_n)}{q(k_i|\widehat{\boldsymbol{\zeta}}_n)}\right)^2 - \left(\frac{1}{n}\sum_{i=1}^{n}\ln\frac{p(k_i|\widehat{\boldsymbol{\xi}}_n)}{q(k_i|\widehat{\boldsymbol{\zeta}}_n)}\right)^2.$$

This provides a very simple test for model selection. Specifically, the actuary chooses a critical value z_ϵ from the $\mathcal{N}or(0, 1)$ distribution for some significance level ϵ. If the value of the test statistic is higher than z_ϵ then he rejects the null hypothesis that the models are equivalent in favour of $p(\cdot|\boldsymbol{\xi})$ being better than $q(\cdot|\boldsymbol{\zeta})$. If the test statistic is smaller than $-z_\epsilon$ then he rejects the null hypothesis in favour of $q(\cdot|\boldsymbol{\zeta})$ being better than $p(\cdot|\boldsymbol{\xi})$. Finally, if the test statistic is between $-z_\epsilon$ and z_ϵ then we cannot discriminate between the two competing models given the data.

The test statistic can be adjusted if the competing models do not have the same number of parameters, i.e. $\dim(\boldsymbol{\xi}) \neq \dim(\boldsymbol{\zeta})$ (which is not the case in this chapter).

1.6 Numerical Illustration

Here, we consider a Belgian motor third party liability insurance portfolio observed during the year 1997 (henceforth referred to as Portfolio A). The observed claim distribution is given in Table 1.1. A thorough description of this portfolio is deferred to Section 2.2.

We see from Table 1.1 that the total exposure is not equal to the number of policies due to the fact that some policies have not been in force during the full observation period (12 months). Some of them have been cancelled before the end of the observation period. Others have been written after the start of the observation period.

Let us now fit the observations to the Poisson, the Negative Binomial, the Poisson-Inverse Gaussian and the Poisson-LogNormal distributions. The results are summarized below:

Table 1.1 Observed claim distribution in Portfolio A.

Number of claims	Number of policies	Total exposure (in years)
0	12962	10 545.94
1	1369	1 187.13
2	157	134.66
3	14	11.08
4	3	2.52
Total	14505	11 881.35

Poisson the maximum likelihood estimate of the Poisson mean is $\widehat{\lambda}=0.1462$. The 95 % confidence interval for λ is (0.1395;0.1532). The log-likelihood of the Poisson model is -5579.339.

Negative Binomial the maximum likelihood estimate of the mean is $\widehat{\lambda}=0.1474$ and the dispersion parameter $\widehat{a}=0.889$. The variance of the random effect is estimated as $\widehat{\mathbb{V}[\Theta]}=1/\widehat{a}=1.1253$. The respective 95 % confidence intervals are (0.1402;0.1551) for λ and (0.8144;1.4361) for $\mathbb{V}[\Theta]$. The log-likelihood of the Negative Binomial model is -5534.36, which is better than the Poisson log-likelihood.

Poisson-Inverse Gaussian the maximum likelihood estimation of the mean is $\widehat{\lambda}=0.1475$, and the variance of the random effect is estimated to $\widehat{\mathbb{V}[\Theta]}=\widehat{\tau}=1.1770$. The respective 95 % confidence intervals are (0.1402;0.1552) for λ and (0.8258;1.5282) for $\mathbb{V}[\Theta]$. The log-likelihood of the Poisson-Inverse Gaussian model is -5534.28, which is better than the Poisson log-likelihood and almost equivalent to the Negative Binomial log-likelihood.

Poisson-LogNormal the maximum likelihood estimation of the mean is $\widehat{\lambda}=0.1476$, and $\widehat{\sigma}^2=0.7964$. The variance of the random effect is estimated to $\widehat{\mathbb{V}[\Theta]}=1.2175$. The respective 95 % confidence intervals are (0.1403;0.1553) for λ and (0.6170;0.9758) for σ^2. The log-likelihood of the Poisson-LogNormal model is -5534.44, which is better than the Poisson log-likelihood and almost equivalent to the Negative Binomial and Poisson-Inverse Gaussian log-likelihoods.

The results have been obtained with the help of the SAS® procedure GENMOD for the Poisson and Negative Binomial distributions (details will be given in the next chapter) and by a direct maximization of the log-likelihood using the Newton–Raphson procedure (coded in the SAS® environment IML) in the Poisson-Inverse Gaussian and Poisson-LogNormal cases.

It is interesting to note that the values of $\widehat{\lambda}$ are different in the Poisson and mixed Poisson models. If all the risk exposures were equal then these values would have been the same in all cases.

Let us now compare the Poisson fit to Portfolio A with each of the mixed Poisson fits. To this end, we use a likelihood ratio test, with an adjusted Chi-square approximation (since the Poisson case is at the border of the mixed Poisson family). Comparing the Poisson fit to any of the three mixed Poisson models leads to a clear rejection of the former one:

Poisson against Negative Binomial likelihood ratio test statistic of 89.95, with a p-value less than 10^{-10}.

Poisson against Poisson-Inverse Gaussian likelihood ratio test statistic of 90.12, with a p-value less than 10^{-10}.

Poisson against Poisson-LogNormal likelihood ratio test statistic of 89.80, with a p-value less than 10^{-10}.

The rejection of the Poisson assumption in favour of a mixed Poisson model is interpreted as a sign that the portfolio is composed of different types of drivers (i.e. the portfolio is heterogeneous).

Now, comparing the three mixed Poisson models with the Vuong test gives:

Negative Binomial against Poisson-Inverse Gaussian Vuong test statistic equal to -0.1086, with p-value 91.36 %.

Poisson-LogNormal against Negative Binomial Vuong test statistic equal to -0.0435, with p-value 96.54 %.

Poisson-LogNormal against Poisson-Inverse Gaussian Vuong test statistic equal to -0.3254, with p-value 74.48 %.

We cannot discriminate between the three competing models given the data, and they all fit the model equally well.

Remark 1.3 (Chi-Square Goodness-of-Fit Tests) In many papers appearing in the actuarial literature devoted to the analysis of claim numbers, as well as in most empirical studies, Chi-square goodness-of-fit tests are performed to select the optimal model. However, this approach neglects the exposures-to-risk (acting as if all the policies were in the portfolio for the whole year). We do not rely on Chi-square goodness-of-fit tests here since they do not allow for unequal risk exposures. Note that the vast majority of papers appearing in the actuarial literature disregard risk exposures (and proceed as if all the risk exposures were equal to 1).

1.7 Further Reading and Bibliographic Notes

1.7.1 Mixed Poisson Distributions

Mixed Poisson distributions are often used to model insurance claim numbers. The statistical analysis of counting random variables is described in much detail in JOHNSON *ET AL.* (1992). An excellent introduction to statistical inference is provided by FRANKLIN (2005). In the actuarial literature, KLUGMAN *ET AL.* (2004) provide a good account of statistical inference applied to insurance data sets, and in particular the analysis of counting random variables. Generating functions are described in KENDALL & STUART (1977) and FELLER (1971).

The axiomatic approach for which the (mixed) Poisson distribution is the counting distribution for a (mixed) Poisson process is presented in GRANDELL (1997). Mixture models are discussed in LINDSAY (1995). See also TITTERINGTON *ET AL.* (1985). Let us mention the work by KARLIS (2005), who applied the EM algorithm for maximum likelihood estimation in mixed Poisson models.

1.7.2 Survey of Empirical Studies Devoted to Claim Frequencies

KESTEMONT & PARIS (1985), using mixtures of Poisson processes, defined a large class of probability distributions and developed an efficient method for estimating their parameters. For the six data sets in GOSSIAUX & LEMAIRE (1981), they proposed a law depending on three parameters and they always obtained extremely good fits. As particular cases of the laws introduced in KESTEMONT & PARIS (1985), we find the ordinary Poisson distribution, the Poisson-Inverse Gaussian distribution, and the Negative Binomial distribution.

TREMBLAY (1992) used the Poisson-Inverse Gaussian distribution. WILLMOT (1987) compared the Poisson-Inverse Gaussian distribution to the Negative Binomial one and concluded that the fits are superior with the Poisson-Inverse Gaussian in all the six cases studied by GOSSIAUX & LEMAIRE (1981). See also the paper by BESSON & PARTRAT (1990). RUOHONEN (1987) considered a model for the claim number process. This model is a mixed Poisson process with a three-parameter Gamma distribution as the structure function and is compared with the two-parameter Gamma model giving the Negative Binomial distribution. He fitted his model to some data that can be found in the actuarial literature and the results were satisfying. PANJER (1987) proposed the Generalized Poisson-Pascal distribution (in fact, the Hofmann distribution), which includes three parameters, for the modelling of the number of automobile claims. The fits obtained were satisfactory, too. Note that the Pólya-Aeppli, the Poisson-Inverse Gaussian and the Negative Binomial are special cases of this distribution. CONSUL (1990) tried to fit the same six data sets by the Generalized Poisson distribution. Although the Generalized Poisson law is not rejected by a Chi-square test, the fits obtained by KESTEMONT & PARIS (1985), for instance, are always better. Furthermore, ELVERS (1991) reported that the Generalized Poisson distribution did not fit the data observed in a motor third party liability insurance portfolio very well. ISLAM & CONSUL (1992) suggested the Consul distribution as a probabilistic model for the distribution of the number of claims in automobile insurance. These authors approximated the chance mechanism which produces vehicle accidents by a branching process. They fit the model to the data sets used by PANJER (1987) and by GOSSIAUX & LEMAIRE (1981). Note that this model deals only with cars in accidents. Consequently, the zero-class has to be excluded. The fitted values seem good. However, this has to be considered cautiously, due to the comments by SHARIF & PANJER (1993) who found serious flaws embedded in the fitting of the Consul model. In particular, the very restricted parameter space and some theoretical problems in the derivation of the maximum likelihood estimators. They refer to other simple probability models, such as the Generalized Poisson-Pascal or the Poisson-Inverse Gaussian, whose fits were found quite satisfying.

DENUIT (1997) demonstrated that the Poisson-Goncharov distribution introduced by LEFÈVRE & PICARD (1996) provides an appropriate probability model to describe the annual number of claims incurred by an insured motorist. Estimation methods were proposed, and the Poisson-Goncharov distribution successfully fitted the six observed claims distributions in GOSSIAUX & LEMAIRE (1981), as well as other insurance data sets.

1.7.3 Semiparametric Approach

Traditionally, actuaries have assumed that the distribution of θ values among all drivers is well approximated by a parametric distribution, be it Gamma, Inverse Gaussian or LogNormal. However, there is no particular reason to believe that F_Θ belongs to some specified parametric family of distributions. Therefore, it seems interesting to resort to a nonparametric estimator for F_Θ.

There have been several attempts to estimate the structure function in a mixed Poisson model nonparametrically. Most of them include the annual claim frequency λ in the random effect, and thus work with $\widetilde{\Theta}=\lambda\Theta$. Assuming that $\widetilde{\Theta}$ has a finite number of support points and that its probability distribution is uniquely determined by its moments, TUCKER (1963) suitably precised by LINDSAY (1989a,b) suggested estimation of the support points of $\widetilde{\Theta}$ and the corresponding probability masses by solving a moment system. This estimator was then smoothed by CARRIÈRE (1993b) using a mixture of LogNormal distribution functions where all the parameters are estimated by a method of moments.

In a seminal paper, SIMAR (1976) gave a detailed description of the nonparametric maximum likelihood estimator of $F_{\tilde{\Theta}}$, as well as an algorithm for its computation. The nonparametric maximum likelihood estimator has a discrete distribution, and SIMAR (1976) obtained an upper bound for the size of its support.

WALHIN & PARIS (1999) showed that, although the nonparametric maximum likelihood estimator is powerful for evaluation of functionals of claim counts, it is not suitable for ratemaking, because it is purely discrete. For this reason, DENUIT & LAMBERT (2001) proposed a smoothed version of the nonparametric maximum likelihood estimator. This approach is somewhat similar to the route followed by CARRIÈRE (1993b), who proposed to smooth the Tucker-Lindsay moment estimator with a LogNormal kernel.

YOUNG (1997) applied nonparametric density estimation techniques to estimate $F_{\tilde{\Theta}}$. Because the actuary only observes claim numbers and not the conditional mean, an estimation of the underlying risk parameter relating to the ith policy of the portfolio is the average claim number $\overline{x}_i$ (i.e. the total number of claims generated by this policy divided by the length of the exposure period). Therefore, given a kernel K, YOUNG (1997) suggested estimating $dF_{\Theta}(\theta)$ by

$$d\widehat{F}_{\tilde{\Theta}}(t) = \sum_{i=1}^{n} \frac{w_i}{h_i} K\left(\frac{t - \overline{x}_i}{h_i}\right)$$

in which h_i is a positive parameter called the bandwidth and w_i is a weight (taken to be the number of years the ith policy is in force divided by the total number of policy-years for the collective). YOUNG (1997) suggested using the Epanechnikov kernel and determined the h_is in order to minimize the mean integrated squared error (by reference to a Normal prior).

2

Risk Classification

2.1 Introduction

2.1.1 Risk Classification, Regression Models and Random Effects

Motor ratemaking is essentially about classifying policies according to their risk characteristics. The classification variables are called *a priori* variables (as their values can be determined before the policyholder starts to drive). In motor insurance, they include the age, gender and occupation of the policyholders, the type and use of their car, the place where they reside and sometimes even the number of cars in the household, marital status, or the colour of the vehicle.

These observable risk characteristics are typically seen as nonrandom covariates. Other risk characteristics are unobservable and must be seen as unknown parameters or, in the vein of credibility theory, latent variables with a common distribution. The literature about premium rating in motor insurance comprises two mainstream approaches: (i) the first one disregards observable covariates altogether and lumps all the individual characteristics into random latent variables and (ii) the second one disregards random individual risk characteristics and tries instead to catch all relevant individual variations by covariates. Chapter 1 adopted the first approach. The present chapter combines both views, employing contemporary, advanced data analysis.

If the data are subdivided into risk classes determined by *a priori* variables, actuaries work with figures which are small in exposure and claim numbers. Therefore, simple averages will be suspect and regression models are needed. Regression analyses the relationship between one variable and another set of variables. This relationship is expressed as an equation that predicts a response variable (the expected number of claims filed by a given policyholder) from a function of explanatory variables and parameters (involving a linear combination of these explanatory variables and parameters, called a linear predictor). The parameters are estimated so that a measure of the goodness-of-fit is optimized (the log-likelihood, in

Actuarial Modelling of Claim Counts: Risk Classification, Credibility and Bonus-Malus Systems M. Denuit, X. Maréchal, S. Pitrebois and J.-F. Walhin

most cases). Actuaries use regression techniques to predict the expected number of claims knowing some information about the policyholders, vehicles and types of contract. It is worth mentioning that even with all the covariates included here, there still remain substantial risk differentials between individual drivers (due to hidden characteristics, like temper and skill, aggressiveness behind the wheel, knowledge of the highway code, etc.). Random effects are added to the linear predictor on the score scale to take this residual heterogeneity into account, reconciling the two approaches (i)–(ii) mentioned above.

In nonlife business, the pure premium is the expected cost of all the claims that policyholders will file during the coverage period (under the assumption of the Law of Large Numbers). The computation of this premium relies on a statistical model incorporating all the available information about the risk. The technical tariff aims to evaluate as accurately as possible the pure premium for each policyholder via regression techniques. It is well-known that market premiums may differ from those computed by actuaries; see, e.g., Coutts (1984) for a discussion. In that respect, the overall market position of the company compared to its competitors with regard to growth and pricing is crucial.

Sometimes, motor ratemaking is performed on panel data. Bringing several observation periods together has some advantages: it increases the sample size and avoids granting too much importance to a single calendar year (during which the particular weather conditions could have increased or decreased the number of traffic accidents, for instance). However, this induces some dependence in the data, since observations relating to the same policyholder across time are expected to be correlated. The analysis of correlated data with Poisson marginals arising from repeated measurements can be performed with the help of Generalized Estimating Equations (GEEs). GEEs provide a practical method with reasonable statistical efficiency to analyse such panel data. GEEs also give initial values for maximum likelihood procedures in models for longitudinal data.

2.1.2 Risk Sharing in Segmented Tariffs

The following discussion is inspired by the paper by De Wit & Van Eeghen (1984). Consider a portfolio of n policies from motor third party liability insurance. The random variable Y_i models a quantity of actuarial interest for policy i (for instance the amount of a claim, the aggregate claim amount in one period or the number of accidents at fault reported by policyholder i during one period). In order to explain the outcomes of Y_i, the actuary has observable covariates $\boldsymbol{X}_i^T = (X_{i1}, X_{i2}, \ldots)$ at his disposal (e.g., age, gender and occupation of policyholder i, the place where he resides, type and use of his car). However, Y_i also depends on a sequence of unknown characteristics $\boldsymbol{Z}_i^T = (Z_{i1}, Z_{i2}, \ldots)$ (e.g., annual mileage, accuracy of judgment, aggressiveness behind the wheel, drinking behaviour, etc.). Some of these quantities are unobservable, others cannot be measured in a cost efficient way.

The 'true' premium for policyholder i is $\mathbb{E}[Y_i|\boldsymbol{X}_i, \boldsymbol{Z}_i]$. It is the function g of $\boldsymbol{X}_i$ and $\boldsymbol{Z}_i$ that is the 'closest' to Y_i, in the sense that $\mathbb{E}[(Y_i - g(\boldsymbol{X}_i, \boldsymbol{Z}_i))^2]$ is minimum for $g(\boldsymbol{X}_i, \boldsymbol{Z}_i) = \mathbb{E}[Y_i|\boldsymbol{X}_i, \boldsymbol{Z}_i]$. If the insurer charges $\mathbb{E}[Y_i|\boldsymbol{X}_i, \boldsymbol{Z}_i]$ to policyholder i, then the policyholders pay premiums that absorb the inter-individual variations (that is, the variations of the premiums due to the modifications in personal characteristics $\boldsymbol{X}_i$ and $\boldsymbol{Z}_i$, the magnitude of which are quantified by $\mathbb{V}\big[\mathbb{E}[Y_i|\boldsymbol{X}_i, \boldsymbol{Z}_i]\big]$). The company covers the purely random intra-individual risks (that is, the random fluctuations of Y_i, which are quantified by the variance $\mathbb{E}\big[\mathbb{V}[Y_i|\boldsymbol{X}_i, \boldsymbol{Z}_i]\big]$ of the outcomes of Y_i once the personal characteristics $\boldsymbol{X}_i$ and $\boldsymbol{Z}_i$ have been fixed). Risk sharing can be summarized as follows: using the variance decomposition formula and then taking expectations gives:

$$\mathbb{V}[Y_i] = \underbrace{\mathbb{E}\big[\mathbb{V}[Y_i|\boldsymbol{X}_i, \boldsymbol{Z}_i]\big]}_{\rightarrow \text{insurer}} + \underbrace{\mathbb{V}\big[\mathbb{E}[Y_i|\boldsymbol{X}_i, \boldsymbol{Z}_i]\big]}_{\rightarrow \text{policyholder}}.$$

Of course, since the elements of $\boldsymbol{Z}_i$ are unknown to the insurer, the situation described above is purely theoretical. Since the company only knows $\boldsymbol{X}_i$, the insurer can only charge $\mathbb{E}[Y_i|\boldsymbol{X}_i]$. The risk sharing is now

$$\mathbb{V}[Y_i] = \underbrace{\mathbb{E}\big[\mathbb{V}[Y_i|\boldsymbol{X}_i]\big]}_{\rightarrow \text{insurer}} + \underbrace{\mathbb{V}\big[\mathbb{E}[Y_i|\boldsymbol{X}_i]\big]}_{\rightarrow \text{policyholder}}.$$

The part of the variance supported by the insurer is now larger, since residual heterogeneity remains with the company. To see this, let us write

$$\mathbb{E}\Big[\mathbb{V}[Y_i|\boldsymbol{X}_i]\Big] = \mathbb{E}\Big[\mathbb{V}[Y_i|\boldsymbol{X}_i, \boldsymbol{Z}_i]\Big] + \mathbb{E}\Big[\mathbb{V}\Big[\mathbb{E}[Y_i|\boldsymbol{X}_i, \boldsymbol{Z}_i]\Big|\boldsymbol{X}_i\Big]\Big].$$

The first term in this sum, i.e. $\mathbb{E}\big[\mathbb{V}[Y_i|\boldsymbol{X}_i, \boldsymbol{Z}_i]\big]$, represents the purely random fluctuations of the risk and is supported by the insurance company in application of the very basic principle of insurance. On the contrary, the second term represents the variations of the expected claims due to the unknown risk characteristics $\boldsymbol{Z}_i$. This quantity should be corrected by an experience rating mechanism (as discussed in Chapters 3 and 4).

We can now clearly see the link existing between *a priori* and *a posteriori* ratemaking. The idea behind experience rating is that past claims experience reveals the hidden features $\boldsymbol{Z}_i$. Let $\boldsymbol{Y}_i^{\leftarrow}$ denote the past claims experience available about Y_i. The idea is that the information contained in $(\boldsymbol{X}_i, \boldsymbol{Y}_i^{\leftarrow})$ becomes comparable to $(\boldsymbol{X}_i, \boldsymbol{Z}_i)$ as time goes on. Therefore, the *a posteriori* premium is $\mathbb{E}[Y_i|\boldsymbol{X}_i, \boldsymbol{Y}_i^{\leftarrow}]$. Experience rating is based on a 'crime and punishment' mechanism: claim-free policyholders are rewarded by premium discounts called *bonuses*, whereas policyholders reporting one or more accidents at fault are penalized by premium surcharges called *maluses*.

In *a priori* ratemaking, the actuary aims to identify the best predictors $\boldsymbol{X}_i$ and to compute the risk premium $\mathbb{E}[Y_i|\boldsymbol{X}_i]$. In *a posteriori* ratemaking, the actuary purposes to compute premium corrections according to past claims history $\boldsymbol{Y}_i^{\leftarrow}$ in order to reflect the unavailable information contained in $\boldsymbol{Z}_i$. A posteriori ratemaking techniques are discussed in the next chapter.

2.1.3 Bonus Hunger and Censoring

In this chapter, we fit statistical models for the number of claims subject to the existing rules adopted by the insurance company to penalize reported claims. Due to bonus-malus mechanisms, some claims are not filed to the company, because policyholders think it is cheaper for them to defray the third party (or to pay for their own costs in first party coverages) to avoid premium surcharges. The data are thus censored in a complicated way, and the conclusions of the actuarial analysis are valid only if the existing rules are kept unchanged. Analysing insurance data, the actuary is able to draw conclusions about the number of claims filed by policyholders subject to a specific *a posteriori* ratemaking mechanism. The actuary is not able to draw any conclusions about the number of accidents caused by these insured drivers. We will come back to this important issue in Chapter 5.

2.1.4 Agenda

This chapter is devoted to *a priori* ratemaking, and focusses on claim frequencies. To make the discussion more concrete, we analyse the statistics from a couple of motor insurance portfolios. We first work with cross-sectional data (i.e. data gathered during one year of observation) from a Belgian motor third party liability insurance portofolio observed during the year 1997 (called Portfolio A, already used in Chapter 1). We also show how it is possible to build a ratemaking on the basis of panel data. To this end, we use another Belgian portfolio (called Portfolio B) for which the data have been collected during 3 years (from 1997 to 1999).

To fix the ideas, in Section 2.2 we present the data observed during the year 1997 and the different explanatory variables available for Portfolio A. We give a detailed description of all the variables and we have a first look at their influence on the risk borne by the insurer. Then, in Section 2.3, we show how it is possible to build an *a priori* ratemaking thanks to a Poisson regression. We illustrate the technique on the data from Portfolio A. Section 2.4 addresses the problem of overdispersion. A random effect is added to the covariates to account for residual heterogeneity (vector $\boldsymbol{Z}_i$ in the preceding discussion). We examine three classical models: the Poisson-Gamma (or Negative Binomial) model, the Poisson-Inverse Gaussian model and the Poisson-LogNormal model. These are extensions of the corresponding models presented in Chapter 1, to incorporate exogenous information about policyholders.

In Section 2.9, we develop ratemaking techniques using panel data. Portfolio B that has been observed during three consecutive years is used for the numerical illustrations. GEE and maximum likelihood are used to estimate the parameters involved in models for longitudinal data. The final Section 2.10 offers an extensive discussion of topics not covered in this chapter, together with appropriate references.

2.2 Descriptive Statistics for Portfolio A

2.2.1 Global Figures

The data relate to a Belgian motor third party liability insurance portfolio observed during the year 1997. The data set (henceforth referred to as Portfolio A, for brevity) comprises 14 505 policies. The observed claim number distribution in the portfolio has been described in Table 1.1. The observed mean claim frequency for Portfolio A is 14.6 %.

2.2.2 Available Information

The following information is available on an individual basis: in addition to the number of claims filed by each policyholder (variable Nclaim) and the exposure-to-risk from which these claims originate (i.e. the number of days the policy has been in force during 1997, variable Expo), we know

Age : Policyholder's age (four categories: 1 = 'between 18 and 24', 2 = 'between 25 and 30', 3 = 'between 31 and 60', 4 = 'more than 60')
Gender : Policyholder's gender (two categories: 1 = 'woman', 2 = 'man')
District : kind of district where the policyholder lives (two categories: 1 = 'urban', 2 = 'rural')
Use : Use of the car (two categories: 1 = 'private use, i.e. leisure and commuting', and 2 = 'professional use')
Split : premium split (two categories: 1 = 'premium paid once a year' and 2 = 'premium split up').

In practice, insurers have at their disposal much more information about their policyholders. Here, we focus on these few explanatory variables for pedagogical purposes, to ease the exposition of ideas.

We see that all the explanatory variables listed above are categorical, i.e. they can be used to partition the portfolio into homogeneous classes with respect to the variables. Such explanatory variables are called factors, each factor having a number of levels. In practice, there are also continuous covariates. We explain in the last section of this chapter how to deal with such explanatory variables.

2.2.3 Exposure-to-Risk

The majority of policies are in force during the whole year. However, in some cases, the observation period does not last for the entire year. This is the case, for instance, for new policyholders entering the portfolio during the observation period, and in case of policy cancellations. It is also common in practice to start a new period if some changes occur in the observable characteristics of the policies (for instance, the policyholder moves from a rural to an urban area and the company uses the rating variable District). The policy is then represented as two different lines in the data base, and observations are recorded separately for the two periods (the policy number allows the actuary to track these changes). Note that independence is lost in this case, and allowance for panel data is preferable (as in Section 2.9). This variety of situations is taken into account in the Poisson process, by multiplying the annual expected claim frequency by the length of the observation period, as explained in Chapter 1.

In Portfolio A, the average coverage period is 298.98 days. Figure 2.1 gives an idea of the distribution of the exposure-to-risk in the portfolio. About 65 % of the policies have been observed during the whole year 1997. Considering the distribution of the exposure-to-risk, we see that policy issuances and lapses are randomly spread over the year. The distribution of the policies in force during less than one year is roughly uniform over [0,365].

It is worth mentioning that the policies that are just issued often differ from those in the portfolio. This is why it may be preferable to conduct a separate analysis for this type of policy.

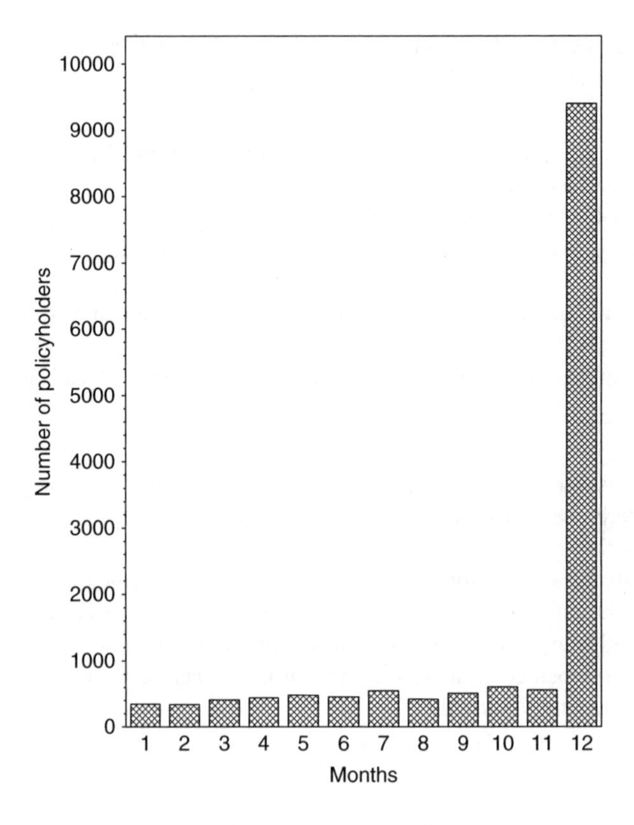

Figure 2.1 Exposure-to-risk in Portfolio A.

2.2.4 One-Way Analyses

Age

The age structure of the portfolio is described in Figure 2.2. Most policyholders are middle-aged as 6722 insured drivers (representing 46.4 % of the portfolio) are between 31 and 60. Only 802 insured drivers (representing 5.8 % of the portfolio) are over 60. The young drivers represent 15.3 % of the portfolio (2295 policyholders) and the remaining 4686 insured drivers (32.5 % of the portfolio) are between 25 and 30.

In the preliminary descriptive analysis, the actuary considers the marginal impact of each rating factor. The possible effect of the other explanatory variables is thus disregarded. Let us assume for a while that the claim frequencies only depend on Age. If the occurrence of the claims filed by the policyholders conforms with a Poisson process, the number N_i of claims reported by policholder i obeys the Poisson distribution with mean $d_i\lambda_{\text{Age}(i)}$, where d_i is the exposure-to-risk (i.e., the length of the coverage period) for policyholder i, Age(i) is the age category to which policyholder i belongs (1, 2, 3 or 4) and the λ_js, $j = 1, 2, 3, 4$, are the annual expected claim frequencies for the 4 age classes.

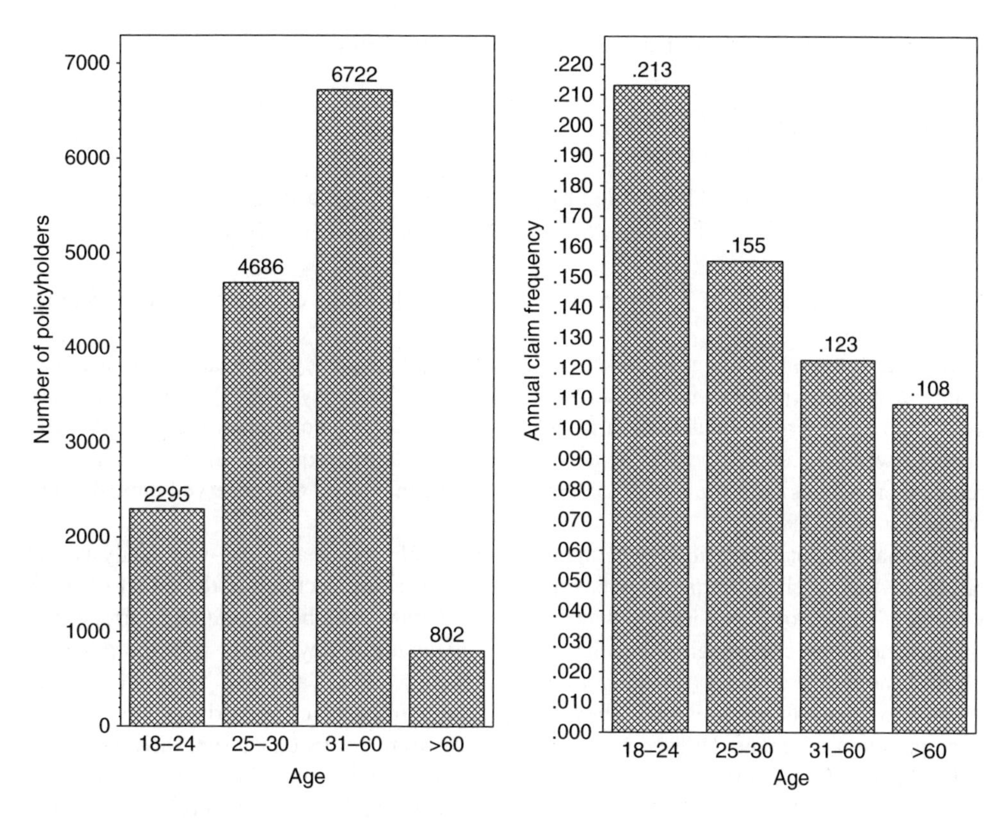

Figure 2.2 Composition of Portfolio A with respect to Age (left panel) and observed annual claim frequencies according to Age (right panel).

Assuming that the numbers of claims filed by the policyholders in the portfolio are independent random variables, the likelihood then becomes

$$
\begin{aligned}
\mathcal{L}(\lambda_1, \lambda_2, \lambda_3, \lambda_4) &= \prod_{i=1}^{n} \exp\Big(-d_i\lambda_{\mathrm{Age}(i)}\Big)\frac{\Big(d_i\lambda_{\mathrm{Age}(i)}\Big)^{k_i}}{k_i!} \\
&\propto \prod_{j=1}^{4} \exp\left(-\lambda_j \sum_{i|\mathrm{Age}(i)=j} d_i\right) \lambda_j^{\sum_{i|\mathrm{Age}(i)=j} k_i}
\end{aligned}
$$

where k_i denotes the observed number of claims for policyholder i, and '$\propto$' reads 'is proportional to'. Differentiating $L(\lambda_1, \lambda_2, \lambda_3, \lambda_4) = \ln \mathcal{L}(\lambda_1, \lambda_2, \lambda_3, \lambda_4)$ with respect to λ_j and setting the derivative equal to 0 gives

$$
-\sum_{i|\mathrm{Age}(i)=j} d_i + \frac{1}{\lambda_j} \sum_{i|\mathrm{Age}(i)=j} k_i = 0.
$$

The maximum likelihood estimator of λ_j is then obtained from

$$\widehat{\lambda}_j = \frac{\sum_{i|\text{Age}(i)=j} k_i}{\sum_{i|\text{Age}(i)=j} d_i}$$

$$= \frac{\text{\# of claims filed by policyholders in age category } j}{\text{total exposure-to-risk (in year) for age category } j}.$$

Of course, the reliability of $\widehat{\lambda}_j$ depends on the magnitude of the exposure-to-risk appearing in the denominator.

We see in Figure 2.2 that the observed annual claim frequency decreases with age. The young drivers are riskier as their observed annual claim frequency is 21.3 %. Old drivers are safer with an observed annual claim frequency of 10.8 %. We notice that the policyholders aged between 25 and 30, with an observed annual claim frequency of 15.5 %, tend to report more claims than the policyholders aged between 31 and 60 (observed annual claim frequency of 12.3 %).

The analysis conducted in this paragraph is often referred to as a one-way analysis: the effect of Age on claim frequencies is studied without taking account of the effect of other variables. The major flaw with one-way analyses is that they can be distorted by correlations. For instance, one can imagine that the majority of young policyholders split the payment of the insurance premiums (for budget reasons). If more claims are filed by young drivers, a one-way analysis of Split may show higher claim frequencies for drivers having split their premium payment. However, this may result from the fact that such drivers are in general the high-risk young policyholders. Premium differentials based on one-way analyses of Split and Age would double-count the effect of Age. Multivariate methods (such as the Poisson regression approach discussed below) adjust for correlations between explanatory variables. The correlations existing between explanatory variables explain why the policies are not uniformly distributed over risk classes but cluster in some specified highly populated classes.

Gender

It is common to include the gender of the main driver in the actuarial ratemaking. Note however that some states have banned the use of this rating factor (as well as of age, for instance). The reason is that age and gender are out of the policyholders' control, in contrast to many other covariates (like the power of the car, or the driving area). The latter may thus be freely used for ratemaking purposes, but some limitations are needed for the former.

In Portfolio A, there are 9358 male policyholders (representing 64.7 % of the portfolio) and 5147 female policyholders (representing 35.3 % of the portfolio). Figure 2.3 suggests a higher annual claim frequency for males (observed annual claim frequency of 15.2 %) than for females (observed annual claim frequency of 13.6 %).

District

Figure 2.4 gives the distribution of the policyholders according to the district where they live. We see that 8664 policyholders (representing 59.8 % of the portfolio) live in an urban area and 5841 policyholders (representing 40.2 % of the portfolio) live in a rural one. The urban policyholders have a larger observed annual claim frequency (15.7 %) than the rural ones (13.0 %).

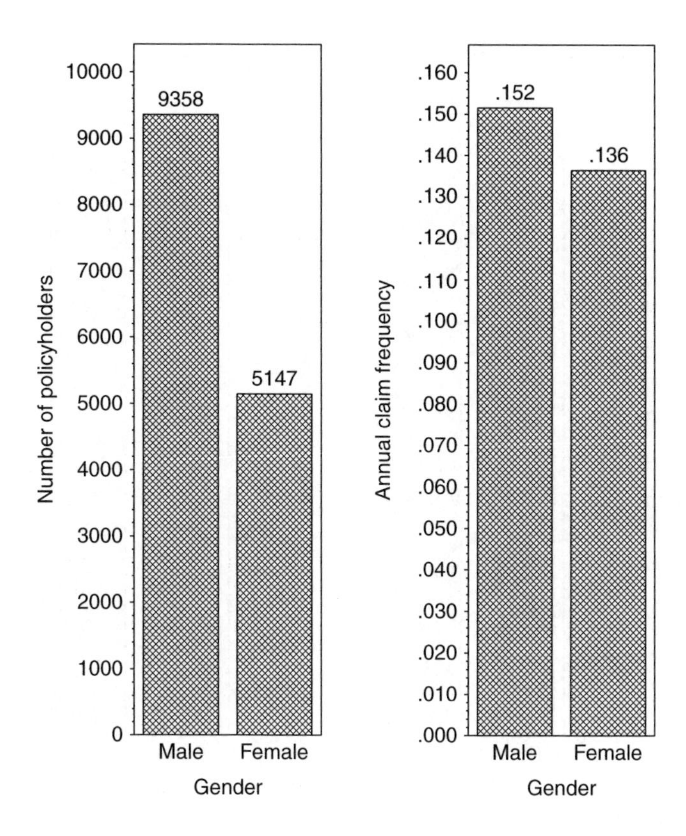

Figure 2.3 Composition of Portfolio A with respect to Gender (left panel) and observed annual claim frequencies according to Gender (right panel).

Use

The majority of the policyholders (12 745 insured drivers, representing 88.0 % of the portfolio) use their vehicle only for leisure and commuting. There are 1760 professional users (representing 12.0 % of the portfolio). Figure 2.5 indicates that the influence of the use of the vehicle on the number of claims is almost negligible, as the annual observed claim frequencies are almost equal for the two categories of insured drivers: 14.6 % for private users and 14.3 % for professional users.

Premium Split

Figure 2.6 indicates that 10 568 policyholders (representing 74.5 % of the portfolio) pay their premium once a year. The remaining 3937 policyholders (representing 25.5 % of the portfolio) pay their premium twice a year, thrice a year or on a monthly basis. Figure 2.6 also shows the influence of the premium split on the number of claims: it can be seen that splitting the premium payment is associated with a considerable increase in the observed annual claim frequency (from 12.4 % to 21.1 %).

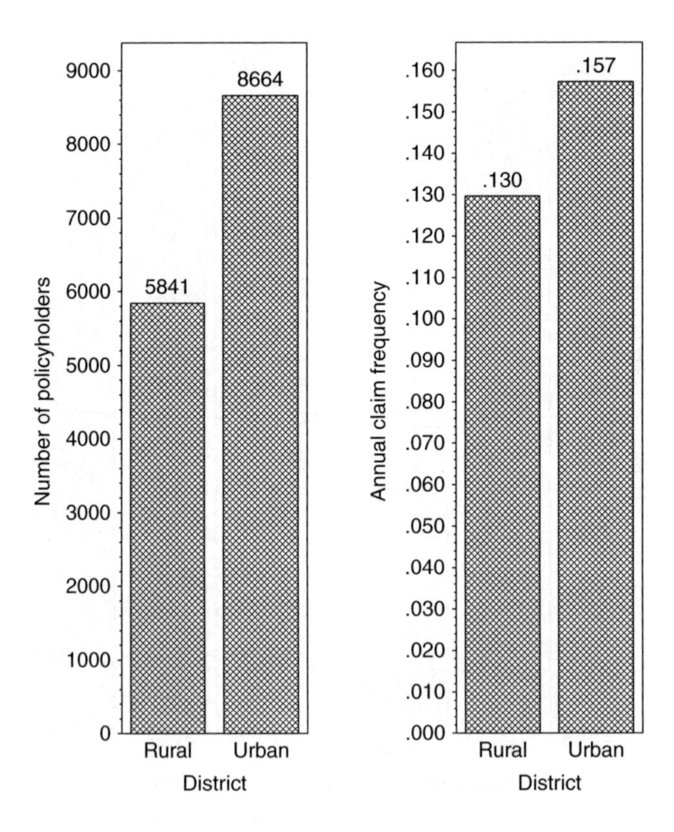

Figure 2.4 Composition of Portfolio A with respect to District (left panel) and observed annual claim frequencies according to District (right panel).

2.2.5 Interactions

So far, only the marginal effect of each observed covariate on the claim frequency has been assessed. Besides these one-way analyses, it is also important to account for possible interactions. Often Gender and Age interact, in the sense that the effect of Age on the average claim frequency is different for males than for females. Typically, young male drivers are more dangerous than young female drivers (but the higher risk associated with young male drivers may be due to higher annual mileage, or to other risk factors correlated to the fact of being a young male). Formally, two explanatory variables are said to interact when the effect of one factor varies depending on the levels of the other factor. Multivariate models allow for investigation into interaction effects.

This phenomenon can be seen from Figure 2.7. The observed annual claim frequency for young males (ages 18–24) peaks at 23.8 %, whereas young females (ages 18–24) have an observed annual claim frequency similar to males aged 25–30. Both genders become more similar for categories 31–60 and over 60. We have thus detected an Age–Gender interaction in Portfolio A.

Note that standard regression models do not automatically account for interaction (in contrast to correlations between covariates, for which the estimated regression coefficients

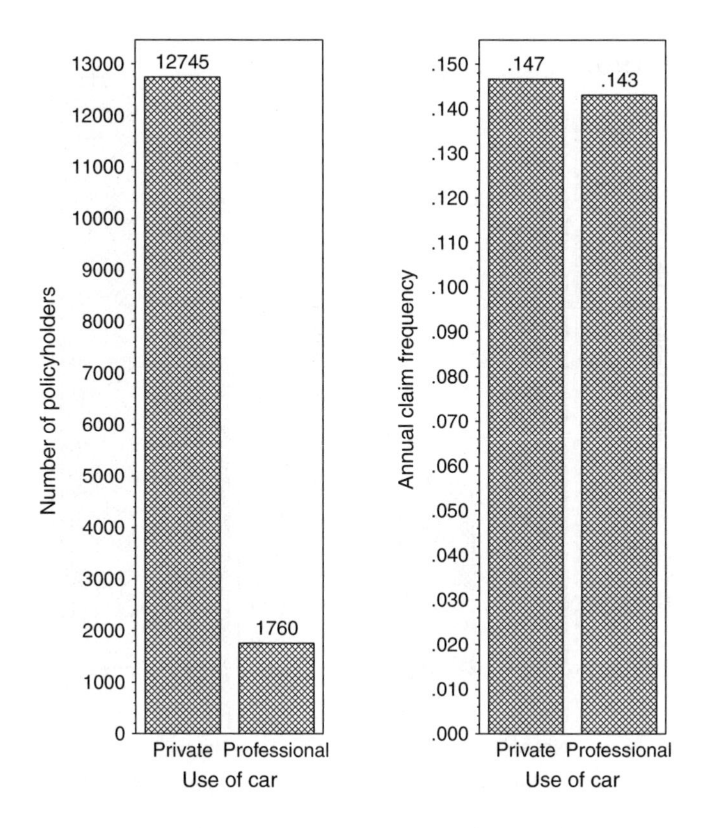

Figure 2.5 Composition of Portfolio A with respect to Use (left panel) and observed annual claim frequencies according to Use (right panel).

are adjusted). The reason is as follows: interactions cannot be rendered by linear combinations of the covariates. To account for interactions, nonlinear functions of the covariates (products) are needed, as explained in Example 2.3 below. In ANOVA terminology, one would speak of interaction when the effects are not just additive. The actuary needs to identify the existing interactions at the preliminary exploratory stage, and then define new ratemaking factors combining the levels of the two interacting variables. Inserting these new factors in the regression model then allows us to account for interaction.

2.2.6 True Versus Apparent Dependence

The descriptive analysis conducted so far suggests that some observed characteristics may influence the number of claims reported to the company. It is nevertheless important to realize the kind of relationship just evidenced: the actuary has to keep in mind that he has not established any causal relationship so far, but only that some correlations seem to exist between the rating factors and the number of claims. Such correlations may have been produced by a causal relationship, but could also result from confounding effects. The following simple example illustrates this situation.

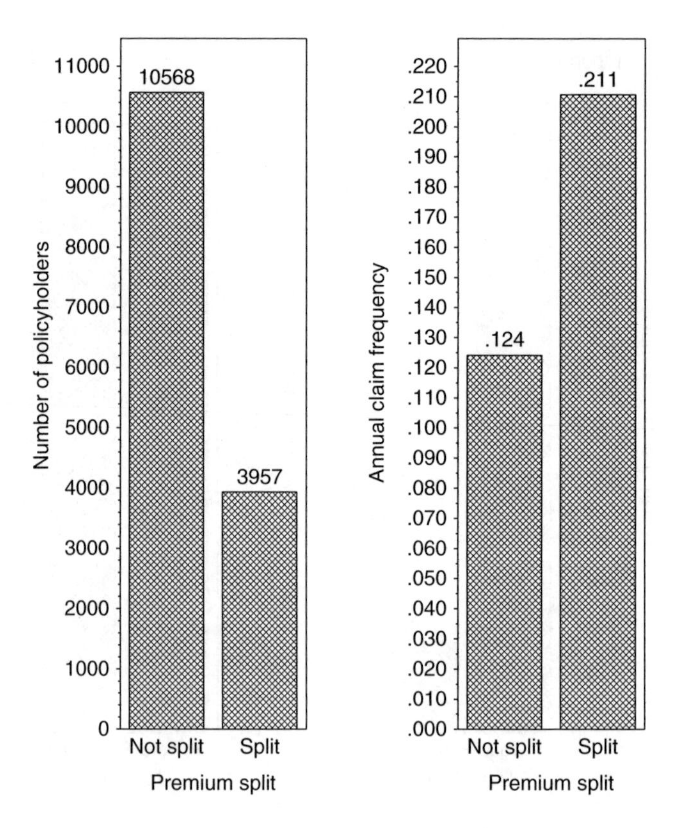

Figure 2.6 Composition of Portfolio A with respect to Split (left panel) and observed annual claim frequencies according to Split (right panel).

Example 2.1 Assume that living in rural or urban areas (observed covariate District) does not influence claim occurrences but the driving experience (hidden characteristic) does. Specifically, let us assume that

$$\begin{aligned}\Pr[N \geq 1|\text{inexperienced, rural}] &= \Pr[N \geq 1|\text{inexperienced, urban}]\\ &= \Pr[N \geq 1|\text{inexperienced}] = 0.15\\ \Pr[N \geq 1|\text{experienced, rural}] &= \Pr[N \geq 1|\text{experienced, urban}]\\ &= \Pr[N \geq 1|\text{experienced}] = 0.05.\end{aligned}$$

The portfolio comprises 50 % inexperienced drivers, but

$$\Pr[\text{inexperienced}|\text{urban}] = 1 - \Pr[\text{experienced}|\text{urban}] = 0.9$$

and

$$\Pr[\text{inexperienced}|\text{rural}] = 1 - \Pr[\text{experienced}|\text{rural}] = 0.1.$$

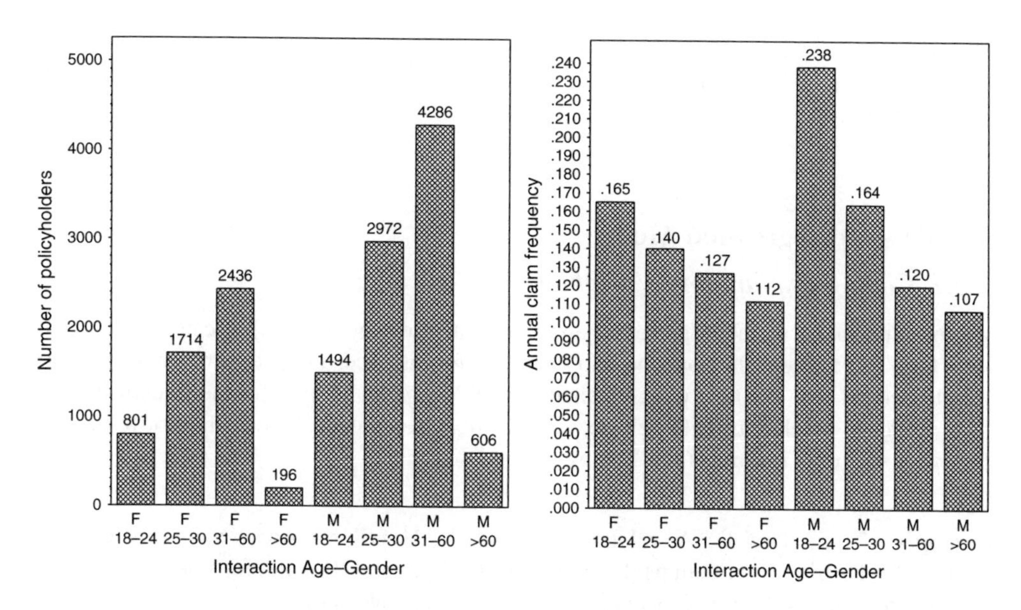

Figure 2.7 Composition of Portfolio A with respect to the Age–Gender interaction (left panel) and observed annual claim frequencies according to the Age–Gender interaction (right panel).

In other words, there is a majority of experienced drivers in rural areas (90 %) whereas in urban areas, the majority of drivers are inexperienced.

Clearly,

$$\begin{aligned}\Pr[N \geq 1|\text{urban}] &= \Pr[N \geq 1|\text{inexperienced, urban}]\Pr[\text{inexperienced}|\text{urban}]\\ &\quad + \Pr[N \geq 1|\text{experienced, urban}]\Pr[\text{experienced}|\text{urban}]\\ &= 0.14\end{aligned}$$

whereas

$$\begin{aligned}\Pr[N \geq 1|\text{rural}] &= \Pr[N \geq 1|\text{inexperienced, rural}]\Pr[\text{inexperienced}|\text{rural}]\\ &\quad + \Pr[N \geq 1|\text{experienced, rural}]\Pr[\text{experienced}|\text{rural}]\\ &= 0.06.\end{aligned}$$

Even if District does not influence the number of claims once the driving experience has been accounted for, unconditionally there is some dependence between District and the number of claims filed by policyholders. The univariate analysis will detect a rural/urban effect, but the latter should disappear in the multivariate analysis (taking experience and the variable District into account). Therefore, we cannot say after a marginal analysis that living in an urban area causes an increase in the average number of claims: both variables are related to driving experience.

The reader has always to keep in mind that it is often not possible to disentangle a true effect of a rating factor from an apparent effect resulting from correlation with unobservable characteristics.

2.3 Poisson Regression Model

2.3.1 Coding Explanatory Variables

All the explanatory variables presented above are categoric (or nominal). A categoric variable with k levels partitions the portfolio into k classes (for instance, 4 classes for Age in Portfolio A). It is coded with the help of $k-1$ binary variables, being all zero for the reference level. The reference level is usually selected as the most populated class in the portfolio. The following example illustrates this coding methodology.

Example 2.2 In Portfolio A, reference levels are '31–60' for Age, 'Male' for Gender, 'Urban' for District, 'Premium paid once a year' for Split and 'Private' for Use. Policyholder i is then represented by a vector of dummies giving the values of

$$x_{i1} = \begin{cases} 1 \text{ if policyholder } i \text{ is less than 24} \\ 0 \text{ otherwise} \end{cases}$$

$$x_{i2} = \begin{cases} 1 \text{ if policyholder } i \text{ is 25–30} \\ 0 \text{ otherwise} \end{cases}$$

$$x_{i3} = \begin{cases} 1 \text{ if policyholder } i \text{ is over 60} \\ 0 \text{ otherwise} \end{cases}$$

$$x_{i4} = \begin{cases} 1 \text{ if policyholder } i \text{ is a female} \\ 0 \text{ otherwise} \end{cases}$$

$$x_{i5} = \begin{cases} 1 \text{ if policyholder } i \text{ lives in a rural area} \\ 0 \text{ otherwise} \end{cases}$$

$$x_{i6} = \begin{cases} 1 \text{ if policyholder } i \text{ splits his premium payment} \\ 0 \text{ otherwise} \end{cases}$$

$$x_{i7} = \begin{cases} 1 \text{ if policyholder } i \text{ uses his car for professional reasons} \\ 0 \text{ otherwise.} \end{cases}$$

The results are interpreted with respect to the reference class (for which all the x_{ij}s are equal to 0) corresponding to a man aged between 31 and 60, living in an urban area, paying the premium once a year and using the car for private purposes only. The sequence (1,0,0,0,0,0,1) represents a man aged less than 24, living in an urban area, paying the premium once a year and using the car for professional reasons.

In case two covariates interact, a new variable is created by combining the levels of each of the interacting covariates, as shown in the following example.

Example 2.3 Figure 2.7 suggests that the variables Age and Gender interact in Portfolio A. It is thus not possible to represent accurately the effect of being a male policyholder (compared with being a female policyholder) in terms of a single multiplier, nor can the effect of Age be represented by a single multiplier. The relevant explanatory variables are not x_{i1}, x_{i2}, x_{i3}, and x_{i4} but rather $x_{i1}x_{i4}$, $x_{i2}x_{i4}$, $x_{i3}x_{i4}$, $(1-x_{i1}x_{i2}x_{i3})x_{i4}$, $x_{i1}(1-x_{i4})$, $x_{i2}(1-x_{i4})$, and $x_{i3}(1-x_{i4})$ (denoted henceforth as x'_{ij}).

To reflect the situation accurately, it is thus necessary to consider multipliers dependent on the combined levels of Age and Gender. To this end, a variable Gender*Age is created, with levels 'Female 18–24', 'Female 25–30', 'Female 31–60', 'Female over 60', 'Male 18–24', 'Male 25–30', 'Male 31–60' and 'Male over 60'. This new variable possesses 8 levels, and is coded by means of 7 dummies, being all 0 for the reference level (taken as 'Male 31–60'). Specifically, the explanatory variables x_{i1} to x_{i4} in Example 2.2 are replaced with

$$x'_{i1} = \begin{cases} 1 \text{ if policyholder } i \text{ is a female less than 24} \\ 0 \text{ otherwise} \end{cases}$$

$$x'_{i2} = \begin{cases} 1 \text{ if policyholder } i \text{ is a female aged 25–30} \\ 0 \text{ otherwise} \end{cases}$$

$$x'_{i3} = \begin{cases} 1 \text{ if policyholder } i \text{ is a female aged 31–60} \\ 0 \text{ otherwise} \end{cases}$$

$$x'_{i4} = \begin{cases} 1 \text{ if policyholder } i \text{ is a female over 60} \\ 0 \text{ otherwise} \end{cases}$$

$$x'_{i5} = \begin{cases} 1 \text{ if policyholder } i \text{ is a male less than 24} \\ 0 \text{ otherwise} \end{cases}$$

$$x'_{i6} = \begin{cases} 1 \text{ if policyholder } i \text{ is a male aged 25–30} \\ 0 \text{ otherwise} \end{cases}$$

$$x'_{i7} = \begin{cases} 1 \text{ if policyholder } i \text{ is a male over 60} \\ 0 \text{ otherwise.} \end{cases}$$

Rather than declaring Age and Gender as two explanatory variables coded with the help of x_{i1} to x_{i4}, a combined Age*Gender variable is declared and is coded with the help of the covariates x'_{i1} to x'_{i7}.

The part of the linear predictor involving Age and Gender is $\beta_1 x_{i1} + \cdots + \beta_4 x_{i4}$ without interaction, and becomes $\beta'_1 x'_{i1} + \cdots + \beta'_7 x'_{i7}$ if allowance is made for interaction. If Age and Gender indeed interact then the values of these two linear combinations differ from each other, whereas they collapse in the absence of interaction.

Allowing for interaction thus dramatically increases the number of explanatory variables. Introducing Age and Gender separately requires 4 binary variables whereas allowing for the Age–Gender interaction requires 7 dummies. As a consequence, the number of parameters to be estimated increases accordingly. We will see below that grouping some levels of the combined variable accounting for interaction is nevertheless often possible.

2.3.2 Loglinear Poisson Regression Model

In Poisson regression, we have a collection of independent Poisson counts whose means are modelled as nonnegative functions of covariates. Specifically, let N_i, $i = 1, 2, \ldots, n$, be the number of claims reported by policyholder i and d_i be the corresponding risk exposure (the N_is are assumed to be independent). All the observable characteristics (the *a priori* variables presented in Section 2.2 for Portfolio A, say) related to this policyholder are summarized into the vector $\boldsymbol{x}_i^T = (x_{i1}, \ldots, x_{ip})$. Poisson regression is a technique analogous to linear regression except that errors no longer follow a Normal distribution but the randomness in the model is described by a Poisson distribution. We first assume that the conditional expectation of N_i given $\boldsymbol{x}_i$ is of the form

$$\mathbb{E}[N_i|\boldsymbol{x}_i] = d_i \exp\left(\beta_0 + \sum_{j=1}^{p} \beta_j x_{ij}\right), \quad i = 1, 2, \ldots, n, \tag{2.1}$$

where $\boldsymbol{\beta}^T = (\beta_0, \beta_1, \ldots, \beta_p)$ is the vector of unknown regression coefficients. The explanatory variables enter the model in the linear combination $\beta_0 + \sum_{j=1}^{p} \beta_j x_{ij}$, where β_0 acts as an intercept and β_j is the coefficient indicating the weight given to the jth covariate. The Poisson regression model consists in stating that N_i is Poisson distributed with mean given by Expression (2.1), that is

$$N_i \sim \mathcal{P}oi\left(d_i \exp\left(\beta_0 + \sum_{j=1}^{p} \beta_j x_{ij}\right)\right), \quad i = 1, 2, \ldots, n.$$

2.3.3 Score

The quantity

$$\text{score}_i = \beta_0 + \sum_{j=1}^{p} \beta_j x_{ij}$$

is called the score (or linear predictor in statistics) because it allows the actuary to rank the policyholders from the least to the most dangerous. The claim frequency for policyholder i is $d_i \exp(\text{score}_i)$; its annual claim frequency being $\exp(\text{score}_i)$. Increasing the score thus means that the associated average annual claim frequency increases. The use of the exponential link function ensures the claim frequency is positive even if the score is negative.

Let us denote as $\widehat{\beta}_0, \widehat{\beta}_1, \ldots, \widehat{\beta}_p$ the estimators of the regression coefficients $\beta_0, \beta_1, \ldots, \beta_p$. In a statistical sense,

$$\widehat{\lambda}_i = d_i \exp\left(\widehat{\text{score}}_i\right) = d_i \exp\left(\widehat{\beta}_0 + \sum_{j=1}^{p} \widehat{\beta}_j x_{ij}\right)$$

is the predicted expected number of claims for policyholder i. Prediction in this sense does not refer to 'predicting the future' (called forecasting by statisticians) but rather to guessing the expected number of claims (i.e., the response) from the values of the regressors in an observation taken under the same circumstances as the sample from which the regression equation was estimated.

2.3.4 Multiplicative Tariff

When the *a priori* variables x_{ij}s are coded by means of binary variables (so that each policyholder is represented by a vector of 0s and 1s), the intercept β_0 represents the risk associated to the reference class. Then, the annual claim frequency λ_i associated with characteristics $\boldsymbol{x}_i$ is of the following multiplicative form

$$\lambda_i = \exp(\beta_0) \prod_{j|x_{ij}=1} \exp(\beta_j),$$

where $\exp(\beta_0)$ is the annual claim frequency corresponding to the reference class and $\exp(\beta_j)$ models the impact of the jth ratemaking variable. The intercept β_0 represents the risk associated with the reference class. If $\beta_j > 0$, being in the class coded by the jth explanatory variable increases the score and thus the annual claim frequency. Conversely, if $\beta_j < 0$, the score and the annual claim frequency decrease when $x_{ij} = 1$. With the exponential link, high (respectively low) scores mean high (respectively low) claim frequencies. The following example makes this clear.

Example 2.4 Let us continue Example 2.2. In Portfolio A, the score for policyholder i is of the form

$$\text{score}_i = \beta_0 + \beta_1 x_{i1} + \beta_2 x_{i2} + \cdots + \beta_7 x_{i7}.$$

Here,

$\exp(\beta_0) =$ annual claim frequency for men aged 31–60, living in an urban area, paying the premium once a year, using the car for private purposes

$\exp(\beta_0 + \beta_1) =$ annual claim frequency for men less than 24, living in an urban area, paying the premium once a year, using the car for private purposes

$\exp(\beta_0 + \beta_2) =$ annual claim frequency for men aged 25–30, living in an urban area, paying the premium once a year, using the car for private purposes

$\exp(\beta_0 + \beta_3) =$ annual claim frequency for men over 60, living in an urban area, paying the premium once a year, using the car for private purposes

$\exp(\beta_0 + \beta_4) =$ annual claim frequency for women aged 31–60, living in an urban area, paying the premium once a year, using the car for private purposes

etc.

2.3.5 Likelihood Equations

Let k_i be the number of claims filed by policyholder i during the observation period. The likelihood associated with these observations equals

$$\mathcal{L}(\boldsymbol{\beta}) = \prod_{i=1}^{n} \Pr[N_i = k_i | \boldsymbol{x}_i] = \prod_{i=1}^{n} \exp(-\lambda_i) \frac{\lambda_i^{k_i}}{k_i!},$$

where

$$\lambda_i = d_i \exp(\text{score}_i) = \exp(\ln d_i + \text{score}_i).$$

The maximum likelihood estimator $\widehat{\boldsymbol{\beta}}$ of $\boldsymbol{\beta}$ maximizes $\mathcal{L}(\boldsymbol{\beta})$: $\widehat{\boldsymbol{\beta}}$ is the value of the regression coefficients that makes the observations the most plausible.

Remark 2.1 (Grouping Data) The maximum likelihood estimators obtained for individual or grouped data are identical. Let us prove it formally. To this end, let $\mathcal{L}_{group}(\boldsymbol{\beta})$ be the likelihood obtained after a grouping in risk classes. We will show below that

$$\mathcal{L}(\boldsymbol{\beta}) \propto \mathcal{L}_{group}(\boldsymbol{\beta}).$$

In words, the likelihood $\mathcal{L}_{group}$ based on grouped data is proportional to the likelihood $\mathcal{L}(\boldsymbol{\beta})$ based on individual data, so that the corresponding maximum likelihood estimates will coincide.

Let $s_1, \ldots, s_q$ be the q possible values for the score, say, and let us define

$$d_{\bullet j} = \sum_{i|\text{score}_i = s_j} d_i \text{ and } k_{\bullet j} = \sum_{i|\text{score}_i = s_j} k_i \text{ for } j = 1, \ldots, q.$$

In words, $d_{\bullet j}$ is the total risk exposure for risk class j (corresponding to the value s_j of the score) and $k_{\bullet j}$ is the total number of claims recorded for the same risk class. Then

$$\begin{aligned}
\mathcal{L}(\boldsymbol{\beta}) &= \prod_{j=1}^{q} \prod_{i|\text{score}_i = s_j} \exp(-\lambda_i) \frac{\lambda_i^{k_i}}{k_i!} \\
&\propto \prod_{j=1}^{q} \exp\left(- \sum_{i|\text{score}_i = s_j} \lambda_i \right) \left(\exp(s_j) d_{\bullet j}\right)^{k_{\bullet j}} \\
&= \prod_{j=1}^{q} \exp\left(- \exp(s_j) \sum_{i|\text{score}_i = s_j} d_i \right) \left(\exp(s_j) d_{\bullet j}\right)^{k_{\bullet j}} \\
&\propto \prod_{j=1}^{q} \exp\left(-\exp(s_j) d_{\bullet j}\right) \frac{\left(\exp(s_j) d_{\bullet j}\right)^{k_{\bullet j}}}{k_{\bullet j}!} = \mathcal{L}_{group}(\boldsymbol{\beta}).
\end{aligned}$$

Maximizing $\mathcal{L}(\boldsymbol{\beta})$ or $\mathcal{L}_{group}(\boldsymbol{\beta})$ then gives the same maximum likelihood estimator $\widehat{\boldsymbol{\beta}}$.

The computation is easier if we change the likelihood to the log-likelihood which is then given by

$$L(\boldsymbol{\beta}) = \ln \mathcal{L}(\boldsymbol{\beta}) = \sum_{i=1}^{n} \Big(-\ln k_i! + k_i \ln \lambda_i - \lambda_i \Big). \tag{2.2}$$

The maximum likelihood estimators $\widehat{\beta}_0$ and the $\widehat{\beta}_j$s are the solutions of the following likelihood equations that are obtained by making the first derivatives of the log-likelihood with respect to the regression coefficients equal to zero:

$$\frac{\partial}{\partial \beta_0} L(\boldsymbol{\beta}) = 0 \Leftrightarrow \sum_{i=1}^{n} (k_i - \lambda_i) = 0, \tag{2.3}$$

$$\frac{\partial}{\partial \beta_j} L(\boldsymbol{\beta}) = 0 \Leftrightarrow \sum_{i=1}^{n} x_{ij}(k_i - \lambda_i) = 0, \quad j = 1, \ldots, p. \tag{2.4}$$

2.3.6 Interpretation of the Likelihood Equations

Equation (2.3) has an obvious interpretation: the fitted total number of claims $\sum_{i=1}^{n} \widehat{\lambda}_i$ is equal to the observed total number of claims $\sum_{i=1}^{n} k_i$. Therefore, provided that an intercept β_0 is included in the score, the total claim number predicted by the regression model equals its observed counterpart. Note that this equality holds for the observation period and not necessarily for the future, when the ratemaking will be implemented in practice. In other words, we cannot be sure that $\sum_{i=1}^{n} \widehat{\lambda}_i$ claims will be filed in the future, just that the actual number of claims should be close to $\sum_{i=1}^{n} \widehat{\lambda}_i$ if the yearly number of claims filed to the company remains stable over time.

The interpretation of the second likelihood Equation (2.4) is as follows: In Example 2.2 with Portfolio A, Equation (2.4) for $j = 4$ gives

$$\sum_{\text{females}} k_i = \sum_{\text{females}} \widehat{\lambda}_i.$$

Therefore, the model fits exactly the total number of claims filed by female policyholders. There is no cross-subsidies between men and women. The conclusion is similar for the other values of j. For $j = 1, 2, 3$ for instance, the Equations (2.4) thus ensure that the sum of all the claims reported for each age category is exactly reproduced by the model.

2.3.7 Solving the Likelihood Equations with the Newton–Raphson Algorithm

The likelihood equations do not admit explicit solutions and must therefore be solved numerically. Let $\boldsymbol{U}(\boldsymbol{\beta})$ be the gradient vector of the log-likelihood $L(\boldsymbol{\beta}) = \ln \mathcal{L}(\boldsymbol{\beta})$ defined in (1.47). Let us define $\widetilde{\boldsymbol{x}}_i$ as the vector $\boldsymbol{x}_i$ of explanatory variables for policyholder i supplemented with a unit first component, that is, $\widetilde{\boldsymbol{x}}_i = (1, \boldsymbol{x}_i^T)^T$. Then, considering (2.3)–(2.4), $\boldsymbol{U}(\boldsymbol{\beta})$ is given by

$$\boldsymbol{U}(\boldsymbol{\beta}) = \sum_{i=1}^{n} \widetilde{\boldsymbol{x}}_i (k_i - \lambda_i) \tag{2.5}$$

in the Poisson regression model. Let $\boldsymbol{H}(\boldsymbol{\beta})$ be the Hessian matrix of $L(\boldsymbol{\beta})$ defined in (1.50). Specifically, $\boldsymbol{H}(\boldsymbol{\beta})$ is given by

$$\boldsymbol{H}(\boldsymbol{\beta}) = -\sum_{i=1}^{n} \widetilde{\boldsymbol{x}}_i \widetilde{\boldsymbol{x}}_i^T \lambda_i \tag{2.6}$$

in the Poisson regression model. The maximum likelihood estimator $\widehat{\beta}_j$ of the parameters β_j then solves $\boldsymbol{U}(\boldsymbol{\beta}) = \boldsymbol{0}$.

The approach used to solve the likelihood equations is the Newton–Raphson algorithm (1.51). Starting from an appropriate $\widehat{\boldsymbol{\beta}}^{(0)}$, the Newton–Raphson algorithm is based on the following iteration

$$\begin{aligned} \widehat{\boldsymbol{\beta}}^{(r+1)} &= \widehat{\boldsymbol{\beta}}^{(r)} - \boldsymbol{H}^{-1}\left(\widehat{\boldsymbol{\beta}}^{(r)}\right) \boldsymbol{U}\left(\widehat{\boldsymbol{\beta}}^{(r)}\right) \\ &= \widehat{\boldsymbol{\beta}}^{(r)} + \left(\sum_{i=1}^{n} \widetilde{\boldsymbol{x}}_i \widetilde{\boldsymbol{x}}_i^T \widehat{\lambda}_i^{(r)}\right)^{-1} \sum_{i=1}^{n} \widetilde{\boldsymbol{x}}_i \left(k_i - \widehat{\lambda}_i^{(r)}\right) \end{aligned} \tag{2.7}$$

for $r = 0, 1, 2, \ldots$, where $\widehat{\lambda}_i^{(r)} = d_i \exp(\widetilde{\boldsymbol{x}}_i^T \widehat{\boldsymbol{\beta}}^{(r)})$. Appropriate starting values are given by

$$\widehat{\beta}_0^{(0)} = \ln \frac{1}{n} \sum_{i=1}^{n} k_i \text{ and } \widehat{\beta}_j^{(0)} = 0 \text{ for } j = 1, \ldots, p.$$

Note that these starting values are equal to the values of the regression coefficients when no segmentation is in force. Therefore, final values close to the starting ones indicate that the portfolio is quite homogeneous.

Remark 2.2 (Iterative Least-Squares) It is possible to interpret the Newton–Raphson approach (2.7) in terms of iterative least-squares. Specifically, it is possible to re-write the iterative algorithm (2.7) in the Poisson model in such a way that $\widehat{\boldsymbol{\beta}}^{(r+1)}$ appears as the maximum likelihood estimator in a linear model with adjusted dependent variables. Fitting the Poisson regression model by maximum likelihood thus boils down to estimating the regression parameter in a sequence of linear models, with adjusted responses and explanatory variables. This is particularly interesting since the numerical aspects of estimation in a linear model are well-known and have been optimized for decades.

2.3.8 Wald Confidence Intervals

The asymptotic variance-covariance matrix $\boldsymbol{\Sigma}_{\widehat{\boldsymbol{\beta}}}$ of the maximum likelihood estimator $\widehat{\boldsymbol{\beta}}$ of the regression coefficients vector $\boldsymbol{\beta}$ is the inverse of the Fisher information matrix. This matrix can be estimated by

$$\widehat{\boldsymbol{\Sigma}}_{\widehat{\boldsymbol{\beta}}} = \left(\sum_{i=1}^{n} \widetilde{\boldsymbol{x}}_i \widetilde{\boldsymbol{x}}_i^T \widehat{\lambda}_i\right)^{-1}, \text{ where } \widehat{\lambda}_i = d_i \exp(\widehat{\text{score}}_i).$$

We know from (1.49) that provided the sample size is large enough $\widehat{\boldsymbol{\beta}} - \boldsymbol{\beta}$ is approximately $\mathcal{N}or(\mathbf{0}, \widehat{\boldsymbol{\Sigma}}_{\widehat{\boldsymbol{\beta}}})$ distributed. It is thus possible to compute confidence intervals at level $1-\alpha$ for each of the β_js. These intervals are of the form

$$\left[\widehat{\beta}_j - z_{\alpha/2}\widehat{\sigma}_{\widehat{\beta}_j}, \widehat{\beta}_j + z_{\alpha/2}\widehat{\sigma}_{\widehat{\beta}_j}\right] \tag{2.8}$$

where $\widehat{\sigma}^2_{\widehat{\beta}_j}$ is the estimated variance of $\widehat{\beta}_j$, given by the element (j, j) of $\widehat{\boldsymbol{\Sigma}}_{\widehat{\boldsymbol{\beta}}}$.

Remark 2.3 (Confidence Intervals for the β_js: the Likelihood Ratio Method) The confidence interval (2.8) is based on the large sample properties of the maximum likelihood estimator $\widehat{\boldsymbol{\beta}}$. Other methods for constructing such a confidence interval are available. One such method is based on the profile likelihood for β_j that is defined as

$$\mathcal{L}_j(\beta_j) = \max_{\beta_0,\ldots,\beta_{j-1},\beta_{j+1},\ldots,\beta_p} \mathcal{L}(\boldsymbol{\beta}).$$

If $\widehat{\boldsymbol{\beta}}$ is the maximum likelihood estimator of $\boldsymbol{\beta}$, we have that $2\big(L(\widehat{\boldsymbol{\beta}}) - L_j(\beta_j)\big)$ is approximately χ^2_1 provided β_j is the true parameter value. A confidence interval at level $1-\alpha$ for β_j is then given by

$$\left\{\beta_j \middle| L_j(\beta_j) \geq L(\widehat{\boldsymbol{\beta}}) - \frac{1}{2}\chi^2_{1,1-\alpha}\right\},$$

where $\chi^2_{1,1-\alpha}$ is the $(1-\alpha)$th quantile of the χ^2_1 distribution.

2.3.9 Testing for Hypothesis on a Single Parameter

It is often interesting to check the validity of the null hypothesis H_0: $\beta_j = 0$ against the alternative H_1: $\beta_j \neq 0$. If the jth explanatory variable is dichotomous (think for instance of gender), then failing to reject H_0 suggests that this variable is not relevant to explaining the expected number of claims. If the jth explanatory variable is coded by means of a set of binary variables, then the nullity of the regression coefficient associated with one of the binary variables means that the corresponding level can be grouped with the reference level. In such a case, equality between the regression coefficients should also be tested to decide about the optimal grouping of the levels. Hypotheses involving a set of regression parameters will be examined in Section 2.3.13.

Considering (1.52), the easiest test statistic for H_0 against H_1 is certainly

$$T = \frac{\widehat{\beta}_j}{\widehat{\sigma}_{\widehat{\beta}_j}}$$

which is approximately $\mathcal{N}or(0, 1)$ under H_0, provided the sample size is large enough. Alternatively, T^2 is approximately Chi-square with one degree of freedom. Rejection of H_0 occurs when T is large in absolute values, or when T^2 is large. In this case, β_j is significantly different from 0, and the associated characteristic has a significant impact in ratemaking. Note that, as always with hypothesis testing, statistical significance is the resultant of two effects: firstly, the distance between the true parameter value and the hypothesized value, and secondly, the number of observations (or more precisely, the amount of information contained in the data). Even a hypothesis that is approximately true (and useful as a working hypothesis) will be rejected with a sufficiently large sample. Conversely, any hypothesis may fail to be rejected (and accepted as a working hypothesis) as long as the actuary has only scanty data at his disposal. The reader should keep this in mind in the numerical illustrations worked out in this book.

If the explanatory variables are correlated (as it is usually the case in actuarial studies), it becomes difficult to disentangle the effects of one explanatory variable from another, and the parameter estimates may be highly dependent on which explanatory variables are used in the model. If the explanatory variables are strongly correlated then the maximum likelihood estimators will have a large variance. The actuary should then reduce the set of regressors.

2.3.10 Confidence Interval for the Expected Annual Claim Frequency

It is possible to build a confidence interval for the annual claim frequency. Recall that the multivariate Normal distribution has the following useful invariance property. Let $\boldsymbol{C}$ be a given $n \times n$ matrix with real entries and let $\boldsymbol{b}$ be a n-dimensional real vector. If $\boldsymbol{X} \sim \mathcal{N}or(\boldsymbol{\mu}, \boldsymbol{M})$ then $\boldsymbol{Y} = \boldsymbol{C}\boldsymbol{X} + \boldsymbol{b}$ is $\mathcal{N}or(\boldsymbol{C}\boldsymbol{\mu} + \boldsymbol{b}, \boldsymbol{C}\boldsymbol{M}\boldsymbol{C}^T)$. The variance of the predicted score, $\widehat{\text{score}_i} = \widetilde{\boldsymbol{x}}_i^T \widehat{\boldsymbol{\beta}}$, is thus given by

$$\mathbb{V}[\widehat{\text{score}_i}] = \widetilde{\boldsymbol{x}}_i^T \boldsymbol{\Sigma}_{\widehat{\boldsymbol{\beta}}} \widetilde{\boldsymbol{x}}_i$$

which is estimated by

$$\widehat{\mathbb{V}[\widehat{\text{score}_i}]} = \widetilde{\boldsymbol{x}}_i^T \widehat{\boldsymbol{\Sigma}}_{\widehat{\boldsymbol{\beta}}} \widetilde{\boldsymbol{x}}_i.$$

As the maximum likelihood estimator $\widehat{\boldsymbol{\beta}}$ is approximately Gaussian when the number of policies is large, $\widehat{\text{score}_i}$ is also Gaussian and an approximate confidence interval at level $1 - \alpha$ for the annual claim frequency can be computed as

$$\left[\exp\left(\widehat{\boldsymbol{\beta}}^T \widetilde{\boldsymbol{x}}_i - z_{\alpha/2}\sqrt{\widetilde{\boldsymbol{x}}_i^T \widehat{\boldsymbol{\Sigma}}_{\widehat{\boldsymbol{\beta}}} \widetilde{\boldsymbol{x}}_i}\right), \exp\left(\widehat{\boldsymbol{\beta}}^T \widetilde{\boldsymbol{x}}_i + z_{\alpha/2}\sqrt{\widetilde{\boldsymbol{x}}_i^T \widehat{\boldsymbol{\Sigma}}_{\widehat{\boldsymbol{\beta}}} \widetilde{\boldsymbol{x}}_i}\right)\right].$$

2.3.11 Deviance

Let $\mathcal{L}(\widehat{\boldsymbol{\lambda}})$ be the model likelihood, i.e.

$$\mathcal{L}(\widehat{\boldsymbol{\lambda}}) = \prod_{i=1}^{n} \exp(-\widehat{\lambda}_i) \frac{\widehat{\lambda}_i^{k_i}}{k_i!}.$$

Note that the maximal value of $\lambda \mapsto \exp(-\lambda)\lambda^k/k!$ is obtained for $\lambda = k$. Therefore, the Poisson likelihood is maximum with expected claim frequencies equal to the observed number of claims. The maximal likelihood possible under the Poisson assumption is then

$$\mathcal{L}(\boldsymbol{k}) = \prod_{i=1}^{n} \exp(-k_i) \frac{k_i^{k_i}}{k_i!}.$$

This is the likelihood of the saturated model, predicting the observed number of claims for each insured driver (there are thus as many parameters as observations). This model just replicates the observed data.

The deviance $D(\boldsymbol{k}, \widehat{\boldsymbol{\lambda}})$ is defined as the likelihood ratio test statistic for the current model against the saturated model, that is,

$$\begin{aligned} D(\boldsymbol{k}, \widehat{\boldsymbol{\lambda}}) &= -2 \ln \frac{\mathcal{L}(\widehat{\boldsymbol{\lambda}})}{\mathcal{L}(\boldsymbol{k})} = 2\Big(\ln \mathcal{L}(\boldsymbol{k}) - \ln \mathcal{L}(\widehat{\boldsymbol{\lambda}}) \Big) \\ &= 2 \ln \left(\prod_{i=1}^{n} \exp(-k_i) \frac{k_i^{k_i}}{k_i!} \right) - 2 \ln \left(\prod_{i=1}^{n} \exp(-\widehat{\lambda}_i) \frac{\widehat{\lambda}_i^{k_i}}{k_i!} \right) \\ &= 2 \sum_{i=1}^{n} \left(k_i \ln \frac{k_i}{\widehat{\lambda}_i} - (k_i - \widehat{\lambda}_i) \right) \end{aligned}$$

where $y \ln y = 0$ for $y = 0$ by convention. It measures the distance of the model likelihood to the saturated model replicating the observed data. The smaller the deviance, the better the current model.

When an intercept β_0 is included in the linear predictor, (2.3) allows us to simplify the deviance as

$$D(\boldsymbol{k}, \widehat{\boldsymbol{\lambda}}) = 2 \sum_{i=1}^{n} k_i \ln \frac{k_i}{\widehat{\lambda}_i}.$$

Provided the data have been grouped as much as possible and the model is correct, D is approximately $\chi^2_{n-\dim(\boldsymbol{\beta})}$ distributed (where n is now the number of classes in the portfolio). The model is considered as inappropriate if D_{obs} is 'too large', that is, if

$$D_{obs} > \chi^2_{n-\dim(\boldsymbol{\beta}); 1-\alpha}.$$

2.3.12 Deviance Residuals

The deviance residuals in the Poisson model are given by:

$$r_i^D = \sqrt{2}\mathrm{sign}(k_i - \widehat{\lambda}_i)\sqrt{k_i \ln \frac{k_i}{\widehat{\lambda}_i} - (k_i - \widehat{\lambda}_i)}.$$

Summing the $(r_i^D)^2$ gives the deviance $D(\boldsymbol{k}, \widehat{\boldsymbol{\lambda}})$. The deviance residual r_i^D is thus the signed square root of the contribution of policyholder i to the deviance $D(\boldsymbol{k}, \widehat{\boldsymbol{\lambda}})$.

A plot for individual data is often uninformative in motor insurance (because of the few observed values for the k_is, the deviance residuals r_i^D are always structured, being concentrated along curves corresponding to 0 claim, 1 claim, 2 claims, etc.). A plot of the residuals against fitted frequencies $\widehat{\lambda}_i$ for grouped data helps to check the adequacy of the model.

2.3.13 Testing a Hypothesis on a Set of Parameters

We would like to test the null hypothesis

$$\begin{cases} H_0 : \boldsymbol{\beta} = \boldsymbol{\beta}_0 = (\beta_0, \beta_1, \beta_2, \ldots, \beta_q)^T \\ H_1 : \boldsymbol{\beta} = \boldsymbol{\beta}_1 = (\beta_0, \beta_1, \beta_2, \ldots, \beta_q, \beta_{q+1}, \ldots, \beta_p)^T. \end{cases}$$

Let D_0 be the deviance of the Poisson regression model under H_0, and D_1 be the deviance under H_1. The test statistic is

$$\begin{aligned} \Delta = D_0 - D_1 &= 2\left(L(\boldsymbol{k}) - L(\widehat{\boldsymbol{\beta}}_0)\right) - 2\left(L(\boldsymbol{k}) - L(\widehat{\boldsymbol{\beta}}_1)\right) \\ &= 2\left(L(\widehat{\boldsymbol{\beta}}_1) - L(\widehat{\boldsymbol{\beta}}_0)\right) \approx_d \chi^2_{p-q}. \end{aligned}$$

Note that Δ is a likelihood ratio test statistic. The null hypothesis H_0 is rejected in favour of H_1 if Δ_{obs} is 'too large', that is, if

$$\Delta_{obs} > \chi^2_{p-q;1-\alpha}.$$

2.3.14 Specification Error and Robust Inference

According to the asymptotic theory related to generalized linear models, the estimator $\widehat{\boldsymbol{\beta}}$ obtained by maximizing the Poisson likelihood remains consistent and efficient provided the mean and variance of the model are correctly specified (even if the underlying data generating process is not Poisson). Moreover, in order to obtain consistency, only the correct specification of the mean function is required, that is, only (2.1) has to be valid. And $\widehat{\boldsymbol{\beta}}$ remains Normally distributed in all cases.

Let us now assume that $\mathbb{E}[N_i|\boldsymbol{x}_i] = d_i \exp(\boldsymbol{\beta}_0^T \boldsymbol{x}_i)$ holds true for some $\boldsymbol{\beta}_0$ (i.e. the conditional mean is correctly specified), but N_i is not Poisson distributed (N_i is in reality Negative Binomial, for instance). In this case, there is a specification error, in that inference conducted with the Poisson likelihood is based on a false distributional assumption. The Poisson maximum likelihood estimator $\widehat{\boldsymbol{\beta}}$ remains nevertheless consistent for the true parameter $\boldsymbol{\beta}_0$, i.e. $\widehat{\boldsymbol{\beta}} \rightarrow_{\text{proba}} \boldsymbol{\beta}_0$ as the sample size $n \rightarrow +\infty$. This explains why the Poisson regression model is so useful: it continues to give reliable estimations for the annual expected claim frequency even if the true model is not Poisson, provided the sample size is large enough.

However, the variances of the $\widehat{\beta}_j$s are mis-estimated. Inference must then be based on the robust information matrix estimate of the variance-covariance matrix of $\widehat{\boldsymbol{\beta}}$ that is based on the empirical estimate of the observed information. Specifically, because of the misspecification, the asymptotic variance-covariance matrix of $\widehat{\boldsymbol{\beta}}$ is now given by

$$\boldsymbol{\Sigma}_{\widehat{\boldsymbol{\beta}}} = \mathcal{I}^{-1}\mathcal{J}\mathcal{I}$$

where

$$\mathcal{I} = \sum_{i=1}^{n} \widetilde{\boldsymbol{x}}_i \widetilde{\boldsymbol{x}}_i^T d_i \exp(\boldsymbol{\beta}_0^T \widetilde{\boldsymbol{x}}_i) \text{ and } \mathcal{J} = \sum_{i=1}^{n} \widetilde{\boldsymbol{x}}_i \widetilde{\boldsymbol{x}}_i^T \mathbb{V}[N_i|\boldsymbol{x}_i].$$

In practice, it will be estimated by

$$\widehat{\boldsymbol{\Sigma}}_{\widehat{\boldsymbol{\beta}}} = \widehat{\mathcal{I}}^{-1}\widehat{\mathcal{J}}\widehat{\mathcal{I}} \tag{2.9}$$

where

$$\widehat{\mathcal{I}} = \sum_{i=1}^{n} \widetilde{\boldsymbol{x}}_i \widetilde{\boldsymbol{x}}_i^T \widehat{\lambda}_i \text{ and } \widehat{\mathcal{J}} = \sum_{i=1}^{n} \widetilde{\boldsymbol{x}}_i \widetilde{\boldsymbol{x}}_i^T (\widehat{\lambda}_i - k_i)^2.$$

with $\widehat{\lambda}_i = d_i \exp(\widehat{\boldsymbol{\beta}}^T \widetilde{\boldsymbol{x}}_i)$. Let us point out that $\widehat{\boldsymbol{\beta}} - \boldsymbol{\beta}$ remains approximately Normally distributed with mean $\mathbf{0}$ and covariance matrix $\boldsymbol{\Sigma}_{\widehat{\boldsymbol{\beta}}}$, provided the sample size is large enough.

2.3.15 Numerical Illustration

Within SAS®, the GENMOD procedure can be used to fit Poisson regression models. This procedure supports the Normal, Binomial, Poisson, Gamma, Inverse Gaussian, Negative Binomial and Multinomial distributions, in the framework of generalized linear models. A typical use of the GENMOD procedure is to perform Poisson regression with a log link function. This type of model is usually called a loglinear model.

The logarithm of the exposure-to-risk is used as an offset, that is, a regression variable with a constant coefficient of 1 for each observation. A log linear relationship between the mean and the explanatory factors is specified by the log link function. The log link function ensures that the mean number of insurance claims predicted from the fitted model is positive.

The results obtained from the Poisson regression for Portfolio A presented in Section 2.2 are shown in Table 2.1. Table 2.1 is similar to the 'Analysis of Parameter Estimates' table produced by the GENMOD procedure. Such a table summarizes the results of the iterative parameter estimation process. For each parameter in the model, the GENMOD procedure displays columns with the parameter name, the degrees of freedom associated with the parameter, the estimated parameter value, the standard error of the parameter estimate, the

Table 2.1 Results of the Poisson regression for the model with the 5 explanatory variables, Portfolio A.

Variable	Level	Coeff β	Std error	Wald 95 %	conf limit	Chi-sq	Pr>Chi-sq
Intercept		−2.2131	0.0582	−2.3271	−2.0991	1447.40	<.0001
Gender*Age	Female 18–24 years	0.3072	0.1117	0.0883	0.5261	7.57	0.0059
Gender*Age	Female 25–30 years	0.1620	0.0876	−0.0098	0.3337	3.42	0.0646
Gender*Age	Female 31–60 years	0.0651	0.0802	−0.0920	0.2222	0.66	0.4166
Gender*Age	Female > 60 years	−0.0010	0.2350	−0.4616	0.4596	0.00	0.9967
Gender*Age	Male 18–24 years	0.6429	0.0797	0.4867	0.7990	65.10	<.0001
Gender*Age	Male 25–30 years	0.2875	0.0713	0.1477	0.4273	16.24	<.0001
Gender*Age	Male > 60 years	−0.0623	0.1425	−0.3416	0.2170	0.19	0.6621
Gender*Age	Male 31–60 years	0	0	0	0	.	.
District	Rural	−0.1828	0.0503	−0.2814	−0.0842	13.19	0.0003
District	Urban	0	0	0	0	.	.
Premium split	Yes	0.4615	0.0515	0.3607	0.5624	80.43	<.0001
Premium split	No	0	0	0	0	.	.
Use of the car	Professional	0.2213	0.0784	0.0677	0.3749	7.98	0.0047
Use of the car	Private	0	0	0	0	.	.

confidence intervals, and the Wald Chi-square statistic and associated p-value for testing the significance of the parameter to the model. The GENMOD procedure assigns zero degrees of freedom to the reference level of each explanatory variable and displays a value of zero for both the parameter estimate and its standard error.

The column 'Coeff β' in Table 2.1 gives the maximum likelihood estimation $\widehat{\beta}_j$ of the parameter β_j associated with each level of the explanatory variables. Recall that positive values of the $\widehat{\beta}_j$s indicate higher risk compared to the reference class, whereas negative values demonstrate lower risk than the reference class. The column 'Std Error' gives the value of $\widehat{\sigma}_{\beta_j}$, which is the square root of the jth diagonal element of $\widehat{\boldsymbol{\Sigma}}_{\boldsymbol{\beta}}$. The columns 'Wald 95 % Conf Limit' give the left and right endpoints of the confidence interval (2.8) at level 95 % (so that $\alpha = 5\,\%$ and the Normal quantile is approximately equal to 2). The columns 'Chi-sq' and 'Pr>Chi-sq' (which is the associated p-value) enable us to test for H_0: $\beta_j = 0$ against H_1: $\beta_j \neq 0$. The test statistic is $\widehat{\beta}_j^2/\widehat{\sigma}^2_{\beta_j}$ that approximately follows the χ^2_1 distribution. The values in the column 'Chi-sq' are obtained by squaring the ratio of the elements of the column 'Coeff β' to the corresponding elements of the column 'Std error'. If the p-value is lower than 5 %, then H_0 is rejected (and β_j is significantly different from 0). In such a case, the characteristic coded by the jth explanatory variable significantly influences the number of claims reported to the company. The reliance on these test results must rest (implicitly) on the amount of information contained in the data.

When the variables Age and Gender are introduced separately in the ratemaking, the analysis leads to the rejection of the variable Gender because it is not relevant (p-value of 9.13 %). However, the Age–Gender interaction is statistically relevant, and is therefore introduced in the model instead of the separated variables Age and Gender. The value of the log-likelihood for this first model is equal to −5482.40 (note that some commercial softwares give the log-likelihood up to a constant term whereas in this chapter, the full expression is used). The ratio of the deviance to the number of degrees of freedom should be close to 1 to indicate goodness-of-fit. Here, we obtain 0.5357 (the deviance is equal to 7764.83). Note however that the data should be grouped to make the Chi-square approximation more reliable, so that we cannot use this statistic at the present stage.

An important aspect of insurance ratemaking with generalized regression models is the selection of explanatory variables in the model. Changes in goodness-of-fit statistics are often used to evaluate the contribution of subsets of explanatory variables to a particular model. One strategy for variable selection is to fit a sequence of models, beginning with a simple model with only an intercept term, and then include one additional explanatory variable in each successive model. The importance of the additional explanatory variable can be measured by the difference in fitted log-likelihoods between successive models. Asymptotic tests computed by the GENMOD procedure enable the actuary to assess the statistical significance of the additional term (this is called Type I analysis in SAS®).

Another strategy (adopted here) consists of starting from a model incorporating all the available information, and then excluding the irrelevant explanatory variables. To this end, the GENMOD procedure generates a Type 3 analysis (analogous to Type III sums of squares in the GLM procedure). A Type 3 analysis does not depend on the order in which the terms for the model are specified (in contrast to the Type 1 analysis).

Type 3 analysis compares the complete model (that is the model which includes all the specified variables) with the different submodels obtained by deleting one of the explanatory variables. It enables the actuary to test the relevance of one variable taking all the others into account. It roughly corresponds to the backward approach: at each step, we exclude the variable with the largest p-value until no more can be excluded (i.e. until all the p-values are smaller than a fixed threshold, generally 5 %). Note that the Type 3 analysis works with the variables and not with the levels of these variables. Indeed, it is possible to obtain a relevant variable for which some levels are not relevant. The results of the Type 3 analysis are as follows:

Source	DF	Chi-square	Pr > Chi-sq
Gender*Age	7	75.16	<.0001
District	1	13.41	0.0003
Premium split	1	77.27	<.0001
Use of the car	1	7.62	0.0058

This analysis allows us to test the contribution of each variable with respect to a model that does not contain this variable. Specifically, the comparison of the model with all the covariates to the model that does not contain the combined Gender*Age variable yields a p-value of less than 10^{-4}, so that the omission of the interaction between Age and Gender significantly degrades the fit to the observed data. This covariate should therefore be kept in the model. All the explanatory variables are statistically relevant. A Type 3 analysis can consume considerable computation time since a constrained model is fitted for each effect. If needed, Wald statistics for Type 3 contrasts can be computed instead of likelihood ratios. Wald statistics for contrasts use less computation time than likelihood ratio statistics but may be less accurate indicators of the significance of the effect of interest.

The option 'Estimate' of GENMOD (which is similar to the option 'Contrast') can be used to assess the relevance of grouping the levels of the interaction Age–Gender 2 by 2. This option computes likelihood ratio statistics for user-defined contrasts, that is, linear functions of the parameters, and p-values based on their asymptotic Chi-square distributions. Here, the option 'Estimate' is used to test for the equality of the regression coefficients (that is, for H_0: $\beta_{j_1} = \beta_{j_2}$ against H_1: $\beta_{j_1} \neq \beta_{j_2}$ for all the combinations $j_1 j_2$ corresponding to binary variables coding Age*Gender). The test statistic is based on the ratio of the likelihoods corresponding to the models under H_0 and under H_1. This grouping process can be summarized as follows:

Step 1 the reference level 'Male 31–60' is first merged with 'Female >60' (p-value of 99.67 %);

Step 2 then 'Male 25–30' and 'Female 18–24' are grouped together (p-value of 85.85 %);

Step 3 'Male 31–60', 'Male > 60' and 'Female > 60' are grouped together (p-value of 66.13 %);

Step 4 'Male 31–60', 'Male > 60', 'Female 31–60' and 'Female > 60' are grouped together (p-value of 35.17 %);

Step 5 finally 'Male 25–30', 'Female 18–24' and 'Female 25–30' are grouped together (p-value of 12.39 %).

The level 'Male 18–24' cannot be grouped with other levels. After grouping, the variable Age*Gender resulting from the interaction of Age with Gender has three levels: 'Males 18–24', 'Females 18–30 and Males 25–30', and 'Males and Females over 30'. As expected, it accounts for the extra risk of young male drivers (aged between 18 and 24), and of the young drivers, whereas the reference level 3 is assigned to drivers over 30.

The fit of the final model is shown in Table 2.2. The log-likelihood is now equal to −5484.2 and the Type 3 analysis gives the following results:

Source	DF	Chi-square	Pr>Chi-sq
Gender*Age	2	71.66	<.0001
District	1	13.15	0.0003
Premium split	1	79.68	<.0001
Use of the car	1	7.34	0.0067

We notice that all the explanatory variables and all the levels of the Age–Gender interaction are now statistically relevant. The deviance is 7768.332, and the ratio of the deviance to the number of degrees of freedom is 0.5358. The estimated variance-covariance matrix of the regression parameters is given by

$$\widehat{\Sigma}_{\widehat{\beta}} = \begin{pmatrix} 0.002170 & -0.001385 & -0.001315 & -0.000907 & -0.000913 & -0.001370 \\ -0.001385 & 0.002893 & 0.001535 & -0.000075 & -0.000262 & 0.000748 \\ -0.001315 & 0.001535 & 0.005167 & -0.000195 & -0.000528 & 0.000967 \\ -0.000907 & -0.000075 & -0.000195 & 0.002529 & 0.000194 & -0.000031 \\ -0.000913 & -0.000262 & -0.000528 & 0.000194 & 0.002633 & 0.000583 \\ -0.001370 & 0.000748 & 0.000967 & -0.000031 & 0.000583 & 0.006014 \end{pmatrix}.$$

Table 2.2 Results of the Poisson regression for the final model, Portfolio A.

Variable	Level	Coeff β	Std error	Wald 95 %	conf limit	Chi-sq	Pr>Chi-sq
Intercept		−2.1975	0.0466	−2.2888	−2.1062	2225.51	<.0001
Gender*Age	Female 18–30 + Male 25–30	0.2351	0.0538	0.1297	0.3405	19.10	<.0001
Gender*Age	Male 18–24	0.6235	0.0719	0.4826	0.7644	75.24	<.0001
Gender*Age	Female > 30+ Male > 30	0	0	0	0	.	.
District	Rural	−0.1809	0.0503	−0.2795	−0.0823	12.94	0.0003
District	Urban	0	0	0	0	.	.
Premium split	Yes	0.4677	0.0513	0.3671	0.5683	83.06	<.0001
Premium split	No	0	0	0	0	.	.
Use of the car	Professional	0.2150	0.0776	0.0630	0.3670	7.69	0.0056
Use of the car	Private	0	0	0	0	.	.

Taking the squares of the values in the column 'Std error' of Table 2.2 gives the diagonal elements of $\widehat{\boldsymbol{\Sigma}}_{\widehat{\beta}}$.

The deviance residuals are displayed in Figure 2.8. A structure inherited from the few observed values is clearly apparent. The individual deviance residuals are concentrated around 5 curves, corresponding to the annual numbers of claims observed in Portfolio A. To get a more informative graph, we compute the deviance residuals for each risk class. The structure then disappears, which supports the regression model. The value of D_{obs} on grouped data is equal to 15.50 and the value of $\chi^2_{17;0.95}$ is 8.67 which leads to the rejection of the Poisson regression model. The reasons for this rejection will be explained below.

Another way to challenge the Poisson assumption is to use the robust version of Poisson regression. The computations of (2.9) can be performed with an appropriate IML program. Alternatively, the 'Repeated' option of the SAS®/STAT procedure GENMOD produces variances and covariances of the maximum likelihood estimators computed from the robust information matrix. We will see below that this option allows us to conduct an analysis of correlated data using the GEE approach. When there is only a single observation per subject, the 'Repeated' statement may still be used, in which case the robust $\widehat{\boldsymbol{\Sigma}}_{\widehat{\beta}}$ given in (2.9) is computed.

The results of the robust Poisson regression are displayed in Table 2.3. The estimated regression coefficients are the same as before, but the standard errors are now computed according to the robust rule. We see that the standard errors are larger than in the Poisson case, but all the variables remain nevertheless significant. A Type 3 analysis gives the following results:

Source	DF	Chi-square	Pr>Chi-sq
Gender*Age	2	54.06	<.0001
District	1	11.95	0.0005
Premium split	1	61.56	<.0001
Use of the car	1	6.04	0.0140

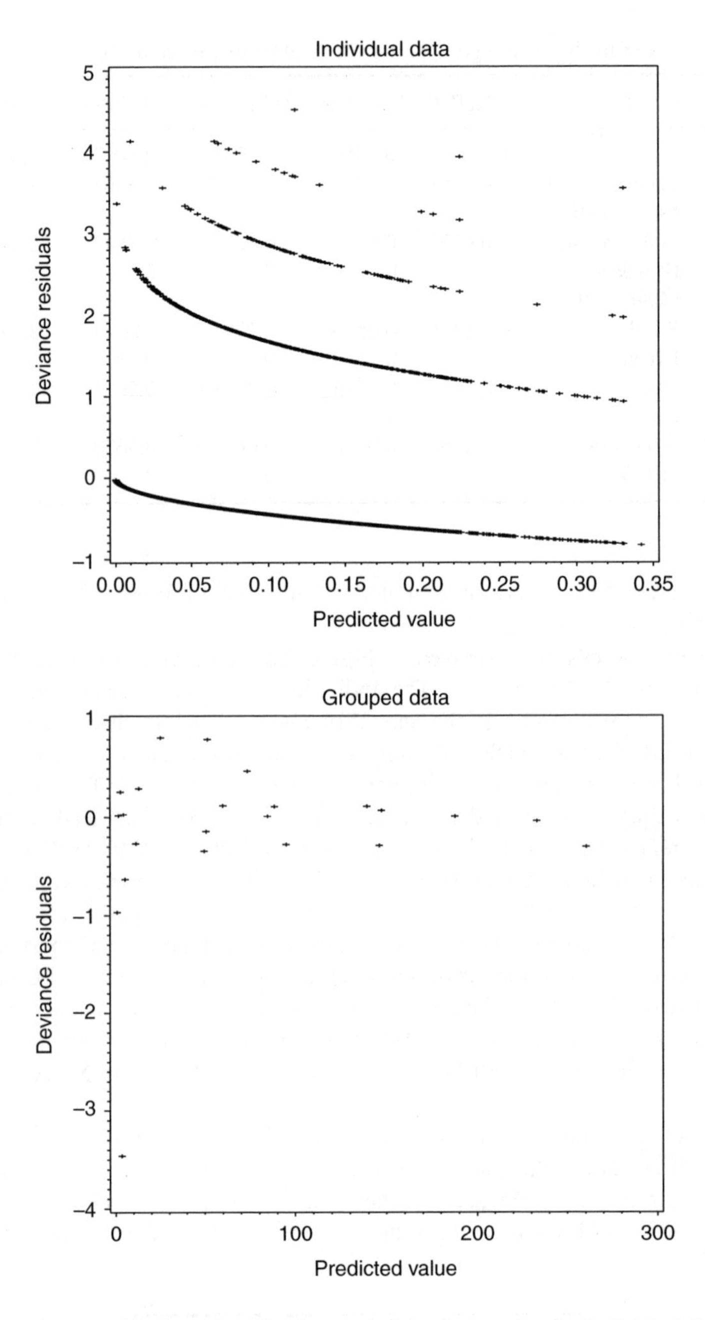

Figure 2.8 Deviance residuals against fitted annual claim frequencies for individual data (top panel) and grouped by risk classes (bottom panel), Portfolio A.

The significance of the explanatory variables is clearly apparent (even if the Chi-square test statistics are smaller than in the regular Poisson case, and the corresponding p-values larger).

Table 2.3 Results of the robust Poisson regression for the final model, Portfolio A.

Variable	Level	Coeff β	Std error	Wald 95 %	conf limit	Z	Pr > $\|Z\|$
Intercept		−2.1975	0.0487	−2.2930	−2.1020	−45.10	<.0001
Gender*Age	Female 18–30 + Male 25–30	0.2351	0.0575	0.1223	0.3479	4.09	<.0001
Gender*Age	Male 18–24	0.6235	0.0762	0.4741	0.7729	8.18	<.0001
Gender*Age	Female > 30+ Male > 30	0	0	0	0	.	.
District	Rural	−0.1809	0.0533	−0.2855	−0.0764	−3.39	0.0007
District	Urban	0	0	0	0	.	.
Premium split	Yes	0.4677	0.0544	0.3611	0.5742	8.60	<.0001
Premium split	No	0	0	0	0	.	.
Use of the car	Professional	0.2150	0.0813	0.0557	0.3743	2.65	0.0082
Use of the car	Private	0	0	0	0	.	.

2.4 Overdispersion

2.4.1 Explanation of the Phenomenon

Despite its prevalence as a starting point in the analysis of count data, the Poisson specification is often inappropriate because of unobserved heterogeneity and failure of the independence assumption if the data consist of repeated observations on the same policyholders (the problem of serial dependence of the data will be addressed in Section 2.9). This has been confirmed by the rejection of the Poisson fit for Portfolio A.

Unobserved heterogeneity and serial dependence will often cause the variance to exceed the mean (a phenomenon termed overdispersion) whereas the Poisson regression model imposes a strong constraint of equidispersion once the explanatory variables have been taken into account, that is,

$$\mathbb{E}[N_i|\boldsymbol{x}_i] = \mathbb{V}[N_i|\boldsymbol{x}_i] = \lambda_i.$$

Failing to account for overdispersion yields underestimated standard errors, thereby conveying erroneously high levels of significance.

The unobserved heterogeneity is due to the fact that important explanatory variables may not have been measured and are consequently incorrectly excluded from the regression relationship. A convenient way to take this phenomenon into account is to introduce a random effect into the model. This is the classical credibility construction (as explained in Chapter 3). We explain, in the following subsections, how to introduce this random effect.

2.4.2 Interpreting Overdispersion

Let us consider two risk classes C_1 and C_2 without overdispersion, i.e. such that $\sigma_1^2 = m_1$ and $\sigma_2^2 = m_2$, where σ_i^2 and m_i denote the variance and the mean of the claim number in C_i, respectively. Now, assume that the actuary is not able to discriminate between those policyholders in C_1 and those in C_2 (and thus works in the class $C_1 \cup C_2$).

In the class $C_1 \cup C_2$, the expected claim number equals

$$m = p_1 m_1 + p_2 m_2$$

where p_1 and p_2 denote the respective weights of C_1 and C_2 (the ratio of the class exposure to the sum of the exposures of both classes, say). Considering the variance of the number of claims in $C_1 \cup C_2$, it can be decomposed as the average of the conditional variances σ_1^2 and σ_2^2 plus the variance of the conditional means m_1 and m_2 (that is, the weighted sum of their squared difference with respect to the grand mean m). The variance thus becomes

$$\underbrace{p_1\sigma_1^2 + p_2\sigma_2^2}_{=m} + p_1\underbrace{(m_1 - m)^2}_{>0} + p_2\underbrace{(m_2 - m)^2}_{>0} > m,$$

which exceeds the mean. Hence, omitting relevant ratemaking variables induces overdispersion.

2.4.3 Consequences of Overdispersion

As mentioned in Section 2.3.14, misspecification of the variance function does not affect the consistency of $\widehat{\boldsymbol{\beta}}$, but leads to misspecification of the asymptotic variance-covariance matrix of $\widehat{\boldsymbol{\beta}}$. As a result, we have a loss of efficiency.

Overdispersion leads to underestimates of standard errors and overestimates of Chi-square statistics (as demonstrated in Table 2.3), which in turn may imply artificial statistical significance for the parameters. Consequently, some explanatory variables may become not significant after overdispersion has been accounted for. In practice, failing to account for overdispersion might produce too many risk classes in the portfolio.

2.4.4 Modelling Overdispersion

Many explanatory variables are unknown to the insurance company or cannot be incorporated in the price list (for legal, moral or economic reasons). There are thus unobservable characteristics $\boldsymbol{Z}_i$ that may influence the number of claims filed by policyholder i as explained in Section 2.1.2. Of course, some Z_{ij}s may be correlated with the observable characteristics $\boldsymbol{X}_i$. To remove these correlations, we could think of first regressing the $\boldsymbol{Z}_i$s on the $\boldsymbol{X}_i$s, with a linear regression model

$$Z_{ij} = \delta_0 + \sum_{k=1}^{p} \delta_k X_{ik} + \epsilon_{ij}.$$

Then, the score becomes

$$\begin{aligned}
\beta_0 + \sum_{j=1}^{p} \beta_j X_{ij} + \sum_{j=1}^{\dim(\boldsymbol{Z}_i)} \gamma_j Z_{ij} &= \beta_0 + \sum_{j=1}^{p} \beta_j X_{ij} + \sum_{j=1}^{\dim(\boldsymbol{Z}_i)} \gamma_j \Big(\delta_0 + \sum_{k=1}^{p} \delta_k X_{ik} + \epsilon_{ij}\Big) \\
&= \widetilde{\beta}_0 + \sum_{j=1}^{p} \widetilde{\beta}_j X_{ij} + \widetilde{\epsilon}_i
\end{aligned}$$

for appropriate $\widetilde{\beta}_0$, $\widetilde{\beta}_j$s and $\widetilde{\epsilon}_i$.

The unobserved heterogeneity (when correlated to observable characteristics) thus modifies the regression coefficients: the true effect $\boldsymbol{\beta}$ of $\boldsymbol{X}_i$ on N_i becomes an apparent effect $\widetilde{\boldsymbol{\beta}}$. Hence, the estimated jth regression coefficient does not only represent the effect of the jth covariate on the number of claims, but also accounts for the effect of all the hidden characteristics $\boldsymbol{Z}_i$ correlated with the jth observable one. This is why the $\widehat{\beta}_j$s may strongly depend on which covariates are included in the model. Moreover, there remains an error term $\widetilde{\epsilon}_i$ representing the influence of the hidden variables on N_i, corrected for the effect of the observed risk factors $\boldsymbol{X}_i$.

For these reasons, we now consider a mixed Poisson model

$$N_i \sim \mathcal{P}oi\left(\exp\left(\beta_0 + \sum_{j=1}^{p} \beta_j x_{ij} + \epsilon_i\right)\right), \quad i = 1, 2, \ldots, n \tag{2.10}$$

where the random variable ϵ_i represents the residual effect of the hidden characteristics. Therefore, the heterogeneity is taken into account by assuming that the number of accidents is Poisson distributed with mean varying from one policyholder to another.

Note that some hidden characteristics are correlated to those in $\boldsymbol{X}_i$ (e.g. the hidden annual mileage and the observable use of the vehicle). The random variable ϵ_i in (2.10) models the effect of hidden characteristics that is not already explained by $\boldsymbol{X}_i$. Since ϵ_i accounts for a residual effect, we will consider in the remainder of this book that ϵ_i is independent from $\boldsymbol{X}_i$. The price to pay is that the estimated regression coefficient β_j does not only express the effect of the jth regressor, but also the effect of all the hidden characteristics correlated with the jth regressor. This is important when the actuary tries to interpret the resulting price list.

The policyholders have different accident proneness because of observable characteristics taken into account in the price list and hidden characteristics to be corrected *a posteriori*. The heterogeneity is taken into account by assuming that the number of accidents is Poisson distributed with mean varying from one policyholder to another. The annual claim frequency becomes a random variable $\lambda_i \Theta_i$ where $\Theta_i = \exp(\epsilon_i)$ models the oscillations around the grand mean λ_i (with $\mathbb{E}[\Theta_i] = 1$). We can now write $N_i \sim \mathcal{MP}oi(\lambda_i, \Theta_i)$ where $\lambda_i = d_i \exp(\widetilde{\boldsymbol{x}}_i^T \boldsymbol{\beta})$. As in (1.29), we have

$$\mathbb{V}[N_i] = \lambda_i + \lambda_i^2 \mathbb{V}[\Theta_i] > \lambda_i = \mathbb{E}[N_i], \tag{2.11}$$

so that any mixed Poisson regression model induces overdispersion.

2.4.5 Detecting Overdispersion

Residual heterogeneity remains considerable in the risk classes despite the use of many *a priori* variables. Indeed, many explanatory variables are unknown to the insurance company and cannot be incorporated in the price list. Let us denote as $\widehat{m}_k$ the empirical mean claim number of risk class k, and $\widehat{\sigma}_k^2$ the associated variance. In order to graphically test the Poisson mixture assumption, we have plotted the points $(\widehat{m}_k, \widehat{\sigma}_k^2)$ and also the bisecting line. The plot is displayed in Figure 2.9. We observe that $\widehat{m}_k < \widehat{\sigma}_k^2$ in numerous risk classes, indicating that the homogeneous Poisson model is inappropriate. We also see that most of the observed pairs $(\widehat{m}_k, \widehat{\sigma}_k^2)$ lie above the bisecting line, thus supporting overdispersion. The three points below the 45-degree line correspond to classes with just a few policies (36, 13 and 12, respectively).

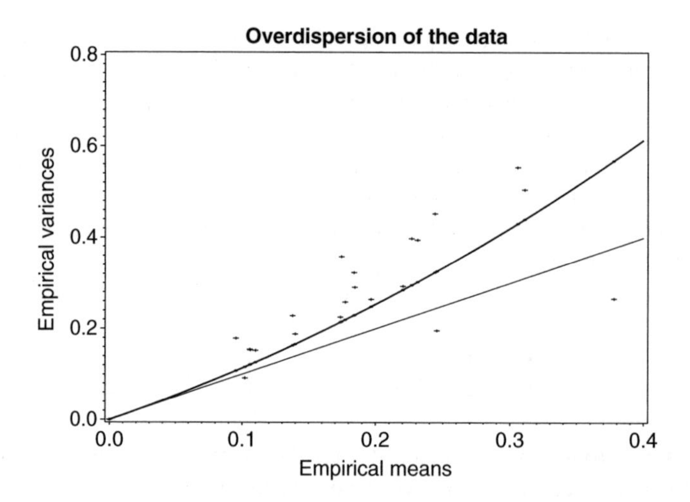

Figure 2.9 Mean–Variance pairs for the risk classes, Portfolio A.

A quadratic curve (without intercept) has been fitted to the mean–variance couples by weighted least-squares (the weights being the exposures of the risk classes). The high value of the R^2 coefficient (86.41 %) supports the quality of the fit. This shows that equation (2.11) is supported by the data, and provides empirical evidence for a mixed Poisson model.

2.4.6 Testing for Overdispersion

The graphical test of the previous section is an easy way of detecting overdispersion. But many statistical tests for the overdispersion assumption have been developed in the literature. Testing for overdispersion can be done by testing for the Poisson distribution against a mixed Poisson model. One problem with standard specification tests (such as likelihood ratio tests) occurs when the null hypothesis is on the boundary of the parameter space, as explained in Chapter 1. When a parameter is bounded by the H_0 hypothesis, the estimate is also bounded and the asymptotic Normality of the maximum likelihood estimator no longer holds under H_0. Consequently, a correction must be made.

Alternatively, testing for overdispersion can be based on the variance function. According to (2.11), the variance function of a heterogeneity model is of the form:

$$\mathbb{V}[N_i] = \lambda_i + \tau\lambda_i^2, \tag{2.12}$$

with $\tau = \mathbb{V}[\Theta_i]$ being the variance of the random effect. Therefore, we have to test the null hypothesis H_0: $\tau = 0$ against H_1: $\tau > 0$. The following score statistics can be used to test the Poisson distribution against heterogeneity models with a variance function of the form (2.12):

$$T_1 = \frac{\sum_{i=1}^{n}\Big((k_i - \widehat{\lambda}_i)^2 - k_i\Big)}{\sqrt{2\sum_{i=1}^{n}\widehat{\lambda}_i^2}},$$

$$T_2 = \frac{\sum_{i=1}^n \left((k_i - \widehat{\lambda}_i)^2 - k_i\right)}{\sqrt{\sum_{i=1}^n \left((k_i - \widehat{\lambda}_i)^2 - k_i\right)^2}},$$

and

$$T_3 = \frac{\sum_{i=1}^n \left((k_i - \widehat{\lambda}_i)^2 - k_i\right)}{\sqrt{\frac{1}{n}\sum_{i=1}^n \widehat{\lambda}_i^{-2}\left((k_i - \widehat{\lambda}_i)^2 - k_i\right)^2}\sqrt{\sum_{i=1}^n \widehat{\lambda}_i^2}}.$$

All these test statistics are $\mathcal{N}or(0, 1)$ distributed. For Portfolio A, $T_1 = 9.18$, $T_2 = 6.13$ and $T_3 = 4.38$. All the p-values are less than 10^{-4} leading to the rejection of the null hypothesis (and thus the Poisson model) in favour of the mixed Poisson model.

2.5 Negative Binomial Regression Model

2.5.1 Likelihood Equations

Overdispersion is taken into account by the inclusion of a random effect, representing an unknown relative risk level. More precisely, assume that $\Theta_1, \ldots, \Theta_n$ are independent $\mathcal{G}am(a, a)$ distributed random variables, i.e. the common probability density function of the Θ_is is given by (1.35). In this case, $\mathbb{E}[\Theta_i] = 1$ and $\mathbb{V}[\Theta_i] = 1/a$. Note that the assumptions made about the Θ_is are rather strong. Their common distribution means that the effect of hidden variables does not depend on observable ones.

Conditional on the observable characteristics summarized in the vector $\boldsymbol{x}_i$ and on the random effect $\Theta_i = \theta$, the annual claim number caused by policyholder i conforms to the $\mathcal{P}oi(\theta\lambda_i)$ law. In other words, λ_i is the expected claim frequency for policyholder i (based on $\boldsymbol{x}_i$) and θ is the relative risk level for this policyholder (if $\theta < 1$, the policyholder is less dangerous than the others in the same risk class). Now, unconditionally,

$$\begin{aligned}\Pr[N_i = k_i | \boldsymbol{x}_i] &= \int_0^\infty \Pr[N_i = k_i | \boldsymbol{x}_i, \Theta_i = \theta] f_\Theta(\theta) d\theta \\ &= \binom{a + k_i - 1}{k_i} \left(\frac{\lambda_i}{a + \lambda_i}\right)^{k_i} \left(\frac{a}{a + \lambda_i}\right)^a,\end{aligned}$$

where we recognize the Negative Binomial distribution.

The likelihood is

$$\mathcal{L}(\boldsymbol{\beta}, a) = \prod_{i=1}^n \frac{\lambda_i^{k_i}}{k_i!} \left(\frac{a}{a + \lambda_i}\right)^a (a + \lambda_i)^{-k_i} \frac{\Gamma(a + k_i)}{\Gamma(a)}.$$

The maximum likelihood estimator for $\boldsymbol{\beta}$ and a solve

$$\frac{\partial}{\partial\boldsymbol{\beta}}L(\boldsymbol{\beta}, a) = \sum_{i=1}^{n} \widetilde{\boldsymbol{x}}_i \left(k_i - \lambda_i \frac{a+k_i}{a+\lambda_i} \right) = \mathbf{0}.$$

Note that these equations are similar to the ones obtained in the Poisson case except that λ_i is now replaced with $\lambda_i(a+k_i)/(a+\lambda_i)$.

Remark 2.4 Let us give an intuitive explanation for the ratio $(a+k_i)/(a+\lambda_i)$ involved in the Negative Binomial likelihood equations. The joint probability density function of (N_i, Θ_i) equals

$$\exp(-\theta_i\lambda_i)\frac{(\theta_i\lambda_i)^{k_i}}{k_i!}\frac{1}{\Gamma(a)}a^a\theta_i^{a-1}\exp(-a\theta_i)$$

$$\propto \exp\left(-\theta_i\lambda_i\right)\theta_i^{k_i+a-1}\exp(-a\theta_i).$$

The probability density function of Θ_i given $N_i = k_i$ is

$$\frac{\exp\left(-\theta_i\left(a+\lambda_i\right)\right)\theta_i^{a+k_i-1}}{\int_0^{+\infty}\exp\left(-\xi\left(a+\lambda_i\right)\right)\xi^{a+k_i-1}d\xi}$$

$$= \exp\left(-\theta_i\left(a+\lambda_i\right)\right)\theta_i^{a+k_i-1}\frac{(a+\lambda_i)^{a+k_i}}{\Gamma\left(a+k_i\right)},$$

so that Θ_i given $N_i = k_i$ follows the $\mathcal{G}am\left(a+k_i, a+\lambda_i\right)$ distribution. Therefore,

$$\mathbb{E}[\Theta_i|N_i = k_i] = \frac{a+k_i}{a+\lambda_i}$$

and the maximum likelihood estimators in the Negative Binomial regression model solve

$$\sum_{i=1}^{n} \boldsymbol{x}_i\Big(k_i - \lambda_i\mathbb{E}[\Theta_i|N_i = k_i]\Big) = \mathbf{0}.$$

Compared to the Poisson likelihood equations, the predicted expected claim number λ_i is replaced with its update $\lambda_i\mathbb{E}[\Theta_i|N_i = k_i]$ based on the information contained in the number k_i of claims filed by policyholder i.

As already shown in Section 2.3.7, it is possible to solve the Negative Binomial likelihood equations with the help of the Newton–Raphson iterative procedure. Starting values for the Newton–Raphson iterative procedure are usually obtained as follows: The Poisson maximum likelihood estimator $\widehat{\boldsymbol{\beta}}$ is known to be consistent, so that we keep it as a reasonable starting value. We still have to find an initial estimate for $\tau = \mathbb{V}[\Theta_i] = 1/a$. So, we first compute the variance of N_i which is given by

$$\mathbb{V}[N_i] = \mathbb{E}[N_i] + \left(d_i\exp(\text{score}_i)\right)^2\tau$$

and then write the empirical analogue of the last relation

$$\sum_{i=1}^{n}\left(\left(k_i - d_i \exp(\text{score}_i)\right)^2 - k_i - \left(d_i \exp(\text{score}_i)\right)^2 \tau\right) = 0.$$

Therefore, the estimator of τ is given by

$$\frac{1}{\widehat{a}} = \widehat{\tau} = \frac{\sum_{i=1}^{n}\left(\left(k_i - d_i \exp(\widehat{\text{score}_i})\right)^2 - k_i\right)}{\sum_{i=1}^{n}\left(d_i \exp(\widehat{\text{score}_i})\right)^2},$$

where $\widehat{\text{score}_i} = \widetilde{\boldsymbol{x}}_i^T \widehat{\boldsymbol{\beta}}$, $\widehat{\boldsymbol{\beta}}$ being the Poisson maximum likelihood estimator for $\boldsymbol{\beta}$. The estimators $\widehat{\boldsymbol{\beta}}$ and $\widehat{\tau}$ are consistent in the Poisson mixture model, and are thus good starting values for finding the maximum likelihood estimators.

2.5.2 *Numerical Illustration*

The Negative Binomial regression with categorical variables can be performed with the SAS®/STAT procedure GENMOD which corrects the estimations for overdispersion. The final model for Portfolio A is shown in Table 2.4. The interpretation of the different columns is the same as in Section 2.3.15. Compared with the Poisson fit, we see that the estimated β_js are very similar, but the standard errors are larger in the Negative Binomial case.

The estimation of the parameter a by the method of moments is given by $\widehat{a} = 1.2401$ whereas the maximum likelihood estimator is equal to $\widehat{a} = 1.065$. The log-likelihood is equal to -5448.5. The variance-covariance matrix of the estimated regression coefficients and the dispersion parameter is

$$\widehat{\boldsymbol{\Sigma}}_{\widehat{\boldsymbol{\beta}}} = \begin{pmatrix} 0.002424 & -0.001537 & -0.001472 & -0.001042 & -0.001033 & -0.001537 & 0.000033 \\ -0.001537 & 0.003249 & 0.001692 & -0.000080 & -0.000286 & 0.000824 & 0.000022 \\ -0.001472 & 0.001692 & 0.006073 & -0.000216 & -0.000537 & 0.001077 & 0.000319 \\ -0.001042 & -0.000080 & -0.000216 & 0.002859 & 0.000217 & -0.000018 & 0.000008 \\ -0.001033 & -0.000286 & -0.000537 & 0.000217 & 0.003038 & 0.000688 & 0.000200 \\ -0.001537 & 0.000824 & 0.001077 & -0.000018 & 0.000688 & 0.006785 & -0.000019 \\ 0.000033 & 0.000022 & 0.000319 & 0.000008 & 0.000200 & -0.000019 & 0.020850 \end{pmatrix}.$$

The Type 3 analysis gives the following results:

Source	DF	Chi-square	Pr>Chi-sq
Gender*Age	2	65.46	<.0001
District	1	11.54	0.0007
Premium split	1	72.73	<.0001
Use of the car	1	6.54	0.0105

All the explanatory variables are still statistically relevant.

Table 2.4 Results of the Negative Binomial regression for the final model, Portfolio A.

Variable	Level	Coeff β	Std error	Wald 95 %	conf limit	Chi-sq	Pr>Chi-sq
Intercept		−2.1963	0.0492	−2.2928	−2.0998	1990.08	<.0001
Gender*Age	Female 18–30 + Male 25–30	0.2363	0.0570	0.1245	0.3480	17.18	<.0001
Gender*Age	Male 18–24	0.6399	0.0779	0.4872	0.7927	67.42	<.0001
Gender*Age	Female > 30+ Male > 30	0	0	0	0	.	.
District	Rural	−0.1805	0.0535	−0.2853	−0.0757	11.40	0.0007
District	Urban	0	0	0	0	.	.
Premium split	Yes	0.4783	0.0551	0.3702	0.5863	75.28	<.0001
Premium split	No	0	0	0	0	.	.
Use of the car	Professional	0.2145	0.0824	0.0531	0.3760	6.78	0.0092
Use of the car	Private	0	0	0	0	.	.

2.6 Poisson-Inverse Gaussian Regression Model

2.6.1 Likelihood Equations

In this section, the random effects $\Theta_1, \ldots, \Theta_n$ are assumed to be independent and to follow the $\mathcal{IG}au(1, \tau)$ distribution with probability density function given by (1.41). In this case, $\mathbb{E}[\Theta_i] = 1$ and $\tau = \mathbb{V}[\Theta_i]$. As in the Negative Binomial case, overdispersion is taken into account with this model.

In the Poisson-Inverse Gaussian regression framework, it is impossible to obtain an analytic expression of $\Pr[N_i = k]$ for large ks. Nevertheless, the recursive formula (1.43) is useful to compute the maximum likelihoood estimator of the regression coefficients (since the observed claim counts correspond to small ks). A Newton–Raphson algorithm can thus be used to get the maximum likelihood estimations.

2.6.2 Numerical Illustration

The Poisson Inverse-Gaussian model is not available in the SAS® procedure GENMOD. A Newton–Raphson algorithm has therefore been implemented in SAS®/IML. The results are presented in Table 2.5 where we can see that the values obtained for the regression coefficients are similar to those obtained in the Negative Binomial case. We observe that the standard deviation in the Poisson-Inverse Gaussian model is greater than the corresponding standard deviation in the Negative Binomial model for each regression coefficient. The estimated variance of the random effect is equal to $\widehat{\tau} = 0.9963$ (the corresponding value in the Negative Binomial model is 0.9388). The log-likelihood is equal to -5448.0 and is slightly better than the log-likelihood of the Negative Binomial regression. The variance-covariance matrix of the estimated regression coefficients and the dispersion parameter is

Table 2.5 Results of the Poisson Inverse-Gaussian regression, Portfolio A.

Variable	Level	Coeff β	Std error	Wald 95 %	conf limit	Chi-sq	Pr>Chi-sq
Intercept		−2.1962	0.0494	−2.2930	−2.0995	1979.4	<.0001
Gender*Age	Female 18–30 + Male 25–30	0.2368	0.0572	0.1248	0.3489	17.17	<.0001
Gender*Age	Male 18–24	0.6435	0.0781	0.4904	0.7966	67.8	<.0001
Gender*Age	Female >30+ Male > 30	0	0	0	0	.	.
District	Rural	−0.1820	0.0537	−0.2871	−0.0768	11.5	0.0007
District	Urban	0	0	0	0	.	.
Premium split	Yes	0.4795	0.0552	0.3712	0.5877	75.3	<.0001
Premium split	No	0	0	0	0	.	.
Use of the car	Professional	0.2159	0.0826	0.0539	0.3778	6.8	0.0090
Use of the car	Private	0	0	0	0	.	.

$$\widehat{\boldsymbol{\Sigma}}_{\widehat{\beta}} = \begin{pmatrix} 0.002437 & -0.001542 & -0.001471 & -0.001050 & -0.001037 & -0.001539 & 0.000246 \\ -0.001542 & 0.003268 & 0.001706 & -0.000085 & -0.000287 & 0.000832 & 0.000243 \\ -0.001471 & 0.001706 & 0.006104 & -0.000228 & -0.000550 & 0.001092 & 0.000532 \\ -0.001050 & -0.000085 & -0.000228 & 0.002878 & 0.000217 & -0.000027 & -0.000399 \\ -0.001037 & -0.000287 & -0.000550 & 0.000217 & 0.003051 & 0.000689 & 0.000067 \\ -0.001539 & 0.000832 & 0.001092 & -0.000027 & 0.000689 & 0.006827 & 0.000423 \\ 0.000246 & 0.000243 & 0.000532 & -0.000399 & 0.000067 & 0.000423 & 0.026767 \end{pmatrix}.$$

The Type 3 analysis for the final model gives

Source	DF	Chi-square	Pr>Chi-sq
Gender*Age	2	65.8	<.0001
District	1	11.6	0.0007
Premium split	1	72.8	<.0001
Use of the car	1	6.6	0.0102

which shows that all the variables are significant.

2.7 Poisson-LogNormal Regression Model

2.7.1 Likelihood Equations

In this section we assume that the random effects $\Theta_1, \ldots, \Theta_n$ are independent and follow the LogNormal distribution $\mathcal{LN}or(-\sigma^2/2, \sigma^2)$. The common probability density function of the Θ_is is then given by (1.45). Then, $\mathbb{E}[\Theta_i] = 1$ and $\mathbb{V}[\Theta_i] = \exp(\sigma^2) - 1$. The probability that the policyholder i files k_i claims is given by

$$\Pr[N_i = k_i | \boldsymbol{x}_i] = \int_0^\infty \exp(-\theta\lambda_i) \frac{(\theta\lambda_i)^{k_i}}{k_i!} \frac{1}{\theta\sigma\sqrt{(2\pi)}} \exp\Big(-\frac{(\ln\theta + \sigma^2/2)^2}{2\sigma^2}\Big) d\theta, \quad (2.13)$$

and does not possess any closed form expression. The likelihood equations must be solved numerically.

2.7.2 Numerical Illustration

The GENMOD procedure of SAS® does not support the Poisson-LogNormal model. We have computed the regression coefficients with the aid of the NLMIXED procedure of SAS®. The results are given in Table 2.6 and the estimation of σ^2 is equal to 0.7064 which leads to

$$\widehat{\tau} = \widehat{\mathbb{V}[\Theta_i]} = 1.027$$

for the variance of the random effect. We observe that this value is greater than those obtained in the Poisson-Inverse Gaussian model and in the Negative Binomial model. The resulting regression coefficients are similar to those obtained previously. The log-likelihood is equal to -5448.1 and is intermediate between the log-likelihoods of the Negative Binomial model and of the Poisson-Inverse Gaussian model (which is the maximum). The variance-covariance matrix of the estimated regression coefficients and $\widehat{\sigma}^2$ is

$$\widehat{\boldsymbol{\Sigma}}_{\widehat{\beta}} = \begin{pmatrix} 0.002438 & -0.001546 & -0.001474 & -0.001047 & -0.001036 & -0.001543 & 0.000075 \\ -0.001546 & 0.003270 & 0.001707 & -0.000082 & -0.000287 & 0.000831 & 0.000064 \\ -0.001474 & 0.001707 & 0.006104 & -0.000224 & -0.000548 & 0.001088 & 0.000354 \\ -0.001047 & -0.000082 & -0.000224 & 0.002875 & 0.000218 & -0.000022 & -0.000050 \\ -0.001036 & -0.000287 & -0.000548 & 0.000218 & 0.003056 & 0.000687 & 0.000188 \\ -0.001543 & 0.000831 & 0.001088 & -0.000022 & 0.000687 & 0.006825 & 0.000058 \\ 0.000075 & 0.000064 & 0.000354 & -0.000050 & 0.000188 & 0.000058 & 0.008056 \end{pmatrix}.$$

The Type 3 analysis for the final model gives

Source	DF	Chi-square	Pr>Chi-sq
Gender*Age	2	66.0	<.0001
District	1	11.6	0.0007
Premium split	1	72.6	<.0001
Use of the car	1	6.6	0.0102

which shows that all the variables included in this model are significant.

Table 2.6 Results of the Poisson Log-Normal regression, Portfolio A.

Variable	Level	Coeff β	Std error	Wald 95 %	conf limit	Chi-sq	Pr>Chi-sq
Intercept		−2.1962	0.0494	−2.2930	−2.0995	1978.1	<.0001
Gender*Age	Female 18–30 + Male 25–30	0.2374	0.0572	0.1253	0.3495	17.2	<.0001
Gender*Age	Male 18–24	0.6443	0.0781	0.4912	0.7974	68.0	<.0001
Gender*Age	Female > 30+ Male > 30	0	0	0	0	.	.
District	Rural	−0.1821	0.0536	−0.2872	−0.0770	11.5	0.0007
District	Urban	0	0	0	0	.	.
Premium split	Yes	0.4793	0.0553	0.3710	0.5877	75.2	<.0001
Premium split	No	0	0	0	0	.	.
Use of the car	Professional	0.2170	0.0826	0.0551	0.3789	6.9	0.0086
Use of the car	Private	0	0	0	0	.	.

2.8 Risk Classification for Portfolio A

So far, we have several competing models for the observed claim frequencies in Portfolio A. This section purposes to compare these models in order to select the optimal one.

2.8.1 Comparison of Competing models with the Vuong Test

Let us consider two non-nested competing models for the number of claims, with respective probability mass functions $p(\cdot|\boldsymbol{x},\boldsymbol{\xi})$ and $q(\cdot|\boldsymbol{x},\boldsymbol{\zeta})$, where $\boldsymbol{x}$ is a vector of explanatory variables, and $\boldsymbol{\xi}$ and $\boldsymbol{\zeta}$ include the regression parameters $\boldsymbol{\beta}$ as well as some dispersion coefficients τ. The corresponding log-likelihoods are

$$L_p(\boldsymbol{\xi}) = \sum_{i=1}^{n} \ln p(k_i|\boldsymbol{x}_i,\boldsymbol{\xi})$$

$$L_q(\boldsymbol{\zeta}) = \sum_{i=1}^{n} \ln q(k_i|\boldsymbol{x}_i,\boldsymbol{\zeta}).$$

In this regression context, the test statistic proposed by VUONG (1989) is

$$T_{LR,NN} = \frac{L_p(\widehat{\boldsymbol{\xi}}) - L_q(\widehat{\boldsymbol{\zeta}})}{\sqrt{n}\omega}, \tag{2.14}$$

where

$$\omega^2 = \frac{1}{n}\sum_{i=1}^{n}\left(\ln\frac{p(k_i|\boldsymbol{x}_i,\widehat{\boldsymbol{\xi}})}{q(k_i|\boldsymbol{x}_i,\widehat{\boldsymbol{\zeta}})}\right)^2 - \left(\frac{1}{n}\sum_{i=1}^{n}\ln\frac{p(k_i|\boldsymbol{x}_i,\widehat{\boldsymbol{\xi}})}{q(k_i|\boldsymbol{x}_i,\widehat{\boldsymbol{\zeta}})}\right)^2 \tag{2.15}$$

is the estimate of the variance of the log-likelihood difference. None of the model has to be true: the test is aimed at selecting the model that is the closer to the true conditional distribution. The null hypothesis of the test is that the two models are equivalent. Under the null hypothesis, the test statistic is asymptotically Normally distributed. Rejection in favour of p happens when $T_{LR,NN} > c$ or in favour of q if $T_{LR,NN} < -c$, where c represents the $\mathcal{N}or\ (0,1)$ critical value for some significance level. If $|T_{LR,NN}| \leq c$ then the null hypothesis is not rejected and the Vuong test cannot discriminate between the two models, given the data.

Now, comparing the three mixed Poisson models with the Vuong test gives:

Negative Binomial against Poisson-Inverse Gaussian the value of the test statistic is equal to -0.754544 leading to a p-value of 45.06 %.

Poisson-LogNormal against Negative Binomial the value of the test statistic is equal to to -0.470894 leading to a p-value of 63.78 %.

Poisson-LogNormal against Poisson-Inverse Gaussian the value of the test statistic is equal to to -0.216702 leading to a p-value of 82.84 %.

These results do not enable us to distinguish between the three models which are therefore statistically equivalent. The Negative Binomial model will be used for the numerical illustrations involving Portfolio A in the next chapters. The reason is that, in this case, explicit expressions are available, providing a deeper insight in the mechanisms behind experience rating systems.

2.8.2 Resulting Risk Classification for Portfolio A

Table 2.7 gives the resulting price list obtained with the Negative Binomial model of Table 2.4. A 'Yes' indicates the presence of the characteristic corresponding to the column. The final *a priori* ratemaking contains 23 classes. Table 2.7 gives the estimated expected annual claim frequencies obtained from the Negative Binomial regression model, and the relative importance of each risk class.

Note that there is another way to present the results displayed in Table 2.7. The idea is to start from the annual expected claim frequency of the reference class, estimated to

$$\exp(\widehat{\beta}_0) = 11.12\,\%$$

according to Table 2.4, and then to apply correction coefficients. Specifically, the annual expected claim frequency of a given policyholder is simply obtained from

$$
\begin{aligned}
& 11.12\,\% \\
& \times \begin{cases} \exp(0.6399) = 1.90, & \text{if the policyholder is a male aged between 18 and 24,} \\ \exp(0.2363) = 1.27, & \text{if the policyholder is a female aged between 18 and 30} \\ & \text{or a male aged between 25 and 30,} \\ 1, & \text{otherwise,} \end{cases} \\
& \times \begin{cases} \exp(-0.1805) = 0.83, & \text{if the policyholder lives in a rural district,} \\ 1, & \text{otherwise,} \end{cases} \\
& \times \begin{cases} \exp(0.4783) = 1.61, & \text{if the policyholder splits the premium payment,} \\ 1, & \text{otherwise,} \end{cases} \\
& \times \begin{cases} \exp(0.2145) = 1.24, & \text{if the policyholder uses the car for professional purposes,} \\ 1, & \text{otherwise.} \end{cases}
\end{aligned}
$$

2.9 Ratemaking using Panel Data

2.9.1 Longitudinal Data

Actuaries often pool several observation periods to determine the price list (the main goal being to increase the size of the data base). The serial dependence arising from the fact that the same individuals are followed and produce correlated claim numbers should prevent the actuaries from using classical statistical techniques (which assume independence).

During the observation period, n policies have been in the portfolio, each one observed during T_i periods. Let N_{it} be the number of claims reported by policyholder i during year t, $i = 1, 2, \ldots, n$, $t = 1, 2, \ldots, T_i$. Such motor insurance data have a panel structure: typically, n is large whereas the T_is are small.

Let d_{it} be the length of observation period t for policyholder i. Usually, $d_{it} = 1$, but there are a variety of situations where this is not the case. Indeed, a new period of observation

Table 2.7 *A priori* risk classification for Portfolio A (Negative Binomial regression).

Gender–age			Use of the car		Premium split		District		Exp. annual claim freq. (%)	Weights (%)
Female 18–30 Male 25–30	Male 18–24	Others	Private	Professional	Annual	Split	Rural	Urban		
Yes	No	No	Yes	No	Yes	No	Yes	No	11.76	10.49
Yes	No	No	Yes	No	Yes	No	No	Yes	14.08	13.96
Yes	No	No	Yes	No	No	Yes	Yes	No	18.97	3.98
Yes	No	No	Yes	No	No	Yes	No	Yes	22.72	7.05
Yes	No	No	No	Yes	Yes	No	Yes	No	14.57	0.76
Yes	No	No	No	Yes	Yes	No	No	Yes	17.46	1.22
Yes	No	No	No	Yes	No	Yes	Yes	No	23.51	0.13
Yes	No	No	No	Yes	No	Yes	No	Yes	28.16	0.14
No	Yes	No	Yes	No	Yes	No	Yes	No	17.61	2.93
No	Yes	No	Yes	No	Yes	No	No	Yes	21.09	2.99
No	Yes	No	Yes	No	No	Yes	Yes	No	28.40	1.52
No	Yes	No	Yes	No	No	Yes	No	Yes	34.02	2.42
No	Yes	No	No	Yes	Yes	No	Yes	No	21.82	0.07
No	Yes	No	No	Yes	Yes	No	No	Yes	26.14	0.09
No	Yes	No	No	Yes	No	Yes	Yes	No	35.20	0.02
No	No	Yes	Yes	No	Yes	No	Yes	No	9.28	13.38
No	No	Yes	Yes	No	Yes	No	No	Yes	11.12	19.73
No	No	Yes	Yes	No	No	Yes	Yes	No	14.98	2.94
No	No	Yes	Yes	No	No	Yes	No	Yes	17.94	6.61
No	No	Yes	No	Yes	Yes	No	Yes	No	11.51	3.72
No	No	Yes	No	Yes	Yes	No	No	Yes	13.78	5.17
No	No	Yes	No	Yes	No	Yes	Yes	No	18.56	0.25
No	No	Yes	No	Yes	No	Yes	No	Yes	22.23	0.44

starts as soon as some policy characteristics are modified (think for instance of a policyholder house moving for a company using postcode as rating factor, a policyholder's wedding for a company using marital status, or simply the policyholder buying a new car). Moreover, in the year the policy is issued and in the one it is possibly cancelled the length of the observation period is generally less than unity.

We face a nested structure: each policyholder generates a sequence $\boldsymbol{N}_i = (N_{i1}, N_{i2}, \ldots, N_{iT_i})^T$ of claim numbers. It is reasonable to assume independence between the series $\boldsymbol{N}_1, \boldsymbol{N}_2, \ldots, \boldsymbol{N}_n$, but this assumption is very questionable inside the $\boldsymbol{N}_i$s. Regarding *a priori* ratemaking, the dependence between the components of each $\boldsymbol{N}_i$ is a nuisance (in the statistical sense). This means that, at this stage, we are not interested in accurately modelling this dependence, but we must take it into account when estimating the regression coefficients. The idea now is to incorporate in the N_{it}s exogenous information (like age, gender, power of the car, and so on) summarized in the vectors $\boldsymbol{x}_{it}$; to this end, we resort to a regression model for longitudinal data.

The distributional assumption for the random component of the regression model has to account for the non-negativity of the data, as well as their integer values. We begin with Poisson regression and assume that the N_{it}s conform to the Poisson distribution with a mean that can be written as an exponential function of a linear combination $\beta_0 + \sum_{j=1}^{p} \beta_j x_{itj}$ of the explanatory variables $\boldsymbol{x}_{it}$, with unknown regression coefficients $\boldsymbol{\beta}$ to be estimated from the data. Despite its prevalence as a starting point in the analysis of count data, the Poisson specification is often inappropriate because of unobserved heterogeneity and failure of the independence assumption if the data consist in repeated observations on the same policyholders. A convenient way to take this phenomenon into account is to introduce a random effect into the model.

Remark 2.5 Before embarking on a panel analysis pooling together the observations relating to several years, it is interesting to first work year by year to assess the stability of the effect of each rating variable on the annual expected claim frequency. Specifically, the vector $\boldsymbol{\beta}$ of the regression coefficients is estimated on the basis of each calendar year and the components are checked for their stability over time. Only stable coefficients are interesting for the purpose of ratemaking. Rating factors with unstable regression coefficients should be excluded from the risk classification scheme. In some cases, a time trend is visible for some estimated regression coefficients (this is typically true for the intercept β_0). A time effect can then be incorporated into the model to account for coefficients with trends.

2.9.2 Descriptive Statistics for Portfolio B

The analysis in this section is based on an insurance portfolio containing 20 354 policies and observed during 3 years (from 1997 to 1999). We have 45 350 observations available as not all the policies have been in force for 3 years. For each policy and for each year, we know the exposure-to-risk, the number of claims filed and some other explanatory variables:

Gender: Policyholder's gender (male–female)
Age: Policyholder's age (18–22, 23–30 and over 30)
Power: The power of the vehicle (less than 66 kW, 66–110 kW and more than 110 kW)
Size: The size of the city where the policyholder lives (large, middle or small), and
Colour: The colour of the vehicle (red or other).

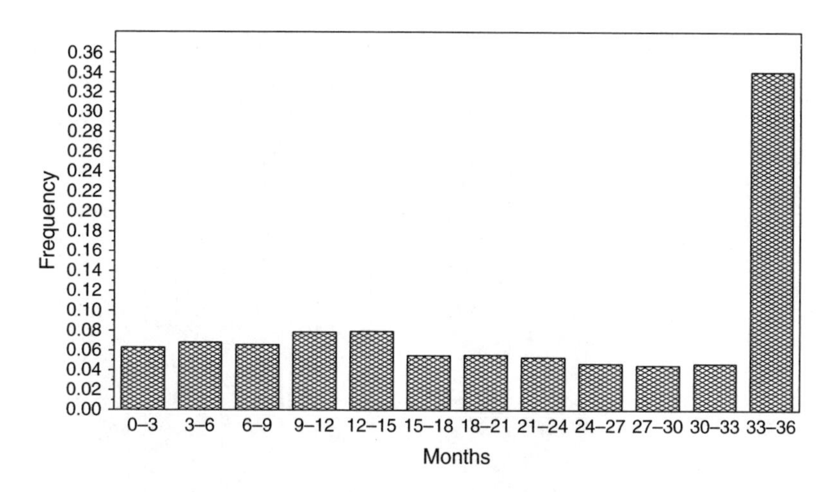

Figure 2.10 Exposure-to-risk in Portfolio B.

The observed mean annual claim frequency is 18.4 %. Figures 2.11 to 2.15 display the histograms giving, for each explanatory variable, the distribution of the portfolio between the different levels of the variable and, for each of these levels, the observed mean annual claim frequency.

Figure 2.10 gives the distribution of the exposure-to-risk in the portfolio. About 34 % of the policies have been in force during 3 years. The exposures for policies in force for less than three years are roughly uniformly distributed accross the triennium.

The age structure of the portfolio is described in Figure 2.11. Most policyholders (60.6 % of the portfolio) are older than 30. Only 3.0 % of the policyholders are less than 22 whereas 36.4 % are between 23 and 30. We see that the annual claim frequency decreases with the age of the policyholders. In this portfolio, the young drivers are rather risky as their observed annual claim frequency is 30.8 %. Drivers over 30 have an observed annual claim frequency of 16.3 %. Finally, the drivers aged between 23 and 30 have an observed annual claim frequency of 20.8 %.

Figure 2.12 suggests a higher annual claim frequency for males (18.8 %) than for females (17.7 %). The portfolio is comprised of 64.0 % male policyholders and 36.0 % female policyholders.

Figure 2.13 gives the distribution of the policyholders according to the size of the city where they live. We see that 31.6 % of the policyholders live in a large city, 35.0 % in a middle-sized city and the remaining 33.4 % in a small city. The annual claim frequency seems to increase with the size of the city (going from 16.6 % for the small cities, to 17.8 % for the middle-sized cities and to 21.0 % for the large cities).

We see, in Figure 2.14, that most of the cars (65.6 % of the portfolio) have a power smaller than 66 kW, 31.2 % have a power included in the 66–110 kW range and only 3.2 % of the cars hold an engine with a power greater than 110 kW. We notice that the most powerful cars are the least risky (with an annual claim frequency of 16.6 %) whereas the cars with an engine power range between 66 kW and 110 kW are the riskiest (with an observed annual claim frequency of 18.7 %). Finally, the least powerful cars have an annual claim frequency of 18.3 %.

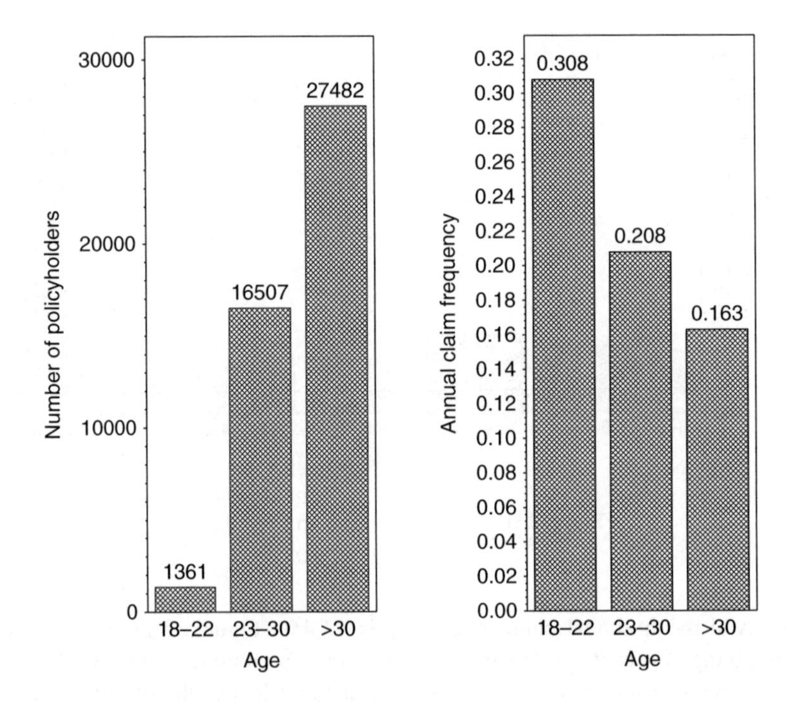

Figure 2.11 Composition of Portfolio B with respect to Age (left panel) and observed annual claim frequencies according to Age (right panel).

Figure 2.15 shows that the colour of the car has nearly no influence on the number of claims. We can notice that 10.1 % of the cars are red.

2.9.3 Poisson Regression with Serial Independence

In this subsection, we assume that the N_{it}s are independent for the different values of i and t. With the help of the SAS®/STAT procedure GENMOD, we have obtained the results displayed in Table 2.8 for the model with the 5 explanatory variables presented in Subsection 2.9.2. The results obtained from a Type 3 analysis are as follows:

Source	DF	Chi-square	Pr>Chi-sq
Gender	1	4.85	0.0276
Age	2	173.56	<.0001
Power	2	4.38	0.1120
Size of the city	2	74.10	<.0001
Colour	1	0.32	0.5698

The log-likelihood associated with this model is $-19\,756.50$ (note that, again, the GENMOD procedure does not incorporate the constant terms in the log-likelihood, so that the value declared by GENMOD is $-19\,282.60$).

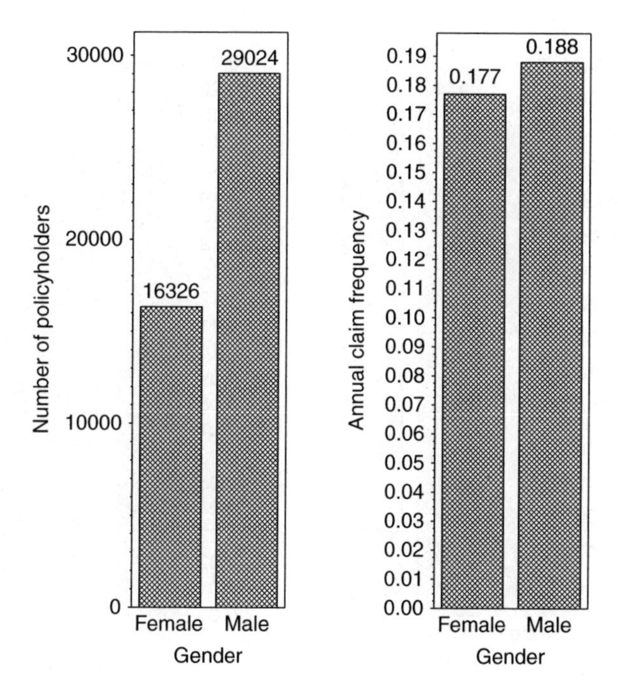

Figure 2.12 Composition of Portfolio B with respect to Gender (left panel) and observed annual claim frequencies according to Gender (right panel).

The variable Colour is not relevant, as the p-value is 56.98 % according to the Type 3 analysis. The variable Power could then also be removed (p-value of 11.21 % after having removed the variable Colour from the regression model) but since this variable is common in the tariff of insurers, we will try an interaction between the variables Age and Power. The interaction between Age and Power is illustrated in Figure 2.16 (where 'LP' stands for large power, that is, over 110 kW, 'MP' stands for medium power, that is, between 66 and 110 kW, and 'SP' for small power, that is, less than 66 kW). The results are given in Table 2.9. The Type 3 analysis is as follows:

Source	DF	Chi-square	Pr>Chi-sq
Gender	1	4.86	0.0275
Age*Power	8	180.01	<.0001
Size of the city	2	74.53	<.0001

We can see that all the variables are now relevant. The log-likelihood is now equal to −19755.10 (as explained above, the value declared by GENMOD is −19281.20, the difference being due to constant terms omitted by the GENMOD procedure) and is slightly better than in the first model.

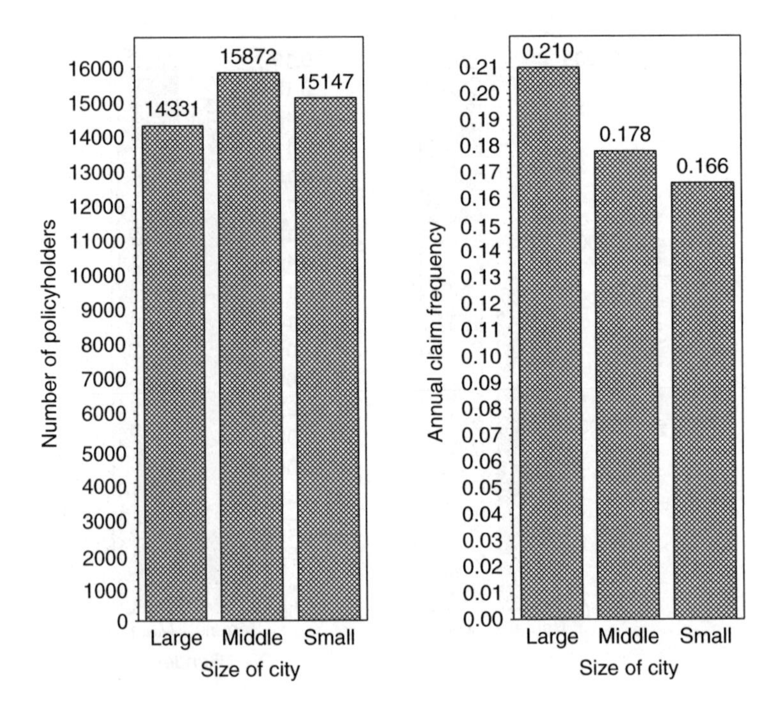

Figure 2.13 Composition of Portfolio B with respect to Size of the city (left panel) and observed annual claim frequencies according to Size of the city (right panel).

Nevertheless, several levels of the interaction between Age and Power can be grouped together. As already explained in Section 2.3.15, the option 'Estimate' of GENMOD can be used to assess the relevance of grouping the levels of the interaction Age–Power 2 by 2. The following groups have been defined:

Group 1 Age > 30 and any power
Group 2 Age 23–30 and any power as well as Age 18–22 and power > 110 kW
Group 3 Age 18–22 and power < 110 kW

The final model is shown in Table 2.10. The variance-covariance matrix of the estimated regression coefficients is

$$\widehat{\boldsymbol{\Sigma}}_{\widehat{\boldsymbol{\beta}}} = \begin{pmatrix} 0.003428 & -0.000244 & -0.003022 & -0.003028 & -0.000441 & -0.000468 \\ -0.000244 & 0.000679 & 0.000012 & 0.000006 & -0.000004 & 0.000007 \\ -0.003022 & 0.000012 & 0.003350 & 0.003064 & -0.000084 & -0.000050 \\ -0.003028 & 0.000006 & 0.003064 & 0.003436 & -0.000059 & -0.000045 \\ -0.000441 & -0.000004 & -0.000084 & -0.000059 & 0.000938 & 0.000511 \\ -0.000468 & 0.000007 & -0.000050 & -0.000045 & 0.000511 & 0.000964 \end{pmatrix}.$$

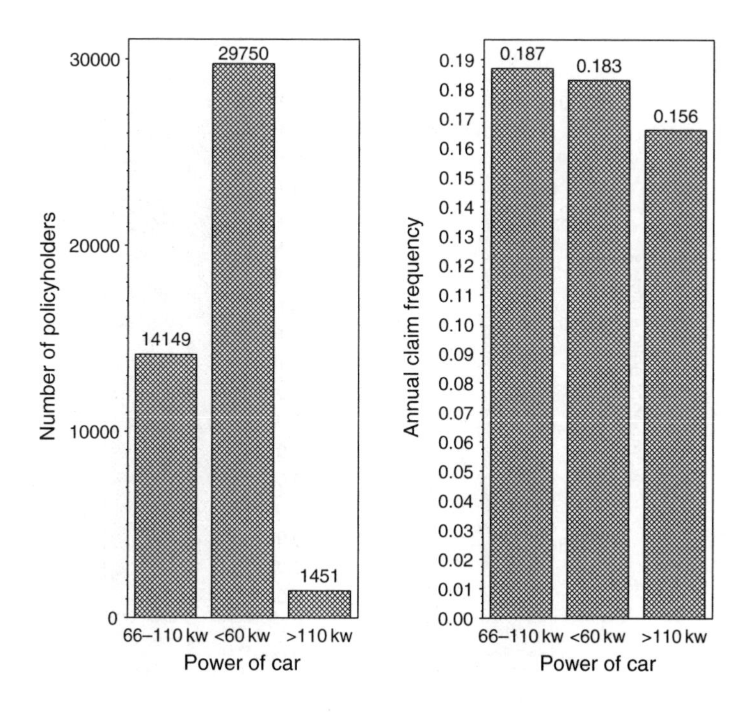

Figure 2.14 Composition of Portfolio B with respect to Power (left panel) and observed annual claim frequencies according to Power (right panel).

Results of the Type 3 analysis are as follows:

Source	DF	Chi-square	Pr>Chi-sq
Gender	1	6.50	0.0108
Age*Power	2	173.14	<.0001
Size of the city	2	74.51	<.0001

2.9.4 Detection of Serial Dependence

Inclusion of the Lagged Claim Numbers as Explanatory Variables

It is interesting, in order to have an idea of the serial dependence between the N_{it}s, to begin by considering the observations N_{it}, $i = 1, 2, \ldots, n$, $t = 2, \ldots T_i$ and to perform a regression of these observations on the corresponding explanatory variables $\boldsymbol{x}_{it}$ and on the number $N_{i,t-1}$ of claims filed during the previous insurance period. Specifically, we start from a model with 4 variables (colour, size of the city, gender and interaction age–power) to be consistent with the previous subsection. We refine it with the help of a Type 3 analysis. Firstly, we must remove the variable Colour which holds a p-value of 28.20 %. Then, the variable Gender could be eliminated since it has a p-value of 10.85 % but we keep it considering its importance in the

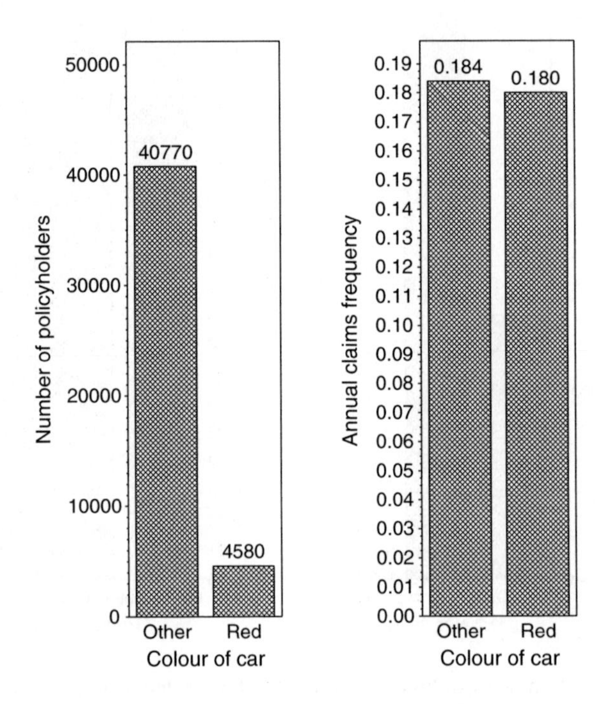

Figure 2.15 Composition of Portfolio B with respect to Colour of the car (left panel) and observed annual claim frequencies according to Colour of the car (right panel).

Table 2.8 Results of the Poisson regression for the model with 5 variables and serial independence, Portfolio B.

Variable	Level	Coeff β	Std error	Wald 95 %	conf limit	Chi-sq	Pr>Chi-sq
Intercept		−1.9242	0.0302	−1.9833	−1.8650	4063.54	<.0001
Gender	Female	−0.0581	0.0265	−0.1100	−0.0063	4.82	0.0281
Gender	Male	0	0	0	0	.	.
Age	17–22	0.6651	0.0583	0.5508	0.7793	130.23	<.0001
Age	23–30	0.2525	0.0261	0.2015	0.3036	93.87	<.0001
Age	> 30	0	0	0	0	.	.
Power	> 110 kW	−0.0116	0.0750	−0.1586	0.1353	0.02	0.8769
Power	66–110 kW	0.0563	0.0275	0.0024	0.1102	4.19	0.0406
Power	< 66 kW	0	0	0	0	.	.
Size of the city	Large	0.2549	0.0306	0.1949	0.3150	69.27	<.0001
Size of the city	Middle	0.0756	0.0311	0.0147	0.1364	5.92	0.0150
Size of the city	Small	0	0	0	0	.	.
Colour	Red	−0.0236	0.0416	−0.1052	0.0580	0.32	0.5710
Colour	Other	0	0	0	0	.	.

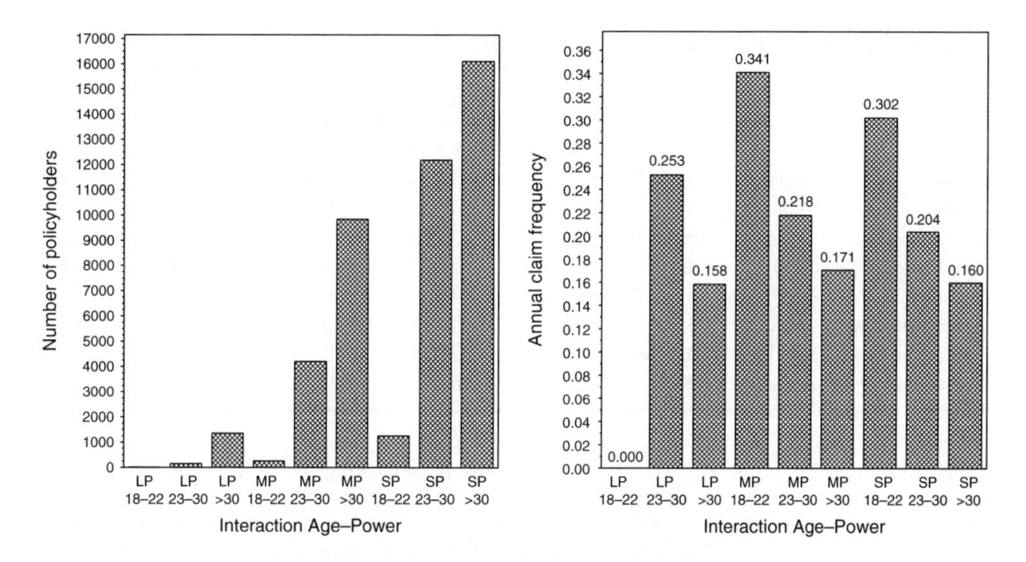

Figure 2.16 Composition of Portfolio B with respect to the Age–Power interaction (left panel) and annual claim frequencies according to the Age–Power interaction (right panel).

Table 2.9 Results of the Poisson regression for model with interactions between Age and Power, Portfolio B.

Variable	Level	Coeff β	Std error	Wald 95 %	conf limit	Chi-sq	Pr>Chi-sq
Intercept		−1.9248	0.0312	−1.9859	−1.8637	3816.03	<.0001
Gender	Female	−0.0582	0.0265	−0.1101	−0.0063	4.83	0.0280
Gender	Male	0	0	0	0	.	.
Age*Power	17–22 > 110 kW	−16.6022	6143.464	−12057.6	12024.37	0.00	0.9978
Age*Power	17–22 66–110 kW	0.7596	0.1333	0.4983	1.0208	32.47	<.0001
Age*Power	17–22 < 66 kW	0.6577	0.0649	0.5304	0.7849	102.60	<.0001
Age*Power	23–30 > 110 kW	0.4605	0.1975	0.0734	0.8475	5.44	0.0197
Age*Power	23–30 66–110 kW	0.3031	0.0441	0.2166	0.3896	47.17	<.0001
Age*Power	23–30 < 66 kW	0.2496	0.0317	0.1875	0.3117	62.11	<.0001
Age*Power	> 30 > 110 kW	−0.0398	0.0812	−0.1989	0.1193	0.24	0.6240
Age*Power	> 30 66–110 kW	0.0547	0.0357	−0.0152	0.1246	2.35	0.1250
Age*Power	> 30 < 66 kW	0	0	0	0	.	.
Size of the city	Large	0.2556	0.0306	0.1956	0.3156	69.62	<.0001
Size of the city	Middle	0.0754	0.0311	0.0145	0.1363	5.90	0.0152
Size of the city	Small	0	0	0	0	.	.

Table 2.10 Results of the Poisson regression for the final model, Portfolio B.

Variable	Level	Coeff β	Std error	Wald 95 %	conf limit	Chi-sq	Pr>Chi-sq
Intercept		−1.2473	0.0585	−1.3620	−1.1325	453.83	<.0001
Gender	Female	−0.0662	0.0260	−0.1173	−0.0152	6.46	0.0110
Gender	Male	0	0	0	0	.	.
Age*Power	Group 1	−0.6571	0.0579	−0.7706	−0.5437	128.88	<.0001
Age*Power	Group 2	−0.4101	0.0586	−0.5250	−0.2953	48.96	<.0001
Age*Power	Group 3	0	0	0	0	.	.
Size of the city	Large	0.2557	0.0306	0.1957	0.3157	69.74	<.0001
Size of the city	Middle	0.0764	0.0311	0.0155	0.1373	6.05	0.0139
Size of the city	Small	0	0	0	0	.	.

Table 2.11 Results of the Poisson regression for the model taking the past claims into account, Portfolio B.

Variable	Level	Coeff β	Std error	Wald 95 %	conf limit	Chi-sq	Pr>Chi-sq
Intercept		−1.4128	0.0994	−1.6076	−1.2179	201.95	<.0001
Gender	Female	−0.0556	0.0347	−0.1237	0.0125	2.56	0.1095
Gender	Male	0	0	0	0	.	.
Age*Power	Group 1	−0.5774	0.0980	−0.7695	−0.3854	34.72	<.0001
Age*Power	Group 2	−0.4023	0.0990	−0.5963	−0.2082	16.51	<.0001
Age*Power	Group 3	0	0	0	0	.	.
Size of the city	Large	0.2214	0.0412	0.1406	0.3021	28.88	<.0001
Size of the city	Middle	0.1055	0.0412	0.0246	0.1863	6.54	0.0106
Size of the city	Small	0	0	0	0	.	.
N_{t-1}		0.3102	0.0370	0.2376	0.3828	70.13	<.0001

tariff of insurers. This leads to the results of Table 2.11. Results of the Type 3 analysis for the model taking the past claims into account are as follows:

Source	DF	Chi-square	Pr>Chi-sq
Gender	1	2.58	0.1085
Age*Power	2	49.31	<.0001
Size of the city	2	29.03	<.0001
N_{t-1}	1	62.98	<.0001

The regression coefficient obtained for the number of past claims shows that this variable is highly relevant, which indicates a serial dependence. Considering the values of the Chi-square statistics, the past number of claims appears as the most important variable to predict the future number of claims.

Inclusion of the Lagged Claim Number as a Correction Factor

In a second approach, we use again the results obtained when assuming independence of the data and without using the number of past claims as an explanatory variable (see Table 2.10).

These results are then corrected with a multiplying factor computed thanks to a Poisson regression with 'number of claims of the previous year' as the single explanatory variable. We use the sum of the logarithm of the estimated annual expected claim frequency obtained when assuming serial independence and of the logarithm of the exposure-to-risk as an offset. Specifically, the expected number of claims for policyholder i during period t, $t = 2, 3$, is now of the form

$$d_{it} \exp\left(\widehat{\beta}_0 + \sum_{j=1}^{p} \widehat{\beta}_j x_{itj}\right) \exp(\widetilde{\beta}_0 + \widetilde{\beta}_1 N_{i,t-1})$$

where the $\widehat{\beta}_j$s are those of Table 2.10 and where the parameters $\widetilde{\beta}_0$ and $\widetilde{\beta}_1$ have to be estimated by Poisson regression. The offset is

$$\ln d_{it} + \widehat{\beta}_0 + \sum_{j=1}^{p} \widehat{\beta}_j x_{itj}.$$

Results of the Poisson regression on the model incorporating the past claims are as follows:

Variable	Coeff $\widetilde{\beta}$	Std error	Wald 95 %	conf limit	Chi-sq	Pr>Chi-sq
Intercept	−0.0412	0.0180	−0.0766	−0.0059	5.23	0.0222
N_{t-1}	0.3241	0.0370	02517	0.3966	76.88	<.0001

Results of the Type 3 analysis for the model incorporating the past claims are as follows:

Source	DF	Chi-sq	Pr>Chi-sq
N_{t-1}	1	68.66	<.0001

The data thus clearly indicate a serial dependence. The results obtained in Section 2.9.3, where the independence of the N_{it}s for different values of i and t was assumed, are thus suspect. However, it can be shown theoretically that the maximum likelihood estimator $\widehat{\boldsymbol{\beta}}$ computed under the hypothesis of serial independence is consistent. So, if the number of policies is large enough, we can expect a small impact on the estimations of the different β_j. However, the variance of $\widehat{\boldsymbol{\beta}}$ can not be computed using the serial independence.

2.9.5 Estimation of the Parameters using GEE

Assuming that the data are independent and Poisson distributed, the likelihood equations are

$$\sum_{i=1}^{n} \boldsymbol{X}_i^T \left(\boldsymbol{n}_i - \mathbb{E}[\boldsymbol{N}_i]\right) = \boldsymbol{0}, \text{ where } \boldsymbol{X}_i = (\widetilde{\boldsymbol{x}}_{i1}, \ldots, \widetilde{\boldsymbol{x}}_{iT_i})^T. \quad (2.16)$$

The variance-covariance matrix of the N_{it}s in the Poisson model with serial independence is given by

$$\boldsymbol{A}_i = \begin{pmatrix} \lambda_{i1} & 0 & \cdots & 0 \\ 0 & \lambda_{i2} & \cdots & 0 \\ \vdots & \vdots & \ddots & \vdots \\ 0 & 0 & \cdots & \lambda_{iT_i} \end{pmatrix}.$$

In fact this matrix does not take the overdispersion and the serial dependence of the data into account. Noting that

$$\frac{\partial}{\partial \boldsymbol{\beta}} \mathbb{E}[\boldsymbol{N}_i] = \boldsymbol{A}_i \boldsymbol{X}_i$$

it is possible to transform (2.16) in order to let $\boldsymbol{A}_i$ explicitly appear in the likelihood equations. This gives

$$\sum_{i=1}^{n} \left(\frac{\partial}{\partial \boldsymbol{\beta}} \mathbb{E}[\boldsymbol{N}_i] \right)^T \boldsymbol{A}_i^{-1} \left(\boldsymbol{n}_i - \mathbb{E}[\boldsymbol{N}_i] \right) = \boldsymbol{0}. \tag{2.17}$$

The principle of Generalized Estimating Equations (GEE) is to find a suitable variance-covariance matrix to insert in Equation (2.17) instead of $\boldsymbol{A}_i$ based on serial independence. This matrix should take the overdispersion and the serial dependence into account. A possible form of this matrix could be

$$\boldsymbol{V}_i = \phi \boldsymbol{A}_i^{1/2} \boldsymbol{R}_i(\boldsymbol{\alpha}) \boldsymbol{A}_i^{1/2}$$

where the 'working' correlation matrix $\boldsymbol{R}_i(\boldsymbol{\alpha})$ takes the serial dependence between the components of $\boldsymbol{N}_i$ into account and depends on a parameter $\boldsymbol{\alpha}$. The overdispersion is also taken into account as $\mathbb{V}[N_{it}] = \phi\lambda_{it}$ exceeds $\mathbb{E}[N_{it}] = \lambda_{it}$ provided $\phi > 1$.

The idea behind the GEE method is then to replace $\boldsymbol{A}_i$ by $\boldsymbol{V}_i$ in (2.17) and to compute the estimator of $\boldsymbol{\beta}$ as the solution of

$$\sum_{i=1}^{n} \left(\frac{\partial}{\partial \boldsymbol{\beta}} \mathbb{E}[\boldsymbol{N}_i] \right)^T \boldsymbol{V}_i^{-1} \left(\boldsymbol{n}_i - \mathbb{E}[\boldsymbol{N}_i] \right) = \boldsymbol{0}. \tag{2.18}$$

The resulting estimator is consistent whatever the choice of the matrix $\boldsymbol{R}_i(\boldsymbol{\alpha})$ but the precision will be much better if $\boldsymbol{R}_i(\boldsymbol{\alpha})$ is close to the true correlation matrix of $\boldsymbol{N}_i$. Equation (2.18) is solved thanks to a modified version of the Fisher scoring method for $\boldsymbol{\beta}$ and a moment estimation for ϕ and $\boldsymbol{\alpha}$. The iterative procedure is as follows:

1. Compute an initial estimate of $\boldsymbol{\beta}$ assuming independence.
2. Compute the current 'working' correlation matrix based on standardized residuals, current $\boldsymbol{\beta}$ and the assumed structure of $\boldsymbol{R}_i(\boldsymbol{\alpha})$.
3. Estimate the covariance matrix $\boldsymbol{V}_i$.
4. Update $\boldsymbol{\beta}$.

Note that GEE is not a likelihood based method of estimation, so that inferences based on likelihoods are not possible in this case.

Modelling Dependence with the 'Working Correlation Matrix'

The 'working' correlation matrix $\boldsymbol{R}_i(\boldsymbol{\alpha})$ takes into account the dependence between the observations corresponding to the same policyholder. The form of this matrix must be specified and depends on the parameter vector $\boldsymbol{\alpha}$.

If $\boldsymbol{R}_i(\boldsymbol{\alpha}) = \boldsymbol{I}$, (2.18) gives exactely the likelihood equations (2.16) under the assumption of independence. The SAS®/STAT procedure GENMOD supports the following structures of the 'working' correlation matrix: fixed (user-specific correlation matrix, not estimated from the data but specified by the actuary), independent ($\boldsymbol{R}_i(\boldsymbol{\alpha}) = \boldsymbol{I}$, giving the estimates of $\boldsymbol{\beta}$ obtained under serial dependence), m-dependent (correlation equal to α_t for lags $t = 1, \ldots, m$, and 0 for higher lags), exchangeable (constant correlation α, whatever the lag), unstructured (each correlation coefficient α_{jk}, between observations made at times j and k, is estimated from the data), and AR1 (autoregressive of order 1, with a correlation coefficient equal to α^t at lag t).

Numerical Example

The GEE approach can be performed thanks to the procedure GENMOD of SAS®. The 'Repeated' statement of GENMOD invokes the GEE method, specifies the correlation structure, and controls the displayed output from the GEE model. Initial parameter estimates for iterative fitting of the GEE model are computed as in a Poisson regression model, as described previously. Results of the initial model fit are displayed as part of the generated output of SAS®. Statistics for the initial model fit such as parameter estimates, standard errors, deviances, and Pearson Chi-squares do not apply to the GEE model, and are only valid for the initial model fit. The SAS® parameter estimates table contains parameter estimates, standard errors, confidence intervals, Z-scores, and p-values for the parameter estimates.

The 'Repeated' statement specifies the covariance structure of multivariate responses for GEE model fitting in the GENMOD procedure. In addition, the 'Repeated' statement controls the iterative fitting algorithm used in GEE and specifies optional output. Other GENMOD procedure statements are used in the same way as they are for ordinary Poisson regression models to specify the regression model for the mean of the responses.

The statement 'SUBJECT = subject-effect' identifies subjects in the input data set. The subject-effect can be a single variable, an interaction effect, a nested effect, or a combination. Each distinct value, or level, of the effect identifies a different subject, or cluster. Responses from different subjects are assumed to be independent, and responses within subjects are assumed to be correlated. A subject-effect must be specified, and variables used in defining the subject-effect must be listed in the CLASS statement. In actuarial applications, the policy number is typically used as subject-effect.

The same variables as for the model where the serial independence was assumed are kept in the final model. The results are given in Table 2.12 that is similar to the SAS® outputs for GEEs. The estimation of the 'working correlation matrix' gives

$$\begin{pmatrix} 1 & 0.0493 & 0.0460 \\ 0.0493 & 1 & 0.0493 \\ 0.0460 & 0.0493 & 1 \end{pmatrix}.$$

and $\widehat{\phi} = 1.3419$.

Since the GEE apprach is not based on a likelihood function, we cannot use the large sample approximations for the estimated variance-covariance matrix $\widehat{\boldsymbol{\Sigma}}_{\widehat{\beta}}$ of the estimated regression coefficients presented in Chapter 1. The model-based estimation of $\boldsymbol{\Sigma}_{\widehat{\boldsymbol{\beta}}}$ in the GEE case is given by $\boldsymbol{\Sigma}_{\widehat{\boldsymbol{\beta}}} = \mathcal{I}_{\text{GEE}}^{-1}$ where

$$\mathcal{I}_{\text{GEE}} = \sum_{i=1}^{n} \left(\frac{\partial}{\partial \boldsymbol{\beta}} \mathbb{E}[\boldsymbol{N}_i] \right)^T \boldsymbol{V}_i^{-1} \left(\frac{\partial}{\partial \boldsymbol{\beta}} \mathbb{E}[\boldsymbol{N}_i] \right).$$

Here, $\mathcal{I}_{\text{GEE}}^{-1}$ is the GEE-equivalent of the inverse of the Fisher information matrix. It is a consistent estimator of the covariance matrix of $\widehat{\boldsymbol{\beta}}$ if the mean structure and the 'working' correlation matrix are correctly specified. Then,

$$\boldsymbol{\Sigma}_{\widehat{\boldsymbol{\beta}}} = \mathcal{I}_{\text{GEE}}^{-1} \mathcal{J}_{\text{GEE}} \mathcal{I}_{\text{GEE}}^{-1}$$

is the robust estimate of the covariance matrix of $\widehat{\boldsymbol{\beta}}$, where

$$\mathcal{J}_{\text{GEE}} = \sum_{i=1}^{n} \left(\frac{\partial}{\partial \boldsymbol{\beta}} \mathbb{E}[\boldsymbol{N}_i] \right)^T \boldsymbol{V}_i^{-1} \mathbb{C}[\boldsymbol{N}_i] \boldsymbol{V}_i^{-1} \left(\frac{\partial}{\partial \boldsymbol{\beta}} \mathbb{E}[\boldsymbol{N}_i] \right).$$

The robust estimate for $\boldsymbol{\Sigma}_{\widehat{\boldsymbol{\beta}}}$ is consistent even if the 'working' correlation matrix is misspecified. In computing, the covariance matrix $\mathbb{C}[\boldsymbol{N}_i]$ of $\boldsymbol{N}_i$ is replaced with

$$\widehat{\mathbb{C}[\boldsymbol{N}_i]} = (\boldsymbol{N}_i - \widehat{\mathbb{E}[\boldsymbol{N}_i]})(\boldsymbol{N}_i - \widehat{\mathbb{E}[\boldsymbol{N}_i]})^T.$$

The robust estimated variance-covariance matrix of the estimated regression coefficients is

$$\widehat{\boldsymbol{\Sigma}}_{\widehat{\boldsymbol{\beta}}} = \begin{pmatrix} 0.003910 & -0.000359 & -0.003418 & -0.003393 & -0.000509 & -0.000559 \\ -0.000359 & 0.000801 & 0.000138 & 0.000081 & -0.000054 & -0.000015 \\ -0.003418 & 0.000138 & 0.003755 & 0.003395 & -0.000089 & -0.000034 \\ -0.003393 & 0.000081 & 0.003395 & 0.003806 & -0.000051 & -0.000029 \\ -0.000509 & -0.000054 & -0.000089 & -0.000051 & 0.001100 & 0.000595 \\ -0.000559 & -0.000015 & -0.000034 & -0.000029 & 0.000595 & 0.001132 \end{pmatrix}.$$

If we compare Tables 2.10 (where the serial independence was assumed) and 2.12 (where the serial dependence is taken into account), we see that the standard errors are systematically larger in the GEE approach as serial dependence induces overdispersion.

Generalized score tests for Type III contrasts are computed for GEE models (Wald tests are also available). Results of the Type 3 analysis are as follows:

Source	DF	Chi-square	Pr>Chi-sq
Gender	1	5.75	0.0165
Age*Power	2	125.78	<.0001
Size of the city	2	61.24	<.0001

This shows that all the explanatory variables are statistically relevant and should be kept in the model.

Table 2.12 Results of the Poisson regression with a GEE approach, Portfolio B.

Variable	Level	Coeff β	Std error	95 %	conf limit	Z	Pr > |Z|
Intercept		−1.2506	0.0625	−1.3731	−1.1280	−20.00	<.0001
Gender	Female	−0.0672	0.0283	−0.1227	−0.0118	−2.38	0.0175
Gender	Male	0	0	0	0	.	.
Age*Power	Group 1	−0.6483	0.0613	−0.7684	−0.5282	−10.58	<.0001
Age*Power	Group 2	−0.4007	0.0617	−0.5216	−0.2798	−6.50	<.0001
Age*Power	Group 3	0	0	0	0	.	.
Size of the city	Large	0.2555	0.0332	0.1905	0.3205	7.71	<.0001
Size of the city	Middle	0.0727	0.0336	0.0068	0.1386	2.16	0.0307
Size of the city	Small	0	0	0	0	.	.

The estimated regression coefficients of Table 2.12 are almost the same as those of Table 2.10. The assumption of independence between the individual data corresponding to different periods of observation thus has little impact on the claim frequencies.

2.9.6 Maximum Likelihood in the Negative Binomial Model for Panel Data

As noted by Liang & Zeger (1986), the GEE method is consistent, but not fully efficient. To achieve full efficiency, the maximum likelihood approach has to be used, which requires complete specification of the model to be estimated. This is why we will now consider models for longitudinal data based on mixed Poisson distributions, and perform maximum likelihood estimations.

Let us assume that, given $\Theta_i = \theta$, the observations $N_{i1}, N_{i2}, \ldots$ are independent and $\mathcal{P}oi(\theta\lambda_{it})$ distributed. If Θ_i follows the Gamma distribution $\mathcal{G}am(a, a)$ with probability density function f_Θ given by (1.35), the joint probability mass function of $N_{i,1}, \ldots, N_{i,T_i}$ is

$$
\begin{aligned}
&\Pr[N_{i1} = k_{i1}, \ldots, N_{iT_i} = k_{iT_i}] \\
&= \int_0^{+\infty} \Pr[N_{i1} = k_{i1}, \ldots, N_{iT_i} = k_{iT_i} | \Theta_i = \theta] f_\Theta(\theta) d\theta \\
&= \int_0^{+\infty} \left(\prod_{t=1}^{T_i} \Pr[N_{it} = k_{it} | \Theta_i = \theta] \right) f_\Theta(\theta) d\theta \\
&= \int_0^{+\infty} \left(\prod_{t=1}^{T_i} \exp(-\theta\lambda_{it}) \frac{(\theta\lambda_{it})^{k_{it}}}{k_{it}!} \right) f_\Theta(\theta) d\theta \\
&= \left(\prod_{t=1}^{T_i} \frac{\lambda_{it}^{k_{it}}}{k_{it}!} \right) \left(\frac{a}{a + \sum_{t=1}^{T_i} \lambda_{it}} \right)^a \left(a + \sum_{t=1}^{T_i} \lambda_{it} \right)^{-\sum_{t=1}^{T_i} k_{it}} \frac{\Gamma\left(a + \sum_{t=1}^{T_i} k_{it}\right)}{\Gamma(a)}. \qquad (2.19)
\end{aligned}
$$

In this case, $\mathbb{E}[N_{i,t}] = \lambda_{it} < \mathbb{V}[N_{i,t}] = \lambda_{it} + (\lambda_{it})^2/a$. Thus we see that the random effect Θ_i induces overdispersion and serial dependence. The multivariate probability mass function

Table 2.13 Results of the Negative Binomial regression model with panel data, Portfolio B.

Variable	Level	Coeff β	Std error	95 %	conf limit	t	Pr > $\|t\|$
Intercept		−1.2277	0.0646	−1.3542	−1.1011	−19.01	<.0001
Gender	Female	−0.0662	0.0287	−0.1223	−0.0101	−2.31	0.0208
Gender	Male	0	0	0	0	.	.
Age*Power	Group 1	−0.6587	0.0639	−0.7839	−0.5336	−10.32	<.0001
Age*Power	Group 2	−0.4069	0.0645	−0.5333	−0.2806	−6.31	<.0001
Age*Power	Group 3	0	0	0	0	.	.
Size of the city	Large	0.2604	0.0338	0.1942	0.3265	7.71	<.0001
Size of the city	Middle	0.0734	0.0340	0.0068	0.1401	2.16	0.0308
Size of the city	Small	0	0	0	0	.	.

(2.19) is often referred to as the Multivariate Negative Binomial, or Negative Multinomial, in the literature. The SAS®/STAT procedure NLMIXED can be used directly to maximize the log-likelihood.

The fit of the Negative Binomial model is described in Table 2.13. The estimated dispersion parameter is $\widehat{a} = 1.9871$. The variance-covariance matrix of the estimated regression coefficients and dispersion parameter $\widehat{a}$ is

$$\widehat{\boldsymbol{\Sigma}}_{\widehat{\beta}} = \begin{pmatrix} 0.004170 & -0.000283 & -0.003685 & -0.003673 & -0.000529 & -0.000561 & -0.000197 \\ -0.000283 & 0.000821 & -0.000004 & -0.000006 & < 10^{-6} & 0.000009 & < 10^{-6} \\ -0.003685 & -0.000004 & 0.004078 & 0.003720 & -0.000096 & -0.000057 & -0.000032 \\ -0.003673 & -0.000006 & 0.003720 & 0.004154 & -0.000069 & -0.000051 & -0.000086 \\ -0.000529 & < 10^{-6} & -0.000096 & -0.000069 & 0.001139 & 0.000610 & -0.000050 \\ -0.000561 & 0.000009 & -0.000057 & -0.000051 & 0.000610 & 0.001156 & 0.000039 \\ -0.000197 & < 10^{-6} & -0.000032 & -0.000086 & -0.000050 & 0.000039 & 0.027629 \end{pmatrix}.$$

The log-likelihood is equal to $-19\,648.5$. The Type 3 analysis gives

Source	DF	Chi-square	Pr>Chi-sq
Gender	1	5.4	0.0201
Age*Power	2	147.8	<.0001
Size of the city	2	64.2	<.0001

All the explanatory variables included in the regression model are thus statistically relevant and must be kept in the model.

2.9.7 Maximum Likelihood in the Poisson-Inverse Gaussian Model for Panel Data

The Inverse Gaussian distribution is another good candidate to model the heterogeneity parameter Θ_i. Let us now assume that the common probability density function of the Θ_is is given by (1.41). Then, the contribution of policyholder i to the likelihood is

$$\Pr[N_{i1}=k_{i1},\ldots,N_{iT_i}=k_{iT_i}]$$
$$=\int_0^{+\infty}\left(\prod_{t=1}^{T_i}\Pr[N_{it}=k_{it}|\Theta_i=\theta]\right)\frac{1}{\sqrt{2\pi\tau\theta^3}}\exp\left(-\frac{1}{2\tau\theta}(\theta-1)^2\right)d\theta.$$

Again, numerical methods are needed to obtain the solutions and the SAS®/STAT procedure NLMIXED can be used to maximize the log-likelihood.

The fit of the Poisson-Inverse Gaussian model is described in Table 2.14. We get $\widehat{\tau}=0.5575$. The variance-covariance matrix of the estimated regression coefficients and dispersion parameter $\widehat{\tau}$ is

$$\widehat{\Sigma}_{\widehat{\beta}}=\begin{pmatrix} 0.004216 & -0.000289 & -0.003723 & -0.003709 & -0.000534 & -0.000567 & 0.000115\\ -0.000289 & 0.000830 & -0.000002 & -0.000004 & 0.000000 & 0.000009 & -0.000002\\ -0.003723 & -0.000002 & 0.004118 & 0.003755 & -0.000097 & -0.000058 & -0.000027\\ -0.003709 & -0.000004 & 0.003755 & 0.004193 & -0.000069 & -0.000052 & 0.000011\\ -0.000534 & 0.000000 & -0.000097 & -0.000069 & 0.001152 & 0.000616 & 0.000030\\ -0.000567 & 0.000009 & -0.000058 & -0.000052 & 0.000616 & 0.001169 & -0.000010\\ 0.000115 & -0.000002 & -0.000027 & 0.000011 & 0.000030 & -0.000010 & 0.002492 \end{pmatrix}.$$

The log-likelihood is equal to $-19\,643.7$ and Type 3 analysis gives

Source	DF	Chi-square	Pr>Chi-sq
Gender	1	5.4	0.0201
Age*Power	2	148.8	<.0001
Size of the city	2	64.2	<.0001

2.9.8 *Maximum Likelihood in the Poisson-LogNormal Model for Panel Data*

In biostatistical circles, the Poisson-LogNormal model is often used, after Hinde (1982). In this case, Θ_i has the probability density function (1.45) and the joint probability mass functions of $N_{i,1},\ldots,N_{i,T_i}$ are

Table 2.14 Results of the Poisson-Inverse Gaussian regression model with panel data, Portfolio B.

Variable	Level	Coeff β	Std error	95 %	conf limit	t	Pr > $\|t\|$
Intercept		−1.2223	0.0649	−1.3496	−1.0951	−18.83	<.0001
Gender	Female	−0.0665	0.0288	−0.1229	−0.0100	−2.31	0.0211
Gender	Male	0	0	0	0	.	.
Age*Power	Group 1	−0.6631	0.0642	−0.7889	−0.5373	−10.33	<.0001
Age*Power	Group 2	−0.4087	0.0648	−0.5356	−0.2817	−6.31	<.0001
Age*Power	Group 3	0	0	0	0	.	.
Size of the city	Large	0.2619	0.0339	0.1954	0.3284	7.72	<.0001
Size of the city	Middle	0.0732	0.0342	0.0062	0.1402	2.14	0.0323
Size of the city	Small	0	0	0	0	.	.

$$\Pr[N_{i1} = k_{i1}, \ldots, N_{iT_i} = k_{iT_i}]$$
$$= \int_0^{+\infty} \left(\prod_{t=1}^{T_i} \exp(-\theta\lambda_{it}) \frac{(\theta\lambda_{it})^{k_{it}}}{k_{it}!} \right) \frac{1}{\theta\sigma\sqrt{(2\pi)}} \exp\left(-\frac{(\ln\theta + \sigma^2/2)^2}{2\sigma^2} \right) d\theta.$$

The integral has no closed-form solution so that it is not possible to derive the log-likelihood equations. Therefore, numerical procedures are needed to solve the integral and to find maximum likelihood estimates (the NLMIXED procedure of SAS®/STAT, for instance). Now,

$$\mathbb{E}[N_{i,t}] = \lambda_{it} \text{ and } \mathbb{V}[N_{i,t}] = \lambda_{it} + \big(\exp(\sigma^2) - 1\big)(\lambda_{it})^2.$$

The fit of the Poisson-LogNormal model is described in Table 2.15. We get $\widehat{\sigma}^2 = 0.4581$. The variance-covariance matrix of the estimated regression coefficients and dispersion parameter $\widehat{\sigma}^2$ is

$$\widehat{\boldsymbol{\Sigma}}_{\widehat{\beta}} = \begin{pmatrix} 0.004225 & -0.000291 & -0.003729 & -0.003714 & -0.000536 & -0.000569 & 0.000091 \\ -0.000291 & 0.000832 & -0.000001 & -0.000004 & -0.000001 & 0.000009 & -0.000003 \\ -0.003729 & -0.000001 & 0.004125 & 0.003760 & -0.000097 & -0.000058 & -0.000022 \\ -0.003714 & -0.000004 & 0.003760 & 0.004199 & -0.000069 & -0.000052 & 0.000006 \\ -0.000536 & -0.000001 & -0.000097 & -0.000069 & 0.001155 & 0.000618 & 0.000022 \\ -0.000569 & 0.000009 & -0.000058 & -0.000052 & 0.000618 & 0.001173 & -0.000007 \\ 0.000091 & -0.000003 & -0.000022 & 0.000006 & 0.000022 & -0.000007 & 0.001193 \end{pmatrix}.$$

The log-likelihood is $-19\,642.3$ and Type 3 analysis gives

Source	DF	Chi-square	Pr>Chi-sq
Gender	1	5.2	0.0226
Age*Power	2	148.8	<.0001
Size of the city	2	64.0	<.0001

Table 2.15 Results of the Poisson-LogNormal regression model with panel data, Portfolio B.

Variable	Level	Coeff β	Std error	95 %	conf limit	t	Pr > $\|t\|$
Intercept		−1.2209	0.0650	−1.3483	−1.0935	−18.78	<.0001
Gender	Female	−0.0669	0.0289	−0.1234	−0.0103	−2.32	0.0205
Gender	Male	0	0	0	0	.	.
Age*Power	Group 1	−0.6640	0.0642	−0.7898	−0.5381	−10.34	<.0001
Age*Power	Group 2	−0.4089	0.0648	−0.5359	−0.2819	−6.31	<.0001
Age*Power	Group 3	0	0	0	0	.	.
Size of the city	Large	0.2620	0.0340	0.1954	0.3287	7.71	<.0001
Size of the city	Middle	0.0731	0.0343	0.0059	0.1402	2.13	0.0329
Size of the city	Small	0	0	0	0	.	.

2.9.9 Vuong Test

In the framework of panel data, the test proposed in Section 2.8.1 is no longer valid. This test can be modified by using the conditional log-likelihood defined, for policyholder i and for the Negative Binomial model, as

$$\begin{aligned} l_i^{NB}(k_{it}|k_{i1}\ldots k_{it-1}) &= \ln \Pr[N_{it}=k_{it}|N_{i1}=k_{i1},\ldots,N_{it-1}=k_{it-1}] \\ &= \ln \frac{\Pr[N_{i1}=k_{i1},\ldots,N_{it}=k_{it}]}{\Pr[N_{i1}=k_{i1},\ldots,N_{it-1}=k_{it-1}]} \\ &= \ln\left(\frac{1}{k_{it}!}\frac{\int_0^\infty \prod_{s=1}^{t}\exp(-\theta\lambda_{is})(\theta\lambda_{is})^{k_{is}} f_\Theta(\theta)d\theta}{\int_0^\infty \prod_{s=1}^{t-1}\exp(-\theta\lambda_{is})(\theta\lambda_{is})^{k_{is}} f_\Theta(\theta)d\theta}\right), \end{aligned}$$

where f_Θ is given by (1.35). Therefore, for panel data, the log-likelihood can be equivalently rewritten as

$$L(\boldsymbol{\beta}) = \sum_{i=1}^{n}\sum_{t=1}^{T_i} l_i^{NB}(k_{it}|k_{i1}\ldots k_{it-1})$$

where for $t=1$, l_i^{NB} is simply the probability $\Pr[N_{i1}=k_{i1}]$. The quantities l_i^{PIG} and l_i^{PLN} are defined equivalently for Θ_is following the Inverse Gaussian and LogNormal distributions, substituting for f_Θ the probability density functions (1.41) and (1.45), respectively.

GOLDEN (2003) extended the test proposed by VUONG (1989) to the serial case. It can be applied here to compare the different mixed Poisson alternatives provided the panel is balanced (that is, the T_is are equal for all the policyholders). In the application to Portfolio B, we keep the policies in the portfolio for three periods (that is, those for which $T_i = T = 3$). There are 9894 such policies out of a total of 20 354 policies in the complete portfolio. The test statistic for comparing the Negative Binomial model to the Poisson-Inverse Gaussian model is then

$$T_{ext} = \frac{\sum_{i=1}^{n}\sum_{t=1}^{T}\left(l_i^{NB}(k_{it}|k_{i1}\ldots k_{it-1}) - l_i^{PIG}(k_{it}|k_{i1}\ldots k_{it-1})\right)}{\sqrt{nT}\sigma},$$

with the variance defined as

$$\begin{aligned} \sigma^2 = &\frac{1}{n}\sum_{i=1}^{n}\sum_{t=1}^{T}\sum_{s=1}^{T}\left(l_i^{NB}(k_{it}|k_{i1}\ldots k_{it-1}) - l_i^{PIG}(k_{it}|k_{i1}\ldots k_{it-1})\right) \\ &\times\left(l_i^{NB}(k_{is}|k_{i1}\ldots k_{is-1}) - l_i^{PIG}(k_{is}|k_{i1}\ldots k_{is-1})\right). \end{aligned}$$

This expression for the variance is close to the one obtained in the independent case, except that covariances between correlated observations appear here. Note that several assumptions have to be checked before the test can be formally performed. We refer the reader to GOLDEN (2003) for more details. Here, we proceed as in FRENCH & JONES (2004) and apply the test directly. The following results are obtained:

- when we compare the Poisson-Gamma model with the Poisson-Inverse Gaussian model, the value of the test statistic is equal to -0.7988 leading to a p-value of 42.44 %.
- when we compare the Poisson-Gamma model with the Poisson-LogNormal model, the value of the test statistic is equal to -0.7688 leading to a p-value of 44.20 %.
- when we compare the Poisson-Inverse Gaussian model with the Poisson-LogNormal model, the value of the test statistic is equal to -0.5922 leading to a p-value of 55.37 %.

Therefore, the three models are not statistically different. Applying the same testing procedure to the policies that are in the portfolio for the first two years (this means 12 202 policies) yields the same conclusion.

2.9.10 Information Criteria

Non-nested models are often compared using likelihood-based criteria, including the well-known AIC, for instance. Since the competing models have the same number of parameters, it is enough to examine the respective log-likelihoods. A commonly used rule-of-thumb consists in considering that two models are significantly different if the difference in the log-likelihoods exceeds five (corresponding to a difference in AICs of more than ten, as discussed in Burnham & Anderson (2002)). This means here that the Poisson-Inverse Gaussian and Poisson-LogNormal models are significantly better than the Negative Binomial model. Considering the maximum of the log-likelihood, we chose the Poisson-LogNormal model for Portfolio B. The same conclusion is obtained using different criteria often used in practice. For instance, Raftery (1995) suggested that a model significantly outperfoms a competitor if the difference in their respective BIC values exceeds 5.

2.9.11 Resulting Classification for Portfolio B

Table 2.16 gives the resulting price list obtained with the Poisson-LogNormal model described in Table 2.15. The final *a priori* ratemaking contains 18 classes. This ratemaking will be used throughout the text for the examples involving Portfolio B.

There is another way to present the results displayed in Table 2.16. The annual expected claim frequency is obtained from the reference class, with

$$\exp(\widehat{\beta}_0) = 29.50\,\%$$

according to Table 2.15. Applying correction coefficients, we then obtain the annual expected claim frequency of any policyholder. Specifically, it is obtained from the multiplicative formula

Table 2.16 *A priori* risk classification for Portfolio B (Poisson-LogNormal regression model).

Age–Power			Size of the city			Gender		Annual claim freq. (%)	Weights (%)
Group 1	Group 2	Group 3	Large	Middle	Small	Female	Male		
No	No	Yes	No	No	Yes	No	Yes	29.50	0.82
No	No	Yes	No	No	Yes	Yes	No	27.59	0.43
No	No	Yes	No	Yes	No	No	Yes	31.73	0.63
No	No	Yes	No	Yes	No	Yes	No	29.68	0.37
No	No	Yes	Yes	No	No	No	Yes	38.33	0.44
No	No	Yes	Yes	No	No	Yes	No	35.86	0.31
No	Yes	No	No	No	Yes	No	Yes	19.60	7.88
No	Yes	No	No	No	Yes	Yes	No	18.33	4.64
No	Yes	No	No	Yes	No	No	Yes	21.08	8.32
No	Yes	No	No	Yes	No	Yes	No	19.72	4.66
No	Yes	No	Yes	No	No	No	Yes	25.47	6.89
No	Yes	No	Yes	No	No	Yes	No	23.82	3.99
Yes	No	No	No	No	Yes	No	Yes	15.19	12.59
Yes	No	No	No	No	Yes	Yes	No	14.20	7.02
Yes	No	No	No	Yes	No	No	Yes	16.34	13.81
Yes	No	No	No	Yes	No	Yes	No	15.28	7.24
Yes	No	No	Yes	No	No	No	Yes	19.74	12.68
Yes	No	No	Yes	No	No	Yes	No	18.46	7.30

$$
\begin{aligned}
&29.50\,\% \\
&\times \begin{cases} \exp(-0.0669) = 0.94, & \text{if the policyholder is a female driver,} \\ 1, & \text{otherwise,} \end{cases} \\
&\times \begin{cases} \exp(-0.6640) = 0.51, & \text{if the policyholder belongs to Group 1 with respect to} \\ & \text{the combined variable Age*Power,} \\ \exp(-0.4089) = 0.66, & \text{if the policyholder belongs to Group 2 with respect to} \\ & \text{the combined variable Age*Power,} \\ 1, & \text{otherwise,} \end{cases} \\
&\times \begin{cases} \exp(0.2620) = 1.30, & \text{if the policyholder resides in a large city,} \\ \exp(0.0731) = 1.08, & \text{if the policyholder resides in a middle-sized city,} \\ 1, & \text{otherwise.} \end{cases}
\end{aligned}
$$

2.10 Further Reading and Bibliographic Notes

2.10.1 Generalized Linear Models

After decades dominated by statistically unsophisticated models, it is now common practice since Mc Cullagh & Nelder (1989) and Brockman & Wright (1992) to achieve *a priori* classification with the help of generalized linear models (GLMs). They are so called because they generalize the classical linear models based on the Normal distribution.

Risk classification techniques for claim counts have been the topic of many papers appearing in the actuarial literature. Early references include TER BERG (1980b) and ALBRECHT (1983a,b,c). DIONNE & VANASSE (1989, 1992) used a Negative Binomial regression model, while DEAN, LAWLESS & WILLMOT (1989) used a Poisson-Inverse Gaussian distribution to fit the number of claims. CUMMINS, DIONNE, MCDONNALD & PRITCHETT (1990) applied the GB2 family of distributions in modelling claim counts. TER BERG (1996) considered the Generalized Poisson distribution of CONSUL (1990) and incorporated explanatory variables with the help of a loglinear model. There are now several textbooks devoted to the statistical analysis of count data. Let us mention CAMERON & TRIVEDI (1998) and WINKELMANN (2003). Before the Poisson regression became popular among actuaries, claims data were often analysed using logistic regression; see, e.g., BEIRLANT, DERVEAUX, DE MEYER, GOOVAERTS, LABIES & MAENHOUDT (1991).

Separate analyses are usually conducted for claim frequencies and costs, including expenses, to arrive at a pure premium. With the noticeable exception of JORGENSEN & PAES DE SOUZA (1994), all the actuarial analyses of the pure premium so far have examined frequencies and severities separately. This approach is particularly relevant in motor insurance, where the risk factors influencing the two components of the pure premium are usually different.

2.10.2 Nonlinear Effects

GLMs however only deal with categorical risk factors in an efficient way. The main drawback of GLMs is that covariate effects are modelled in the form of a linear predictor. GLMs are too restrictive if nonlinear effects of continuous covariates are present. Continuous covariates can efficiently enter GLMs only if they are suitably transformed to reflect their true effect on the score scale. However, it is not always clear how the variables should be transformed.

It has been common practice in insurance companies to model possibly nonlinear effects of a covariate by polynomials. However, it is well known to statisticians that polynomials are often not flexible enough to capture the variability of the data particularly when the polynomial degree is small (see, e.g., FAHRMEIR & TUTZ (2001), Chapter 5). For larger degrees the flexibility of polynomials increases but at the cost of possibly high variability of resulting estimates particularly at the left and right extreme values of the covariate.

A more flexible approach for modelling nonlinear effects can be based on *piecewise polynomials*. More specifically, the unknown functions are approximated by *polynomial splines*, which may be regarded as piecewise polynomials with additional regularity conditions (see, e.g., DE BOOR, 1978). We refer the interested reader to DENUIT & LANG (2004) for an overview of the existing approaches.

2.10.3 Zero-Inflated Models

Insurance data usually include a relatively large number of zeros (no claim). Deductibles and no claim discounts increase the proportion of zeros, since small claims are not reported by insured drivers. Zero-inflated models, including the Zero-Inflated Poisson (ZIP) model, account for this phenomenon.

ZIP models can be considered as a mixture of a zero point mass and a Poisson distribution and were first used to study soldering defects on print wiring boards (LAMBERT, 1992). To account for overdispersion in the Poisson part, generalizations of the model are possible and include the Zero-Inflated Negative Binomial (ZINB) distribution. See Yip & Yau (2005) for an application to insurance claim count data.

Other than the zero-inflated models, parametric methods such as the mixture of distributions can be used to model the claim frequency distribution with extra zeros. Hürlimann (1990) discussed the use of several pseudo compound Poisson distributions in modelling the claim count data. To test for a Poisson mixture, a test statistic was proposed by Carrière (1993a).

2.10.4 Fixed Versus Random Effects

The mixed Poisson distribution is often used to account for unknown characteristics of the driver, influencing the number of accidents reported to the company. When panel data are available, these hidden features can alternatively be captured by an individual heterogeneity term that is constant over time (the standard reference for panel data is Hsiao (2003); the particular case of count variables is treated in Cameron & Trivedi (1998)). Boucher & Denuit (2006) compared the two approaches with emphasis on the actual meaning of the estimated parameters in a mixed Poisson regression when random effects and covariates are correlated. In such a case, parameter estimates should be seen as the apparent effects of the covariates on the frequency. Keeping this in mind allows for a better understanding of the resulting price list.

The results obtained by Boucher & Denuit (2006) legitimate the use of random effects models even if there exists a correlation between the regressors and the heterogeneity. The parameter estimates do not identify the impact of these regressors on the premium but only the apparent effects. Since this is usually the focus for the actuary in ratemaking, there is no problem with this interpretation. However, such a correlation clearly indicates that a correction should be done to obtain a more accurate model. In particular, the apparent high risk of young drivers should deserve some attention. The analysis conducted by Boucher & Denuit (2006) shows that the fixed effects are very heterogeneous for these individuals. Instead of penalizing these insureds in the *a priori* ratemaking, an appropriate bonus-malus scheme could be designed. Merit rating systems improve the fairness of the tariff in that respect. We will come back to this issue in Chapter 8.

2.10.5 Hurdle Models

Boucher, Denuit & Guillén (2006) presented and compared different risk classification models for the annual number of claims reported to the insurer. Generalized heterogeneous, zero-inflated, hurdle and compound frequency models are applied to a sample of an automobile portfolio of a major company operating in Spain.

The hurdle models are widely used in connection with health care demands. An application to credit scoring is proposed in Dionne, Artis & Guillén (1996). With health care demand, it is generally accepted that the demand for certain types of health care services depends on two processes: the decisions of the individual and those of the health care provider. See, e.g. Pohlmeier & Ulrich (1995) or Santos Silva & Windmeijer (2001). The hurdle model also possesses a natural interpretation for the number of reported claims. A reason for the good fit of the zero-inflated models is certainly the reluctance of some insureds to report their accident (since they would then be penalized by some bonus-malus scheme implemented by the insurer). It is reasonable to believe that the behaviour of the insureds is not the same when they have already reported a claim. This suggests that two processes govern the total number of claims, as in the hurdle model.

2.10.6 Geographic Ratemaking

It is common in motor insurance to let the risk premium per unit exposure vary with geographic area when all other risk factors are held constant. Most companies have adopted a risk classification according to the geographical zone where the policyholder lives (urban / nonurban for instance, or a more accurate split of the country according to Zip codes). The spatial variation may be related to geographic factors (e.g. traffic density or proximity to arterial roads) or to socio-demographic factors. In such cases it will be desirable to estimate the spatial variation in risk premium and to price accordingly. Spatial postcode methods for insurance rating attempt to extract information which is in addition to that contained in standard factors (like age or gender for instance). Often, claim characteristics tend to be similar in neighbouring postcode areas (after other factors have been accounted for). The idea of postcode rating models is to exploit this spatial smoothness by allowing for information transfer to and from neighbouring regions.

Following Boskov & Verrall (1994) and Brouhns, Denuit, Masuy & Verrall (2002), the risk associated with each district can be assessed with the help of statistical models for spatial data. The techniques used for geographic ratemaking are closely related to those used in disease mapping by epidemiologists. Figure 2.17 is taken from Brouhns, Denuit, Masuy & Verrall (2002). It displays the raw exposures e_i for each area in Belgium (the number of policy-years, in our case) whilst Figure 2.18 shows crude claim rates (observed number of claims divided by the corresponding expected number of claims given the characteristics of the policyholders living in each area). However, the latter map is at best difficult to interpret and can even be seriously misleading because the crude claim rates tend to be far more extreme in regions with smaller risk exposures (see Figures 2.17–2.18; the high rates in the north-west of Belgium (West Flanders) or in the south of the country correspond to districts for which we have less policyholders). Hence regions with the least reliable data will typically draw the main visual attention. This is one reason why it is difficult in practice to attempt any smoothing or risk assessment 'by eye'.

Whereas epidemiologists and environmetricians have been interested in spatial models for a long time, the actuarial literature is rather poor in respect of ratemaking methods incorporating geographic components. Taylor (1989) used two-dimensional splines on a plane linked to the map of the region by a transformation chosen to match the features of the specific region. He applied this method to a data set from Sydney, Australia. Boskov & Verrall (1994) highlighted some deficiencies in Taylor's model, and provided an alternative treatment which made use of the Gibbs sampler to implement a Bayesian revision of the observation in each area. The main advantage of the Bayesian framework is that it recognizes the magnitudes of sampling error and incorporates the concept of smoothing over neighbouring areas. Taylor (1996) adopted a similar point of view and applied Whittaker graduation (a widely accepted actuarial technique which has also been shown to have a Bayesian interpretation). Dixon, Kelsey & Verrall (2000) proposed an extension of the Boskov & Verrall (1994) model including weighting factors accounting for distances between regions.

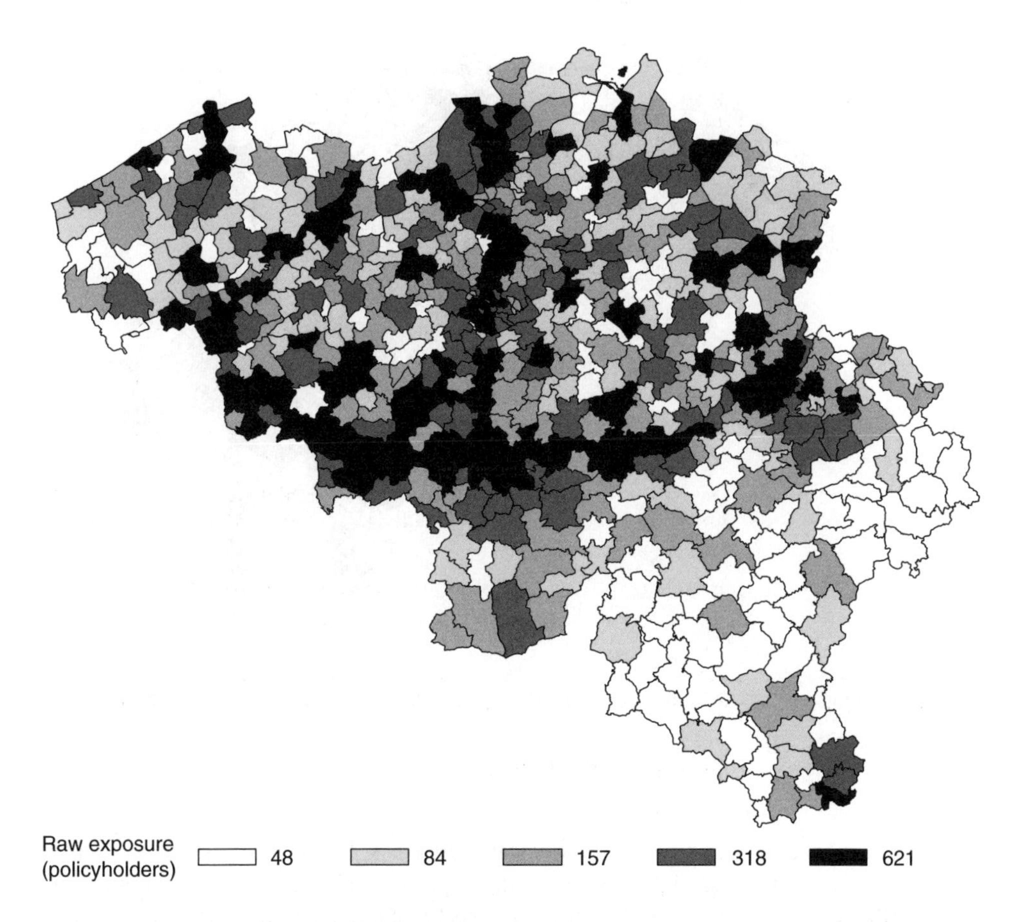

Figure 2.17 Map of Belgium with exposures-to-risk.

A key step in model specification is the definition of neighbours, i.e. those areas whose claim rates are correlated with that of a given area. A traditional definition of neighbours includes all areas contiguous to a given area. The methodology applied in Brouhns, Denuit, Masuy & Verrall (2002) is as follows. In a first stage, available explanatory variables are incorporated to the policyholders' claim frequencies with the help of a (mixed) Poisson regression model. In a second stage, the data are aggregated by districts and overdispersion is accounted for by introduction of a random effect (split into a spatially structured part and a spatially unstructured one). The Boskov and Verrall model is then used to recover the spatial structure of the claims pattern that can be used to design the geographical ratemaking strategy of the company.

In order to figure out the global claim pattern, Figure 2.19 displays the geographical risk variation, and serves as basis for the determination of the rating areas. Clearly, several regions with high, medium and low values of geographic risk emerge. This map can thus serve to design different rating areas, which in turn become a categorical variable that may enter the (mixed) Poisson regression model.

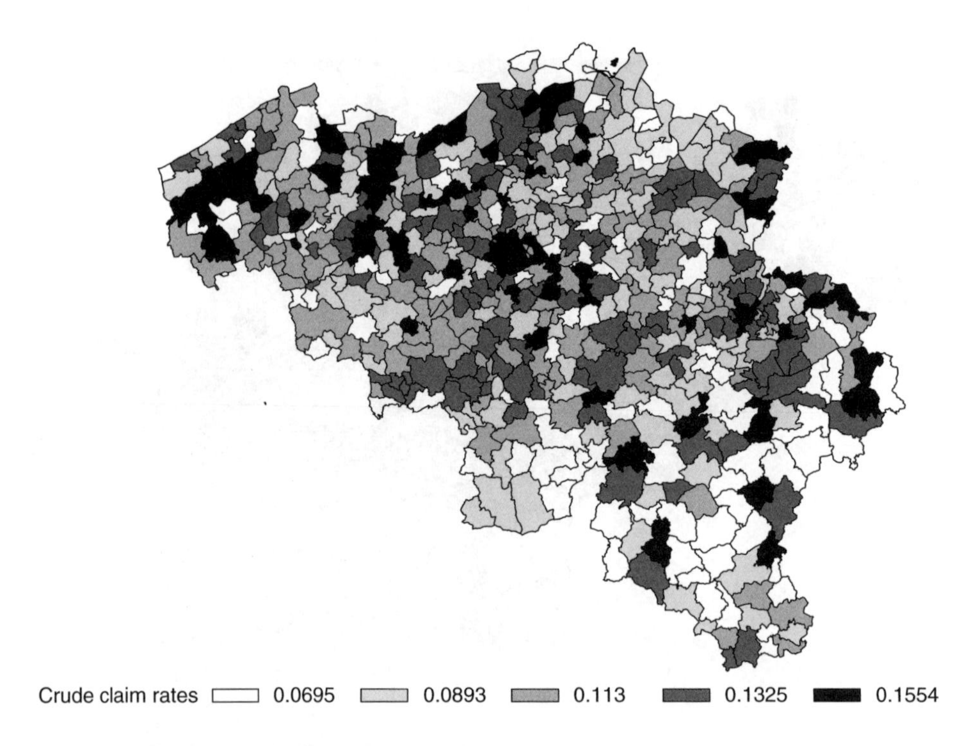

Figure 2.18 Map of Belgium with crude rates.

The approach however proceeds in two steps: first a regression is performed with all the covariates except the spatial ones to get the expected claim number of each area, and then the Boskov and Verrall model recovers the spatial claim pattern. In order to avoid the preprocessing of the data to remove the effect of all risk factors other than the spatial ones, DIMAKOS & RATTALMA (2002) proposed a fully Bayesian approach to nonlife ratemaking. This approach still relies on GLMs and thus suffers from the drawbacks mentioned in Section 2.10.2: continuous covariates such as policyholders' age enter linearly into the model (on the score scale) whereas it is now well established that the effect of some continuous variables is far from linear (typically, convex for policyholders' age).

Statistical modelling tools to perform space-time analysis of insurance data have been proposed by DENUIT & LANG (2004) and FAHRMEIR, LANG & SPIES (2003). This approach enables the actuary to explore spatial and temporal effects simultaneously with the impact of other covariates. Bayesian generalized additive models provide a broad and flexible framework for regression analyses in realistically complex situations with cross-sectional, longitudinal and spatial data. All effects, as well as smoothing parameters, are regarded as random and are assigned appropriate priors.

2.10.7 Software

SAS® has been used throughout this chapter. The readers interested in the use of this software to perform statistical analyses are referred to DER & EVERITT (2002) for a very readable

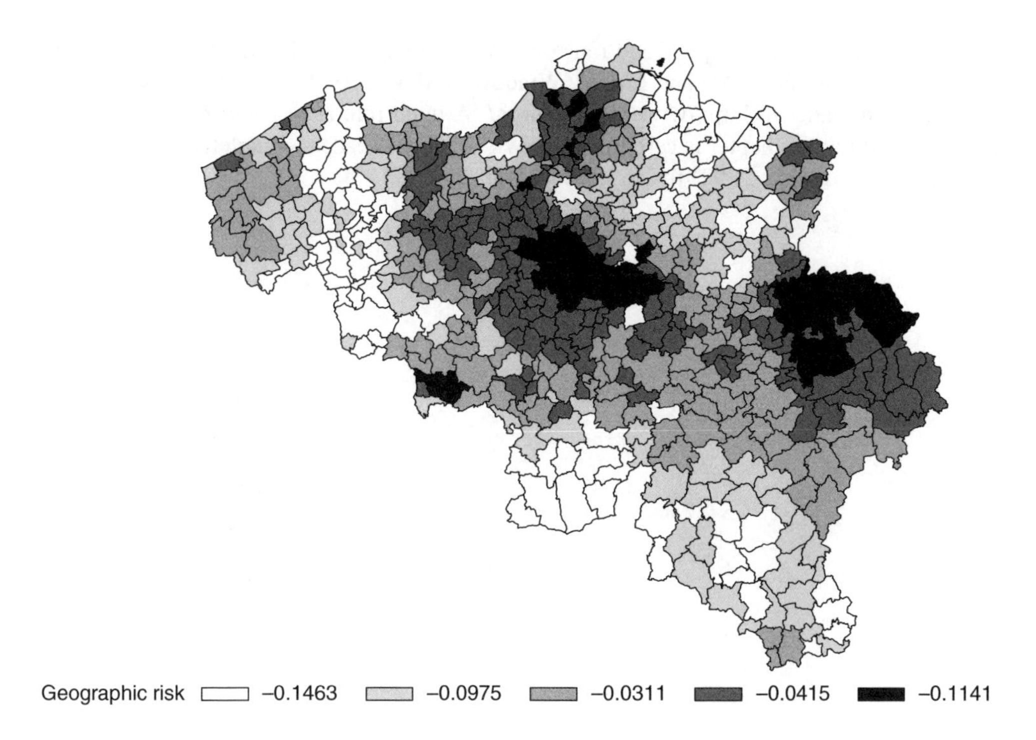

Figure 2.19 Estimation of the geographically structured risk.

introduction. For more information, we refer the interested reader to the SAS® website, http://www.sas.com/.

A number of software packages specific to the insurance industry have been developed, some of them based on SAS®. We mention a few of them hereafter, without being exhaustive. Note also that the authors did not test these tools, which are not available free of charge, so that we could not evaluate their relative performances.

The Tricast suite has been developed using the support software from SAS® and includes actuarial tools to create rating models. For more information, the interested reader may email tricast@tricast-group.com. The consulting firm Watson Wyatt offers the Pretium® system, which integrates with SAS®. For more details, we refer the reader to http://www.watsonwyatt.com/. EMB has developed several computer tools for insurance rating. The Emblem® software can be used to model claims experience and the Classifier™ software allows one to assess and categorize geographically distributed risk. For more details, see http://www.emb.co.uk/.

Academic researchers and research and development actuaries appreciate the software system 'R'. This is a free language and environment for statistical computing and graphics. R is a GNU project which is similar to the S language and environment which was developed at Bell Laboratories (formerly AT&T, now Lucent Technologies) by John Chambers and colleagues. R can be considered as a different implementation of S. There are some important differences, but much code written for S runs unaltered under R. R is available as Free Software under the terms of the Free Software Foundation's GNU General Public License in source code form. It compiles and runs on a wide variety of UNIX platforms and similar systems (including FreeBSD and Linux), Windows and MacOS. For more details, we refer the interested reader to http://www.r-project.org/. A good reference about the use of R for topics related to this chapter is certainly FARAWAY (2006). Readers completely new to R are referred to DALGAARD (2002) for an introduction.

Part II

Basics of Experience Rating

Actuarial Modelling of Claim Counts: Risk Classification, Credibility and Bonus-Malus Systems M. Denuit, X. Maréchal, S. Pitrebois and J.-F. Walhin

3

Credibility Models for Claim Counts

3.1 Introduction

3.1.1 From Risk Classification to Experience Rating

We have seen in Chapter 2 how to partition a heterogeneous portfolio into more homogeneous classes with all policyholders belonging to the same class paying the same premium. However, tariff cells are still quite heterogeneous despite the use of many *a priori* variables. The expected claim frequency for the tariff cell is designed to reflect the average experience of the entire group. If the experience of a policy is consistently better (or worse) than the average experience of the group, the insurance company may consider adapting the amount of premium to be charged for this policy. Of course, this requires a model which can separate random variation from signal in the historical data to indicate whether this policy is of better (or worse) quality compared to the group average.

It is reasonable to believe that the hidden features (unobserved risk characteristics that have been modelled by a random effect in the mixed Poisson regression model) are revealed by the number of claims reported by the policyholders over the successive insurance periods. Hence the adjustment of the premium based on the individual claims experience in order to restore fairness among policyholders. The allowance for the history of the policyholder in a rating model thus derives from interpretation of serial correlation for longitudinal data resulting from hidden features in the risk distribution.

3.1.2 Credibility Theory

Credibility theory is the art of combining different collections of data to obtain an accurate overall estimate. It provides actuaries with techniques to determine insurance premiums for

Actuarial Modelling of Claim Counts: Risk Classification, Credibility and Bonus-Malus Systems M. Denuit, X. Maréchal, S. Pitrebois and J.-F. Walhin

contracts that belong to a (more or less) heterogeneous portfolio, where there is limited or irregular claim experience for each contract but ample claim experience for the portfolio. Credibility theory can be seen as a set of quantitative tools that allows the insurers to perform experience rating, that is, to adjust future premiums based on past experience. In many cases, a compromise estimator is derived from a convex combination of a prior mean and the mean of the current observations. The weight given to the observed mean is called the credibility factor (since it fixes the extent to which the actuary may be confident in the data).

3.1.3 Limited Fluctuation Theory

There are different types of credibility mechanisms: limited fluctuations credibility and greatest accuracy credibility. Limited fluctuation credibility theory was developed in the early part of the 20th century in connection with workers compensation insurance by MOWBRAY (1914). It provides a mechanism for assigning full or partial credibility to a policyholder's experience. In the former case, the policy is rated on the basis of its own claims history, whereas in the latter case, a weighted average of past experience and grand mean is used by the insurer. Although the limited fluctuation approach provides simple solutions to the problem, it suffers from a lack of theoretical justification. We will not consider this approach in this book. Instead, we will consider the greatest accuracy credibility theory formalized by BÜHLMANN (1967,1970).

3.1.4 Greatest Accuracy Credibility

The idea behind greatest accuracy credibility theory can be summarized as follows: Tariff cells include policyholders with similar underwriting characteristics; each of them is viewed as homogeneous with respect to the underwriting characteristics used by the insurance company. Of course, the risks in the cell are not truly homogeneous: there still remains some heterogeneity in each of the tariff cells, as explained in the preceding chapters. To reflect this heterogeneity, the relative risk level of each policyholder in the rating cell is characterized by a risk parameter θ, but the value of θ varies by policyholder. If $\theta = 50\,\%$ then the expected number of claims reported by this policyholder is half of the claim frequency corresponding to the rating cell, whereas if $\theta = 300\,\%$ then the expected number of claims for this individual is three times the claim frequency of the rating cell. Of course, even if assuming the existence of such a θ is reasonable, it is not observable and the actuary can never know its true value for a given policyholder.

Because θ varies by policyholder, there is a distribution function F_Θ giving the proportion of policyholders in the portfolio with relative risk level less than or equal to a certain threshold. Stated another way, $F_\Theta(\theta)$ represents the probability that a policyholder picked at random from the portfolio has a risk parameter that is less than or equal to θ. The connection with the random effect introduced in the statistical models of Chapter 2 to account for the residual heterogeneity is now clear: this random effect Θ becomes the random risk parameter for a policyholder picked at random from the portfolio (the distribution function of Θ is F_Θ). Even if the risk parameter Θ remains unknown, the distribution function F_Θ can be estimated from data, as explained in Chapter 2. Once estimated, the heterogeneity model can be used to perform prediction on longitudinal data and allows for experience rating in motor insurance. In an empirical Bayesian setting, the prediction is derived from the expectation of a random effect with respect to a posterior distribution taking into account the history of the individual.

3.1.5 Linear Credibility

Bayesian statistics offer an intellectually acceptable approach to greatest accuracy credibility theory. Nevertheless, practical applications involve numerical methods to perform integration with respect to *a posteriori* distribution, making more elementary approaches desirable (at least to get an easy-to-compute approximation of the result). In that respect, linear credibility formulas are especially useful. Basically, the actuary still resorts to a quadratic loss function but the shape of the credibility predictor is constrained *ex ante* to be linear.

3.1.6 Financial Equilibrium

When the insurer uses past claims history to reevaluate the amount of premium to be charged to the policyholders, the increases and decreases granted to the policyholders in the portfolio must exactly balance each other. The credibility mechanism indeed has little effect on the number of claims filed to the company. Therefore, the number of claims with and without credibility is very much the same.

For this reason, we expect that the *a posteriori* corrections average to unity. This ensures that the average number of claims without credibility equals the average number of claims with credibility, and that the premium income will be enough to compensate the claims.

3.1.7 Combining a Priori and a Posteriori Ratemaking

The amount of premium paid by the policyholder depends on the rating factors of the current period (think for instance of the type of the car or of the occupation of the policyholder) but also on the claim history. The insurance premium is the product of a base premium and of a credibility coefficient. The base premium is a function of the current rating factors whereas the credibility coefficient usually depends on the history of claims at fault. Clearly, *a priori* and *a posteriori* ratings interact. To the extent that good drivers are rewarded in their base premiums (through other rating variables) the size of the bonus they require for equity is reduced.

The claims history of each policyholder consists of a short integer-valued sequence of yearly claim counts. The basic model used for experience rating is based on the Negative Binomial distribution. This probability law can be seen as a Poisson mixture distribution with Gamma mixing. Therefore, it allows for serial dependence of claim counts, by introducing Gamma-distributed unobserved individual heterogeneity. The serial dependence in claim counts sequences is generated by integrating the unobserved factor, and by updating its prediction when individual information increases. Alternative models with LogNormal or Inverse Gaussian unobserved heterogeneity have also been considered in the actuarial literature (and reviewed in Chapter 2).

3.1.8 Loss Function

Whatever the model selected for the number of claims, the *a posteriori* premium correction is derived from the application of a loss function. The standard choice is a quadratic loss. In this case, the credibility premium is the function of past claim numbers that minimizes the expected squared difference with the next year claim number. It is well known that the solution is given by the *a posteriori* expectation.

The penalties obtained in a credibility system calling upon a quadratic loss function are often so severe that it is almost impossible to implement them in practice, mainly for commercial reasons. In order to avoid this problem, some authors have proposed resorting to an exponential loss function: the hope is that breaking the symmetry between the overcharges and the undercharges leads to reasonable penalties. This reduces the *maluses* and the *bonuses*, and results in a financially balanced system.

3.1.9 Agenda

Section 3.2 introduces the basics of credibility models taking into account *a priori* characteristics. It starts with a simple introductory example that contains all the ideas of credibility theory. Then, the probabilistic tools used in this context are briefly recalled.

Section 3.3 is devoted to credibility formulas based on a quadratic loss function. The optimal predictor is then shown to be the conditional expectation of future claims given past claims history. In the particular case of Gamma distributed risk parameters, explicit expressions are available for *a posteriori* premium corrections. Discrete Poisson mixtures are considered in detail, providing approximate credibility formulas. Also, linear credibility predictions are derived.

In Section 3.4, the quadratic loss function is replaced with an exponential one. Again, the general formulas simplify in the Negative Binomial case. Linear credibility predictions are considered as approximations to exact premium corrections.

Section 3.5 discusses the type of dependence generated by the credibility construction. It is shown that the risk parameter, as well as future claim numbers, increases with the number of claims recorded in the past, and that future and past claim numbers are positively related. This confirms the intuition behind the actuarial credibility model.

The final Section 3.6 gives the references, and discusses further issues.

3.2 Credibility Models

3.2.1 A Simple Introductory Example: the Good Driver / Bad Driver Model

Consider an insurance portfolio where 60 % of the policyholders are good drivers. The probability that a good driver reports k claims during the year is given by the $\mathcal{P}oi(\lambda_G)$ distribution with $\lambda_G = 0.05$. The remaining 40 % of the policyholders are bad drivers. The probability that they report k claims is given by the $\mathcal{P}oi(\lambda_B)$ distribution with $\lambda_B = 0.15$.

A priori (i.e. at time $t = 0$ for a policyholder without any claim record), the actuary is not able to distinguish between good and bad drivers. The expected number of claims is then given by

$$\Pr[\text{Good}]\lambda_G + \Pr[\text{Bad}]\lambda_B = 0.09.$$

Considering a policyholder who reported k claims during the first year, it is nevertheless possible to compute the probability that he is a good driver: calling upon Bayes' Theorem yields

$$\begin{aligned}\Pr[\text{Good}|k \text{ claims}] &= \frac{\Pr[k \text{ claims}|\text{Good}]\Pr[\text{Good}]}{\Pr[k \text{ claims}|\text{Good}]\Pr[\text{Good}] + \Pr[k \text{ claims}|\text{Bad}]\Pr[\text{Bad}]} \\ &= \frac{\exp(-\lambda_G)\lambda_G^k \Pr[\text{Good}]}{\exp(-\lambda_G)\lambda_G^k \Pr[\text{Good}] + \exp(-\lambda_B)\lambda_B^k \Pr[\text{Bad}]}.\end{aligned}$$

We get the values listed in Table 3.1 for increasing ks. Clearly, these probabilities are decreasing with the number of claims reported. If the policyholder does not report any claim during the first year, then the probability that he is a good driver is increased from 60 % *a priori* to 62.37 % *a posteriori*. As soon as one claim is reported during the first year, this probability decreases from 60 % *a priori* to 35.59 % *a posteriori*. The more claims are reported, the less likely that the policyholder is a good driver. The *a posteriori* probability of being a good driver remains nevertheless positive whatever the number of claims reported to the insurance company.

A posteriori (i.e. at time $t = 1$), the actuary knows the number k of claims reported by the policyholder during the year and should incorporate this additional information in the price list. Specifically, if k claims have been reported, the expected number of claims for year two is

$$\Pr[\text{Good}|k \text{ claims}]\lambda_G + \Pr[\text{Bad}|k \text{ claims}]\lambda_B.$$

The claim record of the policyholder thus modifies the weights assigned to λ_G and λ_B. The values of the expected number of claims for year two according to the number k of claims

Table 3.1 Expected numbers of claims for year two in the good driver/bad driver model given the number k of claims reported during the first year.

# of claims k reported during year 1	Pr[Good|k claims] (%)	Pr[Bad|k claims] (%)	Expected number of claims for year 2
0	62.37	37.63	0.0876
1	35.59	64.41	0.1144
2	15.55	84.45	0.1344
3	5.78	94.22	0.1442
4	2.00	98.00	0.1480
5	0.68	99.32	0.1493

reported during the first year are displayed in Table 3.1. This elementary example contains all the ingredients of experience rating.

3.2.2 Credibility Models Incorporating a Priori Risk Classification

This chapter aims to design merit rating plans in accordance with the *a priori* ratemaking structure of the insurance company. Specifically, let us consider a portfolio with n policies, each one observed during T_i periods. Let N_{it} (with mean $\mathbb{E}[N_{it}] = \lambda_{it}$) be the number of claims reported by policyholder i during year t, i.e. during the period $(t-1, t)$, $i = 1, 2, \ldots, n$, $t = 1, 2, \ldots, T_i$. By convention, time 0 corresponds to the issuance of the policy. We thus face a nested structure: each policyholder generates a sequence $\boldsymbol{N}_i = (N_{i1}, N_{i2}, \ldots, N_{iT_i})^T$ of claim numbers. It is reasonable to assume independence between the series $\boldsymbol{N}_1, \boldsymbol{N}_2, \ldots, \boldsymbol{N}_n$ (at least in motor third party liability insurance), but we expect some positive dependence inside the $\boldsymbol{N}_i$s.

The ith policy of the portfolio, $i = 1, 2, \ldots, n$, is represented by a sequence $(\Theta_i, N_{i1}, N_{i2}, N_{i3}, \ldots)$. At the portfolio level, the sequences $(\Theta_i, N_{i1}, N_{i2}, N_{i3}, \ldots)$ are assumed to be independent for $i = 1, 2, \ldots, n$. The risk parameter Θ_i represents the risk proneness of policyholder i, i.e. unknown risk characteristics of the policyholder having a significant impact on the occurrence of claims; it is regarded as a random variable. Given $\Theta_i = \theta$, the random variables $N_{i1}, N_{i2}, N_{i3}, \ldots$ are assumed to be independent. Unconditionally, these random variables are dependent since their behaviour depends on the common Θ_i.

The very basic tenets of a credibility model for claim counts are as follows:

(i) a conditional distribution for the number of claims, that is, for the N_{it}s given $\Theta_i = \theta$;
(ii) a distribution function F_Θ for the risk parameters $\Theta_1, \ldots, \Theta_n$ to describe how the conditional distributions vary accross the portfolio;
(iii) a loss function whose expectation has to be minimized in order to find the optimal experience premium.

Let us briefly comment on these three aspects. In motor third party liability insurance portfolios, the Poisson distribution often provides a good description of the number of claims incurred by an individual policyholder during a given reference period (one year, say): the set of all Poisson assumptions should at least provide (locally in time) a good approximation to the accident generating mechanism. Given $\Theta_i = \theta$, the annual numbers of claims N_{it} for policyholder i are assumed to be independent and to conform to a Poisson distribution with mean $\lambda_{it}\theta$. As in Chapter 2, λ_{it} is a known function of the exposure-to-risk and possibly other covariates.

Let us now consider the choice of F_Θ. Traditionally, actuaries have assumed that the distribution of θ values among all drivers is well approximated by a two-parameter Gamma distribution. The resulting probability distribution for the number of claims is Negative Binomial. Other classical choices for F_Θ include the Inverse-Gaussian and the LogNormal distributions, as explained in Chapters 1–2.

Regarding (iii), quadratic and exponential loss functions will be considered in this chapter. This leads to the following model.

Definition 3.1 In the Poisson credibility model, the ith policy of the portfolio, $i = 1, 2, \ldots, n$, is represented by a sequence $(\Theta_i, \boldsymbol{N}_i)$ where Θ_i is a positive random variable with unit mean representing the unexplained heterogeneity. Moreover,

A1 given $\Theta_i = \theta$, the random variables N_{it}, $t = 1, 2, \ldots$, are independent and conform to the $\mathcal{P}oi(\lambda_{it}\theta)$ distribution;

A2 at the portfolio level, the sequences $(\Theta_i, \boldsymbol{N}_i)$, $i = 1, 2, \ldots, n$, are assumed to be independent.

It is essential to understand the meaning of this classical actuarial construction. In Definition 3.1, dependence between annual claim numbers is a consequence of the heterogeneity of the portfolio (i.e. of Θ_i); the dependence is only apparent. If we had a complete knowledge of policy characteristics then Θ_i would become deterministic and there would be no more dependence between the N_{it}s for fixed i. The unexplained heterogeneity (which has been modelled through the introduction of the risk parameter Θ_i for policyholder i) is then revealed by the claims and premiums histories in a Bayesian way. These histories modify the distribution of Θ_i and hence modify the premium.

Let

$$N_{i\bullet} = \sum_{t=1}^{T_i} N_{it} \text{ and } \lambda_{i\bullet} = \sum_{t=1}^{T_i} \lambda_{it} \tag{3.1}$$

be the total observed and expected claim numbers for policyholder i during the T_i observation periods; the statistic $N_{i\bullet}$ is a convenient summary of past claims history.

Let us prove that in the credibility model of Definition 3.1, $N_{i\bullet}$ is an exhaustive summary of the past claims history.

Property 3.1 *In the Poisson credibility model of Definition 3.1, the predictive distribution of Θ_i only depends on $N_{i\bullet}$, i.e. the equality*

$$\Pr[\Theta_i \leq t | N_{i1}, N_{i2}, \ldots, N_{iT_i}] = \Pr[\Theta_i \leq t | N_{i\bullet}]$$

holds true whatever $t \geq 0$.

Proof Let $f_\Theta(\cdot | k_{i1}, \ldots, k_{iT_i})$ be the conditional probability density function of Θ_i given that $N_{i1} = k_{i1}, \ldots, N_{iT_i} = k_{iT_i}$, and let

$$k_{i\bullet} = \sum_{t=1}^{T_i} k_{it}$$

be the total number of accidents reported by policyholder i to the company. We can then write

$$\begin{aligned} f_\Theta(\theta | k_{i1}, \ldots, k_{iT_i}) &= \frac{\Pr[N_{i1} = k_{i1}, \ldots, N_{iT_i} = k_{iT_i} | \Theta_i = \theta] f_\Theta(\theta)}{\Pr[N_{i1} = k_{i1}, \ldots, N_{iT_i} = k_{iT_i}]} \\ &= \frac{\exp(-\theta \lambda_{i\bullet}) \theta^{k_{i\bullet}} f_\Theta(\theta)}{\int_0^{+\infty} \exp(-\xi \lambda_{i\bullet}) \xi^{k_{i\bullet}} f_\Theta(\xi) d\xi} \end{aligned}$$

which depends only on $k_{i\bullet}$. This ends the proof.

This result has an important practical consequence: the Poisson credibility model A1–A2 disregards the age of the claims. The penalty induced by an old claim is strictly identical to the one induced by a recent claim. This may of course sometimes be undesirable for commercial purposes. We will come back to this issue in the last section of this chapter.

3.3 Credibility Formulas with a Quadratic Loss Function

3.3.1 Optimal Least-Squares Predictor

Often in applied probability, one seeks to approximate an unknown quantity by a function of a set of related variables, by minimizing the expected squared difference between the two items. This leads to the least-squares principle, and to the conditional expectation, as shown in the next result.

Proposition 3.1 *Let us consider a sequence of random variables* $\{X_1, X_2, X_3, \ldots\}$ *and a risk parameter* Θ. *Given* Θ, *the* X_t*s are independent. The first two moments of the* X_t*s are assumed to be finite. Moreover, the conditional mean of the* X_t*s is given by*

$$\mu_t(\Theta) = \mathbb{E}[X_t|\Theta], \quad t = 1, 2, 3, \ldots,$$

and $\mathbb{E}[\mu_t(\Theta)] = \mu_t$.

The minimum of

$$\mathbb{E}\Big[\Big(\mu_{T+1}(\Theta) - \Psi(X_1, X_2, \ldots, X_T)\Big)^2\Big]$$

over all the measurable functions $\Psi: \mathbb{R}^T \to \mathbb{R}$ *is obtained for*

$$\begin{aligned}\Psi^\star(X_1, X_2, \ldots, X_T) &= \mathbb{E}[X_{T+1}|X_1, X_2, \ldots, X_T] \\ &= \mathbb{E}[\mu_{T+1}(\Theta)|X_1, X_2, \ldots, X_T].\end{aligned}$$

Proof An easy way to get the announced result consists in noting that

$$\begin{aligned}&\mathbb{E}\Big[\Big(\mu_{T+1}(\Theta) - \Psi(X_1, X_2, \ldots, X_T)\Big)^2\Big] \\ &= \mathbb{E}\Big[\Big(\mu_{T+1}(\Theta) - \Psi^\star(X_1, X_2, \ldots, X_T) \\ &\quad + \Psi^\star(X_1, X_2, \ldots, X_T) - \Psi(X_1, X_2, \ldots, X_T)\Big)^2\Big] \\ &= \mathbb{E}\Big[\Big(\mu_{T+1}(\Theta) - \Psi^\star(X_1, X_2, \ldots, X_T)\Big)^2\Big] \\ &\quad + \mathbb{E}\Big[\Big(\Psi^\star(X_1, X_2, \ldots, X_T) - \Psi(X_1, X_2, \ldots, X_T)\Big)^2\Big],\end{aligned}$$

which is clearly minimal for $\Psi \equiv \Psi^\star$.

Proposition 3.1 indicates that the best approximation (with respect to the mean squared error) to $\mu_{T+1}(\Theta)$ given $X_1, X_2, \ldots, X_T$ is $\mathbb{E}[X_{T+1}|X_1, X_2, \ldots, X_T]$, that is, the posterior expectation of X_{T+1} given $X_1, X_2, \ldots, X_T$ (also called the predictive mean). The posterior distribution of X_{T+1} is then obtained by conditioning on past claims history.

To calculate the predictive mean one needs a conditional distribution of losses given the parameter of interest (often the conditional mean) and a prior distribution of the parameter of interest.

3.3.2 *Predictive Distribution*

Let us now come back to the credibility model of Definition 3.1. The conditional distribution of N_{i,T_i+1} given $N_{i1} = k_1, \ldots, N_{iT_i} = k_{T_i}$ is called the predictive distribution. It tells the actuary what the next year number of claims might be given the information contained in past claims history. It is the relevant distribution for risk analysis, management and decision making.

In our case, we have

$$\begin{aligned}
&\Pr[N_{i,T_i+1} = k|N_{i1} = k_1, \ldots, N_{iT_i} = k_{T_i}]\\
&= \frac{\int_0^\infty \left(\prod_{t=1}^{T_i} \Pr[N_{it} = k_t|\Theta_i = \theta]\right) \Pr[N_{i,T_i+1} = k|\Theta_i = \theta] dF_\Theta(\theta)}{\int_0^\infty \left(\prod_{t=1}^{T_i} \Pr[N_{it} = k_t|\Theta_i = \xi]\right) dF_\Theta(\xi)}\\
&= \frac{\int_0^\infty \exp(-\lambda_{i\bullet}\theta)\theta^{k_\bullet} \exp(-\lambda_{i,T_i+1}\theta)\dfrac{(\lambda_{i,T_i+1}\theta)^k}{k!} dF_\Theta(\theta)}{\int_0^\infty \exp(-\lambda_{i\bullet}\xi)\xi^{k_\bullet} dF_\Theta(\xi)}.
\end{aligned}$$

Now, the posterior distribution of Θ_i given past claims history $N_{i1} = k_1, \ldots, N_{iT_i} = k_{T_i}$ is given by

$$\begin{aligned}
&\frac{\prod_{t=1}^{T_i}\left(\exp(-\lambda_{it}\theta)\dfrac{(\theta\lambda_{it})^{k_{it}}}{k_{it}!}\right) dF_\Theta(\theta)}{\int_0^\infty \prod_{t=1}^{T_i}\left(\exp(-\lambda_{it}\xi)\dfrac{(\xi\lambda_{it})^{k_{it}}}{k_{it}!}\right) dF_\Theta(\xi)}\\
&= \frac{\exp(-\lambda_{i\bullet}\theta)\theta^{k_\bullet} dF_\Theta(\theta)}{\int_0^\infty \exp(-\lambda_{i\bullet}\xi)\xi^{k_\bullet} dF_\Theta(\xi)}.
\end{aligned}$$

Hence,

$$\begin{aligned}
&\Pr[N_{i,T_i+1} = k|N_{i1} = k_1, \ldots, N_{iT_i} = k_{T_i}]\\
&= \int_0^\infty \exp(-\lambda_{i,T_i+1}\theta)\frac{(\theta\lambda_{i,T_i+1})^k}{k!} dF_\Theta(\theta|k_\bullet)
\end{aligned}$$

where $F_\Theta(\cdot|k_\bullet)$ is the conditional distribution function of Θ_i given $N_{i\bullet} = k$. The predictive distribution thus appears as a Poisson mixture distribution, where the mixing is with respect to the posterior distribution of Θ_i. Past claims history $N_{i1} = k_1, \ldots, N_{iT_i} = k_{T_i}$ modifies the distribution of Θ_i, and this modified distribution is used as a new mixing law for the number of claims N_{i,T_i+1} for year $T_i + 1$.

3.3.3 Bayesian Credibility Premium

We are looking for the function $\Psi^\star$ of $N_{i1}, \ldots, N_{iT_i}$ that is the closest to Θ_i, i.e. minimizing

$$\mathbb{E}\Big[\big(\Theta_i - \Psi(N_{i1}, \ldots, N_{iT_i})\big)^2\Big]$$

over all the measurable functions $\Psi : \mathbb{N}^{T_i} \to \mathbb{R}$. Proposition 3.1 gives the solution of this optimization problem:

$$\Psi^\star(N_{i1}, \ldots, N_{iT_i}) = \mathbb{E}\big[\Theta_i \big| N_{i1}, \ldots, N_{iT_i}\big].$$

In general, the posterior mean of Θ_i given $N_{i1} = k_1, \ldots, N_{i,T_i} = k_{T_i}$ is given by

$$\begin{aligned}
\mathbb{E}\big[\Theta_i \big| N_{i1} = k_1, \ldots, N_{i,T_i} = k_{T_i}\big] &= \frac{\int_0^{+\infty} \theta \left(\prod_{t=1}^{T_i} \Pr[N_{it} = k_t | \Theta_i = \theta]\right) dF_\Theta(\theta)}{\int_0^{+\infty} \left(\prod_{t=1}^{T_i} \Pr[N_{it} = k_t | \Theta_i = \xi]\right) dF_\Theta(\xi)} \\
&= \frac{\int_0^{+\infty} \exp(-\lambda_{i\bullet}\theta)\theta^{k_\bullet+1} dF_\Theta(\theta)}{\int_0^{+\infty} \exp(-\lambda_{i\bullet}\xi)\xi^{k_\bullet} dF_\Theta(\xi)}.
\end{aligned} \tag{3.2}$$

This *a posteriori* expectation thus appears as the ratio of two Mellin transforms $M(k) = \mathbb{E}[\exp(-\xi\Theta_i)\Theta_i^k]$ of Θ_i. It is interesting to note that the *a posteriori* expectation depends only on the total number $k_\bullet$ of accidents caused in the past T_i years of insurance, and not on the history of these claims. This was expected from Property 3.1. This is a characteristic of the credibility models with static random effects.

The Bayesian credibility premium is simply the mean of the predictive distribution. It is given by

$$\begin{aligned}
\mathbb{E}\big[N_{i,T_i+1} \big| N_{i1} = k_1, \ldots, N_{i,T_i} = k_{T_i}\big] &= \int_0^\infty \lambda_{i,T_i+1} \theta dF_\Theta(\theta | k_\bullet) \\
&= \lambda_{i,T_i+1} \mathbb{E}\big[\Theta_i \big| N_{i1} = k_1, \ldots, N_{i,T_i} = k_{T_i}\big]
\end{aligned}$$

where the posterior expectation of Θ_i is given by (3.2). The posterior expected claim number $\mathbb{E}[N_{i,T_i+1} | N_{i1}, \ldots, N_{i,T_i}]$ is obtained by multiplying λ_{i,T_i+1} by the correction coefficient $\mathbb{E}[\Theta_i | N_{i1}, \ldots, N_{i,T_i}]$. This approach always yields financial balance, since

$$\mathbb{E}\Big[\mathbb{E}\big[\Theta_i \big| N_{i1} = k_1, \ldots, N_{i,T_i} = k_{T_i}\big]\Big] = \mathbb{E}[\Theta_i] = 1$$

so that the corrections average to unity.

3.3.4 Poisson-Gamma Credibility Model

Gamma Distribution for the Random Effects

Let us assume that $\Theta_i \sim \mathcal{G}am(a,a)$ with probability density function (1.35). The joint probability mass function of the random vector $\boldsymbol{N}_i = (N_{i1}, N_{i2}, \ldots, N_{iT_i})$ is given by (2.19). The joint probability density of the random vector $(\Theta_i, N_{i1}, N_{i2}, \ldots, N_{iT_i})$ is given by

$$\prod_{t=1}^{T_i} \exp\Big(-\theta_i\lambda_{it}\Big)\frac{(\theta_i\lambda_{it})^{k_{it}}}{k_{it}!}\frac{1}{\Gamma(a)}a^a\theta_i^{a-1}\exp(-a\theta_i)$$

$$\propto \exp\left(-\theta_i\sum_{t=1}^{T_i}\lambda_{it}\right)\theta_i^{\sum_{t=1}^{T_i}k_{it}+a-1}\exp(-a\theta_i). \tag{3.3}$$

A Posteriori Distribution of Random Effects

In Section 2.5.1, we established that in a two-period model, the *a posteriori* distribution of Θ_i remained Gamma, with updated parameters. Let us now extend the result to a multiperiod setting.

Now, the conditional distribution of Θ_i given the past claims frequencies $N_{it} = k_{it}$, $t = 1, 2, \ldots, T_i$, is obtained from (2.19) and (3.3). This gives

$$\frac{\exp\Big(-\theta_i\Big(a+\sum_{t=1}^{T_i}\lambda_{it}\Big)\Big)\theta_i^{a+\sum_{t=1}^{T_i}k_{it}-1}}{\int_0^{+\infty}\exp\Big(-\xi\Big(a+\sum_{t=1}^{T_i}\lambda_{it}\Big)\Big)\xi^{a+\sum_{t=1}^{T_i}k_{it}-1}d\xi}$$

$$=\exp\left(-\theta_i\left(a+\sum_{t=1}^{T_i}\lambda_{it}\right)\right)\theta_i^{a+\sum_{t=1}^{T_i}k_{it}-1}\frac{\Big(a+\sum_{t=1}^{T_i}\lambda_{it}\Big)^{a+\sum_{t=1}^{T_i}k_{it}}}{\Gamma\Big(a+\sum_{t=1}^{T_i}k_{it}\Big)}.$$

Coming back to (1.34), we recognize a Gamma probability density function. Specifically, we thus have that

$$[\Theta_i|N_{i1}, N_{i2}, \ldots, N_{iT_i}] \sim \mathcal{G}am\big(a+N_{i\bullet}, a+\lambda_{i\bullet}\big). \tag{3.4}$$

The correction coefficient is given by

$$\mathbb{E}[\Theta_i|N_{i1}, N_{i2}, \ldots, N_{iT_i}] = \frac{a+N_{i\bullet}}{a+\lambda_{i\bullet}}$$

which clearly increases in the past claims $N_{i\bullet}$. The variance of Θ_i given past claim history is given by

$$\mathbb{V}[\Theta_i|N_{i1}, N_{i2}, \ldots, N_{iT_i}] = \frac{a+N_{i\bullet}}{(a+\lambda_{i\bullet})^2}.$$

The expected number of claims in year T_i+1 given past claims history is given by

$$\mathbb{E}[N_{i,T_i+1}|N_{i1}, N_{i2}, \ldots, N_{iT_i}] = \lambda_{i,T_i+1}\mathbb{E}[\Theta_i|N_{i1}, N_{i2}, \ldots, N_{iT_i}] = \lambda_{i,T_i+1}\frac{a+N_{i\bullet}}{a+\lambda_{i\bullet}}.$$

Let us briefly comment on this *a posteriori* correction:

- We see that the *a posteriori* corrections will be more severe as the residual heterogeneity, measured by $\mathbb{V}[\Theta_i]=1/a$, increases.
- Considering two policyholders (numbered i_1 and i_2) in the portfolio during $T_{i_1}=T_{i_2}$ periods such that i_1 is *a priori* a better driver than i_2, that is, $\lambda_{i_1\bullet}<\lambda_{i_2\bullet}$: if these policyholders do not report any claim (i.e., $N_{i_1\bullet}=N_{i_2\bullet}=0$) then the corrections to be applied to these policyholders satisfy

$$\frac{a}{a+\lambda_{i_1\bullet}}>\frac{a}{a+\lambda_{i_2\bullet}}$$

  so that the *a priori* worse driver receives more discount.
- If these policyholders report $k\geq 1$ claims (i.e., $N_{i_1\bullet}=N_{i_2\bullet}=k$) then the penalties are such that

$$\frac{a+k}{a+\lambda_{i_1\bullet}}>\frac{a+k}{a+\lambda_{i_2\bullet}}.$$

  Hence, the penalty for the *a priori* bad driver is less severe than for the good one.

3.3.5 Predictive Distribution and Bayesian Credibility Premium

From (3.4) we know that the posterior distribution of Θ_i given past claims history is still Gamma, with parameters $a+N_{i\bullet}$ and $a+\lambda_{i\bullet}$. Therefore, the predictive distribution of N_{i,T_i+1} is Negative Binomial, that is,

$$\Pr[N_{i,T_i+1}=k|N_{i\bullet}=k_\bullet]=\binom{a+k_\bullet+k-1}{k}\left(\frac{\lambda_{i,T_i+1}}{a+\lambda_{i\bullet}+\lambda_{i,T_i+1}}\right)^k\left(\frac{a+\lambda_{i\bullet}}{a+\lambda_{i\bullet}+\lambda_{i,T_i+1}}\right)^{a+k_\bullet}.$$

Furthermore, the Bayesian credibility premium is given by

$$\mathbb{E}[N_{i,T_i+1}|N_{i\bullet}=k_\bullet]=\lambda_{i,T_i+1}\mathbb{E}[\Theta_i|N_{i\bullet}=k_\bullet]=\lambda_{i,T_i+1}\frac{a+k_\bullet}{a+\lambda_{i\bullet}}.$$

Remark 3.1 As time goes on, the Bayesian credibility premium tends to 0 for a policyholder reporting no claim. This can be seen as an unrealistic feature. An easy way to avoid this problem is to decompose N_i into two parts: a component $N_i^{(1)}$ distributed according to a pure Poisson distribution that represents the claims occurring purely at random, and a component $N_i^{(2)}$ that is Negative Binomial and influenced by the driver's abilities. This introduces a lower bound on the *a posteriori* claim frequency, which is no more allowed to vanish in the long term.

It is worth mentioning that the Bayesian credibility premium can be cast into

$$\mathbb{E}[N_{i,T_i+1}|N_{i\bullet}=k_\bullet]=\left(\frac{a}{a+\lambda_{i\bullet}}\mathbb{E}[\Theta_i]+\frac{\lambda_{i\bullet}}{a+\lambda_{i\bullet}}\frac{k_\bullet}{\lambda_{i\bullet}}\right)\mathbb{E}[N_{i,T_i+1}].$$

Recall that $\mathbb{E}[\Theta_i] = 1$ and $\mathbb{E}[N_{i,T_i+1}] = \lambda_{i,T_i+1}$. Hence, the Bayesian credibility premium is obtained by multiplying the *a priori* expected number of claims $\mathbb{E}[N_{i,T_i+1}]$ by an appropriate correction factor. This factor appears as a weighted average of the prior expectation of Θ_i, receiving weight $a/(a+\lambda_{i\bullet})$, and the average claim frequency for that policy $k_\bullet/\lambda_{i\bullet}$ receiving a weight $\lambda_{i\bullet}/(a+\lambda_{i\bullet})$.

3.3.6 Numerical Illustration

Let us now compute the coefficients to apply to the pure premium according to the number of claims reported in the past. To this end, let us consider the Negative Binomial fit of Portfolio A given in Table 2.7. Table 3.2 displays the values of $\mathbb{E}[\Theta_i|N_{i\bullet} = k_\bullet]$ for different combinations of observed periods T_i and number of past claims $k_\bullet$ for a good driver (with observable characteristics: man, age 35, rural area, upfront premium, private use, $\lambda_i = 0.0928$). Tables 3.3–3.4 are the analogues for an average driver (with observable characteristics: woman, age 25, urban area, upfront premium, private use, $\lambda_i = 0.1408$) and a bad driver (with observable characteristics: man, age 22, rural area, split premium, private use, $\lambda_i = 0.2840$), respectively.

If the good driver does not report any accident during the first year, we see from Table 3.2 that he will pay 92 % of the base premium to be covered during the second year. If, in addition, he does not file any claim during the second year, the premium decreases to 85.2 % of the base premium. After ten claim-free years, he will have to pay 53.4 % of the base premium to be covered by the insurer.

Considering the average driver, we see from Table 3.3 that he will be awarded more discount than the good driver if he does not file any claim. Indeed he will have to pay 88.3 % (instead of 92 %) of the base premium after one claim-free year, 79.1 % (instead of 85.1 %) of the base premium after two claim-free years, and 43.1 % (instead of 53.4 %) after ten claim-free years.

Table 3.2 Values of $\mathbb{E}[\Theta_i|N_{i\bullet} = k_\bullet]$ for different combinations of observed periods T_i and number of past claims $k_\bullet$ for a good driver from Portfolio A (average annual claim frequency of 9.28 %).

T_i	Number of claims $k_\bullet$					
	0	1	2	3	4	5
1	92.0 %	178.4 %	264.7 %	351.1 %	437.5 %	523.8 %
2	85.2 %	165.1 %	245.1 %	325.0 %	405.0 %	485.0 %
3	79.3 %	153.7 %	228.2 %	302.6 %	377.0 %	451.5 %
4	74.2 %	143.8 %	213.4 %	283.0 %	352.7 %	422.3 %
5	69.7 %	135.1 %	200.5 %	265.9 %	331.3 %	396.7 %
6	65.7 %	127.3 %	189.0 %	250.6 %	312.3 %	374.0 %
7	62.1 %	120.4 %	178.8 %	237.1 %	295.4 %	353.7 %
8	58.9 %	114.3 %	169.6 %	224.9 %	280.2 %	335.6 %
9	56.0 %	108.7 %	161.3 %	213.9 %	266.6 %	319.2 %
10	53.4 %	103.6 %	153.8 %	204.0 %	254.1 %	304.3 %

Table 3.3 Values of $\mathbb{E}[\Theta_i|N_{i\bullet}=k_\bullet]$ for different combinations of observed periods T_i and number of past claims $k_\bullet$ for an average driver from Portfolio A (average annual claim frequency of 14.09 %).

T_i	Number of claims $k_\bullet$					
	0	1	2	3	4	5
1	88.3 %	171.2 %	254.2 %	337.1 %	420.0 %	503.0 %
2	79.1 %	153.3 %	227.6 %	301.8 %	376.1 %	450.4 %
3	71.6 %	138.8 %	206.0 %	273.3 %	340.5 %	407.7 %
4	65.4 %	126.8 %	188.2 %	249.6 %	311.0 %	372.5 %
5	60.2 %	116.7 %	173.2 %	229.8 %	286.3 %	342.8 %
6	55.8 %	108.1 %	160.5 %	212.8 %	265.2 %	317.5 %
7	51.9 %	100.7 %	149.4 %	198.2 %	247.0 %	295.7 %
8	48.6 %	94.2 %	139.8 %	185.5 %	231.1 %	276.7 %
9	45.7 %	88.5 %	131.4 %	174.3 %	217.1 %	260.0 %
10	43.1 %	83.5 %	123.9 %	164.3 %	204.8 %	245.2 %

Table 3.4 Values of $\mathbb{E}[\Theta_i|N_{i\bullet}=k_\bullet]$ for different combinations of observed periods T_i and number of past claims $k_\bullet$ for a bad driver from Portfolio A (average annual claim frequency of 28.40 %).

T_i	Number of claims $k_\bullet$					
	0	1	2	3	4	5
1	78.9 %	153.1 %	227.2 %	301.3 %	375.5 %	449.6 %
2	65.2 %	126.4 %	187.7 %	248.9 %	310.2 %	371.4 %
3	55.5 %	107.7 %	159.9 %	212.0 %	264.2 %	316.4 %
4	48.4 %	93.8 %	139.2 %	184.7 %	230.1 %	275.5 %
5	42.9 %	83.1 %	123.3 %	163.6 %	203.8 %	244.0 %
6	38.5 %	74.6 %	110.7 %	146.8 %	182.9 %	219.0 %
7	34.9 %	67.6 %	100.4 %	133.1 %	165.9 %	198.6 %
8	31.9 %	61.9 %	91.8 %	121.8 %	151.8 %	181.7 %
9	29.4 %	57.0 %	84.6 %	112.3 %	139.9 %	167.5 %
10	27.3 %	52.9 %	78.5 %	104.1 %	129.7 %	155.3 %

Table 3.4 shows higher discounts for the bad driver, with percentages of 78.9 % after one claim-free year, 65.2 % after two claim-free years and 27.3 % after ten claim-free years.

The discounts awarded to policyholders who do not report any accident to the insurance company are thus increasing with the *a priori* annual expected claim frequency. The more claims are expected by the insurance company on the basis of observable characteristics, the higher the discount in case no claims are reported. Note that the *a posteriori* expected claim frequencies remain large for *a priori* bad drivers since reporting no claims happens with a smaller probability.

Now, considering the penalty in case one claim is reported, we see that the good driver who reports one claim during the first year will have to pay 178.4 % of the base premium to be covered by the insurance company during the second year. The average driver in the same situation pays 171.2 % of the base premium, and the bad driver 153.1 % of the base premium. The penalties in the case where an accident is reported to the company are thus decreasing with the *a priori* annual expected claim frequencies.

The system appears rather severe. Reporting a claim entails a penalty of between 50 and 75 %, which seems difficult to implement in practice. This is a consequence of the financial balance property. The weighted averages of all figures (in each infinite row) is equal to 100 %. The modest discounts awarded to the majority of claim-free policyholders have then to be exactly compensated by the penalties supported by the minority of policyholders reporting claims to the company. This causes large penalties. The severity of the credibility corrections also appears in the number of claim free years needed to erase the penalty induced by an accident: 10 years for the good driver, 7 for the average one and 3 for the bad one. These periods make sense compared to the average claim number per policy.

3.3.7 Discrete Poisson Mixture Credibility Model

The good driver / bad driver model presented in Section 3.2.1 assumes that the portfolio is composed of two classes of insured drivers. This can easily be extended to several categories of drivers (for instance, bad, below average, average, above average, excellent). Specifically, let us assume that each risk class of the portfolio is made of q categories of insured drivers. Let us denote as $p_1, p_2, \ldots, p_q$ the proportion of drivers in each of these categories. Then,

$$\Theta_i = \begin{cases} \theta_1 \text{ with probability } p_1 \\ \theta_2 \text{ with probability } p_2 \\ \vdots \\ \theta_q \text{ with probability } p_q \end{cases} \tag{3.5}$$

with $0 < \theta_1 < \theta_2 < \cdots < \theta_q$. The number N_{it} of claims caused by policyholder i during year t is distributed as

$$\Pr[N_{it} = k] = \sum_{j=1}^{q} \exp(-\lambda_{it}\theta_j)\frac{(\lambda_{it}\theta_j)^k}{k!}p_j, \quad k = 0, 1, \ldots$$

Note that the special case $q = 2$ gives the good driver / bad driver model.

The expected number of claims reported by policyholder i in year $T_i + 1$ is given by

$$\mathbb{E}[N_{i,T_i+1}] = \sum_{j=1}^{q} \mathbb{E}[N_{i,T_i+1}|\Theta = \theta_j]p_j = \lambda_{i,T_i+1}\sum_{j=1}^{q}\theta_j p_j.$$

If we know that this policyholder reported k claims during the past T_i years, we expect $\mathbb{E}[N_{i,T_i+1}|N_{i\bullet} = k]$ claims in year $T_i + 1$. The computation of $\mathbb{E}[N_{i,T_i+1}|N_{i\bullet} = k]$ requires the conditional distribution of N_{i,T_i+1} given $N_{i\bullet} = k$. We get it from

$$\Pr[N_{i,T_i+1}=i|N_{i\bullet}=k]=\frac{\Pr[N_{i,T_i+1}=i \text{ and } N_{i\bullet}=k]}{\Pr[N_{i\bullet}=k]}$$
$$=\frac{\sum_{j=1}^{q}\Pr[N_{i,T_i+1}=i|\Theta_i=\theta_j]\Pr[N_{i\bullet}=k|\Theta_i=\theta_j]p_j}{\sum_{j=1}^{q}\Pr[N_{i\bullet}=k|\Theta_i=\theta_j]p_j}$$
$$=\sum_{j=1}^{q}\exp(-\lambda_{i,T_i+1}\theta_j)\frac{(\lambda_{i,T_i+1}\theta_j)^i}{i!}\frac{p_j\exp(-\lambda_{i\bullet}\theta_j)\theta_j^k}{\sum_{l=1}^{q}p_l\exp(-\lambda_{i\bullet}\theta_l)\theta_l^k}.$$

Hence, given $N_{i\bullet}=k$, the law of N_{i,T_i+1} appears as a discrete Poisson mixture with modified weights

$$\widetilde{p}_j(k)=\frac{p_j\exp(-\lambda_{i\bullet}\theta_j)\theta_j^k}{\sum_{l=1}^{q}p_l\exp(-\lambda_{i\bullet}\theta_l)\theta_l^k}.$$

The *a posteriori* expectation of N_{i,T_i+1} is then

$$\mathbb{E}[N_{i,T_i+1}|N_{i\bullet}=k]=\lambda_{i,T_i+1}\sum_{j=1}^{q}\theta_j\widetilde{p}_j(k)=\lambda_{i,T_i+1}\frac{\sum_{j=1}^{q}p_j\exp(-\lambda_{i\bullet}\theta_j)\theta_j^{k+1}}{\sum_{l=1}^{q}p_l\exp(-\lambda_{i\bullet}\theta_l)\theta_l^k}. \tag{3.6}$$

The posterior mean of Θ_i given $N_{i\bullet}=k$ is then given by

$$\mathbb{E}[\Theta_i|N_{\bullet}=k]=\frac{\sum_{j=1}^{q}p_j\exp(-\lambda_{i\bullet}\theta_j)\theta_j^{k+1}}{\sum_{l=1}^{q}p_l\exp(-\lambda_{i\bullet}\theta_l)\theta_l^k}. \tag{3.7}$$

Compared to (3.2), the integrals now reduce to sums over the q components of the discrete mixture.

Even if the reality of the insurance portfolio is a discrete mixture (with a θ_i specific to policyholder i, resulting in $q=n$), this model is not convenient since it involves a large number of parameters. The discrete Poisson mixture nevertheless deserves interest as an approximation of more general Poisson mixtures, as discussed below.

3.3.8 Discrete Approximations for the Heterogeneous Component

Moment Spaces

Apart from the Poisson-Gamma case, the computation of the conditional expectation giving the credibility premium requires numerical integration. A convenient alternative is to approximate the mixing distribution by a suitable discrete analogue sharing the same sequence of moments. We are then back to the discrete Poisson mixture credibility model studied above, and we benefit from the easy-to-compute formulas (3.6)–(3.7) valid in this case.

The discrete approximations to Θ_i are based on the knowledge of its support, $[0, b]$ say, with b possibly infinite, and its first few moments $\mu_1, \mu_2, \ldots$ In general, let us denote by $\mathcal{B}_s([0, b]; \boldsymbol{\mu})$ the class of all the random variables X with support in $[0, b]$ and with prescribed first $s-1$ moments $\mathbb{E}[X^k]=\mu_k$, $k=1, 2, \ldots, s-1$. In the literature, $\mathcal{B}_s([0, b]; \boldsymbol{\mu})$ is referred to as a moment space. Properly speaking, it is a class of distribution functions rather than a class of random variables. Classical problems related to moment spaces are for instance the determination of conditions on $[0, b], \mu_1, \mu_2, \ldots, \mu_{s-1}$ so that $\mathcal{B}_s([0, b]; \boldsymbol{\mu})$ is not void, or the obtention of the elements in $\mathcal{B}_s([0, b]; \boldsymbol{\mu})$ with the minimal number of support points. These elements possess some extremal properties and will be used here in connection with credibility formulas. Henceforth, we tacitly assume that $[0, b]$ and $\mu_1, \mu_2, \ldots, \mu_{s-1}$ are such that the associated moment space $\mathcal{B}_s([0, b]; \boldsymbol{\mu})$ is not void and is not a singleton (i.e., $\mathcal{B}_s([0, b]; \boldsymbol{\mu})$ consists of at least two distinct distribution functions).

De Vylder's Moment Problem

De Vylder (1996, Section 8.3) investigated the following problem: within $\mathcal{B}_s([0, b]; \boldsymbol{\mu})$, determine the random variables $X_{\min}^{(s)}$ and $X_{\max}^{(s)}$ such that the inequalities

$$\mathbb{E}[(X_{\min}^{(s)})^s] \leq \mathbb{E}[X^s] \leq \mathbb{E}[(X_{\max}^{(s)})^s] \quad \text{hold for all} \quad X \in \mathcal{B}_s([0, b]; \boldsymbol{\mu}). \tag{3.8}$$

Explicit solutions to (3.8) are available for s up to five. The supports of the extremal distributions are given in Tables 3.5–3.6.

As shown in Denuit, De Vylder & Lefèvre (1999), the random variables $X_{\min}^{(s)}$ and $X_{\max}^{(s)}$ involved in (3.8) give bounds on $\mathbb{E}[\phi(X)]$ for every function ϕ that is $(s-2)$ times differentiable, with a convex $(s-2)$th derivative (such functions are called s-convex; see Roberts & Varberg (1973) for details). In that context, $X_{\min}^{(s)}$ is called the s-convex minimum, and $X_{\max}^{(s)}$ is called the s-convex maximum.

Table 3.5 Probability distribution of $X_{\min}^{(s)} \in \mathcal{B}_s([0, b]; \boldsymbol{\mu})$ achieving the lower bound in (3.8) for $s = 1$ to 5.

s	Support points	Probability masses
1	0	1
2	μ_1	1
3	0	$\dfrac{\mu_2 - \mu_1^2}{\mu_2}$
	$\dfrac{\mu_2}{\mu_1}$	$\dfrac{\mu_1^2}{\mu_2}$
4	$r_+ = \dfrac{\mu_3 - \mu_1\mu_2 + \sqrt{(\mu_3 - \mu_1\mu_2)^2 - 4(\mu_2 - \mu_1^2)(\mu_1\mu_3 - \mu_2^2)}}{2(\mu_2 - \mu_1^2)}$	$\dfrac{\mu_1 - r_-}{r_+ - r_-}$
	$r_- = \dfrac{\mu_3 - \mu_1\mu_2 - \sqrt{(\mu_3 - \mu_1\mu_2)^2 - 4(\mu_2 - \mu_1^2)(\mu_1\mu_3 - \mu_2^2)}}{2(\mu_2 - \mu_1^2)}$	$1 - \dfrac{\mu_1 - r_-}{r_+ - r_-}$
5	0	$1 - q_+ - q_-$
	$t_+ = \dfrac{\mu_1\mu_4 - \mu_2\mu_3 + \sqrt{(\mu_1\mu_4 - \mu_2\mu_3)^2 - 4(\mu_1\mu_3 - \mu_2^2)(\mu_2\mu_4 - \mu_3^2)}}{2(\mu_1\mu_3 - \mu_2^2)}$	$q_+ = \dfrac{\mu_2 - t_-\mu_1}{t_+(t_+ - t_-)}$
	$t_- = \dfrac{\mu_1\mu_4 - \mu_2\mu_3 - \sqrt{(\mu_1\mu_4 - \mu_2\mu_3)^2 - 4(\mu_1\mu_3 - \mu_2^2)(\mu_2\mu_4 - \mu_3^2)}}{2(\mu_1\mu_3 - \mu_2^2)}$	$q_- = \dfrac{\mu_2 - t_+\mu_1}{t_-(t_- - t_+)}$

Table 3.6 Probability distribution of $X^{(s)}_{\max} \in \mathcal{B}_s([0,b];\boldsymbol{\mu})$ achieving the upper bound in (3.8) for $s=1$ to 5.

s	Support points	Probability masses
1	b	1
2	0	$\dfrac{b-\mu_1}{b}$
	b	$\dfrac{\mu_1}{b}$
3	$\dfrac{b\mu_1-\mu_2}{b-\mu_1}$	$\dfrac{(b-\mu_1)^2}{(b-\mu_1)^2+\mu_2-\mu_1^2}$
	b	$\dfrac{\mu_2-\mu_1^2}{(b-\mu_1)^2+\mu_2-\mu_1^2}$
4	0	$1-p_1-p_2$
	$\dfrac{\mu_3-b\mu_2}{\mu_2-b\mu_1}$	$p_1=\dfrac{(\mu_2-b\mu_1)^3}{(\mu_3-b\mu_2)(\mu_3-2b\mu_2+b^2\mu_1)}$
	b	$p_2=\dfrac{\mu_1\mu_3-\mu_2^2}{b(\mu_3-2b\mu_2+b^2\mu_1)}$
5	$z_+=\dfrac{(\mu_1-b)(\mu_4-b\mu_3)-(\mu_2-b\mu_1)(\mu_3-b\mu_2)+\sqrt{\varrho}}{2\left((\mu_1-b)(\mu_3-b\mu_2)-(\mu_2-b\mu_1)^2\right)}$	$p_+=\dfrac{\mu_2-(b+z_-)\mu_1+bz_-}{(z_+-z_-)(z_+-b)}$
	$z_-=\dfrac{(\mu_1-b)(\mu_4-b\mu_3)-(\mu_2-b\mu_1)(\mu_3-b\mu_2)-\sqrt{\varrho}}{2\left((\mu_1-b)(\mu_3-b\mu_2)-(\mu_2-b\mu_1)^2\right)}$	$p_-=\dfrac{\mu_2-(b+z_+)\mu_1+bz_+}{(z_--z_+)(z_--b)}$
	b	$1-p_+-p_-$

Where $\varrho=((\mu_1-b)(\mu_4-b\mu_3)-(\mu_2-b\mu_1)(\mu_3-b\mu_2))^2-4((\mu_1-b)(\mu_3-b\mu_2)-(\mu_2-b\mu_1)^2)((\mu_2-b\mu_1)(\mu_4-b\mu_3)-(\mu_3-b\mu_2)^2)$.

Approximations Based on First Moments

A given random variable X (think of the risk parameter Θ_i in credibility applications) with known moments μ_k, $k=1,2,\ldots$, can be approximated either by $X^{(s)}_{\min}$ or $X^{(s)}_{\max}$ involved in (3.8). An alternative approximation mixing these two extremal variables is also available. Let us denote as

$$\underline{\mu}_s=\min_{X\in\mathcal{B}_s([0,b];\boldsymbol{\mu})}\mathbb{E}[X^s]=\mathbb{E}\big[(X^{(s)}_{\min})^s\big]$$

and

$$\overline{\mu}_s=\max_{X\in\mathcal{B}_s([0,b];\boldsymbol{\mu})}\mathbb{E}[X^s]=\mathbb{E}\big[(X^{(s)}_{\max})^s\big]$$

the lower and upper bounds involved in (3.8). Explicit expressions of $\underline{\mu}_s$ and of $\overline{\mu}_s$ for s up to five are listed in Table 3.7, in the notation of Tables 3.5–3.6. The quantity $\overline{\mu}_s-\underline{\mu}_s$ can be considered as the 'width' of $\mathcal{B}_s([0,b];\boldsymbol{\mu})$ as explained in Denuit (2002).

Table 3.7 Lower $\underline{\mu}_s$ and upper $\overline{\mu}_s$ bounds in (3.8) for $s = 2$ to 5.

s	$\underline{\mu}_s$	$\overline{\mu}_s$
2	μ_1^2	$b\mu_1$
3	$\dfrac{\mu_2^2}{\mu_1}$	$\left(\dfrac{b\mu_1-\mu_2}{b-\mu_1}\right)^3 \dfrac{(b-\mu_1)^2}{(b-\mu_1)^2+\sigma^2} + b^3 \dfrac{\sigma^2}{(b-\mu_1)^2+\sigma^2}$
4	$\left(1-\dfrac{\mu_1-r_-}{r_+-r_-}\right) r_-^4 + \dfrac{\mu_1-r_-}{r_+-r_-} r_+^4$	$p_1\left(\dfrac{\mu_3-b\mu_2}{\mu_2-b\mu_1}\right)^4 + p_2 b^4$
5	$q_+t_+^5+q_-t_-^5$	$p_+z_+^5+p_-z_-^5+(1-p_+-p_-)b^5$

Now, let $X \in \mathcal{B}_s([0, b]; \boldsymbol{\mu})$. The sth canonical moment of X, denoted as $c_s(X)$, is given by

$$c_s(X) = \frac{\mathbb{E}[X^s] - \underline{\mu}_s}{\overline{\mu}_s - \underline{\mu}_s}.$$

Thus, $c_s(X)$ is simply the position of the sth moment of X relative to its possible range. Now, $c_s(X)$ gives a good indication of the 'position' of X in $\mathcal{B}_s([0, b]; \boldsymbol{\mu})$ with respect to the extrema $X_{\min}^{(s)}$ and $X_{\max}^{(s)}$. Therefore, we might expect that a satisfactory approximation of $X \in \mathcal{B}_s([0, b]; \boldsymbol{\mu})$ is furnished by a convex combination of the stochastic extrema in $\mathcal{B}_{s-1}([0, b]; \boldsymbol{\mu})$ with weights depending on the $(s-1)$th canonical moment $c_{s-1}(X)$, i.e. we use the mixture

$$\widetilde{X}_s = \begin{cases} X_{\min}^{(s-1)} \text{ with probability } 1 - c_{s-1}(X) \\ X_{\max}^{(s-1)} \text{ with probability } c_{s-1}(X) \end{cases} \tag{3.9}$$

in order to approximate X. In the following, we refer to $\widetilde{X}_s$ as the sth canonical approximation of X. It is easily seen that $\widetilde{X}_s \in \mathcal{B}_s([0, b]; \boldsymbol{\mu})$.

The Unimodal Case

A purely discrete approximation to the risk parameter Θ_i causes problems when an experience rating plan has to be designed, as shown in Walhin & Paris (1999). The *a posteriori* corrections obtained with discrete Θ_is exhibit plateaus before and after sudden jumps, which is commercially unacceptable. When F_Θ only has a few support points, a 'block' structure is clearly apparent for the credibility coefficients, each block with almost constant *a posteriori* corrections corresponding to one support point of F_Θ. The policyholder is transferred from smaller to larger mass points as more claims are filed. In order to avoid this, we need a smooth risk distribution. Therefore, we would like to have simple continuous approximations to the risk parameter. This can be done in the unimodal case, as shown below.

A situation of practical interest in actuarial sciences is when the random variables under consideration are known to have a unimodal distribution with a given mode m, together with the fixed moments $\mu_1, \mu_2, \ldots, \mu_{s-1}$. The Gamma, LogNormal and Inverse Gaussian mixing distributions examined in Chapter 2 are all unimodal. Henceforth, we denote by $\mathcal{B}_s([0,b]; m-\text{unim}; \boldsymbol{\mu})$ the *unimodal moment space* of all the random variables of this type.

In the following, we use the notation $\mathcal{U}ni[m,z]$, with $m, z \in \mathbb{R}$, for the Uniform distribution on the interval $[\min(m,z), \max(m,z)]$: if $m < z$, it represents the law with constant probability density function equal to $1/(z-m)$ on $[m,z]$; if $m > z$, it is the law with constant probability density function equal to $1/(m-z)$ on $[z,m]$; if $m = z$, it is the law degenerated at the point m. Given some random variable Z, we denote by $\mathcal{MU}ni[m,Z]$ the mixed Uniform distribution with random extremal point Z as mixing parameter.

A convenient representation of unimodal distributions is provided by Khinchine's theorem (see, e.g., Theorem 1.3 in Dharmadhikari & Joag-Dev (1988)). This theorem states that a random variable Y has a unimodal law with a mode at 0 if, and only if, Y is distributed as UZ where U and Z are two independent random variables and $U \sim \mathcal{U}ni[0,1]$. This condition can be rewritten as $Y \sim \mathcal{MU}ni[0,Z]$ for some random variable Z.

Now, let X be any random variable valued in $[0,b]$ and with a unique mode at m. By Khinchine's theorem,

$$X \sim \mathcal{MU}ni[m,\widetilde{Z}] \text{ where } \widetilde{Z} = m + Z. \tag{3.10}$$

Note that Z is valued in $[-m, b-m]$. Moreover, the moments μ_j of X and $\widetilde{\mu}_j$ of $\widetilde{Z}$ are linked by simple relations. Indeed, we have

$$\begin{aligned}
\mu_j &= \int_{\mathbb{R}} \left(\frac{1}{m-z} \int_{x=z}^{m} x^j dx \right) dF_{\widetilde{Z}}(z) \\
&= \int_{\mathbb{R}} \frac{m^{j+1} - z^{j+1}}{(m-z)(j+1)} dF_{\widetilde{Z}}(z) \\
&= \frac{1}{j+1} \int_{\mathbb{R}} (m^j + m^{j-1} z + m^{j-2} z^2 + \cdots + z^j) dF_{\widetilde{Z}}(z) \\
&= \frac{1}{j+1} (m^j + m^{j-1}\widetilde{\mu}_1 + m^{j-2}\widetilde{\mu}_2 + \cdots + \widetilde{\mu}_j), \quad j = 1, 2, \ldots
\end{aligned} \tag{3.11}$$

From (3.11), we get

$$\widetilde{\mu}_j = (j+1)\mu_j - mj\mu_{j-1}, \quad j = 1, 2, \ldots \tag{3.12}$$

Let us now come back to (3.8) in $\mathcal{B}_s([0,b]; m-\text{unim}; \boldsymbol{\mu})$. Specifically, we would like to determine the random variables $X_{\min}^{(s)\star}$ and $X_{\max}^{(s)\star}$ such that the inequalities

$$\mathbb{E}[(X_{\min}^{(s)\star})^s] \le \mathbb{E}[X^s] \le \mathbb{E}[(X_{\max}^{(s)\star})^s] \quad \text{hold for all} \quad X \in \mathcal{B}_s([0,b]; m-\text{unim}; \boldsymbol{\mu}). \tag{3.13}$$

As shown in Denuit, De Vylder & Lefèvre (1999), the random variables $X_{\min}^{(s)\star}$ and $X_{\max}^{(s)\star}$ involved in (3.13) give bounds on $\mathbb{E}[\phi(X)]$ for every s-convex function ϕ. Finding the solution of (3.13) for X in $\mathcal{B}_s([0,b]; m-\text{unim}; \boldsymbol{\mu})$ thus amounts to solving the corresponding problem in $\mathcal{B}_s([0,b]; \widetilde{\boldsymbol{\mu}})$.

Tables 3.8–3.9 give explicit expressions for the improved extremal distributions, also for values of s up to five. In these tables $\sum_{i=1}^{k} p_i \mathcal{U}ni[\alpha_i, \beta_i], 0 \le p_i \le 1, \alpha_i, \beta_i \in \mathbb{R}, i = 1, 2, \ldots, k$, represents a mixture of the distributions $\mathcal{U}ni[\alpha_i, \beta_i]$, with respective weights p_i.

Table 3.8 Probability distribution of $X_{\min}^{(s)\star} \in \mathcal{B}_s([0,b]; m-\text{unim}; \boldsymbol{\mu})$ achieving the lower bound in (3.13), $s=1,\ldots,5$.

s	Distributions
1	$\mathcal{U}ni[0,m]$
2	$\mathcal{U}ni[m,\tilde{\mu}_1]$
3	$\frac{\tilde{\mu}_2-\tilde{\mu}_1^2}{\tilde{\mu}_2}\mathcal{U}ni[0,m]+\frac{\tilde{\mu}_1^2}{\tilde{\mu}_2}\mathcal{U}ni[m,\tilde{\mu}_2/\tilde{\mu}_1]$
4	$\frac{\tilde{\mu}_1-\tilde{r}_-}{\tilde{r}_+-\tilde{r}_-}\mathcal{U}ni[m,\tilde{r}_+]+(1-\frac{\tilde{\mu}_1-\tilde{r}_-}{\tilde{r}_+-\tilde{r}_-})\mathcal{U}ni[m,\tilde{r}_-]$
5	$(1-\tilde{q}_+-\tilde{q}_-)\mathcal{U}ni[0,m]+\tilde{q}_+\mathcal{U}ni[m,\tilde{t}_+]+\tilde{q}_-\mathcal{U}ni[m,\tilde{t}_-]$

Here $\tilde{r}_\pm$, $\tilde{t}_\pm$ and $\tilde{q}_\pm$ are those from Table 3.5, with the $\tilde{\mu}_j$s substituted for the μ_js.

Table 3.9 Probability distribution of $X_{\max}^{(s)\star} \in \mathcal{B}_s([0,b]; m-\text{unim}; \boldsymbol{\mu})$ achieving the upper bound in (3.13), $s=1,\ldots,5$.

s	Distribution
1	$\mathcal{U}ni[m,b]$
2	$\frac{b-\tilde{\mu}_1}{b}\mathcal{U}ni[0,m]+\frac{\tilde{\mu}_1}{b}\mathcal{U}ni[m,b]$
3	$\frac{(b-\tilde{\mu}_1)^2}{(b-\tilde{\mu}_1)^2+\tilde{\mu}_2-\tilde{\mu}_1^2}\mathcal{U}ni\big[m,\frac{b\tilde{\mu}_1-\tilde{\mu}_2}{b-\tilde{\mu}_1}\big]+\frac{\tilde{\mu}_2-\tilde{\mu}_1^2}{(b-\tilde{\mu}_1)^2+\tilde{\mu}_2-\tilde{\mu}_1^2}\mathcal{U}ni[m,b]$
4	$(1-\tilde{p}_1-\tilde{p}_2)\mathcal{U}ni[0,m]+\tilde{p}_1\mathcal{U}ni\big[m,\frac{\tilde{\mu}_3-b\tilde{\mu}_2}{\tilde{\mu}_2-b\tilde{\mu}_1}\big]+\tilde{p}_2\mathcal{U}ni[m,b]$
5	$\tilde{p}_+\mathcal{U}ni[m,\tilde{z}_+]+\tilde{p}_-\mathcal{U}ni[m,\tilde{z}_-]+(1-\tilde{p}_+-\tilde{p}_-)\mathcal{U}ni[m,b]$

Here $\tilde{p}_1$, $\tilde{p}_2$, $\tilde{z}_\pm$ and $\tilde{p}_\pm$ are those from Table 3.6, with the $\tilde{\mu}_j$s substituted for the μ_js.

Application to Credibility Formulas

Considering a risk parameter Θ_i with moments $\mu_1,\ldots,\mu_{s-1}$ and support $[0,b]$, b possibly infinite, the idea is now to approximate Θ_i by either $\Theta_i^{\min}$ or $\Theta_i^{\max}$ according to the form of the support of Θ_i. This amounts to replacing the general formulas (3.2) with the simpler (3.6)–(3.7) where the support points $\theta_1,\ldots,\theta_q$ are those given in Tables 3.5–3.6. Note that (3.2) can be rewritten as

$$\mathbb{E}[\Theta_i|N_{i1}=k_1,\ldots,N_{iT_i}=k_{T_i}]=\frac{\mathbb{E}[\exp(-\lambda_{i\bullet}\Theta_i)\Theta_i^{k_\bullet+1}]}{\mathbb{E}[\exp(-\lambda_{i\bullet}\Theta_i)\Theta_i^{k_\bullet}]}$$

so that we need to approximate the Mellin transform

$$M(k)=\mathbb{E}[\exp(-\xi\Theta_i)\Theta_i^k]$$

for any integer k and real $\xi>0$.

When the risk parameter Θ_i is known to be unimodal (with mode m, say) then the improved extremal distributions of Tables 3.8–3.9 can be used in lieu of those coming from Tables 3.5–3.6. This amounts to evaluating the numerator and denominator of (3.2), taking for the structure function F_Θ a discrete mixture of uniform distributions. This provides an easy-to-compute approximation for (3.2) based on incomplete Gamma functions, as can be seen from the following example.

Example 3.1 Assume that Θ_i has support $\mathbb{R}^+$, mode m and moments μ_1 and μ_2. Then, we can use the approximation $M(k) \approx M^{(3)}_{\min}(k)$ where

$$
\begin{aligned}
M^{(3)}_{\min}(k) &= \frac{3\mu_2+2\mu_1 m - m^2 - 4\mu_1^2}{(m-2\mu_1)^2+3\mu_2+2\mu_1 m - m^2 - 4\mu_1^2}\frac{1}{m}\int_0^m \exp(-\xi\theta)\theta^k d\theta \\
&+ \frac{(m-2\mu_1)^2}{(m-2\mu_1)^2+3\mu_2+2\mu_1 m - m^2 - 4\mu_1^2}\frac{1}{\overline{c}-\underline{c}}\int_{\underline{c}}^{\overline{c}} \exp(-\xi\theta)\theta^k d\theta
\end{aligned}
$$

with $\underline{c}$ and $\overline{c}$ defined as

$$\underline{c} = \min\left(m, \frac{3\mu_2 - 2m\mu_1}{2\mu_1 - m}\right)$$

and

$$\overline{c} = \max\left(m, \frac{3\mu_2 - 2m\mu_1}{2\mu_1 - m}\right).$$

The value of $M^{(3)}_{\min}(k)$ is easily obtained from the incomplete Gamma function. Specifically,

$$
\begin{aligned}
M^{(3)}_{\min}(k) &= \frac{3\mu_2+2\mu_1 m - m^2 - 4\mu_1^2}{(m-2\mu_1)^2+3\mu_2+2\mu_1 m - m^2 - 4\mu_1^2}\frac{k!}{m\xi^{k+1}}\Gamma(k+1, m\xi) \\
&+ \frac{(m-2\mu_1)^2}{(m-2\mu_1)^2+3\mu_2+2\mu_1 m - m^2 - 4\mu_1^2}\frac{k!}{(\overline{c}-\underline{c})\xi^{k+1}}\big(\Gamma(k+1, \overline{c}\xi) - \Gamma(k+1, \underline{c}\xi)\big).
\end{aligned}
$$

Example 3.2 Now, if the support of Θ_i is known to be contained in $[0, b]$ then we can use the approximation $M(k) \approx M^{(3)}_{\max}(k)$ where

$$
\begin{aligned}
M^{(3)}_{\max}(k) &= \frac{3\mu_2+2\mu_1 m - m^2 - 4\mu_1^2}{(b+m-2\mu_1)^2+3\mu_2+2\mu_1 m - m^2 - 4\mu_1^2}\frac{1}{b-m}\int_m^b \exp(-\xi\theta)\theta^k d\theta \\
&+ \frac{(b+m-2\mu_1)^2}{(b+m-2\mu_1)^2+3\mu_2+2\mu_1 m - m^2 - 4\mu_1^2}\frac{1}{\overline{d}-\underline{d}}\int_{\underline{d}}^{\overline{d}} \exp(-\xi\theta)\theta^k d\theta
\end{aligned}
$$

with $\underline{d}$ and $\overline{d}$ defined as

$$\underline{d} = \min\left(m, \frac{2b\mu_1 + 2\mu_1 m - bm - 3\mu_2}{b+m-2\mu_1}\right)$$

and

$$\overline{d} = \max\left(m, \frac{2b\mu_1 + 2\mu_1 m - bm - 3\mu_2}{b + m - 2\mu_1}\right).$$

As above, this expression can be simplified with the help of the incomplete Gamma function as

$$\begin{aligned}
M_{\max}^{(3)}(k) = & \frac{3\mu_2 + 2\mu_1 m - m^2 - 4\mu_1^2}{(b+m-2\mu_1)^2 + 3\mu_2 + 2\mu_1 m - m^2 - 4\mu_1^2} \\
& \times \frac{k!}{(b-m)\xi^{k+1}}\big(\Gamma(k+1, b\xi) - \Gamma(k+1, m\xi)\big) \\
& + \frac{(b+m-2\mu_1)^2}{(b+m-2\mu_1)^2 + 3\mu_2 + 2\mu_1 m - m^2 - 4\mu_1^2} \\
& \times \frac{k!}{(\overline{d} - \underline{d})\xi^{k+1}}\big(\Gamma(k+1, \overline{d}\xi) - \Gamma(k+1, \underline{d}\xi)\big).
\end{aligned}$$

Example 3.3 If the third moment of Θ_i, μ_3 say, is known then we can use a mixture of the two preceding approximations, with weights defined by the canonical moments as suggested by (3.9). To this end, note that

$$\begin{aligned}
\underline{\mu}_3^\star = & \min_{X \in \mathcal{B}_3([0,b]; m-unim; \mu_1, \mu_2)} \mathbb{E}[X^3] \\
= & \frac{3\mu_2 + 2\mu_1 m - m^2 - 4\mu_1^2}{(m-2\mu_1)^2 + 3\mu_2 + 2\mu_1 m - m^2 - 4\mu_1^2} \frac{m^3}{3} \\
& + \frac{(m-2\mu_1)^2}{(m-2\mu_1)^2 + 3\mu_2 + 2\mu_1 m - m^2 - 4\mu_1^2} \frac{\overline{c}^4 - \underline{c}^4}{\overline{c} - \underline{c}}
\end{aligned}$$

and

$$\begin{aligned}
\overline{\mu_3^\star} = & \max_{X \in \mathcal{B}_3([0,b]; m-unim; \mu_1, \mu_2)} \mathbb{E}[X^3] \\
= & \frac{3\mu_2 + 2\mu_1 m - m^2 - 4\mu_1^2}{(b+m-2\mu_1)^2 + 3\mu_2 + 2\mu_1 m - m^2 - 4\mu_1^2} \frac{b^4 - m^4}{b - m} \\
& + \frac{(b+m-2\mu_1)^2}{(b+m-2\mu_1)^2 + 3\mu_2 + 2\mu_1 m - m^2 - 4\mu_1^2} \frac{\overline{d}^4 - \underline{d}^4}{\overline{d} - \underline{d}}.
\end{aligned}$$

The approximation to the Mellin transform is then as follows

$$M(k) \approx \frac{\overline{\mu_3^\star} - \mu_3}{\overline{\mu_3^\star} - \underline{\mu}_3^\star} M_{\min}^{(3)}(k) + \frac{\mu_3 - \underline{\mu}_3^\star}{\overline{\mu_3^\star} - \underline{\mu}_3^\star} M_{\max}^{(3)}(k).$$

Numerical Illustration

Since we worked with mixing distributions with unbounded support, we cannot use the maximal variables described above. In practice, setting b equal to a high quantile (99.99 %, say) of the mixing distribution is expected to give good results.

Let us consider the Negative Binomial model for the annual claim number in Portfolio A. In this case, $\Theta_i \sim \mathcal{G}am(a, a)$, with estimated parameter $\widehat{a} = 1.065$. The estimated moments of Θ_i are

$$\widehat{\mu}_1 = 1$$
$$\widehat{\mu}_2 = \frac{\widehat{a}+1}{\widehat{a}} = 1.939$$
$$\widehat{\mu}_3 = \frac{(\widehat{a}+1)(\widehat{a}+2)}{\widehat{a}^2} = 5.580$$
$$\widehat{\mu}_4 = \frac{(\widehat{a}+1)(\widehat{a}+2)(\widehat{a}+3)}{\widehat{a}^3} = 21.299.$$

Applying formula (3.7) with discrete approximation given by Table 3.5, we can approximate Θ_i with the help of the 4-convex minimum with two support points: $r_+ = 3.2883$ with probability 0.1521 and $r_- = 0.5897$ with probability 0.8479, or with the 5-convex minimum with three support points: $t_+ = 4.5218$ with probability 0.0474 and $t_- = 1.2341$ with probability 0.6366 and 0 with probability 0.3160. The accuracy of these approximations decreases with T_i and $k_\bullet$. Discrete approximations are useful for short claim history and rather good driving record. Since the Gamma distribution (playing the role of the mixing distribution in the Negative Binomial model) has a unimodal probability density function, we can also use a mixture of uniform distributions to approximate Θ_i in Portfolio A. It turns out that the approximation is satisfactory, except for high values of $k_\bullet$. Again, this is due to the fact that the approximation uses s-convex minima that underestimate the riskiness of the worse drivers.

To get accurate approximations we need to use a mixture (3.9) of the improved 4-convex extrema in the unimodal case taking for b the 99.99th quantile of the Gamma distribution. This gives the results displayed in Table 3.10 for a good driver, in Table 3.11 for an average driver and in Table 3.12 for a bad driver. We can see that the approximations are now very satisfactory for the vast majority of the combinations T_i–$k_\bullet$.

3.3.9 Linear Credibility

Bayesian statistics offer an intellectually acceptable approach to credibility theory. Bayes revision $\mathbb{E}[\Theta_i|N_{i\bullet}]$ of the heterogeneity component is theoretically very satisfying but is often difficult to compute (except for conjugate distributions or discrete approximations). Practical applications involve numerical methods to perform integration with respect to *a posteriori* distributions, making more elementary approaches desirable (at least to get a first easy-to-compute approximation of the result). Because we have observed $N_{i1}, \ldots, N_{iT_i}$, one suggestion is to approximate $\mathbb{E}[\Theta_i|N_{i1}, \ldots, N_{iT_i}]$ by a linear function of the N_{it}s. Basically, the actuary still resorts to a quadratic loss function but the shape of the credibility predictor

Table 3.10 Values of $\mathbb{E}[\Theta_i|N_{i\bullet}=k_\bullet]$ for different combinations of observed periods T_i and number of past claims $k_\bullet$ for a good driver from Portfolio A (expected annual claim frequency of 9.28 %) and for a mixture of mixed uniform approximations of Θ_i (improved 4-convex extrema).

T_i	Number of claims $k_\bullet$					
	0	1	2	3	4	5
1	91.98 %	178.36 %	264.74 %	351.04 %	437.76 %	524.92 %
2	85.16 %	165.12 %	245.13 %	325.04 %	404.57 %	486.88 %
3	79.28 %	153.70 %	228.23 %	302.76 %	376.26 %	451.61 %
4	74.16 %	143.77 %	213.51 %	283.40 %	352.14 %	420.74 %
5	69.66 %	135.06 %	200.54 %	266.38 %	331.27 %	394.52 %
6	65.67 %	127.37 %	189.00 %	251.28 %	312.83 %	372.29 %
7	62.18 %	120.55 %	178.65 %	237.81 %	296.27 %	353.06 %
8	58.93 %	114.47 %	169.29 %	225.73 %	281.26 %	335.95 %
9	56.05 %	109.04 %	160.80 %	214.82 %	267.62 %	320.35 %
10	53.43 %	104.17 %	153.05 %	204.90 %	255.23 %	305.90 %

Table 3.11 Values of $\mathbb{E}[\Theta_i|N_{i\bullet}=k_\bullet]$ for different combinations of observed periods T_i and number of past claims $k_\bullet$ for an average driver from Portfolio A (expected annual claim frequency of 14.09 %) and for a mixture of mixed uniform approximations of Θ_i (improved 4-convex extrema).

T_i	Number of claims $k_\bullet$					
	0	1	2	3	4	5
1	88.32 %	171.25 %	254.21 %	337.07 %	419.97 %	505.05 %
2	79.09 %	153.34 %	227.69 %	302.05 %	375.37 %	450.47 %
3	71.60 %	138.83 %	206.16 %	273.75 %	340.28 %	405.73 %
4	65.41 %	126.87 %	188.25 %	250.30 %	311.63 %	370.88 %
5	60.21 %	116.90 %	173.05 %	230.58 %	287.30 %	342.82 %
6	55.76 %	108.51 %	159.96 %	213.75 %	266.28 %	318.81 %
7	51.93 %	101.38 %	148.58 %	199.16 %	248.12 %	297.44 %
8	48.58 %	95.27 %	138.64 %	186.29 %	232.40 %	278.28 %
9	45.63 %	89.96 %	129.95 %	174.75 %	218.71 %	261.30 %
10	43.01 %	85.30 %	122.35 %	164.32 %	206.58 %	246.42 %

is constrained *ex ante* to be linear in past observations, i.e. the predictor $\widehat{N}_{iT_i+1}$ of N_{iT_i+1} is of the form

$$\widehat{N}_{iT_i+1} = c_{i0} + \sum_{t=1}^{T_i} c_{it} N_{it}$$

The coefficients c_{i0} and the c_{it}s involved in $\widehat{N}_{iT_i+1}$ are obtained from the minimization of an expected squared difference.

Table 3.12 Values of $\mathbb{E}[\Theta_i|N_{i\bullet}=k_\bullet]$ for different combinations of observed periods T_i and number of past claims $k_\bullet$ for a bad driver from Portfolio A (expected annual claim frequency of 28.40 %) and for a mixture of mixed uniform approximations of Θ_i (improved 4-convex extrema).

T_i	Number of claims $k_\bullet$					
	0	1	2	3	4	5
1	78.95 %	153.07 %	227.29 %	301.52 %	374.70 %	449.61 %
2	65.22 %	126.50 %	187.69 %	249.57 %	310.74 %	369.83 %
3	55.55 %	108.12 %	159.34 %	212.96 %	265.29 %	317.66 %
4	48.36 %	94.88 %	138.01 %	185.46 %	231.41 %	277.06 %
5	42.80 %	84.93 %	121.75 %	163.48 %	205.61 %	245.24 %
6	38.39 %	77.09 %	109.33 %	145.45 %	184.61 %	220.87 %
7	34.79 %	70.65 %	99.70 %	130.79 %	166.57 %	201.26 %
8	31.83 %	65.18 %	92.05 %	119.03 %	150.92 %	184.27 %
9	29.35 %	60.42 %	85.75 %	109.64 %	137.59 %	168.95 %
10	27.24 %	56.22 %	80.40 %	102.06 %	126.49 %	155.15 %

Specifically, we look for c_{i0} and c_{it}s, $t=1,\ldots,T_i$, such that the expected square difference between N_{iT_i+1} and its prediction $\hat{N}_{iT_i+1}$ is minimum, i.e. such that

$$\boldsymbol{c}=\arg\min_{\boldsymbol{c}} \mathcal{Q}_1,$$

where

$$\mathcal{Q}_1=\mathbb{E}\left[\left(N_{i,T_i+1}-c_{i0}-\sum_{t=1}^{T_i}c_{it}N_{it}\right)^2\right].$$

Alternatively, it can be shown that $\boldsymbol{c}$ also solves

$$\boldsymbol{c}=\arg\min_{\boldsymbol{c}} \mathcal{Q}_j,\quad j=2,3,$$

where

$$\mathcal{Q}_2=\mathbb{E}\left[\left(\mathbb{E}[N_{i,T_i+1}|\Theta_i]-c_{i0}-\sum_{t=1}^{T_i}c_{it}N_{it}\right)^2\right]$$

$$\mathcal{Q}_3=\mathbb{E}\left[\left(\mathbb{E}[N_{i,T_i+1}|N_{i1},\ldots,N_{iT_i}]-c_{i0}-\sum_{t=1}^{T_i}c_{it}N_{it}\right)^2\right].$$

Let us show for instance that $\arg\min_{\boldsymbol{c}}\mathcal{Q}_1=\arg\min_{\boldsymbol{c}}\mathcal{Q}_2$. To this end, let us write

$$\mathcal{Q}_1=\mathbb{E}\left[\left(\Big(N_{i,T_i+1}-\mathbb{E}[N_{i,T_i+1}|\Theta_i]\Big)+\left(\mathbb{E}[N_{i,T_i+1}|\Theta_i]-c_{i0}-\sum_{t=1}^{T_i}c_{it}N_{it}\right)\right)^2\right]$$

and let us expand the squared sum to get

$$\mathcal{Q}_1 = \mathbb{E}\left[\left(N_{i,T_i+1} - \mathbb{E}[N_{i,T_i+1}|\Theta_i]\right)^2\right]$$
$$+2\mathbb{E}\left[\left(N_{i,T_i+1} - \mathbb{E}[N_{i,T_i+1}|\Theta_i]\right)\left(\mathbb{E}[N_{i,T_i+1}|\Theta_i] - c_{i0} - \sum_{t=1}^{T_i} c_{it}N_{it}\right)\right] + \mathcal{Q}_2.$$

The second term in the expansion of $\mathcal{Q}_1$ vanishes, and the first one does not depend on the c_{it}s.

Let us determine $\boldsymbol{c}$ as $\arg\min \mathcal{Q}_2$. Recall that $\mathbb{E}[N_{i,T_i+1}|\Theta_i] = \lambda_{i,T_i+1}\Theta_i$. Setting equal to 0 the partial derivative of $\mathcal{Q}_2$ with respect to c_{i0} allows us to write

$$0 = \lambda_{i,T_i+1}\mathbb{E}[\Theta_i] - c_{i0} - \sum_{t=1}^{T_i} c_{it}\mathbb{E}[N_{it}]$$

which gives

$$c_{i0} = \lambda_{i,T_i+1} - \sum_{t=1}^{T_i} c_{it}\lambda_{it}. \tag{3.14}$$

Now, setting equal to 0 the partial derivatives of $\mathcal{Q}_2$ with respect to c_{is} for $s = 1, 2, \ldots, T_i$, yields

$$0 = \lambda_{i,T_i+1}\mathbb{E}[N_{is}\Theta_i] - c_{i0}\mathbb{E}[N_{is}] - \sum_{t=1}^{T_i} c_{it}\mathbb{E}[N_{is}N_{it}]. \tag{3.15}$$

Noting that

$$\begin{aligned}
\mathbb{E}[N_{it}N_{is}] &= \mathbb{C}[N_{is}, N_{it}] + \mathbb{E}[N_{is}]\mathbb{E}[N_{it}] \\
&= \mathbb{E}\Big[\mathbb{C}[N_{is}, N_{it}|\Theta_i]\Big] + \mathbb{C}\Big[\mathbb{E}[N_{is}|\Theta_i], \mathbb{E}[N_{it}|\Theta_i]\Big] + \lambda_{is}\lambda_{it} \\
&= \begin{cases} \lambda_{is} + (\lambda_{is})^2(1+\mathbb{V}[\Theta_i]) \text{ if } s = t \\ \lambda_{is}\lambda_{it}(1+\mathbb{V}[\Theta_i]) \text{ otherwise} \end{cases}
\end{aligned}$$

and that

$$\begin{aligned}
\mathbb{E}[N_{is}\Theta_i] &= \mathbb{E}\Big[\mathbb{C}[N_{is}, \Theta_i|\Theta_i]\Big] + \mathbb{C}\Big[\mathbb{E}[N_{is}|\Theta_i], \Theta_i\Big] + \lambda_{is} \\
&= \lambda_{is}(1+\mathbb{V}[\Theta_i]).
\end{aligned}$$

allows us to cast equation (3.15) into the form

$$c_{is} = \mathbb{V}[\Theta_i]\left(\lambda_{i,T_i+1} - \sum_{t=1}^{T_i} c_{it}\lambda_{it}\right) = \mathbb{V}[\Theta_i]c_{i0}.$$

Hence, the value of c_{is} does not depend on s and the same weight is given to all the past annual claim numbers. Now, inserting the value of c_{is} that we just obtained in (3.14) finally gives

$$c_{i0} = \frac{\lambda_{i,T_i+1}}{1+\mathbb{V}[\Theta_i]\sum_{t=1}^{T_i}\lambda_{it}}$$

which in turn yields.

$$c_{is} = \frac{\lambda_{i,T_i+1}\mathbb{V}[\Theta_i]}{1+\mathbb{V}[\Theta_i]\sum_{t=1}^{T_i}\lambda_{it}}.$$

The expected claim frequency for year T_i+1 given past claims history is

$$\widehat{N}_{i,T_i+1} = \lambda_{i,T_i+1}\frac{1+\mathbb{V}[\Theta_i]N_{i\bullet}}{1+\mathbb{V}[\Theta_i]\lambda_{i\bullet}} \tag{3.16}$$

where $N_{i\bullet}$ and $\lambda_{i\bullet}$ have been defined in (3.1). Note that $\widehat{N}_{i,T_i+1}$ is the best linear predictor of each of the true unknown means $\lambda_{i,T_i+1}\Theta_i$, of the Bayesian credibility premium $\mathbb{E}[N_{i,T_i+1}|N_{i1},\ldots,N_{iT_i}]$ and of the number of claims N_{i,T_i+1} for year T_i+1.

The linear predictor for year T_i+1 thus appears as the product of the *a priori* expected claim frequency, λ_{i,T_i+1}, times an approximation of the theoretical correction $\mathbb{E}[\Theta_i|N_{i1},\ldots,N_{iT_i}]$. This approximation possesses a particularly simple interpretation since it entails a malus when $N_{i\bullet} > \lambda_{i\bullet}$, that is, if the policyholder reported more claims than expected *a priori*.

Remark 3.2 Note that (3.16) agrees with the result obtained in the Poisson-Gamma case. This is because the Bayesian credibility premium is linear in the past observations in the Poisson-Gamma case. The term exact credibility is used to describe the situation where the linear credibility premium equals the Bayesian one. Intuitively speaking, using linear credibility formulas in the mixed Poisson model boils down to approximating the mixing distribution with the Gamma distribution (or, equivalently, the distribution of the claim numbers with the Negative Binomial one).

Remark 3.3 Note that $\widehat{N}_{i,T_i+1}$ could have been obtained by a direct application of the Bühlmann–Straub formula. Let us consider a sequence of random variables $\{X_1, X_2, X_3, \ldots\}$ such that, given a random variable Θ, the X_ts are independent, with finite first and second moments

$$\mu(\Theta) = \mathbb{E}[X_t|\Theta] \text{ and } \mathbb{E}[X_t] = \mu = \mathbb{E}[\mu(\Theta)].$$

Now, define M^2 and Σ^2 as

$$\mathbb{E}\big[\mathbb{V}[X_t|\Theta]\big] = \frac{\Sigma^2}{w_t} \text{ and } \mathbb{V}\big[\mathbb{E}[X_t|\Theta]\big] = M^2,$$

and put

$$w_\bullet = \sum_{t=1}^{n} w_t \text{ and } \widetilde{X}^{(n)} = \frac{1}{w_\bullet}\sum_{t=1}^{n} w_j X_j.$$

Clearly, $\widetilde{X}^{(n)}$ is the weighted average of the X_js. The minimum of

$$\mathbb{E}\Big[\big(\mu(\Theta) - a - b\widetilde{X}^{(n)}\big)^2\Big]$$

on all the couples (a, b) is obtained for

$$a = \frac{\Sigma^2}{\Sigma^2 + w_\bullet M^2}\mu \text{ and } b = \frac{w_\bullet M^2}{\Sigma^2 + w_\bullet M^2}.$$

To apply this general result to the Poisson case, we define $X_j = N_{ij}/\lambda_{ij}$ which gives $\mu(\Theta_i) = \Theta_i$, $\mu = 1$, $w_j = \lambda_{ij}$, $\Sigma^2 = 1$, $M^2 = \mathbb{V}[\Theta_i]$. The best linear approximation to Θ_i is then given by $\widehat{N}_{i,T_i+1}/\lambda_{i,T_i+1}$.

3.4 Credibility Formulas with an Exponential Loss Function

3.4.1 Optimal Predictor

This section purposes to describe an alternative approach based on an exponential loss function. The exponential loss function is asymmetric and possesses one parameter that reflects the severity of the credibility correction. This allows us to soften the *a posteriori* corrections in case of claims, keeping the financial balance.

When the new premium amount is fixed by the insurer, two kinds of errors may arise: either the policyholder is undercharged and the insurance company loses its money or the insured is overcharged and the insurer is at risk of losing the policy. In order to penalize large mistakes to a greater extent, it is usually assumed that the loss function is a nonnegative convex function of the error. The loss is zero when no error is made and strictly positive otherwise. The loss function is generally taken to be quadratic as in the preceding section. Among other choices we find also the absolute loss and the 4-degree loss; see, e.g., Lemaire & Vandermeulen (1983). The problem with these two last losses is that the resulting bonus-malus systems are unbalanced.

We give here a technical result involving an exponential loss function. It is the analogue of Proposition 3.1.

Proposition 3.2 *Under the conditions of Proposition* 3.1, *the minimum of*

$$\mathbb{E}\Big[\exp\Big(-c\big(\mu_{T+1}(\Theta) - \Psi(X_1, X_2, \ldots, X_T)\big)\Big)\Big]$$

on all the measurable functions $\Psi : \mathbb{R}^T \to \mathbb{R}$ *satisfying the constraint* $\mathbb{E}[\Psi(X_1, X_2, \ldots, X_T)] = \mu_{T+1}$ *is obtained for*

$$\Psi^{\star\star}(X_1, X_2, \ldots, X_T) = \mu_{T+1} + \frac{1}{c}\Big(\mathbb{E}\big[\ln \mathbb{E}\big[\exp(-c\mu_{T+1}(\Theta))|X_1, X_2, \ldots, X_T\big]\big]$$
$$- \ln \mathbb{E}\big[\exp(-c\mu_{T+1}(\Theta))|X_1, X_2, \ldots, X_T\big]\Big).$$

Proof Starting from

$$\mathbb{E}\Big[\exp\Big(-c\big(\mu_{T+1}(\Theta) - \Psi(X_1, X_2, \ldots, X_T)\big)\Big)\Big]$$
$$= \mathbb{E}\Big[\exp\big(c\Psi(X_1, X_2, \ldots, X_T)\big)\mathbb{E}\big[\exp(-c\mu_{T+1}(\Theta))|X_1, X_2, \ldots, X_T\big]\Big]$$
$$= \mathbb{E}\Big[\exp\Big(c\big(\Psi(X_1, X_2, \ldots, X_T) - \Psi^{\star\star}(X_1, X_2, \ldots, X_T)\big)\Big)\Big]$$
$$\times \exp(c\mu_{T+1})\exp\Big(\mathbb{E}\Big[\ln \mathbb{E}\big[\exp(-c\mu_{T+1}(\Theta))|X_1, X_2, \ldots, X_T\big]\Big]\Big).$$

Now, let us apply Jensen's inequality to get

$$\mathbb{E}\Big[\exp\Big(-c\big(\mu_{T+1}(\Theta) - \Psi(X_1, X_2, \ldots, X_T)\big)\Big)\Big]$$
$$\geq \exp\Big(c\mathbb{E}\Big[\Psi(X_1, X_2, \ldots, X_T) - \Psi^{\star\star}(X_1, X_2, \ldots, X_T)\Big]\Big)$$
$$\exp(c\mu_{T+1})\exp\Big(\mathbb{E}\Big[\ln \mathbb{E}\big[\exp(-c\mu_{T+1}(\Theta))|X_1, X_2, \ldots, X_T\big]\Big]\Big).$$

Because of the constraint on the expectation of the Ψs, the first exponential is 1, yielding

$$\mathbb{E}\Big[\exp\Big(-c\big(\mu_{T+1}(\Theta) - \Psi(X_1, X_2, \ldots, X_T)\big)\Big)\Big]$$
$$\geq \exp(c\mu_{T+1})\exp\Big(\mathbb{E}\Big[\ln \mathbb{E}\big[\exp(-c\mu_{T+1}(\Theta))|X_1, X_2, \ldots, X_T\big]\Big]\Big)$$
$$= \mathbb{E}\Big[\exp\Big(-c\big(\mu_{T+1}(\Theta) - \Psi^{\star\star}(X_1, X_2, \ldots, X_T)\big)\Big)\Big]$$

which is the expected result. □

Remark that in Proposition 3.2 the constraint is made in order to guarantee the financial equilibrium.

Let us now apply the result contained in Proposition 3.2 to the credibility problem. In this case,

$$X_t = N_{it} \text{ and } \mu_t(\Theta_i) = \lambda_{it}\Theta_i$$

so that the optimal predictor of N_{i,T_i+1} for the exponential loss function is of the form

$$\Psi^{\star\star}(N_{i1}, \ldots, N_{iT_i}) = \lambda_{i,T_i+1} + \frac{1}{c}\Big(\mathbb{E}\big[\ln \mathbb{E}\big[\exp(-c\lambda_{i,T_i+1}\Theta_i)|N_{i1}, \ldots, N_{iT_i}\big]\big]$$
$$- \ln \mathbb{E}\big[\exp(-c\lambda_{i,T_i+1}\Theta_i)|N_{i1}, \ldots, N_{iT_i}\big]\Big).$$

3.4.2 Poisson-Gamma Credibility Model

Let us now apply the result contained in Proposition 3.2 to the Poisson-Gamma credibility model. To this end, assume that $\Theta_i \sim \mathcal{G}am(a, a)$. Given $N_{i\bullet}$, we know from (3.4) that Θ_i follows the $\mathcal{G}am(a+N_{i\bullet}, a+\lambda_{i\bullet})$ distribution, so that we know from (1.36) that

$$\mathbb{E}\Big[\exp(-c\lambda_{i,T_i+1}\Theta_i)|N_{i1},\ldots,N_{iT_i}\Big] = \left(\frac{a+\lambda_{i\bullet}}{a+\lambda_{i\bullet}+c\lambda_{i,T_i+1}}\right)^{a+N_{i\bullet}}.$$

It follows that

$$\begin{aligned}&\ln\mathbb{E}\Big[\exp(-c\lambda_{i,T_i+1}\Theta_i)|N_{i1},\ldots,N_{iT_i}\Big]\\ &= -(a+N_{i\bullet})\ln\left(1+\frac{c\lambda_{i,T_i+1}}{a+\lambda_{i\bullet}}\right)\end{aligned}$$

and

$$\begin{aligned}&\mathbb{E}\Big[\ln\mathbb{E}\big[\exp(-c\lambda_{i,T_i+1}\Theta_i)|N_{i1},\ldots,N_{iT_i}\big]\Big]\\ &= -(a+\lambda_{i\bullet})\ln\left(1+\frac{c\lambda_{i,T_i+1}}{a+\lambda_{i\bullet}}\right).\end{aligned}$$

Proposition 3.2 then gives

$$\Psi^{\star\star}(k_{i1},\ldots,k_{iT_i}) = \lambda_{i,T_i+1} + \frac{k_{i\bullet}-\lambda_{i\bullet}}{c}\ln\left(1+\frac{c\lambda_{i,T_i+1}}{a+\lambda_{i\bullet}}\right). \tag{3.17}$$

Considering (3.17), we see that $\Psi^{\star\star}(k_{i1},\ldots,k_{iT_i})$ is equal to the *a priori* expectation $\lambda_{i,T_i+1} = \mathbb{E}[N_{i,T_i+1}]$ plus a correction term. This correction is positive, so that

$$\Psi^{\star\star}(k_{i1},\ldots,k_{iT_i}) > \lambda_{i,T_i+1}$$

if $k_{i\bullet} > \lambda_{i\bullet}$, that is, if the policyholder reported more claims than expected. Otherwise, the correction is negative. As with the quadratic loss function, the penalty is caused by an excess of observed claims over expected ones.

Let us now compare the credibility formulas obtained with a quadratic and exponential loss function. Since for any $c \geq 0$,

$$\ln\left(1+\frac{c\lambda_{i,T_i+1}}{a+\lambda_{i\bullet}}\right) \leq \frac{c\lambda_{i,T_i+1}}{a+\lambda_{i\bullet}},$$

it is easily seen that we have

$$\Psi^{\star\star}(k_{i1},\ldots,k_{iT_i}) \leq \Psi^{\star}(k_{i1},\ldots,k_{iT_i}) \text{ if } k_{i\bullet} > \lambda_{i\bullet},$$

and

$$\Psi^{**}(k_{i1},\ldots,k_{iT_i}) \geq \Psi^{*}(k_{i1},\ldots,k_{iT_i}) \text{ if } k_{i\bullet} < \lambda_{i\bullet}.$$

Let us define

$$\rho_{exp}(c) = \frac{1}{c}\ln\left(1+\frac{c\lambda_{i,T_i+1}}{a+\lambda_{i\bullet}}\right)$$

to be the weight given to $k_{i\bullet}$ in the *a posteriori* evaluation (3.17) in the Poisson-Gamma model. We have that $\lim_{c\to+\infty}\rho_{exp}(c)=0$. Moreover, routine calculations show that $d/dc\rho_{exp}(c)<0$, so that the weight given to the observed average claim number decreases as c increases. This provides an intuitive meaning of the parameter c: if c increases, then the *a posteriori* merit-rating scheme becomes less severe, and at the limit for $c\to+\infty$, the premium no longer depends on the incurred claims. If the asymmetry factor c tends to $+\infty$ then all the risks within the same tariff class pay the same premium: there is no longer an experience rating. Conversely, the weight given to past claims under an exponential loss function tends to the weight under a quadratic loss function as $c\to 0$.

3.4.3 Linear Credibility

Another possibility is to determine a linear credibility premium based on an exponential loss function, by considering predictors of the form $b_0+\sum_{t=1}^{T_i} b_j N_{it}$. The b_js minimize the Lagrangian function

$$\begin{aligned}\mathcal{L}(\psi,b_0,b_1,\cdots,b_t) &= \mathbb{E}\Big[\exp\big(-c(\lambda_{i,T_i+1}\Theta_i-b_1N_{i1}-b_2N_{i2}-\cdots-b_{T_i}N_{iT_i}-b_0)\big)\Big]\\ &\quad -\psi\mathbb{E}\Big[b_0+b_1N_{i1}+b_2N_{i2}+\cdots+b_{T_i}N_{iT_i}-\lambda_{i,T_i+1}\Big].\end{aligned}$$

Setting to 0 the derivatives of $\mathcal{L}$ with respect to $\psi, b_0, b_1, \ldots, b_{T_i}$ yields the same result as in the Poisson-Gamma case, i.e. (3.17).

3.4.4 Numerical Illustration

Let us now illustrate the use of the exponential loss function in credibility. To this end, let us consider the Negative Binomial fit to Portfolio A described in Table 2.7. Thus, Θ_i is taken to be $\mathcal{G}am(a,a)$ distributed, with estimated parameter $\widehat{a}=1.065$.

Formula (3.17) allows us to compute the *a posteriori* correction as a function of the number T_i of coverage periods and of the total number of claims $k_\bullet$ filed to the company. The results obtained with $c=1$ are displayed in Table 3.13 for a good driver, in Table 3.14 for an average driver, and in Table 3.15 for a bad driver. Tables 3.16–3.18 are the analogues for $c=5$.

Let us first compare the *a posteriori* corrections listed in Table 3.13 with those of Table 3.2 corresponding to a quadratic loss function (i.e. to $c=0$). We see that the application of an exponential loss function slightly reduces the penalties in case of claims

Table 3.13 Values of the *a posteriori* corrections obtained from (3.17) for different combinations of observed periods T_i and number of past claims $k_\bullet$ for a good driver (expected annual claim frequency equal to 9.28 %) from Portfolio A, with $c = 1$.

T_i	Number of claims $k_\bullet$					
	0	1	2	3	4	5
1	92.3 %	175.4 %	258.5 %	341.5 %	424.6 %	507.7 %
2	85.7 %	162.8 %	240.0 %	317.1 %	394.2 %	471.4 %
3	80.0 %	151.9 %	223.9 %	295.9 %	367.9 %	439.9 %
4	75.0 %	142.4 %	209.9 %	277.4 %	344.8 %	412.3 %
5	70.5 %	134.0 %	197.5 %	261.0 %	324.5 %	388.0 %
6	66.6 %	126.6 %	186.5 %	246.5 %	306.5 %	366.4 %
7	63.1 %	119.9 %	176.7 %	233.5 %	290.3 %	347.1 %
8	59.9 %	113.9 %	167.9 %	221.8 %	275.8 %	329.7 %
9	57.1 %	108.5 %	159.8 %	211.2 %	262.6 %	314.0 %
10	54.5 %	103.5 %	152.6 %	201.6 %	250.7 %	299.7 %

Table 3.14 Values of the *a posteriori* corrections obtained from (3.17) for different combinations of observed periods T_i and number of past claims $k_\bullet$ for an average driver (expected annual claim frequency equal to 14.09 %) from Portfolio A, with $c = 1$.

T_i	Number of claims $k_\bullet$					
	0	1	2	3	4	5
1	89.0 %	167.4 %	245.8 %	324.3 %	402.7 %	481.1 %
2	80.1 %	150.7 %	221.4 %	292.0 %	362.6 %	433.2 %
3	72.9 %	137.1 %	201.3 %	265.5 %	329.8 %	394.0 %
4	66.8 %	125.7 %	184.6 %	243.5 %	302.4 %	361.3 %
5	61.7 %	116.1 %	170.5 %	224.9 %	279.2 %	333.6 %
6	57.3 %	107.8 %	158.3 %	208.9 %	259.4 %	309.9 %
7	53.5 %	100.7 %	147.8 %	195.0 %	242.1 %	289.3 %
8	50.2 %	94.4 %	138.6 %	182.8 %	227.0 %	271.3 %
9	47.2 %	88.9 %	130.5 %	172.1 %	213.7 %	255.4 %
10	44.6 %	83.9 %	123.3 %	162.6 %	201.9 %	241.2 %

(the values listed in the columns entitled $k_\bullet = 1$ to 5 are smaller in Table 3.13 compared to Table 3.2). Since the financial balance is fulfilled by the credibility premiums obtained with an exponential loss function, the discounts for claim-free policyholders are also reduced (the values in the column entitled $k_\bullet = 0$ are higher in Table 3.13 compared to Table 3.2).

Note however that the *a posteriori* corrections obtained with an exponential loss function with $c = 1$ are very similar to those coming from a quadratic loss function. To see this, let us now increase the value of c to 5 in Table 3.16. Increasing c results in reduced discounts and also in reduced penalties.

Table 3.15 Values of the *a posteriori* corrections obtained from (3.17) for different combinations of observed periods T_i and number of past claims $k_\bullet$ for a bad driver (expected annual claim frequency equal to 28.40 %) from Portfolio A, with $c = 1$.

T_i	Number of claims $k_\bullet$					
	0	1	2	3	4	5
1	80.9 %	148.2 %	215.4 %	282.7 %	350.0 %	417.2 %
2	67.9 %	124.4 %	180.8 %	237.3 %	293.8 %	350.2 %
3	58.6 %	107.2 %	155.8 %	204.5 %	253.1 %	301.8 %
4	51.5 %	94.2 %	136.9 %	179.6 %	222.4 %	265.1 %
5	45.9 %	84.0 %	122.1 %	160.2 %	198.3 %	236.4 %
6	41.4 %	75.8 %	110.2 %	144.5 %	178.9 %	213.3 %
7	37.7 %	69.1 %	100.4 %	131.7 %	163.0 %	194.3 %
8	34.7 %	63.4 %	92.2 %	120.9 %	149.7 %	178.4 %
9	32.0 %	58.6 %	85.2 %	111.8 %	138.4 %	165.0 %
10	29.8 %	54.5 %	79.2 %	103.9 %	128.7 %	153.4 %

Table 3.16 Values of the *a posteriori* corrections obtained from (3.17) for different combinations of observed periods T_i and number of past claims $k_\bullet$ for a good driver (expected annual claim frequency equal to 9.28 %) from Portfolio A, with $c = 5$.

T_i	Number of claims $k_\bullet$					
	0	1	2	3	4	5
1	93.3 %	165.9 %	238.5 %	311.1 %	383.8 %	456.4 %
2	87.4 %	155.4 %	223.4 %	291.4 %	359.4 %	427.4 %
3	82.2 %	146.1 %	210.1 %	274.0 %	338.0 %	401.9 %
4	77.6 %	137.9 %	198.3 %	258.6 %	318.9 %	379.3 %
5	73.5 %	130.6 %	187.7 %	244.9 %	302.0 %	359.1 %
6	69.8 %	124.0 %	178.3 %	232.5 %	286.7 %	340.9 %
7	66.5 %	118.1 %	169.7 %	221.3 %	272.9 %	324.6 %
8	63.4 %	112.7 %	161.9 %	211.2 %	260.4 %	309.7 %
9	60.7 %	107.8 %	154.8 %	201.9 %	249.0 %	296.1 %
10	58.1 %	103.2 %	148.4 %	193.5 %	238.6 %	283.7 %

If we compare the *a posteriori* corrections for the different types of drivers, we see that the discounts increase with the average annual claim frequency, as was the case with the quadratic loss function. Also, the penalties appear to decrease with the average annual claim frequencies. The fact that *a priori* bad drivers need a greater premium reduction when no claim is filed to the insurance company thus remains with exponential loss functions.

Table 3.17 Values of the *a posteriori* corrections obtained from (3.17) for different combinations of observed periods T_i and number of past claims $k_\bullet$ for an average driver (expected annual claim frequency equal to 14.09 %) from Portfolio A, with $c = 5$.

T_i	Number of claims $k_\bullet$					
	0	1	2	3	4	5
1	90.8 %	156.1 %	221.4 %	286.7 %	352.0 %	417.4 %
2	83.2 %	142.9 %	202.6 %	262.4 %	322.1 %	381.8 %
3	76.7 %	131.8 %	186.8 %	241.9 %	296.9 %	351.9 %
4	71.2 %	122.3 %	173.3 %	224.3 %	275.4 %	326.4 %
5	66.5 %	114.1 %	161.7 %	209.2 %	256.8 %	304.4 %
6	62.3 %	106.9 %	151.5 %	196.0 %	240.6 %	285.2 %
7	58.7 %	100.6 %	142.5 %	184.4 %	226.3 %	268.2 %
8	55.4 %	95.0 %	134.5 %	174.1 %	213.7 %	253.2 %
9	52.5 %	90.0 %	127.4 %	164.9 %	202.4 %	239.8 %
10	49.9 %	85.5 %	121.0 %	156.6 %	192.2 %	227.8 %

Table 3.18 Values of the *a posteriori* corrections obtained from (3.17) for different combinations of observed periods T_i and number of past claims $k_\bullet$ for a bad driver (expected annual claim frequency equal to 28.40 %) from Portfolio A, with $c = 5$.

T_i	Number of claims $k_\bullet$					
	0	1	2	3	4	5
1	85.6 %	136.3 %	186.9 %	237.5 %	288.2 %	338.8 %
2	75.0 %	119.0 %	163.1 %	207.2 %	251.2 %	295.3 %
3	66.7 %	105.8 %	144.8 %	183.8 %	222.9 %	261.9 %
4	60.2 %	95.2 %	130.3 %	165.3 %	200.4 %	235.5 %
5	54.8 %	86.6 %	118.5 %	150.3 %	182.1 %	213.9 %
6	50.3 %	79.5 %	108.6 %	137.8 %	166.9 %	196.1 %
7	46.5 %	73.4 %	100.3 %	127.2 %	154.1 %	181.0 %
8	43.3 %	68.2 %	93.2 %	118.2 %	143.1 %	168.1 %
9	40.4 %	63.7 %	87.0 %	110.3 %	133.6 %	156.9 %
10	38.0 %	59.8 %	81.6 %	103.5 %	125.3 %	147.2 %

3.5 Dependence in the Mixed Poisson Credibility Model

3.5.1 Intuitive Ideas

The main focus of this section is to formalize intuitive ideas with the help of stochastic orderings. Every actuary intuitively feels that the *a posteriori* claim frequency distribution must become more dangerous as more claims are reported. Here we precisely define 'more dangerous' and explain that the *a posteriori* premium must increase with the total claim number in the mixed Poisson model.

In the model A1–A2 of Definition 3.1, we intuitively feel that the following statements are true:

Statement S1 Θ_i 'increases' in the past claims $N_{i\bullet}$
Statement S2 N_{i,T_i+1} 'increases' in the past claims $N_{i\bullet}$
Statement S3 N_{i,T_i+1} and $N_{i\bullet}$ are 'positively dependent'.

This section aims to precisely define the meaning of 'increases' in statements S1 and S2, as well as the nature of the 'positive dependence' involved in statement S3. The proofs will be omitted because of their technical nature; for a detailed study, we refer the reader to DENUIT ET AL. (2005, Chapter 7). Note that we present the results in terms of stochastic dominance whereas in fact the stronger (but less intuitive) likelihood ratio order applies.

3.5.2 Stochastic Order Relations

In order to formalize the increasingness involved in statements S1–S2, our study will extensively resort to stochastic orderings. Therefore, we recall in this section the definition of stochastic dominance, as well as some intuitive intepretations. Given two random variables X and Y, X is said to be smaller than Y in the stochastic dominance, written as $X \preceq_{\text{ST}} Y$, if

$$\Pr[X > t] \leq \Pr[Y > t] \text{ for all } t \in \mathbb{R}.$$

We see that a ranking in the $\preceq_{\text{ST}}$-sense translates the intuitive meaning of 'being smaller than' in probability models: indeed, we compare the probability that both random variables exceed some given threshold t, and the smallest one in the $\preceq_{\text{ST}}$-sense has the smallest probability of exceeding the threshold. If M and N are two counting random variables then

$$M \preceq_{\text{ST}} N \Leftrightarrow \sum_{j=k}^{+\infty} \Pr[M = j] \leq \sum_{j=k}^{+\infty} \Pr[N = j] \text{ for all } k = 0, 1, \ldots$$

One intuitively feels that a random variable N_θ following the Poisson distribution with mean θ gets bigger as θ increases. The next implication formalizes this intuitive statement:

$$\theta \leq \theta' \Rightarrow N_\theta \preceq_{\text{ST}} N_{\theta'}.$$

The $\mathcal{P}oi(\theta)$ family thus increases in its parameter θ in the $\preceq_{\text{ST}}$-sense.

3.5.3 Comparisons of Predictive Distributions

We then have the following results that formalize statements S1 and S2. First, Θ_i increases in the past claims $N_{i\bullet}$ in the $\preceq_{\text{ST}}$-sense, that is

$$[\Theta_i | N_{i\bullet} = k] \preceq_{\text{ST}} [\Theta_i | N_{i\bullet} = k'] \text{ for } k \leq k' \tag{3.18}$$

$$\Leftrightarrow \Pr[\Theta_i > t | N_{i\bullet} = k] \leq \Pr[\Theta_i > t | N_{i\bullet} = k'] \text{ whatever } t, \text{ provided } k \leq k'.$$

This relation is transmitted to the number of claims, in the sense that

$$[N_{i,T_i+1}|N_{i\bullet}=k] \preceq_{\text{ST}} [N_{i,T_i+1}|N_{i\bullet}=k'] \text{ for } k \leq k' \tag{3.19}$$

$$\Leftrightarrow \Pr[N_{i,T_i+1}>j|N_{i\bullet}=k] \leq [N_{i,T_i+1}>j|N_{i\bullet}=k'] \text{ whatever } j\text{, provided } k \leq k'.$$

3.5.4 Positive Dependence Notions

In order to formalize the positive dependence involved in statement S3, we will present some concepts of dependence related to $\preceq_{\text{ST}}$. The study of concepts of positive dependence for random variables, started in the late 1960s, has yielded numerous useful results in both statistical theory and applications. Applications of these concepts in actuarial science recently received increased interest.

Let us formalize the positive dependence existing between the two components of a random couple (i.e. the fact that large values of one component tend to be associated with large values for the other). Formally, let $\boldsymbol{X}=(X_1,X_2)$ be a bivariate random vector. Then, $\boldsymbol{X}$ is positive regression dependent (PRD, for short) if $[X_2|X_1=x_1] \preceq_{\text{ST}} [X_2|X_1=x_1']$ for all $x_1 \leq x_1'$ and $[X_1|X_2=x_2] \preceq_{\text{ST}} [X_1|X_2=x_2']$ for all $x_2 \leq x_2'$. PRD imposes stochastic increasingness of one component of the random couple in the value assumed by the other component in the $\preceq_{\text{ST}}$-sense. This dependence notion is thus rather intuitive.

PRD naturally extends to higher dimension. Specifically, let $\boldsymbol{X}=(X_1,\ldots,X_n)$ be a n-dimensional random vector. Then,

(i) $\boldsymbol{X}$ is conditionally increasing (CI, for short) if

$$[X_i|X_j=x_j, \quad j\in J] \preceq_{\text{ST}} [X_i|X_j=x_j', \quad j\in J]$$

whenever $x_j \leq x_j'$, $j\in J$, $J\subset\{1,2,\ldots,n\}$ and $i\notin J$.

(ii) $\boldsymbol{X}$ is conditionally increasing in sequence (CIS, for short) if X_i is stochastically increasing in $X_1,\ldots,X_{i-1}$, for $i\in\{2,\ldots,n\}$ i.e.

$$[X_i|X_1=x_1,\ldots,X_{i-1}=x_{i-1}] \preceq_{\text{ST}} [X_i|X_1=x_1',\ldots,X_{i-1}=x_{i-1}'],$$

whenever $x_j \leq x_j'$, $j\in\{1,\ldots,i-1\}$.

The conditional increasingness in sequence is interesting when there is a natural order in the components of $\boldsymbol{X}$, induced by obervation times for instance.

3.5.5 Dependence Between Annual Claim Numbers

The total claim number $N_{i\bullet}$ reported in the past periods and the claim number N_{i,T_i+1} for the next coverage period are PRD. The fact that $N_{i\bullet}$ and N_{i,T_i+1} are PRD completes the statement (3.19). This provides a host of useful inequalities. In particular, whatever the distribution of Θ_i, the credibility coefficient $\mathbb{E}[\Theta_i|N_{i\bullet}=k]$ is increasing in k, which is easily deduced from (3.18).

Considering the dependence existing between the components of $\boldsymbol{N}_i$, i.e. between the N_{it}s, $t=1,2,\ldots,T_i$, we can prove that $\boldsymbol{N}_i$ is CI.

3.5.6 Increasingness in the Linear Credibility Model

Let $\widehat{N}_{i,T_i+1}$ be the predictor (3.16) of N_{i,T_i+1}. It can be shown that N_{iT_i+1} is indeed increasing in $\widehat{N}_{i,T_i+1}$, in the sense that

$$[N_{iT_i+1}|\widehat{N}_{i,T_i+1}=p] \preceq_{\mathrm{ST}} [N_{iT_i+1}|\widehat{N}_{i,T_i+1}=p'] \text{ whenever } p \leq p'$$

$$\Leftrightarrow \Pr[N_{iT_i+1}>k|\widehat{N}_{i,T_i+1}=p] \leq \Pr[N_{iT_i+1}>k|\widehat{N}_{i,T_i+1}=p'] \text{ whatever } k, \text{ provided } p \leq p'.$$

This means that the linear credibility premium is indeed a good predictor of the future claim number in model A1–A2. Basically, we prove that increasing the linear credibility premium (i.e. degrading the claim record of the policyholder) makes the probability of observing more claims in the future greater.

3.6 Further Reading and Bibliographic Notes

3.6.1 Credibility Models

Credibility theory began with the papers by Mowbray (1914) and Whitney (1918). These papers purposed to derive a premium which was a balance between the experience of an individual risk and of a class of risks. An excellent introduction to credibility theory can be found, e.g., in Goovaerts & Hoogstad (1987), Herzog (1994), Dannenburg, Kaas & Goovaerts (1996), Klugman, Panjer & Willmot (2004, Chapter 16) and Bühlmann & Gisler (2005). See also Norberg (2004) for an overview with useful references and links to Bayesian statistics and linear estimation. The underlying assumption of credibility theory which sets it apart from formulas based on the individual risk alone is that the risk parameter is regarded as a random variable. This naturally leads to a Bayesian approach to credibility theory. The book by Klugman (1992) provides an in-depth treatment of the question. See also the review papers by Makov *et al.* (1996) and Makov (2002). The connection between credibility formulas and Mellin transforms in the Poisson case has been established by Albrecht (1984).

In a couple of seminal papers, Dionne & Vanasse (1989, 1992) proposed a credibility model which integrates *a priori* and *a posteriori* information on an individual basis. The unexplained heterogeneity was then modelled by the introduction of a latent variable representing the influence of hidden policy characteristics. Taking this random effect to be Gamma distributed yields the Negative Binomial model for the claim number. An excellent summary of the statistical models that may lead to experience rating in insurance can be found in Pinquet (2000). The nature of serial correlation (endogeneous or exogeneous) is discussed there.

There are many applications of credibility techniques to various branches of insurance. Let us mention a nonstandard one, by Rejesus *et al.* (2006). These authors examined the feasibility of implementing an experience-based premium rate discount in crop insurance.

3.6.2 Claim Count Distributions

Other credibility models for claim counts can be found in the literature, going beyond the mixed Poisson model studied in this chapter. The model suggested by SHENGWANG, WEI & WHITMORE (1999) employs the Negative Binomial distribution for the conditional distribution of the annual claim numbers together with a Pareto structure function. Some credibility models are designed for stratified portfolios. DESJARDINS, DIONNE & PINQUET (2001) considered fleets of vehicles, and used individual characteristics of both the vehicles and the carriers. See also ANGERS, DESJARDINS, DIONNE & GUERTIN (2006).

An interesting alternative to the Negative Binomial model can be obtained using the conditional specification technique introduced by ARNOLD, CASTILLO & SARABIA (1999). The idea is to specify the joint distribution of (N_t, Θ) through its conditionals. More precisely, the conditional distribution of N_t given $\Theta = \theta$ is $\mathcal{P}oi(\gamma(\theta))$ for some function $\gamma : \mathbb{R}^+ \to \mathbb{R}^+$, and the conditional distribution of Θ given $N_t = k$ is $\mathcal{G}am(\alpha(k), \beta(k))$ where $\alpha(\cdot)$ and $\beta(\cdot)$ are two functions mapping $\mathbb{N}$ to $\mathbb{R}^+$. For an application of the model to experience rating, see SARABIA, GOMEZ-DENIZ & VAZQUEZ-POLO (2004).

3.6.3 Loss Functions

The quadratic loss function is by far the most widely used in practice. The results with the exponential loss function are taken from BERMÚDEZ, DENUIT & DHAENE (2000). Early references about the use of this kind of loss function include FERREIRA (1977) and LEMAIRE (1979). MORILLO & BERMUDEZ (2003) used an exponential loss function in connection with the Poisson-Inverse Gaussian model.

Other loss functions can be envisaged. YOUNG (1998a) uses a loss function that is a linear combination of a squared-error term and a second-derivative term. The squared-error term measures the accuracy of the estimator, while the second-derivative term constrains the estimator to be close to linear. See also YOUNG & DE VYLDER (2000), where the loss function is a linear combination of a squared-error term and a term that encourages the estimator to be close to constant, especially in the tails of the distribution of claims, where YOUNG (1997) noted the difficulty with her semiparametric approach. YOUNG (2000) resorts to a loss function that can be decomposed into a squared-error term and a term that encourages the credibility premium to be constant. This author shows that by using this loss function, the problem of upward divergence noted in YOUNG (1997) is reduced. See also YOUNG (1998b). YOUNG (2000) also provides a simple routine for minimizing the loss function, based on the discussion of De Vylder in YOUNG (1998a).

Adopting the semiparametric model proposed in YOUNG (1997, 2000) but considering that the piecewise linear function has better characteristics in simplicity and intuition than the kernel, HUANG, SONG & LIANG (2003) used the piecewise linear function as the estimate of the prior distribution and to obtain the estimates for the credibility formula.

3.6.4 Credibility and Regression Models

HACHEMEISTER (1975) contributed to the credibility context by introducing a regression model. Since DE VYLDER (1985), it has been known that credibility formulas can be recovered from appropriate (non-)linear statistical regression models. More recently, NELDER

& VERRALL (1997) recognized that credibility theory can be encompassed within the theory of hierarchical generalized linear models developed by LEE & NELDER (1996). This extends to the GLM family the pioneering work by NORBERG (1986). The likelihood-based approach can be carried out using standard statistical packages. All the assumptions underlying the model can be checked (e.g., using appropriate residual analyses), avoiding dogmatic application of risk theory models. The mean random effects are estimated for each policyholder by maximizing the hierarchical likelihood. With an appropriate choice for the distribution of the random effects and using the canonical link function, the estimate is in the form of a linear credibility premium. See also BÜHLMANN & BÜHLMANN (1999) and LUO, YOUNG & FREES (2004). FREES, YOUNG & LUO (1999) developed links between credibility theory and statistical models for longitudinal (or panel) data, as explained below.

FREES & WANG (2006) used a longitudinal data set-up, so that the experience from several risk classes are observed over time, whereas FREES *ET AL.* (1999) focussed on credibility predictors that are linear combinations of the data and/or that are minimizers of a squared-error loss function. In contrast, FREES & WANG (2006) considered severity distributions that may be long-tailed so that averaging or using squared-error loss do not yield appropriate prediction tools.

ANTONIO & BEIRLANT (2007) suggested the use of the generalized linear mixed models (where a transformation of the mean is expressed as a linear function of both fixed and random effects) in credibility theory. Many actuarial credibility models appear to be particular cases of generalized linear mixed models. YEO & VALDEZ (2006) addressed a simultaneous dependence of claims across individuals for a fixed time period and across time periods for a fixed individual. This is accomplished by introducing the notion of a common effect affecting all individuals and another common effect affecting a fixed individual over time. This construction falls within the broader framework of generalized linear mixed models.

LO, FUNG & ZHU (2006) considered a regression credibility model with random regression coefficients. The variance components represented by the uncertainty about the regression coefficients then account for the heterogeneity in risks borne by policyholders across contracts. From a different perspective, the dependence between contracts has been introduced by treating the contract-specific regression coefficients as being generated by the same random mechanism such that they are random deviations from the collective mean.

Autoregressive specifications of the error structure in the credibility context have been proposed by BOLANCÉ *ET AL.* (2003). In SUNDT (1983), the generalized Bühlmann–Straub model was proposed with consecutive error terms assumed to follow AR(1) dependences.

QIAN (2000) used the nonparametric regression method to establish estimators for credibility premiums under some principles of premium calculation. The asymptotic properties of the estimators are studied in this paper.

3.6.5 Credibility and Copulas

Copulas are a powerful tool to model dependencies between multivariate outcomes. See, e.g., DENUIT *ET AL.* (2005) for an introduction. Several works have successfully applied copulas to solve credibility problems. Let us mention FREES & WANG (2005) who handled serial (time) dependence through a t-copula, and FREES & WANG (2006) who extended that formulation by introducing elliptical copulas for serial dependencies. Like the t-copula, the elliptical copulas turn out to have an analytically tractable form for predictive distributions.

Multivariate credibility models may be considered for several lines of business, or several types of claims. Multivariate credibility models are discussed, e.g., in Frees (2003). This topic will be treated in Chapter 6.

3.6.6 Time Dependent Random Effects

The vast majority of the papers which have appeared in the actuarial literature considered time-independent heterogeneous models. This chapter is restricted to the case of static random effects: in the classical credibility construction A1–A2 of Definition 3.1, the risk parameter Θ_i relating to policyholder i is assumed to be constant over time. This is of course rather unrealistic since driving ability may vary during the driving career (because of the learning effect, or modification in the risk characteristics). In automobile insurance, an unknown underlying random parameter that develops over time expresses the fact that the abilities of a driver are not constant. Moreover, the hidden exogeneous variables revealed by claims experience may vary with time, as do observable ones.

Another reason to allow for random effects that vary with time relates to moral hazard. Indeed, individual efforts to prevent accidents are unobserved and feature temporal dependence. The policyholders may adjust their efforts for loss prevention according to their experience with past claims, the amount of premium and awareness of future consequences of an accident (due to experience rating schemes). The effort variable determines the moral hazard and is modelled by a dynamic unobserved factor.

Of course, it is hopeless in practice to discriminate between residual heterogeneity due to unobservable characteristics of drivers that significantly affect the risk of accident, and their individual efforts to prevent such accidents. Both effects get mixed in the latent process. Since the observed contagion between annual claim numbers is always positive, the effect of omitted explanatory variables seems to dominate moral hazard. Anyway, this issue has no practical implication since predictions depend on observed contagion, but not on its nature.

Hence, instead of assuming that the risk characteristics are given once and for all by a single risk parameter, we might suppose that the unknown risk characteristics of each policy are described by dynamic random effects. In the terminology of Jewell (1975), evolutionary credibility models allow for random effects to vary in successive periods. Now, the ith policy of the portfolio, $i = 1, 2, \ldots, n$, is represented by a double sequence $(\boldsymbol{\Theta}_i, \boldsymbol{N}_i)$ where $\boldsymbol{\Theta}_i$ is a positive random vector with unit mean representing the unexplained heterogeneity. Specifically, the model is based on the following assumptions:

B1 Given $\boldsymbol{\Theta}_i = \boldsymbol{\theta}_i$, the random variables N_{it}, $t = 1, 2, \ldots, T_i$, are independent and conform to the Poisson distribution with mean $\lambda_{it}\theta_{it}$, i.e.

$$\Pr[N_{it} = k|\Theta_{it} = \theta_{it}] = \exp(-\lambda_{it}\theta_{it})\frac{(\lambda_{it}\theta_{it})^k}{k!}, \quad k = 0, 1, \ldots$$

with $\lambda_{it} = d_{it}\exp(\boldsymbol{\beta}^T\widetilde{\boldsymbol{x}}_{it})$.

B2 At the portfolio level, the $\boldsymbol{\Theta}_i$s are assumed to be independent. Moreover, $(\Theta_{i1}, \ldots, \Theta_{iT_i})$ is distributed as $(\Theta_1, \ldots, \Theta_{T_i})$ where $\boldsymbol{\Theta} = (\Theta_1, \ldots, \Theta_{T_{\max}})^T$ is a stationary random vector (with $T_{\max} = \max T_i$). It is further assumed that $\mathbb{E}[\Theta_{it}] = 1$ for all i, t. The unit mean condition is imposed for identification (otherwise, the mean could be absorbed into the intercept term of the Poisson regression). This condition means that the *a priori* ratemaking is correct on average.

B3 The sequences (Θ_i, N_i), $i = 1, 2, \ldots, n$, are assumed to be independent.

PURCARU & DENUIT (2003) studied the kind of dependence arising in these credibility models for claim counts. ALBRECHT (1985) studied such credibility models for claim counts, whereas GERBER & JONES (1975) and SUNDT (1981,1988) dealt with general random variables.

A fundamental difference between static A1–A3 and dynamic B1–B3 credibility models is that the latter incorporate the age of the claims in the risk prediction, whereas the former neglect this information. Since we intuitively feel that the predictive ability of a claim should decrease with its age, dynamic specification seems more acceptable. As pointed out by PINQUET, GUILLÉN & BOLANCÉ (2001), dynamic credibility models agree with economic analysis of multiperiod optimal insurance under moral hazard. In this optic, the stationarity of the Θ_is implies that the predictive ability of claims depends mainly on the lag between the date of prediction and the date of occurrence (because of time translation invariance of the marginals of the Θ_is).

Empirical studies performed on panel data, as in PINQUET, GUILLÉN & BOLANCÉ (2001) and BOLANCÉ, GUILLÉN & PINQUET (2003), support time-varying (or dynamic) random effects. An interesting feature of credibility premium derived from stationary random effects with a decreasing correlogram is that the age of the claims are taken into account in the *a posteriori* correction: a recent claim will be more penalized than an old one (whereas the age of the claim is not taken into account with static random effects).

This kind of *a posteriori* correction reconciles actuaries' and economists' approaches to experience rating. HENRIET & ROCHET (1986) distinguished two roles played by *a posteriori* corrections, showing that these two roles involve different rating structures. The first role deals with the problem of adverse selection, where the very aim is to evaluate as accurately as possible the true distribution of reported accidents. This is the classical actuarial perspective. The second role is linked to moral hazard and implies that the distribution of reported accidents over time must be taken into account to maintain incentives to drive carefully. This means that more weight must be given to recent information in order to maintain such incentives. This is the economic point of view. The credibility model B1–B3 with dynamic random effects, although theoretically more intricate, takes these two objectives into account.

3.6.7 Credibility and Panel Data Models

FREES, YOUNG & LUO (1999) developed links between credibility theory and longitudinal (or panel) data models. They demonstrated how longitudinal data models can be applied to the credibility ratemaking problem. As pointed out by these authors, by expressing credibility ratemaking applications in the framework of longitudinal data models, actuaries can realize several benefits: (1) Longitudinal data models provide a wide variety of models from which to choose. (2) Standard statistical software makes analysing data relatively easy. (3) Actuaries have another method for explaining the ratemaking process. (4) Actuaries can use graphical and diagnostic tools to select a model and assess its usefulness.

3.6.8 Credibility and Empirical Bayes Methods

Credibility theory has an empirical Bayes flavour, as pointed out by NORBERG (1980). Analyses commonly used in highway safety include the widely applied empirical Bayes method. According to LORD (2006), this method has become increasingly popular since it corrects for the regression-to-the-mean bias, refines the predicted mean of an entity, and is relatively simple to manipulate compared to the full Bayes approach. The empirical Bayes method combines information obtained from a reference group having similar characteristics with the information specific to the individual under study. A weight factor is assigned to both the reference population and the individual under study, and a credibility formula is obtained.

Credibility theory has thus a clear empirical Bayes flavour. We refer the interested reader to QUIGLEY, BEDFORD & WALLS (2006) for a case study involving the rate of occurrence of train derailments within the United Kingdom.

4

Bonus-Malus Scales

4.1 Introduction

4.1.1 From Credibility to Bonus-Malus Scales

One of the main tasks of the actuary is to design a tariff structure that will fairly distribute the burden of claims among policyholders. To this end, he often has to partition all policies into risk classes with all policyholders belonging to the same class paying the same premium. It is convenient to achieve *a priori* classification by resorting to generalized linear models (e.g. Poisson regression for claim counts), as explained in Chapter 2. However, many important factors cannot be taken into account at this stage. Consequently, risk classes are still quite heterogeneous despite the use of many *a priori* variables.

Rating systems penalizing insureds responsible for one or more accidents by premium surcharges (or *maluses*), and rewarding claim-free policyholders by awarding them discounts (or *bonuses*) are now in force in many developed countries. Besides encouraging policyholders to drive carefully (i.e. counteracting moral hazard), they aim to better assess individual risks. The amount of premium is adjusted each year on the basis of the individual claims experience using techniques from credibility theory, as shown in Chapter 3.

However, credibility formulas are difficult to implement in practice, because of their mathematical complexity (complexity refers here to the sphere of commercial relations, where customers are often reluctant in using mechanisms that they consider to be complex, especially in connection with insurance products). For this reason, bonus-malus scales have been proposed by insurance companies. Such scales have to be seen as commercial versions of credibility formulas. The typical customer can figure out what the premium will be for any given claims history.

4.1.2 The Nature of Bonus-Malus Scales

When a merit rating plan is in force, the amount of premium paid by the policyholder depends on the rating factors of the current period but also on claim history. In practice, a bonus-malus scale consists of a finite number of levels, each with its own relative premium. New policyholders have access to a specified level. After each year, the policy moves up or down according to transition rules and to the number of claims at fault. The premium charged to a policyholder is obtained by applying the relative premium associated to his current level in the scale to a base premium depending on his observable characteristics incorporated into the price list.

4.1.3 Relativities

The problem addressed in this chapter is the determination of the relative premiums attached to each of the levels of the scale when *a priori* classification is used by the company. The relativity associated to level ℓ is denoted as r_ℓ. The meaning is that a policyholder occupying level ℓ in the bonus-malus scale has to pay r_ℓ times the base premium to be covered by the insurance company.

The severity of the *a posteriori* corrections must depend on the extent to which amounts of premiums vary according to observable characteristics of policyholders. The key idea is that both *a priori* classification and *a posteriori* corrections aim to create tariff cells as homogeneous as possible. The residual heterogeneity inside each of these cells being smaller for insurers incorporating more variables in their *a priori* ratemaking, the *a posteriori* corrections must be softer for those insurers.

The framework of credibility theory, with its fundamental notion of randomly distributed risk parameters, was employed in analysis of bonus-malus systems by Pesonen as early as 1963. In this chapter, we will keep the framework of Definition 3.1. According to NORBERG (1976), once the number of classes, the starting level and the transition rules have been fixed, the optimal relativity associated with level ℓ is determined by maximizing the asymptotic predictive accuracy. Formally, the relativities minimize the mean squared deviation between a policy's expected claim frequency and its premium in the year t as $t \to +\infty$. The optimal relativity for level ℓ is thus equal to the conditional expected risk parameter for an infinitely old policy, given that the policy is in level ℓ.

4.1.4 Bonus-Malus Scales and Markov Chains

In most of the commercial bonus-malus systems, the knowledge of the current level and the number of claims during the current period suffice to determine the next level in the scale. So the future (the level for year $t+1$) depends only on the present (the level for year t and the number of accidents reported during year t) and not on the past. This is closely related to the memoryless property of the Markov chains. If the claim numbers in different years are (conditionally) independent then the trajectory of a given policyholder in the bonus-malus scale will be a (conditional) Markov chain. Sometimes, fictitious levels have to be introduced to recover the memoryless property.

The treatment of bonus-malus scales is best performed in the framework of Markov chains. This chapter is nevertheless self-contained, and does not require any prior knowledge of this topic. All the useful results have been derived in an elementary way (the readers having acquaintance with the theory of Markov chains will rapidly recognize all the classical machinery taught in textbooks devoted to stochastic processes).

4.1.5 *Financial Equilibrium*

Exactly as for credibility mechanisms, it is important that the relativities average to 100 %, resulting in financial equilibrium. This fundamental property is highly desirable: it guarantees that the introduction of a bonus-malus system has no impact on the yearly premium collection. The distribution of the amounts paid by the policyholders is modified according to the reported claims but on the whole, the company gets the same amount of money.

Throughout this chapter, we work with the long-run equilibrium distribution of policyholders in the bonus-malus levels. We will see that in the long run, the way the relativities are computed in this chapter ensures that the bonus-malus system is financially stable. Things are however more complicated in practice. Specifically, some undesirable phenomena can arise in a transient regime. These issues will be addressed in Chapters 8–9.

4.1.6 *Agenda*

In Section 4.2, the trajectory of the policyholder accross the bonus-malus levels is modelled as a Markov chain. Section 4.3 is devoted to transition probabilities, that is, the probability that the policyholder moves from one level to another over a given time horizon. The long-term behaviour of the scale is studied in Section 4.4. It is shown there that the proportions of policyholders in each level of the scale tend to stabilize over time. Various methods to compute the stationary probabilities are described.

Section 4.5 explains how to compute the relativities using a quadratic loss function. As for credibility formulas, relativities that are linear in the bonus-malus level are also derived. Section 4.5.3 examines the interaction between the bonus-malus scale and *a priori* risk classification. It is shown there that creating several scales decreases the rating inadequacies.

In Section 4.6, the quadratic loss function is replaced with an exponential one. A comparison with quadratic relativities is performed, and the influence of the severity parameter is carefully assessed.

In Section 4.7, we will consider the so-called special bonus rule. According to this rule, a policyholder who did not report any claim for a certain number of years, and is still in the malus zone (i.e. in a level with a relativity above 100 %) is automatically sent to the initial level (i.e. to the level with relativity equal to 100 %). Many compulsory systems formerly imposed by governments included such a rule. Because of this special rule, the stochastic process describing the trajectory of the drivers accross the levels is no longer Markovian. The memoryless property can nevertheless be re-obtained by adding fictitious levels in the scale. To fix the ideas, we will study the special bonus rule in the former compulsory Belgian system.

In a competitive market, it can be expected that some policyholders switch from one insurance company to the other. In a regulated framework, with a unique compulsory bonus-malus system imposed on all the insurance companies, the drivers will be subject to the same *a posteriori* corrections whatever the insurer. If a driver decides to switch to another insurer, he must first obtain a certificate from the former insurer stating his attained bonus-malus level and whether pending claims could affect this level. The new insurer must then award the same discount or apply the same surcharges. The competition between insurance companies is limited to the services offered and the *a priori* premiums.

Things become more complicated in a deregulated market, where each insurer is free to design its own bonus-malus system. Then, insurers compete also on the basis of *a posteriori* corrections. It rapidly becomes extremely difficult for policyholders to determine the optimal insurance provider, since companies apply different penalties when claims are reported. In Section 4.8, we consider a policyholder switching from insurer A to insurer B. He occupies the level ℓ_1 in the bonus-malus scale used by company A, and the question is where to place him in the bonus-malus scale used by company B.

Section 4.9 examines the dependence properties existing between the successive levels occupied by the policyholders and the random risk parameter. It is argued that contrarily to the results obtained with credibility models, the risk parameters do not necessarily increase with the level occupied in the scale.

The final Section 4.10 gives references and addresses further issues.

4.2 Modelling Bonus-Malus Systems

4.2.1 Typical Bonus-Malus Scales

Before embarking on an abstract definition of bonus-malus systems, let us discuss a couple of examples that will be used throughout this chapter.

Example 4.1 (−1/Top Scale) This bonus-malus scale has 6 levels (numbered 0 to 5). Policyholders are classified according to the number of claim-free years since their last claim (0, 1, 2, 3, 4 or at least 5). After a claim all premium reductions are lost. The transition rules are described in Table 4.1. Specifically, the starting class is the highest level 5. Each claim-free year is rewarded by one bonus class. When an accident is reported, all the discounts are lost and the policyholder is transferred to level 5.

Note that the philosophy behind such a bonus-malus system is different from credibility theory. Indeed, this bonus-malus scale only aims to counteract moral hazard: it is in fact more or less equivalent to a deductible which is not paid at once but smoothed over the time

Table 4.1 Transition rules for the scale −1/top.

Starting level	Level occupied if 0 claim is reported	Level occupied if ≥ 1 claim is reported
0	0	5
1	0	5
2	1	5
3	2	5
4	3	5
5	4	5

needed to go back to the lowest class. Note however that this 'smoothed' deductible only applies to the first claim: subsequent claims are 'for free'.

Example 4.2 (−1/+2 Scale) There are 6 levels. Level 5 is the starting level. A higher level number indicates a higher premium. The discount per claim-free year is one level: if no claims have been reported by the policyholder then he moves one level down. The penalty per claim is two levels. If a number of claims, $n_t > 0$, has been reported during year t then the policyholder moves $2n_t$ levels up. The transition rules are described in Table 4.2.

In the subsequent sections, we will also make the $-1/+2$ scale more severe, by penalizing each claim by 3 levels instead of 2. This alternative bonus-malus system will be henceforth referred to as the $-1/+3$ system.

4.2.2 Characteristics of Bonus-Malus Scales

The bonus-malus scales investigated in this book are assumed to possess $s+1$ levels, numbered from 0 to s. A specified level is assigned to a new driver. In practice, the initial level may depend upon the use of the vehicle (or upon another observable risk characteristic). Each claim-free year is rewarded by a bonus point (i.e. the driver goes one level down). Claims are penalized by malus points (i.e. the driver goes up a certain number of levels each time he files a claim). We assume that the penalty is a given number of classes per claim. This facilitates the mathematical treatment of the problem. More general systems can nevertheless be considered, with higher penalties for subsequent claims. After sufficiently many claim-free years, the driver enters level 0 where he enjoys the maximal bonus.

In Chapter 3, updating premiums with credibility formulas only uses the total number of claims reported by the policyholder in the past. The new premium does not depend on the way the accidents are distributed over the years. This property is never satisfied by bonus-malus systems, where it would be the policyholder's interest to concentrate all the claims during a single year.

In commercial bonus-malus systems, the knowledge of the present level and of the number of claims of the present year suffices to determine the next level. Together with the (conditional) independence of annual claim numbers, this ensures that the trajectory accross the bonus-malus levels may be represented by a (conditional) Markov chain: the

Table 4.2 Transition rules for the scale $-1/+2$.

Starting level	Level occupied if 0	1	2	≥3
	claim(s) is/are reported			
5	4	5	5	5
4	3	5	5	5
3	2	5	5	5
2	1	4	5	5
1	0	3	5	5
0	0	2	4	5

future (the level for year $t+1$) depends on the present (the level for year t and the number of accidents reported during year t) and not on the past (the complete claim history and the levels occupied during years $1, 2, \ldots, t-1$). Sometimes, fictitious levels have to be introduced in order to meet this memoryless property. Indeed, in some bonus-malus systems, policyholders occupying high levels are sent to the starting class after a few years without claims. This issue will be addressed in Section 4.7.

4.2.3 Trajectory

New drivers start in level ℓ_0 of the scale. Note that experienced drivers arriving in the portfolio are not necessarily placed in level ℓ_0, but in a level corresponding to their claim history or to the level occupied in the bonus-malus scale used by a competitor. This problem will be dealt with in Section 4.8.

The trajectory of the policyholder in the bonus-malus scale is modelled by a sequence $\{L_1, L_2, \ldots\}$ of random variables valued in $\{0, 1, \ldots, s\}$, such that L_k is the level occupied during the $(k+1)$th year, i.e. during the time interval $(k, k+1)$. Since movements in the scale occur once a year (at policy anniversary), the policyholder occupies level L_k from time k until time $k+1$. Once the number N_k of claims reported by the policyholder during $(k-1, k)$ is known, this information is used to reevaluate the position of the driver in the scale. We supplement the sequence of the L_ks with $L_0 = \ell_0$.

The L_ks obviously depend on the past numbers of claims $N_1, N_2, \ldots, N_k$ reported by the policyholder. If we denote as 'pen' the penalty induced by each claim (expressed as a number of levels), then the L_ks obey the recursion

$$
\begin{aligned}
L_k &= \begin{cases} \max\{L_{k-1} - 1, 0\} \text{ if } N_k = 0 \\ \min\{L_{k-1} + N_k \times \text{pen}, s\} \text{ if } N_k \geq 1 \end{cases} \\
&= \max\Big\{ \min\big\{L_{k-1} - (1 - I_k) + N_k \times \text{pen}, s\big\}, 0 \Big\}
\end{aligned}
$$

where

$$
I_k = \begin{cases} 1 \text{ if } N_k \geq 1, \\ 0 \text{ otherwise,} \end{cases}
$$

indicates whether at least one claim has been reported in year k. This is an example of a stochastic recursive equation. This representation of the L_ks clearly shows that the future trajectory of the policyholder in the scale is independent of the levels occupied in the past, provided that the present level is given. This conditional independence property is at the heart of Markov models.

The stochastic recursive equations given above assume that the bonus is lost in case at least one claim is filed with the company. In some cases (like the former compulsory Belgian bonus-malus scale), the bonus is granted in any case. The L_ks then obey the recursion

$$
L_k = \max\Big\{ \min\big\{L_{k-1} - 1 + N_k \times \text{pen}, s\big\}, 0 \Big\}.
$$

This means that the first claim is penalized by pen−1 levels, and the subsequent ones by pen levels.

4.2.4 Transition Rules

The probability of moving from one level to another depends on the number of claims reported during the current year. Therefore, we can introduce more formally the transition rules which impose the transfer from one level to another level once the number of claims is known. If k claims are reported,

$$t_{ij}(k) = \begin{cases} 1, & \text{if the policy gets transferred from level } i \text{ to level } j, \\ 0, & \text{otherwise.} \end{cases}$$

The $t_{ij}(k)$s are put in matrix form $\boldsymbol{T}(k)$, i.e.

$$\boldsymbol{T}(k) = \begin{pmatrix} t_{00}(k) & t_{01}(k) & \cdots & t_{0s}(k) \\ t_{10}(k) & t_{11}(k) & \cdots & t_{1s}(k) \\ \vdots & \vdots & \ddots & \vdots \\ t_{s0}(k) & t_{s1}(k) & \cdots & t_{ss}(k) \end{pmatrix}.$$

Then, $\boldsymbol{T}(k)$ is a 0-1 matrix having in each row exactly one 1.

Example 4.3 (−1/Top Scale) In this case, we have

$$\boldsymbol{T}(0) = \begin{pmatrix} 1 & 0 & 0 & 0 & 0 & 0 \\ 1 & 0 & 0 & 0 & 0 & 0 \\ 0 & 1 & 0 & 0 & 0 & 0 \\ 0 & 0 & 1 & 0 & 0 & 0 \\ 0 & 0 & 0 & 1 & 0 & 0 \\ 0 & 0 & 0 & 0 & 1 & 0 \end{pmatrix}, \quad \boldsymbol{T}(1) = \begin{pmatrix} 0 & 0 & 0 & 0 & 0 & 1 \\ 0 & 0 & 0 & 0 & 0 & 1 \\ 0 & 0 & 0 & 0 & 0 & 1 \\ 0 & 0 & 0 & 0 & 0 & 1 \\ 0 & 0 & 0 & 0 & 0 & 1 \\ 0 & 0 & 0 & 0 & 0 & 1 \end{pmatrix}$$

and $\boldsymbol{T}(k) = \boldsymbol{T}(1)$ for all $k \geq 2$.

Example 4.4 (−1/+2 Scale) In this case, we have

$$\boldsymbol{T}(0) = \begin{pmatrix} 1 & 0 & 0 & 0 & 0 & 0 \\ 1 & 0 & 0 & 0 & 0 & 0 \\ 0 & 1 & 0 & 0 & 0 & 0 \\ 0 & 0 & 1 & 0 & 0 & 0 \\ 0 & 0 & 0 & 1 & 0 & 0 \\ 0 & 0 & 0 & 0 & 1 & 0 \end{pmatrix}, \quad \boldsymbol{T}(1) = \begin{pmatrix} 0 & 0 & 1 & 0 & 0 & 0 \\ 0 & 0 & 0 & 1 & 0 & 0 \\ 0 & 0 & 0 & 0 & 1 & 0 \\ 0 & 0 & 0 & 0 & 0 & 1 \\ 0 & 0 & 0 & 0 & 0 & 1 \\ 0 & 0 & 0 & 0 & 0 & 1 \end{pmatrix}$$

$$\boldsymbol{T}(2) = \begin{pmatrix} 0 & 0 & 0 & 0 & 1 & 0 \\ 0 & 0 & 0 & 0 & 0 & 1 \\ 0 & 0 & 0 & 0 & 0 & 1 \\ 0 & 0 & 0 & 0 & 0 & 1 \\ 0 & 0 & 0 & 0 & 0 & 1 \\ 0 & 0 & 0 & 0 & 0 & 1 \end{pmatrix} \text{ and } \boldsymbol{T}(k) = \begin{pmatrix} 0 & 0 & 0 & 0 & 0 & 1 \\ 0 & 0 & 0 & 0 & 0 & 1 \\ 0 & 0 & 0 & 0 & 0 & 1 \\ 0 & 0 & 0 & 0 & 0 & 1 \\ 0 & 0 & 0 & 0 & 0 & 1 \\ 0 & 0 & 0 & 0 & 0 & 1 \end{pmatrix} \text{ for all } k \geq 3.$$

4.3 Transition Probabilities

4.3.1 Definition

Let us now assume that $N_1, N_2, \ldots$ are independent and $\mathcal{P}oi(\vartheta)$ distributed. The trajectory will be denoted as $\{L_1(\vartheta), L_2(\vartheta), \ldots\}$ to emphasize the dependence upon the annual expected claim frequency ϑ. Note however that the argument ϑ in $L_k(\vartheta)$ does not mean that the $L_k(\vartheta)$s are functions of the parameter ϑ, but only that their distribution depends on ϑ.

Let $p_{\ell_1\ell_2}(\vartheta)$ be the probability of moving from level ℓ_1 to level ℓ_2 for a policyholder with annual mean claim frequency ϑ, that is,

$$p_{\ell_1\ell_2}(\vartheta) = \Pr[L_{k+1}(\vartheta) = \ell_2 | L_k(\vartheta) = \ell_1]$$

with $\ell_1, \ell_2 \in \{0, 1, \ldots, s\}$. Clearly, the $p_{\ell_1\ell_2}(\vartheta)$s satisfy

$$p_{\ell_1\ell_2}(\vartheta) \geq 0, \text{ for all } \ell_1 \text{ and } \ell_2, \text{ and } \sum_{\ell_2=0}^{s} p_{\ell_1\ell_2}(\vartheta) = 1. \tag{4.1}$$

Moreover, the transition probabilities can be expressed using the $t_{ij}(\cdot)$s introduced above. To see this, it suffices to write

$$\begin{aligned} p_{\ell_1\ell_2}(\vartheta) &= \sum_{n=0}^{+\infty} \Pr[L_{k+1}(\vartheta) = \ell_2 | N_{k+1} = n, L_k(\vartheta) = \ell_1] \Pr[N_{k+1} = n | L_k(\vartheta) = \ell_1] \\ &= \sum_{n=0}^{\infty} \frac{\vartheta^n}{n!} \exp(-\vartheta) t_{\ell_1\ell_2}(n). \end{aligned}$$

Note that we have used the fact that N_{k+1} and $L_k(\vartheta)$ are independent (since $L_k(\vartheta)$ depends on $N_1, \ldots, N_k$), so that

$$\Pr[N_{k+1} = n | L_k(\vartheta) = \ell_1] = \Pr[N_{k+1} = n] = \frac{\vartheta^n}{n!} \exp(-\vartheta).$$

The transition probabilities allow the actuary to compute the probability of any trajectory in the scale. Specifically, since the probability that a certain policyholder with expected annual claim frequency ϑ is in level $\ell_1, \ldots, \ell_n$ at time $1, \ldots, n$ is simply the probability of going from ℓ_0 to ℓ_n via the intermediate levels $\ell_1, \ldots, \ell_{n-1}$, we have

$$\Pr[L_1(\vartheta) = \ell_1, \ldots, L_n(\vartheta) = \ell_n | L_0(\vartheta) = \ell_0] = p_{\ell_0\ell_1}(\vartheta) \cdots p_{\ell_{n-1}\ell_n}(\vartheta). \tag{4.2}$$

Furthermore, it is enough to know the current position in the scale to determine the probability of being transferred to any other level in the bonus-malus scale. Formally,

$$\Pr[L_n(\vartheta) = \ell_n \mid L_{n-1}(\vartheta) = \ell_{n-1}, \ldots, L_0(\vartheta) = \ell_0] = p_{\ell_{n-1}\ell_n}(\vartheta),$$

whenever $\Pr[L_{n-1}(\vartheta) = \ell_{n-1}, \ldots, L_0(\vartheta) = \ell_0] > 0$.

4.3.2 Transition Matrix

Further, $\boldsymbol{P}(\vartheta)$ is the one-step transition matrix, i.e.

$$\boldsymbol{P}(\vartheta) = \begin{pmatrix} p_{00}(\vartheta) & p_{01}(\vartheta) & \cdots & p_{0s}(\vartheta) \\ p_{10}(\vartheta) & p_{11}(\vartheta) & \cdots & p_{1s}(\vartheta) \\ \vdots & \vdots & \ddots & \vdots \\ p_{s0}(\vartheta) & p_{s1}(\vartheta) & \cdots & p_{ss}(\vartheta) \end{pmatrix}.$$

From (4.1), we see that the matrix $\boldsymbol{P}(\vartheta)$ is a stochastic matrix. As already mentioned, the future level of a policyholder is independent of its past levels and only depends on its present level (and also on the number of claims reported during the present year).

In matrix form, we can write $\boldsymbol{P}(\vartheta)$ as

$$\boldsymbol{P}(\vartheta) = \sum_{k=0}^{\infty} \frac{\vartheta^k}{k!} \exp(-\vartheta) \boldsymbol{T}(k)$$

provided the N_ts are independent and $\mathcal{P}oi(\vartheta)$ distributed.

Example 4.5 (−1/Top Scale) The transition matrix $\boldsymbol{P}(\vartheta)$ associated with this bonus-malus system is given by

$$\boldsymbol{P}(\vartheta) = \begin{pmatrix} \exp(-\vartheta) & 0 & 0 & 0 & 0 & 1-\exp(-\vartheta) \\ \exp(-\vartheta) & 0 & 0 & 0 & 0 & 1-\exp(-\vartheta) \\ 0 & \exp(-\vartheta) & 0 & 0 & 0 & 1-\exp(-\vartheta) \\ 0 & 0 & \exp(-\vartheta) & 0 & 0 & 1-\exp(-\vartheta) \\ 0 & 0 & 0 & \exp(-\vartheta) & 0 & 1-\exp(-\vartheta) \\ 0 & 0 & 0 & 0 & \exp(-\vartheta) & 1-\exp(-\vartheta) \end{pmatrix}.$$

Example 4.6 (−1/+2 Scale) The transition matrix $\boldsymbol{P}(\vartheta)$ associated with this bonus-malus system is given by

$$\boldsymbol{P}(\vartheta) = \begin{pmatrix} \exp(-\vartheta) & 0 & \vartheta\exp(-\vartheta) & 0 & \frac{\vartheta^2}{2}\exp(-\vartheta) & 1-\Sigma_1 \\ \exp(-\vartheta) & 0 & 0 & \vartheta\exp(-\vartheta) & 0 & 1-\Sigma_2 \\ 0 & \exp(-\vartheta) & 0 & 0 & \vartheta\exp(-\vartheta) & 1-\Sigma_3 \\ 0 & 0 & \exp(-\vartheta) & 0 & 0 & 1-\exp(-\vartheta) \\ 0 & 0 & 0 & \exp(-\vartheta) & 0 & 1-\exp(-\vartheta) \\ 0 & 0 & 0 & 0 & \exp(-\vartheta) & 1-\exp(-\vartheta) \end{pmatrix}.$$

where Σ_i represents the sum of the elements in columns 1 to 5 in row i, $i = 1, 2, 3$, that is,

$$\Sigma_1 = \exp(-\vartheta)\left(1 + \vartheta + \frac{\vartheta^2}{2}\right)$$

$$\Sigma_2 = \Sigma_3 = \exp(-\vartheta)(1 + \vartheta).$$

4.3.3 Multi-Step Transition Probabilities

The probability

$$p_{ij}^{(n)}(\vartheta) = \Pr[L_{k+n}(\vartheta) = j | L_k(\vartheta) = i]$$

evaluates the likelihood of being transferred from level i to level j in n steps. Note that this is the probability that $L_{k+n}(\vartheta) = j$ given $L_k(\vartheta) = i$ for any k. The process describing the trajectory of the policyholder accross the levels is thus stationary. From

$$p_{ij}^{(n)}(\vartheta) = \sum_{i_1=0}^{s} \sum_{i_2=0}^{s} \cdots \sum_{i_{n-1}=0}^{s} p_{ii_1}(\vartheta) p_{i_1 i_2}(\vartheta) \cdots p_{i_{n-1}j}(\vartheta),$$

we clearly see that it includes all the possible paths from i to j and the probability of their occurrence. This is the n-step transition probability $p_{ij}^{(n)}(\vartheta)$. Therefore, the matrix

$$\boldsymbol{P}^{(n)}(\vartheta) = \begin{pmatrix} p_{00}^{(n)}(\vartheta) & p_{01}^{(n)}(\vartheta) & \cdots & p_{0s}^{(n)}(\vartheta) \\ p_{10}^{(n)}(\vartheta) & p_{11}^{(n)}(\vartheta) & \cdots & p_{1s}^{(n)}(\vartheta) \\ \vdots & \vdots & \ddots & \vdots \\ p_{s0}^{(n)}(\vartheta) & p_{s1}^{(n)}(\vartheta) & \cdots & p_{ss}^{(n)}(\vartheta) \end{pmatrix}$$

is called the n-step transition matrix corresponding to $\boldsymbol{P}(\vartheta)$.

The following result shows that $\boldsymbol{P}^{(n)}(\vartheta)$ is a stochastic matrix, being the nth power of the one-step transition matrix $\boldsymbol{P}(\vartheta)$.

Property 4.1 *For all $n, m = 0, 1, \ldots,$*

$$\boldsymbol{P}^{(n)}(\vartheta) = \boldsymbol{P}^{n}(\vartheta) \tag{4.3}$$

and hence,

$$\boldsymbol{P}^{(n+m)}(\vartheta) = \boldsymbol{P}^{(n)}(\vartheta)\boldsymbol{P}^{(m)}(\vartheta). \tag{4.4}$$

Proof The proof is by induction on n. The result is obviously true for $n = 1$. Assume it holds for n and let us show that it is still true for $n+1$. Clearly, by conditioning on the level ℓ occupied at time n we get

$$p_{ij}^{(n+1)}(\vartheta) = \sum_{\ell=0}^{s} p_{i\ell}^{(n)}(\vartheta) p_{\ell j}(\vartheta) \tag{4.5}$$

which corresponds to matrix multiplication. This proves (4.3), from which (4.4) readily follows.

□

The matrix identity (4.4) is usually called the *Chapman Kolmogorov equation*. Taking the nth power of $\boldsymbol{P}(\vartheta)$ yields the n-step transition matrix whose element $(\ell_1 \ell_2)$, denoted as $p_{\ell_1\ell_2}^{(n)}(\vartheta)$, is the probability of moving from level ℓ_1 to level ℓ_2 in n transitions. Using Property 4.1, we get the following representation for the distribution of the state variable L_n: denoting as

$$p^{(k)}(\vartheta) = \big(\Pr[L_k(\vartheta)=0], \ldots, \Pr[L_k(\vartheta)=s]\big)^T,$$

we have

$$p^{(k+n)}(\vartheta) = p^{(k)}(\vartheta)P^n(\vartheta). \tag{4.6}$$

Remark 4.1 (Numerical Aspects) The computation of the probability distribution of $L_n(\vartheta)$ amounts to calculating the nth power of the transition matrix $P(\vartheta)$. For large values of n, this may pose some computational difficulties. This is why we now discuss an algebraic method which makes use of the concept of eigenvalues and eigenvectors (that will be encountered further in this chapter).

The vector v with at least one component different from 0 is a right eigenvector of $P(\vartheta)$ if $P(\vartheta)v = \lambda v$ for some $\lambda \in \mathbb{R}$. In this case, λ is said to be an eigenvalue of $P(\vartheta)$. Finding the eigenvalues of $P(\vartheta)$ amounts to solving the characteristic equation $\det(P(\vartheta) - \lambda I) = 0$. A nonzero vector u that is a solution of $u^T P(\vartheta) = \lambda u^T$ is called a left eigenvector corresponding to λ.

In general, the characteristic equation possesses $s+1$ solutions $\lambda_0, \ldots, \lambda_s$ which can be complex and some of them can coincide (we assume that the eigenvalues are numbered so that $|\lambda_0| \geq |\lambda_1| \geq \cdots \geq |\lambda_s|$). The Perron-Froebenius theorem for regular matrices ensures that provided the transition matrix $P(\vartheta)$ is regular then $\lambda_0 = 1$, $v_0 = e$, $u_0 \geq \mathbf{0}$ with $u^T e = \sum_{j=0}^{s} u_{0j} = 1$. Moreover, all the other eigenvalues of $P(\vartheta)$ lie inside the unit circle of the complex plane, that is $|\lambda_j| < 1$ for $j = 1, \ldots, s$.

Let $V = (v_0, \ldots, v_s)$ be an $(s+1)\times(s+1)$ matrix consisting of right column eigenvectors and

$$U = \begin{pmatrix} u_0^T \\ \vdots \\ u_{s+1}^T \end{pmatrix}$$

an $(s+1)\times(s+1)$ matrix consisting of left column eigenvectors. Let us assume that the eigenvalues $\lambda_0, \ldots, \lambda_s$ are distinct. This ensures that $v_0, \ldots, v_s$ are linearly independent so that V is invertible. Moreover, $U = V^{-1}$. In this case, $P(\vartheta)$ can be represented as

$$P(\vartheta) = V \begin{pmatrix} \lambda_0 & \cdots & 0 \\ \vdots & \ddots & \vdots \\ 0 & \cdots & \lambda_s \end{pmatrix} U = \sum_{j=0}^{s} \lambda_j v_j u_j^T.$$

This representation is useful for computing the nth power of $P(\vartheta)$ in that

$$P^n(\vartheta) = V \begin{pmatrix} \lambda_0^n & \cdots & 0 \\ \vdots & \ddots & \vdots \\ 0 & \cdots & \lambda_s^n \end{pmatrix} U = \sum_{j=0}^{s} \lambda_j^n v_j u_j^T.$$

4.3.4 Ergodicity and Regular Transition Matrix

A Markov chain with transition matrix $\boldsymbol{P}$ is said to be ergodic if $\boldsymbol{P}$ is regular, that is, if there exists some $n_0 \geq 1$ such that all entries of $\boldsymbol{P}^{n_0}$ are strictly positive. This condition means that it is possible, with a strictly positive probability, to go from one level i to another level j in a finite number of transitions or, in other words, that all states of the Markov chain are accessible from any initial state in a finite number of steps.

All bonus-malus scales in practical use have a 'best' level, with the property that a policy in that level remains in the same level after a claim-free period. In our framework, the best level is level 0 and, for any level, it is possible to reach the superbonus level 0 after a sufficiently large number of claim-free years, resulting in $p_{\ell 0}^{(n)}(\vartheta) > 0$ for all sufficiently large n. In the following, we restrict our attention to such non-periodic bonus rules. The transition matrix $\boldsymbol{P}(\vartheta)$ associated with such a bonus-malus scale is regular, i.e. there exists some integer $n_0 \geq 1$ such that all entries of the n_0th power $\boldsymbol{P}^{n_0}(\vartheta)$ of the one-step transition matrix are strictly positive.

4.4 Long-Term Behaviour of Bonus-Malus Systems

4.4.1 Stationary Distribution

A natural question that arises concerns the long term behaviour of a bonus-malus system. Intuitively, we expect that the system will stabilize in the long run. Since the annual claim numbers have been assumed to be independent and identically distributed, each policyholder will ultimately stabilize around an equilibrium level correponding to the expected annual claim frequency ϑ, and will gravitate around this level.

To formalize this intuitive idea, let us compute the powers of the transition matrix $\boldsymbol{P}(\vartheta)$ for $\vartheta = 0.1$ in the -1/top and $-1/+2$ bonus-malus scales. This is done in the following examples.

Example 4.7 (−1/Top Scale) Starting from

$$\boldsymbol{P}(0.1) = \begin{pmatrix} 0.904837 & 0 & 0 & 0 & 0 & 0.095163 \\ 0.904837 & 0 & 0 & 0 & 0 & 0.095163 \\ 0 & 0.904837 & 0 & 0 & 0 & 0.095163 \\ 0 & 0 & 0.904837 & 0 & 0 & 0.095163 \\ 0 & 0 & 0 & 0.904837 & 0 & 0.095163 \\ 0 & 0 & 0 & 0 & 0.904837 & 0.095163 \end{pmatrix}$$

we get

$$\boldsymbol{P}^2(0.1) = \begin{pmatrix} 0.818731 & 0 & 0 & 0 & 0.086107 & 0.095163 \\ 0.818731 & 0 & 0 & 0 & 0.086107 & 0.095163 \\ 0.818731 & 0 & 0 & 0 & 0.086107 & 0.095163 \\ 0 & 0.818731 & 0 & 0 & 0.086107 & 0.095163 \\ 0 & 0 & 0.818731 & 0 & 0.086107 & 0.095163 \\ 0 & 0 & 0 & 0.818731 & 0.086107 & 0.095163 \end{pmatrix},$$

$$
\boldsymbol{P}^3(0.1) = \begin{pmatrix} 0.740818 & 0 & 0 & 0.077913 & 0.086107 & 0.095163 \\ 0.740818 & 0 & 0 & 0.077913 & 0.086107 & 0.095163 \\ 0.740818 & 0 & 0 & 0.077913 & 0.086107 & 0.095163 \\ 0.740818 & 0 & 0 & 0.077913 & 0.086107 & 0.095163 \\ 0 & 0.740818 & 0 & 0.077913 & 0.086107 & 0.095163 \\ 0 & 0 & 0.740818 & 0.077913 & 0.086107 & 0.095163 \end{pmatrix},
$$

$$
\boldsymbol{P}^4(0.1) = \begin{pmatrix} 0.67032 & 0.00000 & 0.070498 & 0.077913 & 0.086107 & 0.095163 \\ 0.67032 & 0.00000 & 0.070498 & 0.077913 & 0.086107 & 0.095163 \\ 0.67032 & 0.00000 & 0.070498 & 0.077913 & 0.086107 & 0.095163 \\ 0.67032 & 0.00000 & 0.070498 & 0.077913 & 0.086107 & 0.095163 \\ 0.67032 & 0.00000 & 0.070498 & 0.077913 & 0.086107 & 0.095163 \\ 0.00000 & 0.67032 & 0.070498 & 0.077913 & 0.086107 & 0.095163 \end{pmatrix},
$$

and

$$
\boldsymbol{P}^5(0.1) = \begin{pmatrix} 0.606531 & 0.063789 & 0.070498 & 0.077913 & 0.086107 & 0.095163 \\ 0.606531 & 0.063789 & 0.070498 & 0.077913 & 0.086107 & 0.095163 \\ 0.606531 & 0.063789 & 0.070498 & 0.077913 & 0.086107 & 0.095163 \\ 0.606531 & 0.063789 & 0.070498 & 0.077913 & 0.086107 & 0.095163 \\ 0.606531 & 0.063789 & 0.070498 & 0.077913 & 0.086107 & 0.095163 \\ 0.606531 & 0.063789 & 0.070498 & 0.077913 & 0.086107 & 0.095163 \end{pmatrix},
$$

where all the rows are identical. Of course,

$$
\boldsymbol{P}^k(0.1) = \boldsymbol{P}^5(0.1) \text{ for any integer } k \geq 6.
$$

This means that, whatever the initial distribution,

$$
\boldsymbol{p}^{(k)}(0.1) = (0.606531, 0.063789, 0.070498, 0.077913, 0.086107, 0.095163)^T
$$

for any $k \geq 5$. The proportion of policyholders occupying each of the levels of the -1/top scale thus remains unchanged after 5 years.

Example 4.8 (−1/+2 Scale) In this case,

$$
\boldsymbol{P}(0.1) = \begin{pmatrix} 0.904837 & 0 & 0.090484 & 0 & 0.004524 & 0.000155 \\ 0.904837 & 0 & 0 & 0.090484 & 0 & 0.004679 \\ 0 & 0.904837 & 0 & 0 & 0.090484 & 0.004679 \\ 0 & 0 & 0.904837 & 0 & 0 & 0.095163 \\ 0 & 0 & 0 & 0.904837 & 0 & 0.095163 \\ 0 & 0 & 0 & 0 & 0.904837 & 0.095163 \end{pmatrix}.
$$

The convergence is now much slower:

$$\boldsymbol{P}^5(0.1) = \begin{pmatrix} 0.791523 & 0.081985 & 0.088694 & 0.018169 & 0.015456 & 0.004173 \\ 0.788490 & 0.066822 & 0.106890 & 0.017259 & 0.014546 & 0.005992 \\ 0.788490 & 0.063789 & 0.073531 & 0.053651 & 0.013637 & 0.006902 \\ 0.606531 & 0.245749 & 0.070498 & 0.020292 & 0.050028 & 0.006902 \\ 0.606531 & 0.063789 & 0.252457 & 0.017259 & 0.034865 & 0.025098 \\ 0.606531 & 0.063789 & 0.070498 & 0.199219 & 0.031833 & 0.028131 \end{pmatrix}$$

$$\boldsymbol{P}^{10}(0.1) = \begin{pmatrix} 0.784013 & 0.081747 & 0.090966 & 0.022022 & 0.016217 & 0.005037 \\ 0.784003 & 0.081480 & 0.090248 & 0.023009 & 0.016178 & 0.005081 \\ 0.777382 & 0.088092 & 0.089871 & 0.022071 & 0.017497 & 0.005087 \\ 0.776278 & 0.079263 & 0.099795 & 0.021694 & 0.016890 & 0.006080 \\ 0.776278 & 0.078160 & 0.090966 & 0.031618 & 0.016623 & 0.006356 \\ 0.743169 & 0.111269 & 0.089862 & 0.026100 & 0.023236 & 0.006365 \end{pmatrix}$$

$$\boldsymbol{P}^{20}(0.1) = \begin{pmatrix} 0.782907 & 0.082338 & 0.090996 & 0.022276 & 0.016387 & 0.005096 \\ 0.782903 & 0.082332 & 0.091006 & 0.022275 & 0.016387 & 0.005097 \\ 0.782902 & 0.082326 & 0.090993 & 0.022295 & 0.016386 & 0.005098 \\ 0.782803 & 0.082424 & 0.090984 & 0.022285 & 0.016406 & 0.005098 \\ 0.782776 & 0.082352 & 0.091082 & 0.022278 & 0.016403 & 0.005108 \\ 0.782774 & 0.082327 & 0.091011 & 0.022376 & 0.016399 & 0.005113 \end{pmatrix}$$

which slowly converges to

$$\boldsymbol{\Pi}(0.1) = \begin{pmatrix} 0.782901 & 0.082338 & 0.090998 & 0.022278 & 0.016387 & 0.005097 \\ 0.782901 & 0.082338 & 0.090998 & 0.022278 & 0.016387 & 0.005097 \\ 0.782901 & 0.082338 & 0.090998 & 0.022278 & 0.016387 & 0.005097 \\ 0.782901 & 0.082338 & 0.090998 & 0.022278 & 0.016387 & 0.005097 \\ 0.782901 & 0.082338 & 0.090998 & 0.022278 & 0.016387 & 0.005097 \\ 0.782901 & 0.082338 & 0.090998 & 0.022278 & 0.016387 & 0.005097 \end{pmatrix}.$$

In this case, the system is not stable after 20 years.

Let us consider the trajectory of a policyholder with expected claim frequency ϑ accross the levels of the bonus-malus scale. We define the stationary distribution $\boldsymbol{\pi}(\vartheta) = (\pi_0(\vartheta), \pi_1(\vartheta), \ldots, \pi_s(\vartheta))^T$ as follows: $\pi_\ell(\vartheta)$ is the stationary probability for a policyholder with mean frequency ϑ to be in level ℓ i.e.

$$\pi_{\ell_2}(\vartheta) = \lim_{n\to+\infty} p^{(n)}_{\ell_1\ell_2}(\vartheta).$$

The term $\pi_\ell(\vartheta)$ is the limit value of the probability that the policyholder is in level ℓ, when the number of periods tends to $+\infty$. It is also the fraction of the time a policyholder with claim frequency ϑ spends in level ℓ, once the steady state has been reached. Note that $\boldsymbol{\pi}(\vartheta)$

does not depend on the starting class. This means that the nth power $\boldsymbol{P}^n(\vartheta)$ of the one-step transition matrix $\boldsymbol{P}(\vartheta)$ converges to a matrix $\boldsymbol{\Pi}(\vartheta)$ with all the same rows $\boldsymbol{\pi}^T(\vartheta)$, that is

$$\lim_{n\to+\infty} \boldsymbol{P}^{(n)}(\vartheta) = \boldsymbol{\Pi}(\vartheta) = \begin{pmatrix} \boldsymbol{\pi}^T(\vartheta) \\ \boldsymbol{\pi}^T(\vartheta) \\ \vdots \\ \boldsymbol{\pi}^T(\vartheta) \end{pmatrix},$$

exactly as we saw in the introductory examples.

Let us now explain how to compute the $\pi_\ell(\vartheta)$s. Taking the limit in both sides of (4.5) for $n \to +\infty$, we see that the vector $\boldsymbol{\pi}(\vartheta)$ is the unique probabilistic solution to the system of linear equations

$$\pi_j(\vartheta) = \sum_{\ell=0}^{s} \pi_\ell(\vartheta) p_{\ell j}(\vartheta), \qquad j \in \{0, \ldots, s\}. \tag{4.7}$$

In matrix notation, (4.7) can be written as

$$\begin{cases} \boldsymbol{\pi}^T(\vartheta) = \boldsymbol{\pi}^T(\vartheta)\boldsymbol{P}(\vartheta), \\ \boldsymbol{\pi}^T(\vartheta)\boldsymbol{e} = 1 \end{cases} \tag{4.8}$$

where $\boldsymbol{e}$ is a column vector of 1s. This means that $\boldsymbol{\pi}(\vartheta)$ is the left eigenvector $\boldsymbol{u}_0$ of $\boldsymbol{P}(\vartheta)$ encountered above. Thus we see that if the initial distribution in the scale is $\boldsymbol{\pi}(\vartheta)$, then the probability distribution remains equal to $\boldsymbol{\pi}(\vartheta)$.

4.4.2 Rolski–Schmidli–Schmidt–Teugels Formula

Let $\boldsymbol{E}$ be the $(s+1)\times(s+1)$ matrix all of whose entries are 1, i.e. consisting of $s+1$ column vectors $\boldsymbol{e}$. Then, the following result provides a direct method to get $\boldsymbol{\pi}(\vartheta)$.

Property 4.2 *Assume that the stochastic matrix* $\boldsymbol{P}(\vartheta)$ *is regular. Then the matrix* $\boldsymbol{I}-\boldsymbol{P}(\vartheta)+\boldsymbol{E}$ *is invertible and the solution of (4.8) is given by*

$$\boldsymbol{\pi}^T(\vartheta) = \boldsymbol{e}^T(\boldsymbol{I}-\boldsymbol{P}(\vartheta)+\boldsymbol{E})^{-1}. \tag{4.9}$$

Proof Let us first check that $\boldsymbol{I}-\boldsymbol{P}(\vartheta)+\boldsymbol{E}$ is invertible. We must show that

$$(\boldsymbol{I}-\boldsymbol{P}(\vartheta)+\boldsymbol{E})\boldsymbol{x} = \boldsymbol{0} \Rightarrow \boldsymbol{x} = \boldsymbol{0}.$$

From (4.8), we have $\boldsymbol{\pi}^T(\vartheta)(\boldsymbol{I}-\boldsymbol{P}(\vartheta)) = \boldsymbol{0}$. Thus,

$$(\boldsymbol{I}-\boldsymbol{P}(\vartheta)+\boldsymbol{E})\boldsymbol{x} = \boldsymbol{0} \Rightarrow 0 = \boldsymbol{\pi}^T(\vartheta)(\boldsymbol{I}-\boldsymbol{P}(\vartheta)+\boldsymbol{E})\boldsymbol{x} = 0 + \boldsymbol{\pi}^T(\vartheta)\boldsymbol{E}\boldsymbol{x}$$

$$\Leftrightarrow \boldsymbol{\pi}^T(\vartheta)\boldsymbol{E}\boldsymbol{x} = \boldsymbol{0}.$$

On the other hand, $\boldsymbol{\pi}^T(\vartheta)\boldsymbol{E} = \boldsymbol{e}^T$. Thus, $\boldsymbol{e}^T\boldsymbol{x} = 0$, which implies $\boldsymbol{E}\boldsymbol{x} = \boldsymbol{0}$. Consequently,

$$(\boldsymbol{I} - \boldsymbol{P}(\vartheta))\boldsymbol{x} = \boldsymbol{0} \Leftrightarrow \boldsymbol{P}(\vartheta)\boldsymbol{x} = \boldsymbol{x}.$$

This implies for any $n \geq 1$ that

$$\boldsymbol{x} = \boldsymbol{P}^n(\vartheta)\boldsymbol{x} \to \boldsymbol{\Pi}(\vartheta)\boldsymbol{x},$$

i.e. $x_i = \sum_{j=0}^{s} \pi_j(\vartheta)x_j$ for all $i = 0, \ldots, s$. Because the right-hand side of these equations does not depend on i, we have $\boldsymbol{x} = c\boldsymbol{e}$ for some $c \in \mathbb{R}$. Since we also have

$$0 = \boldsymbol{e}^T\boldsymbol{x} = c\boldsymbol{e}^T\boldsymbol{e} = c(s+1) \Rightarrow c = 0.$$

Thus, $\boldsymbol{I} - \boldsymbol{P}(\vartheta) + \boldsymbol{E}$ is invertible. Furthermore, since $\boldsymbol{\pi}^T(\vartheta)(\boldsymbol{I} - \boldsymbol{P}(\vartheta)) = \boldsymbol{0}$, we have

$$\boldsymbol{\pi}^T(\vartheta)(\boldsymbol{I} - \boldsymbol{P}(\vartheta) + \boldsymbol{E}) = \boldsymbol{\pi}^T(\vartheta)\boldsymbol{E} = \boldsymbol{e}^T.$$

This proves (4.9).

□

If the number $s+1$ of states is small, the matrix $\boldsymbol{I} - \boldsymbol{P}(\vartheta) + \boldsymbol{E}$ can easily be inverted. For larger $s+1$, numerical methods have to be used, like the Gaussian elimination algorithm.

Example 4.9 (−1/top scale) We have seen above that $p_{5\ell}^{(5)}(\vartheta) = \pi_\ell(\vartheta)$ for $\ell = 0, 1, \ldots, 5$, so that the system needs 5 years to reach stationarity (i.e. the time needed by the best policyholders starting from level 5 to arrive in level 0). Formula (4.9) gives here

$$\boldsymbol{\pi}^T(\vartheta) = (1, 1, 1, 1, 1, 1)$$

$$\times \begin{pmatrix} 2-\exp(-\vartheta) & 1 & 1 & 1 & 1 & \exp(-\vartheta) \\ 1-\exp(-\vartheta) & 2 & 1 & 1 & 1 & \exp(-\vartheta) \\ 1 & 1-\exp(-\vartheta) & 2 & 1 & 1 & \exp(-\vartheta) \\ 1 & 1 & 1-\exp(-\vartheta) & 2 & 1 & \exp(-\vartheta) \\ 1 & 1 & 1 & 1-\exp(-\vartheta) & 2 & \exp(-\vartheta) \\ 1 & 1 & 1 & 1 & 1-\exp(-\vartheta) & 1+\exp(-\vartheta) \end{pmatrix}^{-1}.$$

Applying this formula in the particular case $\vartheta = 0.1$, we get

$$\boldsymbol{\pi}^T(0.1) = (1, 1, 1, 1, 1, 1)$$

$$\times \begin{pmatrix} 1.095163 & 1 & 1 & 1 & 1 & 0.904837 \\ 0.095163 & 2 & 1 & 1 & 1 & 0.904837 \\ 1 & 0.095163 & 2 & 1 & 1 & 0.904837 \\ 1 & 1 & 0.095163 & 2 & 1 & 0.904837 \\ 1 & 1 & 1 & 0.095163 & 2 & 0.904837 \\ 1 & 1 & 1 & 1 & 0.095163 & 1.904837 \end{pmatrix}^{-1}.$$

$$
= (1, 1, 1, 1, 1, 1)
$$

$$
\times \begin{pmatrix} 2.305672 & -0.678486 & -0.565648 & -0.440943 & -0.303122 & -0.150806 \\ 1.305672 & 0.321514 & -0.565648 & -0.440943 & -0.303122 & -0.150806 \\ 0.400834 & 0.226351 & 0.434352 & -0.440943 & -0.303122 & -0.150806 \\ -0.417897 & 0.140245 & 0.339189 & 0.559057 & -0.303122 & -0.150806 \\ -1.158715 & 0.062332 & 0.253083 & 0.463895 & 0.696878 & -0.150806 \\ -1.829035 & -0.008166 & 0.175170 & 0.377788 & 0.601716 & 0.849194 \end{pmatrix}.
$$

$$
= (0.606531, 0.063789, 0.070498, 0.077913, 0.086107, 0.095163).
$$

Coming back to Example 4.7, we see that the matrices $\boldsymbol{P}^k(0.1)$, $k \geq 5$, have all their rows equal to $\boldsymbol{\pi}^T(0.1)$.

Example 4.10 (−1/+2 scale) Formula (4.9) gives here

$$
\boldsymbol{\pi}^T(\vartheta) = (1, 1, 1, 1, 1, 1)
$$

$$
\times \begin{pmatrix} 2-\exp(-\vartheta) & 1 & 1-\vartheta\exp(-\vartheta) & 1 & 1-\frac{\vartheta^2}{2}\exp(-\vartheta) & \exp(-\vartheta)\left(1+\vartheta+\frac{\vartheta^2}{2}\right) \\ 1-\exp(-\vartheta) & 2 & 1 & 1-\vartheta\exp(-\vartheta) & 1 & \exp(-\vartheta)(1+\vartheta) \\ 1 & 1-\exp(-\vartheta) & 2 & 1 & 1-\vartheta\exp(-\vartheta) & \exp(-\vartheta)(1+\vartheta) \\ 1 & 1 & 1-\exp(-\vartheta) & 2 & 1 & \exp(-\vartheta) \\ 1 & 1 & 1 & 1-\exp(-\vartheta) & 2 & \exp(-\vartheta) \\ 1 & 1 & 1 & 1 & 1-\exp(-\vartheta) & 1+\exp(-\vartheta) \end{pmatrix}^{-1}.
$$

Applying this formula in the particular case $\vartheta = 0.1$, we get

$$
\boldsymbol{\pi}^T(0.1) = (1, 1, 1, 1, 1, 1)
$$

$$
\times \begin{pmatrix} 1.095163 & 1 & 0.909516 & 1 & 0.995476 & 0.999845 \\ 0.095163 & 2 & 1 & 0.909516 & 1 & 0.995321 \\ 1 & 0.095163 & 2 & 1 & 0.909516 & 0.995321 \\ 1 & 1 & 0.095163 & 2 & 1 & 0.904837 \\ 1 & 1 & 1 & 0.095163 & 2 & 0.904837 \\ 1 & 1 & 1 & 1 & 0.095163 & 1.904837 \end{pmatrix}^{-1}
$$

$$
= (1, 1, 1, 1, 1, 1)
$$

$$
\times \begin{pmatrix} 2.759072 & -0.630802 & -0.512948 & -0.658608 & -0.480599 & -0.309449 \\ 1.659534 & 0.358730 & -0.524518 & -0.561440 & -0.472165 & -0.293474 \\ 0.578107 & 0.244995 & 0.454957 & -0.475982 & -0.366345 & -0.269065 \\ -0.478699 & 0.133850 & 0.332122 & 0.599117 & -0.272234 & -0.147489 \\ -1.434937 & 0.033282 & 0.220977 & 0.571906 & 0.812921 & -0.037482 \\ -2.300176 & -0.057717 & 0.120409 & 0.547285 & 0.794810 & 1.062056 \end{pmatrix}
$$

$$
= (0.782901, 0.082338, 0.090998, 0.022278, 0.016387, 0.005097).
$$

4.4.3 Dufresne Algorithm

Dufresne (1988,1995) proposed a simple and efficient iterative algorithm for deriving $\boldsymbol{\pi}(\vartheta)$ provided that the driver goes one level down if no claims are filed to the company, and goes $n\times$pen levels up if n claims are reported to the insurer. Then, the move at the end of year k can be modelled as

$$\Delta_{k+1}(\vartheta)=\begin{cases}-1 \text{ if no claims}\\ n\times \text{pen if } n \text{ claims.}\end{cases}$$

If the annual numbers of claims $N_1, N_2, \ldots$ are independent and $\mathcal{P}oi(\vartheta)$ distributed then the sequence $\Delta_1(\vartheta), \Delta_2(\vartheta), \ldots$ is made up of independent and identically distributed random variables, with common probability mass function

$$\Pr[\Delta_{k+1}(\vartheta)=-1]=\Pr[N_{k+1}=0]=\exp(-\vartheta)$$

$$\Pr[\Delta_{k+1}(\vartheta)=n\times\text{pen}]=\Pr[N_{k+1}=n]=\exp(-\vartheta)\frac{\vartheta^n}{n!} \text{ for } n=1,2,\ldots$$

$$\Pr[\Delta_{k+1}(\vartheta)=\xi]=0 \text{ otherwise.}$$

The level $L_{k+1}(\vartheta)$ can then be represented as

$$L_{k+1}(\vartheta)=\begin{cases}L_k(\vartheta)+\Delta_{k+1}(\vartheta) \text{ if } 0\leq L_k(\vartheta)+\Delta_{k+1}(\vartheta)\leq s\\ 0 \text{ if } L_k(\vartheta)+\Delta_{k+1}(\vartheta)=-1\\ s \text{ if } L_k(\vartheta)+\Delta_{k+1}(\vartheta)>s.\end{cases}$$

Let us denote as $F_k(\cdot|\vartheta)$ the distribution function of $L_k(\vartheta)$, that is

$$F_k(\ell|\vartheta)=\Pr[L_k(\vartheta)\leq\ell],\quad \ell=0,1,\ldots,s.$$

Furthermore, let us denote as $p_\Delta(\cdot)$ the common probability mass function of the Δ_ks. We then have

$$F_{k+1}(\ell|\vartheta)=\sum_{y=-1}^{\ell}F_k(\ell-y|\vartheta)p_\Delta(y),$$

with $F_{k+1}(s|\vartheta)=1$. The stationary distribution $F_\infty(\cdot|\vartheta)$ is then obtained as

$$F_\infty(\ell|\vartheta)=\lim_{k\to+\infty}F_k(\ell|\vartheta)=\sum_{y=-1}^{\ell}F_\infty(\ell-y|\vartheta)p_\Delta(y),$$

with $F_\infty(s|\vartheta)=1$. Obviously, the $\pi_\ell(\vartheta)$s are then recovered from $\pi_0(\vartheta)=F_\infty(0|\vartheta)$ and

$$\pi_\ell(\vartheta)=F_\infty(\ell|\vartheta)-F_\infty(\ell-1|\vartheta) \text{ for } \ell=1,\ldots,s.$$

The values of $F_\infty(\ell|\vartheta)$ can be computed recursively from the following algorithm:

(i) set $A(0)=1$
(ii) compute for $\ell=0,1,\ldots,s-1$

$$A(\ell+1)=\frac{1}{p_{\Delta}(-1)}\left(A(\ell)-\sum_{y=0}^{\ell}A(\ell-y)p_{\Delta}(y)\right)$$

(iii) set

$$F_{\infty}(\ell|\vartheta)=\frac{A(\ell)}{A(s)}\text{ for }\ell=0,1,\ldots,s.$$

This recursive formula is computationally efficient, and easy to implement.

4.4.4 Convergence to the Stationary Distribution

Geometric Bound for the Speed of Convergence

What matters for a unique limit $\boldsymbol{\pi}(\vartheta)$ to exist is that one can find a positive integer n_0 such that

$$\alpha(\vartheta)=\min_{i,j\in\{0,\ldots,s\}}p_{ij}^{(n_0)}(\vartheta)>0.$$

In other words, n_0 is such that all the entries of $\boldsymbol{P}^{n_0}(\vartheta)$ are positive, and thus it is possible to reach any level starting from any other level in n_0 periods. Various inequalities indicate the speed of convergence to the limit distribution $\boldsymbol{\pi}(\vartheta)$. It can be shown that

$$|p_{ij}^{(n)}(\vartheta)-\pi_j(\vartheta)|\leq\max_{i\in\{0,\ldots,s\}}p_{ij}^{(n)}(\vartheta)-\min_{i\in\{0,\ldots,s\}}p_{ij}^{(n)}(\vartheta)\leq\big(1-\alpha(\vartheta)\big)^{\left\lfloor\frac{n}{n_0}\right\rfloor-1}.$$

This inequality provides us with a geometric bound for the rate of convergence to the limit distribution. Further bounds can be determined using concepts of matrix algebra.

Total Variation Distance

The total variation metric is often used to measure the distance to the stationary distribution $\boldsymbol{\pi}(\vartheta)$. Recall that the total variation distance between two random variables X and Y, denoted as $d_{TV}(X,Y)$, is given by

$$d_{TV}(X,Y)=\int_{-\infty}^{+\infty}|dF_X(t)-dF_Y(t)|. \tag{4.10}$$

For counting random variables M and N, (4.10) obviously reduces to

$$d_{TV}(M,N)=\sum_{k=0}^{+\infty}\big|\Pr[M=k]-\Pr[N=k]\big|.$$

There is a close connection between d_{TV} and the standard variation distance, which considers the supremum of the difference between the probability masses given to some random events. Specifically, given two random variables X and Y, $d_{TV}(X, Y)$ can be represented as

$$d_{TV}(X, Y) = 2 \sup_A \Big| \Pr[X \in A] - \Pr[Y \in A] \Big|. \tag{4.11}$$

Selection of the Initial Level

The main objective of a bonus-malus system is to correct the inadequacies of *a priori* rating by separating the good from the bad drivers. This separation process should proceed as fast as possible; the time needed to achieve this operation is the time needed to reach stationarity. A convenient way to select the initial level has been suggested by BONSDORFF (1992). The idea is to select it in order to minimize the time needed to reach stationarity. It relies on the total variation distance d_{TV} between the nth transient distribution starting from level ℓ_1, i.e. $\{p^{(n)}_{\ell_1\ell_2}(\vartheta),\ \ \ell_2 = 0, 1, \ldots, s\}$, and the stationary distribution $\{\pi_{\ell_2}(\vartheta),\ \ \ell_2 = 0, 1, \ldots, s\}$, computed as

$$d_{TV}(\ell_1, n, \vartheta) = \sum_{\ell_2=0}^{s} |p^{(n)}_{\ell_1\ell_2}(\vartheta) - \pi_{\ell_2}(\vartheta)|;$$

it measures the degree of convergence of the system after n transitions. Of course,

$$\lim_{n\to+\infty} d_{TV}(\ell_1, n, \vartheta) = 0 \text{ for all } \ell_1 \text{ and } \vartheta.$$

The convergence to $\boldsymbol{\pi}(\vartheta)$ is essentially controlled by the second largest eigenvalue λ_1. For any $\varrho > |\lambda_1|$, there exists an $a < +\infty$ such that

$$\max_{\ell_1=0,1,\ldots,s} d_{TV}(\ell_1, n, \vartheta) \leq a\varrho^n$$

for all n. BONSDORFF (1992) referred to $|\lambda_1|$ as the rate of convergence of the bonus-malus system to equilibrium at intensity ϑ. The function $\vartheta \mapsto |\lambda_1(\vartheta)|$ is then called the convergence rate function of the system.

The initial level could be selected in order to minimize $d_{TV}(\ell_1, n, \vartheta)$ for some fixed value of n. In fact, the further the starting level is from the stationary average level, the slower the convergence.

4.5 Relativities with a Quadratic Loss Function

4.5.1 Relativities

The relativity associated with level ℓ is denoted as r_ℓ; the meaning is that an insured occupying that level pays an amount of premium equal to r_ℓ % of the *a priori* premium determined on the basis of his observable characteristics.

The determination of the r_ℓs given the *a priori* classification implemented by the insurer is the main task of the actuary. The idea is to make r_ℓ 'as close as possible' to the risk factor Θ of a policyholder picked at random from the portfolio. The closeness is usually measured by the expected square difference between Θ and r_ℓ, but other loss functions can be used, too.

4.5.2 Bayesian Relativities

Predictive accuracy is a useful measure of the efficiency of a bonus-malus scale. The idea behind this notion is as follows: A bonus-malus scale is good at discriminating among the good and the bad drivers if the premium they pay is close to their 'true' premium. According to NORBERG (1976), once the number of classes and the transition rules have been fixed, the optimal relativity r_ℓ associated with level ℓ is determined by maximizing the asymptotic predictive accuracy.

Let us pick at random a policyholder from the portfolio. Both the *a priori* expected claim frequency and relative risk parameter are random in this case. Let us denote as Λ the (random) *a priori* expected claim frequency of this randomly selected policyholder, and as Θ the residual effect of the risk factors not included in the ratemaking. The actual (unknown) annual expected claim frequency of this policyholder is then $\Lambda\Theta$. Since the random effect Θ represents residual effects of hidden covariates, the random variables Λ and Θ may reasonably be assumed to be mutually independent. Let w_k be the weight of the kth risk class whose annual expected claim frequency is λ_k. Clearly, $\Pr[\Lambda = \lambda_k] = w_k$.

Let L be the level occupied by this randomly selected policyholder once the steady state has been reached. The distribution of L can be written as

$$\Pr[L = \ell] = \sum_k w_k \int_0^{+\infty} \pi_\ell(\lambda_k \theta) dF_\Theta(\theta). \tag{4.12}$$

Here, $\Pr[L = \ell]$ represents the proportion of the policyholders in level ℓ.

Our aim is to minimize the expected squared difference between the 'true' relative premium Θ and the relative premium r_L applicable to this policyholder (after the steady state has been reached), i.e. the goal is to minimize

$$\begin{aligned}
\mathbb{E}\Big[(\Theta - r_L)^2\Big] &= \sum_{\ell=0}^{s} \mathbb{E}\Big[(\Theta - r_\ell)^2 \Big| L = \ell\Big] \Pr[L = \ell] \\
&= \sum_{\ell=0}^{s} \int_0^{+\infty} (\theta - r_\ell)^2 \Pr[L = \ell | \Theta = \theta] dF_\Theta(\theta) \\
&= \sum_k w_k \int_0^{+\infty} \sum_{\ell=0}^{s} (\theta - r_\ell)^2 \pi_\ell(\lambda_k \theta) dF_\Theta(\theta).
\end{aligned}$$

The solution is given by

$$\begin{aligned}
r_\ell &= \mathbb{E}[\Theta | L = \ell] \\
&= \mathbb{E}\Big[\mathbb{E}[\Theta | L = \ell, \Lambda] \Big| L = \ell\Big] \\
&= \sum_k \mathbb{E}[\Theta | L = \ell, \Lambda = \lambda_k] \Pr[\Lambda = \lambda_k | L = \ell] \\
&= \sum_k \int_0^{+\infty} \theta \frac{\Pr[L = \ell | \Theta = \theta, \Lambda = \lambda_k] w_k}{\Pr[L = \ell, \Lambda = \lambda_k]} dF_\Theta(\theta) \frac{\Pr[\Lambda = \lambda_k, L = \ell]}{\Pr[L = \ell]}
\end{aligned}$$

$$= \frac{\sum_k w_k \int_0^{+\infty} \theta \pi_\ell(\lambda_k \theta) dF_\Theta(\theta)}{\sum_k w_k \int_0^{+\infty} \pi_\ell(\lambda_k \theta) dF_\Theta(\theta)}. \tag{4.13}$$

It is easily seen that

$$\mathbb{E}[r_L] = \mathbb{E}\Big[\mathbb{E}[\Theta|L]\Big] = \mathbb{E}[\Theta] = 1,$$

resulting in financial equilibrium once steady state is reached.

Remark 4.2 If the insurance company does not enforce any *a priori* ratemaking system, all the λ_ks are equal to $\mathbb{E}[\Lambda] = \overline{\lambda}$ and (4.13) reduces to the formula

$$r_\ell = \frac{\int_0^{+\infty} \theta \pi_\ell(\overline{\lambda}\theta) dF_\Theta(\theta)}{\int_0^{+\infty} \pi_\ell(\overline{\lambda}\theta) dF_\Theta(\theta)} \tag{4.14}$$

that has been derived in Norberg (1976).

The way *a priori* and *a posteriori* ratemakings interact is described by

$$\begin{aligned}
\mathbb{E}[\Lambda|L=\ell] &= \sum_k \lambda_k \Pr[\Lambda = \lambda_k | L = \ell] \\
&= \sum_k \lambda_k \frac{\Pr[L=\ell|\Lambda=\lambda_k] w_k}{\Pr[L=\ell]} \\
&= \frac{\sum_k \lambda_k w_k \int_0^{+\infty} \pi_\ell(\lambda_k \theta) dF_\Theta(\theta)}{\sum_k w_k \int_0^{+\infty} \pi_\ell(\lambda_k \theta) dF_\Theta(\theta)}.
\end{aligned} \tag{4.15}$$

If $\mathbb{E}[\Lambda|L=\ell]$ is indeed increasing in the level ℓ, those policyholders who have been granted premium discounts at policy issuance (on the basis of their observable characteristics) will also be rewarded *a posteriori* (because they occupy the lowest levels of the bonus-malus scale). Conversely, the policyholders who have been penalized at policy issuance (because of their observable characteristics) will cluster in the highest bonus-malus levels and will consequently be penalized again.

Example 4.11 (−1/Top Scale, Portfolio A) The results for the bonus-malus scale −1/top are displayed in Table 4.3. Specifically, the values in the third column are computed with the help of (4.14) with $\widehat{a} = 0.889$ and $\widehat{\overline{\lambda}} = 0.1474$. Those values were obtained in Section 1.6 by fitting a Negative Binomial distribution to the portfolio observed claim frequencies given in Table 1.1. Integrations have been performed numerically with the QUAD procedure of SAS®/IML. The fourth column is based on (4.13) with $\widehat{a} = 1.065$ and the $\widehat{\lambda}_k$s listed in Table 2.7.

Once the steady state has been reached, the majority of the policies (58.5 % of the portfolio) occupy level 0 and enjoy the maximum discount. The remaining 41.5 % of the portfolio are distributed over levels 1–5, with about 13 % in level 5 (those policyholders who just claimed). Concerning the relativities, the minimum percentage of 54.7 % when the *a priori* ratemaking is not recognized becomes 61.2 % when the relativities are adapted to the

Table 4.3 Numerical characteristics for the system -1/top and for Portfolio A.

Level ℓ	$\Pr[L=\ell]$	$r_\ell=\mathbb{E}[\Theta\|L=\ell]$ without *a priori* ratemaking	$r_\ell=\mathbb{E}[\Theta\|L=\ell]$ with *a priori* ratemaking	$\mathbb{E}[\Lambda\|L=\ell]$
5	12.8 %	197.3 %	181.2 %	16.3 %
4	9.7 %	170.9 %	159.9 %	15.8 %
3	7.7 %	150.7 %	143.9 %	15.5 %
2	6.2 %	134.8 %	131.3 %	15.2 %
1	5.2 %	122.0 %	120.9 %	15.0 %
0	58.5 %	54.7 %	61.2 %	14.1 %

a priori risk classification. Similarly, the relativity attached to the highest level of 197.3 % gets reduced to 181.2 %. The severity of the *a posteriori* corrections is thus weaker once the *a priori* ratemaking is taken into account in the determination of the r_ℓs. The last column of Table 4.3 indicates the extent to which *a priori* and *a posteriori* ratemakings interact. The numbers in this column are computed as (4.15). The average *a priori* expected claim frequency clearly increases with the level ℓ occupied by the policyholder.

Example 4.12 (−1/Top Scale, Portfolio B) The results for the bonus-malus scale -1/top are displayed in Table 4.4. We only give the relativities computed by taking into account the *a priori* risk classification that is taken from Table 2.16 with $\widehat{\sigma}=0.677$. We see that in Portfolio B, only 46.6 % of the policyholders occupy level 0. The relativities are now less dispersed, ranging from 70.6 % to 146.9 % (instead of 61.2 % to 181.2 %). Again, the last column indicates that *a priori* risk classification and *a posteriori* premium corrections interact.

Example 4.13 (−1/+2 Scale, Portfolio A) Results are displayed in Table 4.5 which is the analogue of Table 4.3 for the bonus-malus scale $-1/+2$. The bonus-malus system is perhaps too soft since the vast majority of the portfolio (about 71 %) clusters in the super bonus level 0. The higher levels are occupied by a very small minority of drivers. Such a system does not really discriminate between good and bad drivers. Consequently, only those

Table 4.4 Numerical characteristics for the system -1/top and for portfolio B.

Level ℓ	$\Pr[L=\ell]$	$r_\ell=\mathbb{E}[\Theta\|L=\ell]$ with *a priori* ratemaking	$\mathbb{E}[\Lambda\|L=\ell]$
5	16.3 %	146.9 %	19.5 %
4	12.5 %	129.3 %	19.3 %
3	9.9 %	117.2 %	19.1 %
2	8.0 %	108.0 %	19.0 %
1	6.6 %	100.7 %	18.9 %
0	46.6 %	70.6 %	18.4 %

Table 4.5 Numerical characteristics for the system $-1/+2$ and for Portfolio A.

Level ℓ	$\Pr[L=\ell]$	$r_\ell = \mathbb{E}[\Theta \mid L=\ell]$ without *a priori* ratemaking	$r_\ell = \mathbb{E}[\Theta \mid L=\ell]$ with *a priori* ratemaking	$\mathbb{E}[\Lambda \mid L=\ell]$
5	4.4 %	309.1 %	271.4 %	18.5 %
4	4.7 %	241.4 %	218.5 %	17.1 %
3	4.4 %	207.7 %	192.5 %	16.4 %
2	8.7 %	142.9 %	138.8 %	15.3 %
1	7.1 %	130.2 %	128.6 %	15.1 %
0	70.6 %	62.4 %	68.5 %	14.2 %

policyholders in level 0 get some discount whereas occupancy of any level 1–5 implies some penalty. Again, the *a posteriori* corrections are softened when *a priori* risk classification is taken into account in the determination of the r_ℓs. The comments made for the scale -1/top still apply to this bonus-malus scale.

Example 4.14 (−1/+2 Scale, Portfolio B) Results are displayed in Table 4.6 which is the analogue of Table 4.4 for the bonus-malus scale $-1/+2$. The comparison with Portfolio A yields the same comments as before.

Example 4.15 (−1/+3 Scale, Portfolio A) Let us now make the $-1/+2$ bonus-malus scale more severe: to this end, each claim is now penalized by 3 levels (instead of 2 in the $-1/+2$ system). The numerical results are displayed in Table 4.7.

We see that less policyholders occupy level 0 (64.5 % compared with 70.6 % with the $-1/+2$ system), and that the upper levels are now more populated. Drivers in level 0 deserve more bonus compared to the $-1/+2$ system (they pay 57.8 % of the base premium compared to 62.4% in the non-segmented case, and 64.2 % compared to 68.5 % in the segmented case). Also, the maximal penalties get reduced when claims are more severely penalized.

Example 4.16 (−1/+3 Scale, Portfolio B) Let us now consider Portfolio B where each claim is penalized by 3 levels (instead of 2 in the $-1/+2$ system). The numerical results

Table 4.6 Numerical characteristics for the system $-1/+2$ and for Portfolio B.

Level ℓ	$\Pr[L=\ell]$	$r_\ell = \mathbb{E}[\Theta \mid L=\ell]$ with *a priori* ratemaking	$\mathbb{E}[\Lambda \mid L=\ell]$
5	5.4 %	223.2 %	20.4 %
4	6.0 %	171.3 %	19.9 %
3	5.8 %	148.9 %	19.6 %
2	11.5 %	112.1 %	19.1 %
1	9.3 %	105.0 %	19.0 %
0	62.0 %	74.7 %	18.5 %

Table 4.7 Numerical characteristics for the system $-1/+3$ and for Portfolio A.

Level ℓ	$\Pr[L=\ell]$	$r_\ell=\mathbb{E}[\Theta\|L=\ell]$ without *a priori* ratemaking	$r_\ell=\mathbb{E}[\Theta\|L=\ell]$ with *a priori* ratemaking	$\mathbb{E}[\Lambda\|L=\ell]$
5	7.3 %	257.1 %	230.8 %	17.4 %
4	5.9 %	219.4 %	200.9 %	16.7 %
3	9.0 %	151.7 %	145.2 %	15.5 %
2	7.3 %	136.5 %	133.1 %	15.2 %
1	6.0 %	124.0 %	123.0 %	15.0 %
0	64.5 %	57.8 %	64.2 %	14.1 %

Table 4.8 Numerical characteristics for the system $-1/+3$ and for Portfolio B.

Level ℓ	$\Pr[L=\ell]$	$r_\ell=\mathbb{E}[\Theta\|L=\ell]$ with *a priori* ratemaking	$\mathbb{E}[\Lambda\|L=\ell]$
5	9.2 %	184.0 %	20.0 %
4	7.8 %	156.6 %	19.7 %
3	11.7 %	117.4 %	19.2 %
2	9.5 %	108.6 %	19.0 %
1	7.8 %	101.6 %	18.9 %
0	54.0 %	72.0 %	18.4 %

are displayed in Table 4.8. The comparison with Portfolio A shows that the relativities are much less dispersed here.

4.5.3 Interaction between Bonus-Malus Systems and a Priori Ratemaking

Since the relativities attached to the different levels are the same whatever the risk class to which the policyholders belong, those scales overpenalize *a priori* bad risks. Let us explain this phenomenon, put in evidence by Taylor (1997). Over time, policyholders will be distributed over the levels of the bonus-malus scale. Since their trajectory is a function of past claims history, policyholders with low *a priori* expected claim frequencies will tend to gravitate to the lowest levels of the scale. Conversely for individuals with high *a priori* expected claim frequencies. Consider for instance a policyholder with a high *a priori* expected claim frequency, a young male driver living in a urban area, say. This driver is expected to report many claims (this is precisely why he has been penalized *a priori*) and so to be transferred to the highest levels of the bonus-malus scale. On the contrary, a policyholder with a low *a priori* expected claim frequency, a middle-aged lady living in a rural area, say, is expected to report few claims and so to gravitate to the lowest levels of the scale. The level occupied by the policyholders in the bonus-malus scale can thus be partly explained by their observable characteristics included in the price list. It is thus fair to isolate that part of the information contained in the level occupied by the policyholder that does not reflect observables characteristics. *A posteriori* corrections should be driven only by this part of the bonus-malus information.

In credibility theory, we have seen that to the extent that good drivers are rewarded in their base premiums (through other rating variables) the size of the bonus they require for equity is reduced. This can be summarized as follows:

- when *a priori* segmentation is used, then the severity of the *a posteriori* differentiation has to be lowered.
- when both *a priori* and *a posteriori* ratemakings are used, the size of the bonus is always smaller for good drivers than for bad ones.

If a single bonus-malus scale is applied to the entire portfolio, even if the relativites take *a priori* risk classification into account, the resulting ratemaking is unfair to *a priori* bad drivers. The bonuses and the maluses need to be functions of the drivers' *a priori* characteristics, used in the price list.

We know from credibility theory that the *a posteriori* corrections are functions of the *a priori* characteristics; see for instance (3.16). On the contrary, when a bonus-malus system is in force, the same *a posteriori* corrections apply to all policyholders, whatever their *a priori* expected claim frequency. This of course induces unfairness in the portfolio.

In order to reduce the unfairness of the tariff, we could propose several bonus-malus scales, according to the *a priori* characteristics. The idea is to select a few *a priori* characteristics inducing large differences in expected claim frequencies (typically, those associated with the largest regression coefficients), and to build separate scales according to these characteristics.

Example 4.17 (−1/+2 System, Portfolio A, with a Dichotomy Rural–Urban) Table 4.9 describes such a system where the company differentiates policyholders according to the type of district where they live (urban or rural). People living in urban areas have higher *a priori* expected claim frequencies. Thus, they should be better rewarded when they do not file any claim and less penalized when they report accidents compared to people living in rural zones. This is indeed what we observe when we compare the relative premiums obtained for the system $-1/+2$: the maximal discount is 66.6 % for urban policyholders, compared to 71.3 % for rural ones. Similarly, the highest penalty is 267.2 % for urbans against 280.4 % for rurals.

Table 4.9 Relativities obtained by differentiating policyholders according to the type of district for the system $-1/+2$ and for Portfolio A.

Level ℓ	Rural	Urban
5	280.4 %	267.2 %
4	226.4 %	214.3 %
3	200.8 %	187.9 %
2	144.4 %	135.3 %
1	134.3 %	124.9 %
0	71.3 %	66.6 %

4.5.4 Linear Relativities

As demonstrated in the numerical examples with the -1/top, $-1/+2$ and $-1/+3$ scales, the relativities obtained above may exhibit a rather irregular pattern, and this may be undesirable for commercial purposes. It may therefore be interesting to smooth this scale in order to obtain relativities which are regularly increasing according to the level. As suggested by GILDE & SUNDT (1989), a linear scale of the form $r_\ell^{lin} = \alpha + \beta\ell, \ell = 0, 1, \ldots, s$ could then be desirable. Then, Norberg's maximum accuracy criterion becomes a constrained minimization:

$$\min \mathbb{E}\Big[(\Theta - r_L^{lin})^2\Big] = \min \mathbb{E}\Big[(\Theta - \alpha - \beta L)^2\Big]. \tag{4.16}$$

Setting the derivative of the objective function with respect to α equal to 0 yields

$$\alpha = \mathbb{E}[\Theta] - \beta\mathbb{E}[L].$$

Doing the same with the derivative of the objective function with respect to β gives

$$\begin{aligned} 0 &= \mathbb{E}\Big[L(\Theta - \alpha - \beta L)\Big] \\ &= \mathbb{E}[L\Theta] - \alpha\mathbb{E}[L] - \beta\mathbb{E}[L^2]. \end{aligned}$$

Replacing the value of α with the expression found above, we get

$$\begin{aligned} 0 &= \mathbb{E}[L\Theta] - \mathbb{E}[L]\mathbb{E}[\Theta] + \beta(\mathbb{E}[L])^2 - \beta\mathbb{E}[L^2] \\ &= \mathbb{C}[L, \Theta] - \beta\mathbb{V}[L]. \end{aligned}$$

The solution of the optimization problem (4.16) is thus given by:

$$\beta = \frac{\mathbb{C}[L, \Theta]}{\mathbb{V}[L]} \text{ and } \alpha = \mathbb{E}[\Theta] - \frac{\mathbb{C}[L, \Theta]}{\mathbb{V}[L]}\mathbb{E}[L]. \tag{4.17}$$

The linear relative premium scale is thus of the form

$$r_\ell^{lin} = 1 + \frac{\mathbb{C}[L, \Theta]}{\mathbb{V}[L]}(\ell - \mathbb{E}[L]),$$

where

$$\begin{aligned} \mathbb{C}[L, \Theta] &= \mathbb{E}[L\Theta] - \mathbb{E}[L] \\ &= \sum_k w_k \sum_{\ell=0}^{s} \ell\mathbb{E}[\Theta\pi_\ell(\lambda_k\Theta)] - \mathbb{E}[L] \\ &= \sum_k w_k \sum_{\ell=0}^{s} \ell \int_0^{+\infty} \theta\pi_\ell(\lambda_k\theta)dF_\Theta(\theta) - \mathbb{E}[L], \end{aligned}$$

$$\Pr[L = \ell] = \sum_k w_k \int_0^{+\infty} \pi_\ell(\lambda_k \theta) dF_\Theta(\theta),$$

$$\mathbb{E}[L] = \sum_{\ell=0}^{s} \ell \Pr[L = \ell],$$

$$\mathbb{V}[L] = \sum_{\ell=0}^{s} (\ell - \mathbb{E}[L])^2 \Pr[L = \ell].$$

Remark 4.3 It is interesting to note that the optimal α and β solving (4.16) also minimize

$$\mathbb{E}\Big[\big(\mathbb{E}[\Theta|L] - r_L^{lin}\big)^2\Big] = \mathbb{E}\Big[\big(\mathbb{E}[\Theta|L] - \alpha - \beta L\big)^2\Big]$$

so that the linear relativities provide the best linear fit to the Bayesian ones. To check this assertion, note that

$$\begin{aligned}
&\mathbb{E}\Big[\big(\Theta - \alpha - \beta L\big)^2\Big] \\
&= \mathbb{E}\Big[\Big(\big(\Theta - \mathbb{E}[\Theta|L]\big) + \big(\mathbb{E}[\Theta|L] - \alpha - \beta L\big)\Big)^2\Big] \\
&= \mathbb{E}\Big[\big(\Theta - \mathbb{E}[\Theta|L]\big)^2\Big] \\
&\quad + 2\mathbb{E}\Big[\big(\Theta - \mathbb{E}[\Theta|L]\big)\big(\mathbb{E}[\Theta|L] - \alpha - \beta L\big)\Big] \\
&\quad + \mathbb{E}\Big[\big(\mathbb{E}[\Theta|L] - \alpha - \beta L\big)^2\Big]
\end{aligned}$$

and the second term vanishes by definition of the conditional expectation $\mathbb{E}[\Theta|L]$, since $\Theta - \mathbb{E}[\Theta|L]$ is orthogonal to any function of L.

Example 4.18 (−1/+2 Scale, Portfolio A) Table 4.10 allows comparison of the relativities of the two scales in the segmented case. As we can see, the values are close to each other. The constant step between two levels in the linear scale is equal to 39.3 %.

The values of the expected errors $Q_1 = \mathbb{E}\left[(\Theta - r_L)^2\right]$ and $Q_2 = \mathbb{E}\left[(\Theta - r_L^{lin})^2\right]$ are respectively given by

$$Q_1 = 0.6219 \text{ and } Q_2 = 0.6261.$$

The small difference between Q_1 and Q_2 indicates that the additional linear restriction does not really produce any deterioration in the fit. Note that the large differences in levels 1–5 are given low weights in the computation of Q_1 and Q_2.

Table 4.11 displays the relativities without *a priori* segmentation. As can be observed, the scale without *a priori* segmentation is more elastic, which is logical since it has to take into account the full heterogeneity. The mean square errors are now given by

$$Q_1 = 0.6630 \text{ and } Q_2 = 0.6688$$

Table 4.10 Linear relativities with *a priori* risk classification for the system $-1/+2$ and for Portfolio A.

Level ℓ	Unconstrained relativities r_ℓ with *a priori* ratemaking	Linear relativities r_ℓ^{lin} with *a priori* ratemaking
5	271.4 %	266.2 %
4	218.5 %	226.9 %
3	192.5 %	187.5 %
2	138.8 %	148.2 %
1	128.6 %	108.8 %
0	68.5 %	69.5 %

Table 4.11 Linear relativities without *a priori* risk classification for the system $-1/+2$ and for Portfolio A.

Level ℓ	Unconstrained relativities r_ℓ without *a priori* ratemaking	Linear relativities r_ℓ^{lin} without *a priori* ratemaking
5	309.1 %	298.3 %
4	241.4 %	251.2 %
3	207.7 %	204.1 %
2	142.9 %	157.1 %
1	130.2 %	110.0 %
0	62.4 %	62.9 %

so that again the additional linear restriction does not really produce any deterioration in the fit. Obviously, Q_1 and Q_2 are higher than before (when *a priori* risk classification was in force). This was expected as Θ is now more variable.

Example 4.19 (−1/+2 Scale, Portfolio B) Table 4.12 gives the linear relativities for Portfolio B. The mean square errors are

$$Q_1 = 0.4134 \text{ and } Q_2 = 0.4182.$$

We observe large discrepancies between r_ℓ and r_ℓ^{lin}, especially in the higher levels. The constant penalty by step in the linear scale is 26.2 %.

4.5.5 Approximations

In Chapter 3, several discrete approximations to Θ allowed us to derive simplified versions of the credibility formulas (replacing integrals with sums). The same idea can be applied here. Considering (4.13), the expression for r_ℓ when Θ has support points $\{\theta_1, \ldots, \theta_k\}$ with respective probability masses $p_1, \ldots, p_q$ as in (3.5) becomes

$$r_\ell = \frac{\sum_k w_k \sum_{j=1}^{q} \theta_j \pi_\ell(\lambda_k \theta_j) p_j}{\sum_k w_k \sum_{j=1}^{q} \pi_\ell(\lambda_k \theta_j) p_j}.$$

Table 4.12 Linear relativities with risk classification for the system $-1/+2$ and for Portfolio B.

Level ℓ	Unconstrained relativities r_ℓ with *a priori* ratemaking	Linear relativities r_ℓ^{lin} with *a priori* ratemaking
5	223.2 %	204.7 %
4	171.3 %	178.5 %
3	148.9 %	152.2 %
2	112.1 %	126.0 %
1	105.0 %	99.8 %
0	74.7 %	73.6 %

The discrete approximations listed in Tables 3.5–3.6 can then be used in this formula.

In the case where Θ has a unimodal probability density function, the mixed uniform approximations of Tables 3.8–3.9 can also be used.

4.6 Relativities with an Exponential Loss Function

4.6.1 Bayesian Relativities

This section proposes an asymmetric loss function with one parameter that reflects the severity of the bonus-malus system. In order to reduce the maluses obtained with a quadratic loss, keeping a financially balanced system, we resort on an exponential loss function. Such loss functions have been applied in Section 3.4 in the classical credibility setting. Our purpose here is to apply exponential loss functions to determine the optimal relativities.

When using the exponential loss function, the goal is now to minimize

$$Q^{exp} = \mathbb{E}\Big[\exp(-c(\Theta - r_L))\Big] \tag{4.18}$$

under the financial balance constraint $\mathbb{E}[r_L] = 1$. The parameter $c > 0$ determines the 'severity' of the bonus-malus scale. The loss (4.18) puts more weight on the errors resulting in an overestimation of the premium (i.e. $r_L > \Theta$), than on those coming from an underestimation. Consequently, the maluses are reduced, as well as the bonuses since financial stability has been imposed.

Let us derive the general solution of (4.18).

Proposition 4.1 *The solution of the constrained optimization problem (4.18) is*

$$r_L^{exp} = 1 + \frac{1}{c}\Big(\mathbb{E}\Big[\ln \mathbb{E}[\exp(-c\Theta)|L]\Big] - \ln \mathbb{E}[\exp(-c\Theta)|L]\Big). \tag{4.19}$$

Proof First, note that

$$\exp\left(cr_L^{exp}\right) = \frac{\exp(c)\exp\left(\mathbb{E}\Big[\ln \mathbb{E}\left[\exp(-c\Theta)|L\right]\Big]\right)}{\mathbb{E}\Big[\exp(-c\Theta)|L\Big]}.$$

Now, we have to minimize (4.18) that can be rewritten as

$$\mathbb{E}\Big[\exp(-c(\Theta - r_L))\Big] = \mathbb{E}\Big[\exp(c(r_L - r_L^{exp}))\Big] \times \exp(c)\exp\big(\mathbb{E}\big[\ln\mathbb{E}\left[\exp(-c\Theta)|L\right]\big]\big).$$

Invoking Jensen's inequality yields

$$\begin{aligned}\mathbb{E}\Big[\exp(-c(\Theta - r_L))\Big] &\geq \underbrace{\exp\Big(c\mathbb{E}[r_L - r_L^{exp}]\Big)}_{=1} \\ &\quad \times \exp(c)\exp\Big(\mathbb{E}\Big[\ln\mathbb{E}\left[\exp(-c\Theta)|L\right]\Big]\Big) \\ &= \mathbb{E}\Big[\exp(-c(\Theta - r_L^{exp}))\Big],\end{aligned}$$

which ends the proof.

□

Let us now compute the quantities in (4.19). Firstly,

$$\begin{aligned}\mathbb{E}[\exp(-c\Theta)|L=\ell] &= \mathbb{E}\Big[\mathbb{E}[\exp(-c\Theta)|L=\ell,\Lambda]\Big|L=\ell\Big] \\ &= \sum_k \mathbb{E}[\exp(-c\Theta)|L=\ell,\Lambda=\lambda_k]\Pr[\Lambda=\lambda_k|L=\ell] \\ &= \sum_k \int_0^{+\infty}\exp(-c\theta)\frac{\Pr[L=\ell|\Theta=\theta,\Lambda=\lambda_k]w_k}{\Pr[L=\ell,\Lambda=\lambda_k]}dF_\Theta(\theta) \\ &\quad \times \frac{\Pr[\Lambda=\lambda_k,L=\ell]}{\Pr[L=\ell]} \\ &= \frac{\sum_k w_k\int_0^{+\infty}\exp(-c\theta)\pi_\ell(\lambda_k\theta)dF_\Theta(\theta)}{\sum_k w_k\int_0^{+\infty}\pi_\ell(\lambda_k\theta)dF_\Theta(\theta)}. \end{aligned} \tag{4.20}$$

and secondly

$$\mathbb{E}\Big[\ln\mathbb{E}[\exp(-c\Theta)|L]\Big] \tag{4.21}$$

$$= \sum_{\ell=0}^{s}\Pr[L=\ell]\ln\mathbb{E}[\exp(-c\theta)|L=\ell]$$

$$= \sum_{\ell=0}^{s}\Pr[L=\ell]\ln\left(\frac{\sum_k w_k\int_0^{+\infty}\exp(-c\theta)\pi_\ell(\lambda_k\theta)dF_\Theta(\theta)}{\Pr[L=\ell]}\right). \tag{4.22}$$

Remark 4.4 If no *a priori* ratemaking is in force, the expressions (4.20) and (4.22) are equal to those derived in DENUIT & DHAENE (2001), that is

$$\mathbb{E}[\exp(-c\Theta)|L=\ell] = \frac{\int_0^{+\infty}\exp(-c\theta)\pi_\ell(\theta)dF_\Theta(\theta)}{\Pr[L=\ell]}$$

and

$$\begin{aligned}\mathbb{E}\Big[\ln\mathbb{E}[\exp(-c\Theta)|L]\Big] &= \sum_{\ell=0}^{s}\Pr[L=\ell]\ln\mathbb{E}[\exp(-c\Theta)|L=\ell]\\ &= \sum_{\ell=0}^{s}\Pr[L=\ell]\ln\left(\frac{\int_0^{+\infty}\exp(-c\theta)\pi_\ell(\theta)dF_\Theta(\theta)}{\Pr[L=\ell]}\right).\end{aligned}$$

4.6.2 Fixing the Value of the Severity Parameter

Let us briefly explain a possible criterion to fix the value of the parameter c. First, note that

$$\lim_{c\to 0} r_\ell^{exp} = \mathbb{E}[\Theta|L=\ell] = r_\ell^{quad},$$

where r_ℓ^{quad} is the relativity obtained with a quadratic loss function. Letting c tend to 0 thus yields Norberg's approach. In other words, the bonus-malus scale becomes more severe as c decreases. Now, the ratio of the variances of the premiums obtained with an exponential and a quadratic loss is given by

$$\frac{\mathbb{V}[r_L^{exp}]}{\mathbb{V}[r_L^{quad}]} = \frac{1}{c^2}\,\frac{\mathbb{V}\big[\ln\mathbb{E}[\exp(-c\Theta)|L]\big]}{\mathbb{V}\big[\mathbb{E}[\Theta|L]\big]} = \eta\,\% \leq 100\,\%.$$

The fact that the ratio of the variances is less than unity comes from the Jensen inequality. The idea is then to select the variance of the premium in the new system as a fraction of the corresponding variance under a quadratic loss (for instance $\eta = 25$, 50 or 75 %). Of course, other procedures can be applied. For instance, the actuary could select the value of r_0, or of r_s, and then compute c in order to match this value.

4.6.3 Linear Relativities

In practice, a linear scale of the form $r_\ell^{lin} = \alpha + \beta\ell$, $\ell = 0, 1, \ldots, s$, could be desirable. Let us now indicate how GILDE & SUNDT's (1989) approach can be extended using exponential loss functions. The aim is now to minimize the objective function

$$\mathcal{O}(\alpha,\beta) = \mathbb{E}\Big[\exp\big(-c(\Theta-\alpha-\beta L)\big)\Big]$$

under the financial balance constraint

$$\mathbb{E}[r_L^{lin}] = \mathbb{E}[\Theta] = 1$$

$$\Leftrightarrow \mathbb{E}[\Theta] = \alpha + \beta\mathbb{E}[L] \Leftrightarrow \alpha = \mathbb{E}[\Theta] - \beta\mathbb{E}[L].$$

It suffices to minimize

$$\widetilde{\mathcal{O}}(\beta) = \mathbb{E}\Big[\exp\Big(-c\Big(\Theta - \mathbb{E}[\Theta] - \beta(L - \mathbb{E}[L])\Big)\Big)\Big].$$

Differentiating $\widetilde{\mathcal{O}}$ with respect to β and equating to zero yields

$$\mathbb{E}\Big[(L - \mathbb{E}[L])\exp\Big(-c(\Theta - \mathbb{E}[\Theta] - \beta(L - \mathbb{E}[L]))\Big)\Big] = 0$$

$$\Leftrightarrow \quad \int_0^{+\infty}\sum_{\ell=0}^{s}(\ell - \mathbb{E}[L])\exp\Big(-c\Big(\theta - \mathbb{E}[\Theta] - \beta(\ell - \mathbb{E}[L])\Big)\Big)\pi_\ell(\theta)dF_\Theta(\theta) = 0$$

which has to be solved numerically to get the value of β (and hence of α). Convenient starting values for the numerical search are provided by (4.17).

4.6.4 Numerical Illustration

In this section, we give numerical examples of computation of bonus-malus scales when using an exponential loss function. We compare the results with those obtained previously.

In order to be able to compare the results, we have computed the relativities associated with the same severity factor $c = 1$. These results are given in Table 4.13 for the -1/top, $-1/+2$ and $-1/+3$ systems and Portfolio A. Specifically, Table 4.13 gives the relativities obtained with an exponential loss function with severity parameter $c = 1$, with and without *a priori* risk classification, as well as the analogues for the $-1/+2$ and $-1/+3$ systems. As was the case with the quadratic loss function, we see that *a priori* risk classification reduces the dispersion of the relativities.

We observe that the relativities computed when no *a priori* ratemaking is in force give bigger bonuses when the policyholders are in level 0 but also impose bigger maluses in the other levels. Indeed, when no *a priori* ratemaking is in force, the *a posteriori* correction must be more severe in order to distinguish between good and bad drivers. On the contrary, when an *a priori* ratemaking is in force, the correction applied with the *a posteriori* tariff must be softer because a greater part of the risk is already taken into account in the *a priori* ratemaking.

Influence of the Loss Function

We can also compare the results obtained when using the quadratic loss function and when using the exponential loss function (with different values of c). These are given in

Table 4.13 Numerical characteristics for the systems −1/top, −1/+2 and −1/+3 with a severity $c = 1$ and Portofolio A.

−1/top		
Level ℓ	Relativity r_ℓ^{exp} without *a priori* ratemaking	Relativity r_ℓ^{exp} with *a priori* ratemaking
5	161.3 %	154.1 %
4	148.0 %	142.3 %
3	137.1 %	132.9 %
2	128.1 %	125.2 %
1	120.4 %	118.7 %
0	69.0 %	72.3 %
−1/+2		
Level ℓ	Relativity r_ℓ^{exp} without *a priori* ratemaking	Relativity r_ℓ^{exp} with *a priori* ratemaking
5	253.4 %	229.3 %
4	204.4 %	189.8 %
3	183.4 %	172.6 %
2	132.8 %	130.0 %
1	124.9 %	123.4 %
0	71.6 %	75.8 %
−1/+3		
Level ℓ	Relativity r_ℓ^{exp} without *a priori* ratemaking	Relativity r_ℓ^{exp} with *a priori* ratemaking
5	209.3 %	194.7 %
4	187.8 %	176.3 %
3	137.2 %	133.4 %
2	128.3 %	125.9 %
1	120.7 %	119.4 %
0	69.3 %	73.2 %

Tables 4.14, 4.15 and 4.16 respectively for the −1/top, −1/+2 and −1/+3 systems. The exponential relativities have been computed for a severity coefficient ranging from 0 to 5. The limit value of $c = 0$ provides the same result as with the quadratic loss function. We observe in Table 4.14 that an increasing value of c leads to less dispersed relativities. The maximal penalty decreases as c increases: keeping financial balance, increasing c tends to soften *a posteriori* corrections.

Table 4.14 Numerical characteristics for the system −1/top and Portofolio A.

Level ℓ	Relativity r_ℓ^{quad} with *a priori* ratemaking ($c=0$)	Relativity r_ℓ^{exp} with *a priori* ratemaking ($c=1$)	Relativity r_ℓ^{exp} with *a priori* ratemaking ($c=2$)	Relativity r_ℓ^{exp} with *a priori* ratemaking ($c=5$)
5	181.2 %	154.1 %	141.8 %	126.2 %
4	159.9 %	142.3 %	133.6 %	121.9 %
3	143.9 %	132.9 %	127.0 %	118.4 %
2	131.2 %	125.2 %	121.4 %	115.3 %
1	120.9 %	118.7 %	116.6 %	112.6 %
0	61.2 %	72.3 %	77.9 %	85.4 %

Table 4.15 Numerical characteristics for the system −1/+2 and Portofolio A.

Level ℓ	Relativity r_ℓ^{quad} with *a priori* ratemaking ($c=0$)	Relativity r_ℓ^{exp} with *a priori* ratemaking ($c=1$)	Relativity r_ℓ^{exp} with *a priori* ratemaking ($c=2$)	Relativity r_ℓ^{exp} with *a priori* ratemaking ($c=5$)
5	271.4 %	229.3 %	207.3 %	174.9 %
4	218.5 %	189.8 %	174.3 %	151.1 %
3	192.5 %	172.6 %	161.2 %	143.3 %
2	138.8 %	130.0 %	124.9 %	117.2 %
1	128.6 %	123.4 %	120.0 %	114.3 %
0	68.5 %	75.8 %	79.8 %	85.9 %

Table 4.16 Numerical characteristics for the system −1/+3 and Portofolio A.

Level ℓ	Relativity r_ℓ^{quad} with *a priori* ratemaking ($c=0$)	Relativity r_ℓ^{exp} with *a priori* ratemaking ($c=1$)	Relativity r_ℓ^{exp} with *a priori* ratemaking ($c=2$)	Relativity r_ℓ^{exp} with *a priori* ratemaking ($c=5$)
5	230.8 %	194.7 %	176.7 %	151.6 %
4	200.9 %	176.3 %	163.2 %	143.9 %
3	145.2 %	133.4 %	127.1 %	118.2 %
2	133.1 %	125.9 %	121.6 %	115.1 %
1	123.0 %	119.4 %	116.9 %	112.4 %
0	64.2 %	73.2 %	78.0 %	84.9 %

4.7 Special Bonus Rule

4.7.1 The Former Belgian Compulsory System

In this section we will concentrate on the former compulsory Belgian bonus-malus system, which all companies operating in Belgium have been obliged to use from 1992 to 2002. The Belgian system consists of a scale of 23 levels (numbered from 0 to 22). A new driver starts in class 11 if he uses his vehicle for pleasure and commuting and in class 14 if he uses his vehicle for business. Each claim-free year is rewarded by a bonus point. The first claim is penalized by four malus points and the subsequent ones by five malus points each.

According to the special bonus rule, a policyholder with four claim-free years cannot be in a class above 14. This restriction is a concession to insureds with many claims in a few years and who suddenly improve; very few policyholders are ever able to take advantage of this rule.

Actually, the Belgian bonus-malus system is not Markovian due to the special bonus rule (i.e. due to the fact that policyholders occupying high levels are sent to level 14 after four claim-free years). Fortunately, it is possible to introduce fictitious classes in order to meet the memoryless property by splitting the levels 16 to 21 into subclasses, depending on the number of consecutive years without accident.

4.7.2 Fictitious Levels

Splitting the levels 16 to 21 into sub-levels, depending on the number of consecutive years without accident, allows us to account for the special bonus rule. Let n_j be the number of sub-levels to be associated with bonus level j. A level $j.i$ is to be understood as level j and i consecutive years without accidents. The transition rules are completely defined in Table 4.17 and the different values for n_j are given in Table 4.18. We take some liberty with the notation by using the value 0 for the subscript i and by not using a subscript when $n_j = 1$.

4.7.3 Determination of the Relativities

The relativities $r_{j.i}$ are obtained by minimizing the squared difference between the true relative premium Θ and the relative premium r_L applicable to the policyholder when stationary state has been reached.

The current situation is more complicated because some levels have to be constrained to have the same relativity. Indeed the artificial levels $j.i$ have the property that

$$r_j = r_{j.1} = \cdots = r_{j.n_j}, \quad j = 0, \ldots, s.$$

We have to minimize $\mathbb{E}\left[(\Theta - r_L)^2\right]$ under these constraints.

The solution is given by

$$r_j = \frac{\sum_k w_k \int_0^\infty \sum_{i=1}^{n_j} \theta \pi_{j.i}(\lambda_k \theta) dF_\Theta(\theta)}{\sum_k w_k \int_0^\infty \sum_{i=1}^{n_j} \pi_{j.i}(\lambda_k \theta) dF_\Theta(\theta)}. \tag{4.23}$$

Table 4.17 Transition rules of the Belgian bonus-malus system with fictitious levels accounting for the special bonus rule.

Class	Class after k accidents					
	k = 0	1	2	3	4	5
22	21.1	22	22	22	22	22
21.0	20.1	22	22	22	22	22
21.1	20.2	22	22	22	22	22
20.0	19.1	22	22	22	22	22
20.1	19.2	22	22	22	22	22
20.2	19.3	22	22	22	22	22
19.0	18.1	22	22	22	22	22
19.1	18.2	22	22	22	22	22
19.2	18.3	22	22	22	22	22
19.3	14	22	22	22	22	22
18.0	17	22	22	22	22	22
18.1	17.2	22	22	22	22	22
18.2	17.3	22	22	22	22	22
18.3	14	22	22	22	22	22
17	16	21.0	22	22	22	22
17.2	16.3	21.0	22	22	22	22
17.3	14	21.0	22	22	22	22
16	15	20.0	22	22	22	22
16.3	14	20.0	22	22	22	22
15	14	19.0	22	22	22	22
14	13	18.0	22	22	22	22
13	12	17	22	22	22	22
12	11	16	21.0	22	22	22
11	10	15	20.0	22	22	22
10	9	14	19.0	22	22	22
9	8	13	18.0	22	22	22
8	7	12	17	22	22	22
7	6	11	16	21.0	22	22
6	5	10	15	20.0	22	22
5	4	9	14	19.0	22	22
4	3	8	13	18.0	22	22
3	2	7	12	17	22	22
2	1	6	11	16	21.0	22
1	0	5	10	15	20.0	22
0	0	4	9	14	19.0	22

Note that it is easily seen that

$$r_j = \sum_{i=1}^{n_j} \frac{\Pr[L = j.i]}{\sum_{i=1}^{n_j} \Pr[L = j.i]} r_{j.i},$$

where the $r_{j.i}$s represent the non-constrained solution of the minimization of $\mathbb{E}\left[(\Theta - r_L)^2\right]$.

Table 4.18 Sub-levels of the Belgian bonus-malus system.

j	n_j
22	1
21	2
20	3
19	4
18	4
17	3
16	2
15	1
14	1
13	1
12	1
11	1
10	1
9	1
8	1
7	1
6	1
5	1
4	1
3	1
2	1
1	1
0	1

We also immediately verify that

$$\sum_{j=0}^{s} r_j \Pr[L = j] = 1$$

which ensures that the bonus-malus scale is financially balanced at stationary state.

4.7.4 Numerical Illustration

The numerical results for the Belgian bonus-malus system and for Portfolio A are displayed in Table 4.19. In this table, the special bonus rule has not been taken into account. Specifically, the values in the third column are computed with the help of (4.14) with $\widehat{a} = 0.889$ and $\widehat{\lambda} = 0.1474$. Those values were obtained by fitting a Negative Binomial distribution to the portfolio's observed claims frequencies. The fourth column is based on (4.13) with $\widehat{a} = 1.065$ and the $\widehat{\lambda}_k$s obtained from the *a priori* risk classification described in Table 2.7. The last column is computed with the help of (4.15).

Once the stationary state has been reached, more or less half of the policies occupy level 0 and enjoy the maximum discount. This is due to the fact that the transition rules of the

Table 4.19 Numerical characteristics for the Belgian bonus-malus system without the special bonus rule, computed on the basis of Portfolio A.

Level ℓ	$\Pr[L=\ell]$	$r_\ell=\mathbb{E}[\Theta \mid L=\ell]$ without *a priori* ratemaking	$r_\ell=\mathbb{E}[\Theta \mid L=\ell]$ with *a priori* ratemaking	$\mathbb{E}[\Lambda \mid L=\ell]$
22	5.4 %	306.0 %	271.5 %	18.4 %
21	3.8 %	273.9 %	247.5 %	17.7 %
20	2.9 %	248.7 %	229.1 %	17.2 %
19	2.3 %	228.3 %	214.1 %	16.8 %
18	1.9 %	211.2 %	201.4 %	16.5 %
17	1.6 %	196.5 %	190.3 %	16.2 %
16	1.5 %	183.8 %	180.3 %	16.0 %
15	1.3 %	172.6 %	171.4 %	15.8 %
14	1.3 %	162.5 %	163.1 %	15.6 %
13	1.2 %	152.8 %	155.0 %	15.5 %
12	1.2 %	143.9 %	147.2 %	15.3 %
11	1.2 %	135.9 %	140.2 %	15.2 %
10	1.2 %	128.9 %	134.0 %	15.1 %
9	1.4 %	119.8 %	125.5 %	14.9 %
8	1.6 %	111.1 %	117.3 %	14.8 %
7	1.7 %	104.8 %	111.4 %	14.7 %
6	1.8 %	99.8 %	106.7 %	14.6 %
5	1.8 %	95.6 %	102.8 %	14.6 %
4	4.3 %	75.2 %	82.6 %	14.3 %
3	3.8 %	72.7 %	80.2 %	14.3 %
2	3.4 %	70.3 %	77.8 %	14.2 %
1	3.1 %	68.1 %	75.5 %	14.2 %
0	50.3 %	37.6 %	45.1 %	13.8 %

Belgian system are not severe enough in comparison with the average claims frequency. And the situation is even more serious when looking at the actual figures on the market. Indeed the market average claims frequency is even smaller than the one of the analysed portfolio.

If we compute the relativities without taking into account the *a priori* ratemaking system, the relativities vary between 37.6 % for level 0 and 306.0 % for level 22. On the other hand, if we adapt the relativities to the *a priori* risk classification, these relativities vary between 45.1 % and 271.5 %; the severity of the *a posteriori* corrections is thus weaker in this case.

Tables 4.20, 4.21 and 4.22 display the results for the Belgian bonus-malus system when the special bonus rule is taken into account. Table 4.20 compares the probability mass functions of L when the special bonus rule is taken into account, with respect to the case without special bonus rule. This rule decreases the probabilities associated with upper levels. The decrease for level 18 is even more apparent, since policyholders in that level benefit from the special bonus rule.

Table 4.21 gives the relativities computed with and without the special bonus rule. We see that the relativities are larger with the rule, since it boils down to softening the penalty in the case where claims are filed to the insurance company.

Table 4.20 Distribution of L for the Belgian bonus-malus system, computed on the basis of Portfolio A.

Level ℓ	$\Pr[L=\ell]$	
	Without special bonus rule	With special bonus rule
22	5.4 %	4.3 %
21	3.8 %	3.0 %
20	2.9 %	2.2 %
19	2.3 %	1.7 %
18	1.9 %	0.9 %
17	1.6 %	1.0 %
16	1.5 %	1.0 %
15	1.3 %	1.0 %
14	1.3 %	2.1 %
13	1.2 %	1.9 %
12	1.2 %	1.8 %
11	1.2 %	1.7 %
10	1.2 %	1.6 %
9	1.4 %	1.7 %
8	1.6 %	1.9 %
7	1.7 %	2.0 %
6	1.8 %	2.0 %
5	1.8 %	2.0 %
4	4.3 %	4.5 %
3	3.8 %	4.0 %
2	3.4 %	3.6 %
1	3.1 %	3.2 %
0	50.3 %	50.9 %

We observe in Table 4.22 that the average *a priori* expected claim frequency $\mathbb{E}[\Lambda|L=\ell]$ in level ℓ is always higher with the special bonus rule than without that rule. The effect is more pronounced in the highest levels of the scale and less pronounced in the lowest levels of the scale. This fact is obvious from the definition of the special bonus rule. The policyholders attaining the highest classes of the scale benefit from the special bonus rule. Those staying in these highest classes show therefore a higher expected frequency. Even below level 14 the effect remains true because the policyholders have benefitted from it before attaining the lowest levels. Obviously the effect is less and less pronounced at the bottom of the scale.

Some insurance companies use the bonus-malus scale as an underwriting tool. For instance, they systematically refuse drivers with a bonus-malus level > 14. Our calculations show that this is unreasonable because drivers at level 15 are on average less risky than drivers at level 14.

We see from Table 4.19 that without the special bonus rule, the relativities are always increasing from level 0 to level 22. The same increasing pattern is observed for $\mathbb{E}[\Lambda|L=\ell]$. When looking at the results for the bonus-malus system with special bonus rule (Table 4.21), we observe that the relativities at levels 13–16 are not ordered any more. This can be

Table 4.21 Relativities $r_\ell = \mathbb{E}[\Theta|L = \ell]$ for the Belgian bonus-malus system computed on the basis of Portfolio A.

Level ℓ	Relativities	
	Without special bonus rule	With special bonus rule
22	271.5 %	284.3 %
21	247.5 %	258.9 %
20	229.1 %	238.9 %
19	214.1 %	222.5 %
18	201.4 %	199.4 %
17	190.3 %	194.2 %
16	180.3 %	186.9 %
15	171.4 %	179.1 %
14	163.1 %	192.4 %
13	155.0 %	179.8 %
12	147.2 %	168.4 %
11	140.2 %	158.3 %
10	134.0 %	149.5 %
9	125.5 %	138.6 %
8	117.3 %	128.3 %
7	111.4 %	120.7 %
6	106.7 %	114.7 %
5	102.8 %	109.6 %
4	82.6 %	87.0 %
3	80.2 %	84.1 %
2	77.8 %	81.3 %
1	75.5 %	78.6 %
0	45.1 %	46.2 %

explained by the fact that many drivers at level 14 have benefitted from the special bonus rule. It is clear that such a situation is not acceptable from a commercial point of view. We may constrain the scale to be linear in the spirit of GILDE & SUNDT (1989). However we propose a local adjustment to the scale in order to keep $\mathbb{E}[(\Theta - r_L)^2]$ as small as possible.

Let us constrain the scale to be linear between levels 13 and 16. We are looking for updated values for r_j', $j = 13, \ldots, 16$. They are such that $r_j' = r_{j-1}' + a$, $j = 14, 15, 16$ where $a = (r_{16}' - r_{13}')/3$. We also want to keep the financial equilibrium of the system. Therefore we constrain a local equilibrium :

$$\sum_{j=13}^{16} r_j \pi_j = \sum_{j=13}^{16} r_j' \pi_j.$$

Choosing $r_{13}' = 179.8\,\%$, we obtain $r_{16}' = 193.7\,\%$.

Table 4.22 Average *a priori* claim frequency $\mathbb{E}[\Lambda|L=\ell]$ in level ℓ for the Belgian bonus-malus system, computed on the basis of Portfolio A.

Level ℓ	$\mathbb{E}[\Lambda\|L=\ell]$	
	Without special bonus rule	With special bonus rule
22	18.4 %	18.7 %
21	17.7 %	18.0 %
20	17.2 %	17.4 %
19	16.8 %	17.0 %
18	16.5 %	16.4 %
17	16.2 %	16.3 %
16	16.0 %	16.1 %
15	15.8 %	16.0 %
14	15.6 %	16.3 %
13	15.5 %	16.0 %
12	15.3 %	15.8 %
11	15.2 %	15.6 %
10	15.1 %	15.4 %
9	14.9 %	15.2 %
8	14.8 %	15.0 %
7	14.7 %	14.9 %
6	14.6 %	14.8 %
5	14.6 %	14.7 %
4	14.3 %	14.4 %
3	14.3 %	14.3 %
2	14.2 %	14.3 %
1	14.2 %	14.2 %
0	13.8 %	13.8 %

Now let us compare the value of the expected error $Q=\mathbb{E}\left[(\Theta-r_L)^2\right]$ with the original model, Q_1, and with the constrained model, Q_2: we get

$$Q_1=0.41121 \text{ and } Q_2=0.41150.$$

This shows that the error induced by the commercial constraint is really small. So we may adapt the scale locally without resorting to a full linear scale constraint.

We can perform the local minimization numerically without imposing a linear scale between levels 13 and 16. We use the following constraints :

$$r'_{13}\leq r'_{14},$$

$$r'_{14}\leq r'_{15},$$

$$r'_{15}\leq r'_{16},$$

$$\sum_{j=13}^{16} r_j \pi_j = \sum_{j=13}^{16} r_j^{'} \pi_j,$$

$$r_j^{'} \geq 169\,\% \quad j = 13, \ldots, 16,$$

$$r_j^{'} \leq 194\,\% \quad j = 13, \ldots, 16.$$

And we obtain $r_{13}^{'} = 179.8\,\%$ and $r_{14}^{'} = r_{15}^{'} = r_{16}^{'} = 187.8\,\%$. The value of Q is now 0.41135.

4.7.5 Linear Relativities for the Belgian Scale

We shall now demonstrate how to get linear relativities for the Belgian scale. In addition to the constraint of building a linear scale, we add a new constraint: the artificial states must have the same relativity. The minimization problem then becomes :

$$\min \mathbb{E}\Big[(\Theta - r_L^{lin})^2\Big] = \min \mathbb{E}\Big[(\Theta - \alpha - \beta L)^2\Big]$$

Table 4.23 Relativities r_ℓ and r_ℓ^{lin} for the Belgian bonus-malus system, taking into account the special bonus rule and computed on the basis of Portfolio A.

Level ℓ	Relativities	
	General scale	Linear scale
22	284.3 %	267.4 %
21	258.9 %	257.4 %
20	238.9 %	247.5 %
19	222.5 %	237.6 %
18	199.4 %	227.6 %
17	194.2 %	217.7 %
16	186.9 %	207.7 %
15	179.1 %	197.8 %
14	192.4 %	187.9 %
13	179.8 %	177.9 %
12	168.4 %	168.0 %
11	158.3 %	158.0 %
10	149.5 %	148.1 %
9	138.6 %	138.2 %
8	128.3 %	128.2 %
7	120.7 %	118.3 %
6	114.7 %	108.3 %
5	109.6 %	98.4 %
4	87.0 %	88.5 %
3	84.1 %	78.5 %
2	81.3 %	68.6 %
1	78.6 %	58.6 %
0	46.2 %	48.7 %

so that

$$r_\ell = r_{\ell.1} = r_{\ell.2} = \cdots = r_{\ell.n_\ell}, \quad \ell = 0, 1, \ldots, s.$$

The solution to this problem is the same as above. We merely have to replace π_ℓ by $\sum_{i=1}^{n_\ell} \pi_{\ell.i}$. The results are displayed in Table 4.23. The constant step between two levels of the linear scale is equal to 9.9 % and the value of the mean square error is $Q = 0.41779$. The major advantage of the linear scale is that it gives a system that is commercially more acceptable.

4.8 Change of Scale

4.8.1 Migration from One Scale to Another

Since the 90s insurance markets in the EU have been deregulated. More competition is allowed. Two related problems arise with the deregulation of the bonus-malus systems. The first one consists of transferring the policyholders to the new scales. The second one is more difficult: it consists of transferring a new policyholder to the scale of the company knowing his level in the scale of his previous insurer.

The aim of the present section is to show how to develop rules allowing the transfer of a policyholder to a bonus-malus scale knowing his level in his previous bonus-malus scale. The *a posteriori* probability density function of Θ given the level L occupied in the bonus-malus scale is given by

$$\begin{aligned} f_{\Theta|L=\ell}(\theta) &= \frac{\Pr[L=\ell|\Theta=\theta] f_\theta(\theta)}{\Pr[L=\ell]} \\ &= \frac{\sum_k \omega_k \pi_\ell(\lambda_k \theta) dF_\Theta(\theta)}{\sum_k \omega_k \int_0^\infty \pi_\ell(\lambda_k \theta) dF_\Theta(\theta)}. \end{aligned}$$

Because we want to move a policyholder from one scale to the other, we should try to put the policyholder at a level which is as close as possible to his level in his original bonus-malus scale. By 'close', we mean here having the *a posteriori* random effect as close as possible.

4.8.2 Kolmogorov Distance

In addition to the total variation distance d_{TV} used previously, we also need the Kolmogorov distance. The Kolmogorov (or uniform) metric based on the well-known Kolmogorov-Smirnov statistic (associated with the goodness-of-fit test with that name), is defined as follows: The Kolmogorov distance d_K between the random variables X and Y is given by

$$d_K(X, Y) = \sup_{t \in \mathbb{R}} |\overline{F}_X(t) - \overline{F}_Y(t)|. \tag{4.24}$$

Given two random variables X and Y, we have that $d_K(X, Y) \leq d_{TV}(X, Y)$. This result is an immediate consequence of (4.11) since $\overline{F}_X(t) = \Pr[X \in (t, +\infty)]$.

4.8.3 Distances between the Random Effects

A first measure may be to compare the expected *a posteriori* random effect, which actually is the relative premium at level ℓ in the bonus-malus scale :

$$r_{L=\ell} = \mathbb{E}[\Theta|L=\ell].$$

So the closest level ℓ_j in scale 2 to the level ℓ_i occupied in scale 1 is given by

$$\text{argmin}_{\ell_j}(r_{L_1=\ell_i} - r_{L_2=\ell_j})^2.$$

This rule simply amounts to placing the policyholder in the new scale at the level with the closest relativity to the one applicable in the previous scale. Because most commercial scales are normalized (to associate a unit relativity with the entry level), this means that the insurer has to compute the relativities for both scales. The implicit assumption is that the new entrant has the same characteristics as the policyholders in the portfolio (no adverse selection being allowed).

Another measure of discrepancy consists of comparing the distribution functions of the *a posteriori* random effects. This can be done by using the Kolmogorov distance d_K or the total variation distance d_{TV}:

$$d_K(\Theta|L_1=\ell_i, \Theta|L_2=\ell_j) = \max_\theta |\Pr[\Theta \leq \theta|L_1=\ell_i] - \Pr[\Theta \leq \theta|L_2=\ell_j]|,$$

$$d_{TV}(\Theta|L_1=\ell_i, \Theta|L_2=\ell_j) = \int_0^\infty |f_{\Theta|L_1=\ell_i}(\theta) - f_{\Theta|L_2=\ell_j}(\theta)|d\theta.$$

Summarizing, a policyholder being at level ℓ_i in scale 1 and moving to scale 2 will be put in the level ℓ_j of scale 2 that minimizes one of the following distances:

(i) $d_{\mathbb{E}}(\Theta|L_1=\ell_i, \Theta|L_2=\ell_j) = (r_{L_1=\ell_i} - r_{L_2=\ell_j})^2$

(ii) $d_K(\Theta|L_1=\ell_i, \Theta|L_2=\ell_j)$

(iii) $d_{TV}(\Theta|L_1=\ell_i, \Theta|L_2=\ell_j)$.

4.8.4 Numerical Illustration

In this section we use Portfolio A and its risk classification described in Table 2.7. Let us assume that we want to move policyholders between the following two bonus-malus scales:

- The -1/top scale with 6 levels, numbered 0 to 5.
- The $-1/+2$ scale of Taylor (1997) with 9 levels, having transition rules given in Table 4.24.

Table 4.24 Transition rules for the scale $-1/+2$ scale of Taylor (1997) with 9 levels.

Starting level	Level occupied if 0	1	2	3	≥ 4 claim is/are reported
8	7	8	8	8	8
7	6	8	8	8	8
6	5	8	8	8	8
5	4	7	8	8	8
4	3	6	8	8	8
3	2	5	7	8	8
2	1	4	6	8	8
1	0	3	5	7	8
0	0	2	4	6	8

Table 4.25 Distances between $\Theta|L_1 = \ell_1$ and $\Theta|L_2 = \ell_2$ for the three metrics.

$\ell_2 \backslash \ell_1$	0	1	2	3	d_K 4	5	6	7	8
0	0.042	0.377	0.404	0.574	0.613	0.701	0.740	0.790	0.825
1	0.323	0.016	0.044	0.245	0.299	0.421	0.483	0.562	0.625
2	0.357	0.061	0.036	0.200	0.254	0.376	0.439	0.520	0.585
3	0.396	0.110	0.080	0.153	0.204	0.325	0.388	0.471	0.539
4	0.439	0.168	0.137	0.105	0.151	0.267	0.330	0.414	0.483
5	0.488	0.237	0.207	0.061	0.097	0.202	0.262	0.344	0.414

$\ell_2 \backslash \ell_1$	0	1	2	3	d_{TV} 4	5	6	7	8
0	0.089	0.754	0.808	1.148	1.225	1.402	1.481	1.580	1.651
1	0.646	0.081	0.107	0.490	0.599	0.841	0.966	1.125	1.249
2	0.715	0.123	0.094	0.402	0.508	0.751	0.878	1.041	1.172
3	0.791	0.220	0.160	0.320	0.411	0.649	0.777	0.943	1.078
4	0.877	0.337	0.274	0.270	0.324	0.535	0.660	0.827	0.966
5	0.976	0.475	0.413	0.281	0.286	0.426	0.527	0.688	0.834

$\ell_2 \backslash \ell_1$	0	1	2	3	$d_{\mathbb{E}}$ 4	5	6	7	8
0	0.001	0.308	0.391	1.077	1.415	2.329	3.093	4.349	5.908
1	0.314	0.002	0.001	0.194	0.351	0.863	1.350	2.215	3.362
2	0.440	0.021	0.006	0.114	0.239	0.682	1.120	1.917	2.993
3	0.624	0.074	0.041	0.044	0.132	0.489	0.868	1.584	2.572
4	0.902	0.186	0.131	0.003	0.041	0.291	0.596	1.207	2.085
5	1.377	0.430	0.343	0.030	0.000	0.100	0.301	0.765	1.489

Table 4.25 provides the distances between the conditional random effect for our three metrics. On the basis of these distances, we see that the minima are attained for the following transition rules. Transition rules from scale 1 to scale 2 are

	d_K	d_{TV}	$d_{\mathbb{E}}$
ℓ_1	ℓ_2	ℓ_2	ℓ_2
0	0	0	0
1	1	1	2
2	2	2	2
3	2	2	2
4	3	3	3
5	3	3	4

and transition rules from scale 2 to scale 1 are

	d_K	d_{TV}	$d_{\mathbb{E}}$
ℓ_2	ℓ_1	ℓ_1	ℓ_1
0	0	0	0
1	1	1	1
2	2	2	1
3	5	4	4
4	5	5	5
5	5	5	5
6	5	5	5
7	5	5	5
8	5	5	5

Let us now take into account the *a priori* characteristics of the driver. It will be seen that good and bad drivers are not placed in the same way when they are transferred from one scale to the other. The transition rules for an *a priori* bad driver (specifically, a male driver aged more than 30 years, with premium split, having private use of the car and living in an urban environment, whose *a priori* expected frequency is 0.1794) are given in the next tables: from scale 1 to scale 2

	d_K	d_{TV}	$d_{\mathbb{E}}$
ℓ_1	ℓ_2	ℓ_2	ℓ_2
0	0	0	0
1	2	1	2
2	2	2	2
3	2	2	3
4	3	3	3
5	3	4	4

and from scale 2 to scale 1

ℓ_2	d_K ℓ_1	d_{TV} ℓ_1	$d_{\mathbb{E}}$ ℓ_1
0	0	0	0
1	1	1	1
2	1	1	1
3	5	4	4
4	5	4	4
5	5	5	5
6	5	5	5
7	5	5	5
8	5	5	5

We here compare the distance between $\Theta|L_1 = \ell_i, \Lambda = 0.1794$ and $\Theta|L_2 = \ell_j, \Lambda = 0.1794$. We have

$$f_{\Theta|L=\ell,\Lambda=\lambda}(\theta) = \frac{\pi_\ell(\lambda\theta) f_\Theta(\theta)}{\int_0^\infty \pi_\ell(\lambda\theta) f_\Theta(\theta) d\theta}.$$

The transition rules for a good policyholder (specifically, a male driver aged more than 30 years, with upfront premium, having private use of the car and living in a rural environment, whose *a priori* expected frequency is 0.0928) are given in the next tables: from scale 1 to scale 2

ℓ_1	d_K ℓ_2	d_{TV} ℓ_2	$d_{\mathbb{E}}$ ℓ_2
0	0	0	0
1	1	1	1
2	1	1	2
3	2	2	2
4	2	2	2
5	2	2	3

and from scale 2 to scale 1

ℓ_2	d_K ℓ_1	d_{TV} ℓ_1	$d_{\mathbb{E}}$ ℓ_1
0	0	0	0
1	2	2	1
2	2	3	2
3	5	5	5
4	5	5	5
5	5	5	5
6	5	5	5
7	5	5	5
8	5	5	5

We here compare the distance between $\Theta|L_1 = \ell_i, \Lambda = 0.0928$ and $\Theta|L_2 = \ell_j, \Lambda = 0.0928$.

We observe that the different metrics we have chosen do not provide very different results. Because the expected posterior random effect has a financial meaning (i.e. it is the multiplier of the average cost to get the *a posteriori* premium), we may be tempted to choose its corresponding metric to transfer a policyholder from one scale to the other. Note also that the *a priori* characteristics of the driver influence the way the policy is transferred from one scale to the other. This results in a number of rules according to the risk classification scheme applied by the insurance company.

Remark 4.5 Transferring a policyholder from the bonus-malus scale of a given insurer to the bonus-malus scale of another insurer remains a more complicated task. Indeed, the actuary needs to know the *a posteriori* random effect in both situations. However the *a priori* random effect may be different due to another type of *a priori* tariff or due to adverse selection. The distance to minimize then extends to $d(\Theta_1|L_1=\ell_1, \Theta_2|L_2=\ell_2)$.

4.9 Dependence in Bonus-Malus Scales

It is clear that the sequence $(L_1, L_2, \ldots)$ is CIS, but we do not have in general that Θ given $L=\ell$ increases with ℓ. The reason is as follows: Because of the finite number of levels, it does not automatically follow that a policyholder in a higher level filed more claims in the past. To show this, consider the scale -1/top. A policyholder in level 3 in year 3 may occupy that level having filed two claims in year 1 and no claim after. On the contrary, a policyholder in level 4 could have filed one claim in year 2. From (3.18) we conclude that Θ should be larger (in the $\preceq_{\rm ST}$-sense) for policyholder 1 than for policyholder 2 despite the fact that policyholder 1 is in an inferior level.

As a consequence, we cannot be sure that the Bayesian relativities obtained with a quadratic loss function are increasing with the level occupied in the scale. This is why linear relativities are so useful in practice.

Remark 4.6 This counterintuitive fact becomes reasonable if we allow for random effects to vary in time. Provided the autocorrelogram is decreasing, old claims have less predictive power than recent ones. Then, we have to compare two old claims to one recent claim, and it becomes less obvious that policyholder 1 is more dangerous than policyholder 2.

4.10 Further Reading and Bibliographic Notes

Chapter 7 in Rolski *et al.* (1999) offers an excellent introduction to Markov chains, with applications to bonus-malus systems. Several parts of this chapter are directly inspired from this source. For the most part, this chapter is based on Pitrebois, Denuit & Walhin (2003b), following on from Taylor (1997) for the extension of Norberg's (1976) pioneering work on segmented tariffs; on Pitrebois, Denuit & Walhin (2004) for the linear relativities; on Pitrebois, Walhin & Denuit (2006c) for Section 4.8; and on Pitrebois, Denuit & Walhin (2003a) for the Belgian bonus-malus scale and its special bonus rule.

Norberg (1976), Borgan, Hoem & Norberg (1981) and Gilde & Sundt (1989) assumed that the bonus-malus system forms a first order Markov chain. Centeno & Andrade e Silva (2002) considered bonus-malus systems that are not first order Markovian processes, but that can be made Markovian by increasing the number of states as we did in Section 4.7.

The notion of distance found a second life in probability in the form of metrics in spaces of random variables and their probability distributions. The study of limit theorems (among other questions) made it necessary to introduce functionals evaluating the nearness of probability distributions in some probabilistic sense. In this chapter, the total variance and Kolmogorov distances have been used in connection with bonus-malus systems. We refer the interested reader to Chapter 9 of DENUIT *ET AL.* (2005) for a detailed account of probability metrics and their applications in risk theory.

The premium relativities are traditionally computed with the help of a quadratic loss function, in the vein of NORBERG (1976). Other loss functions nevertheless also deserve consideration, such as the exponential loss function applied by DENUIT & DHAENE (2001) to the computation of the relativities. For the sake of completeness, let us mention that the absolute value loss function has been successfully applied to the determination of the relativities by HERAS, VILAR & GIL (2002) and HERAS, GIL, GARCIA-PINEDA & VILAR (2004).

If in a given market, companies start to compete on the basis of bonus-malus systems, many policyholders could leave the portfolio after the occurrence of an accident, in order to avoid the resulting penalties. Those attritions can be incorporated in the model by adding a supplementary level to the Markov chain (in the spirit of CENTENO & ANDRADE E SILVA (2001)). Transitions from a level of the bonus-malus scale to this state represent a policyholder leaving the portfolio whereas transitions from this state to any level of the bonus-malus scale mean that a new policy enters the portfolio.

It has been assumed throughout this chapter that the unknown expected claim frequencies were constant and that the random effects representing hidden characteristics were time-invariant. Dropping these assumptions makes the determination of the relativities much harder. We refer the interested reader to BROUHNS, GUILLÉN, DENUIT & PINQUET (2003) for a thorough study of this general situation. A fundamental difference with the traditional approaches is that we lose the homogeneity of the chain in a dynamic segmented environment. Indeed, if the observable characteristics of the policyholders are allowed to vary in time, the claim frequencies are no longer constant and the trajectory of the policyholder in the bonus-malus scale is no longer described by a homogeneous Markov chain, but well described by a non-homogeneous one. Consequently, the classical techniques based on stationary distributions cannot be applied to the problem of determining the relativities. BROUHNS, GUILLÉN, DENUIT & PINQUET (2003) propose a computer-intensive method to calibrate bonus-malus scales. Their paper clearly illustrates the strong complementarity of *a priori* and *a posteriori* ratemakings. The main originality of their approach is to compare on the basis of real data four different credibility models: static versus dynamic heterogeneity, with and without recognizing *a priori* risk classification. The impact of the different assumptions becomes clear in the numerical illustrations.

ANDRADE E SILVA & CENTENO (2005) suggested the use of geometric relativities instead of linear ones. Specifically, under a quadratic loss function, the relativity associated with level ℓ becomes $r_\ell^{geo} = \alpha\beta^\ell$, with α and β positive. As pointed out in Remark 4.3 for linear relativities, finding the geometric relativities amounts to finding the best approximation $\alpha\beta^\ell$ to the Bayesian relativities, that is, α and β solve

$$\min \mathbb{E}\Big[\big(\mathbb{E}[\Theta|L] - \alpha\beta^L\big)^2\Big].$$

No explicit expressions are available for the optimal α and β, which must be determined numerically.

As pointed out by SUBRAMANIAN (1998), the market shares of competitors in a given market can be strongly affected when some insurers adopt an agressive competitive behaviour by modifying the bonus-malus systems. In a deregulated market, insurers have an incentive to innovate in their pricing decisions by partitioning their portfolios (*a priori* ratemaking) and by designing new bonus-malus systems (*a posteriori* ratemaking). VISWANATHAN & LEMAIRE (2005) examined the evolution of market shares and claim frequencies in a two-company market, when one insurer breaks off the existing stability by introducing a super discount class in its bonus-malus system.

To end with, let us point out a final remark of primary importance. Merit-rating structures in automobile insurance require the insured to decide whether to file a claim for an accident when he is at fault. Since the penalties are independent of the claim amounts, one could imagine that some policyholders prefer to carry the cost of the accident themselves in order to avoid the premium increase in the future. Therefore, the data the actuary has at his disposal are contingent on the actual bonus-malus system and are 'censored' in a very complicated way. Hence, the policyholder might modify their behaviour when a new bonus-malus system is introduced, resulting in an increasing (or decreasing) number of reported claims. This is a particularly important side-effect when a bonus-malus system is modified. We will come back to this issue in Chapter 5.

The numerical results presented in this chapter can be obtained using software such as SAS® or R. There is also a specific software package, called BM-builder, which works in the SAS® environment and allows for actuarial computations related to bonus-malus scales. It has been developed by ReacFin SA, which is a spin-off of the *Université Catholique de Louvain*, Louvain-la-Neuve, Belgium, created in January 2004 by the authors. Reacfin's aim is to provide actuarial solutions to its clients. Its strong link with the university guarantees the use of up-to-date techniques. BM-builder aims to compute the relativities attached to the different levels of a bonus-malus scale, taking into account the actual structure of the insurance portfolio. In that respect, it properly integrates the interactions between *a priori* and *a posteriori* ratemakings and allows for an efficient ratemaking. For more details, see http://www.reacfin.com.

Part III

Advances in Experience Rating

5

Efficiency and Bonus Hunger

5.1 Introduction

5.1.1 Pure Premium

In nonlife business, the pure premium is the expected cost of all the claims that the policyholder will file during the coverage period (under the assumption of the Law of Large Numbers, i.e. a large portfolio comprising independent and identically distributed risks). The actuarial ratemaking is based on a claim frequency distribution and a loss distribution. The average claim frequency is defined as the number of incurred claims per unit of earned exposure (the exposure is usually measured in car-year for motor insurance). The average loss severity is the average payment per incurred claim.

In this chapter, as well as in the next one, we need to consider the claim severities. Even if the premium updates induced by bonus-malus systems only depend on the number of claims at fault filed with the insurance company, the design of efficiency measures, as well as the study of the bonus-hunger phenomenon (i.e. the tendency for the policyholder to self-defray minor accidents to avoid premium surcharges), require an accurate modelling for the cost of the claims. This topic is dealt with in Section 5.2.

5.1.2 Statistical Analysis of Claim Costs

The computation of the pure premium relies on a statistical model incorporating all the available information about the risk. The technical tariff aims to evaluate as accurately as possible the pure premium for each policyholder via regression techniques. It is well known that market premiums may differ from those computed by actuaries. In that respect, the overall market position of the company compared to its competitors with regard to growth and pricing is crucial. This chapter is devoted to technical tariff only.

Actuarial Modelling of Claim Counts: Risk Classification, Credibility and Bonus-Malus Systems M. Denuit, X. Maréchal, S. Pitrebois and J.-F. Walhin

The first step of any actuarial analysis consists in displaying descriptive statistics in order to figure out the composition of the portfolio and the marginal impact of the rating factors. In a second stage, available explanatory variables are incorporated to the policyholders' expected claim frequencies and severities with the help of generalized regression models.

5.1.3 Large Claims and Extreme Value Theory

Large claims generally affect liability coverages. These major accidents require a separate analysis. The reason for a separate analysis of small (or moderate) and large losses is that no standard parametric model seems to emerge as providing an acceptable fit to both small and large claims. The main goal is then to determine an optimal threshold separating the two types of losses.

Extreme Value Theory and Generalized Pareto distributions can be used to set the value of this threshold. Specifically, graphical tools including the Pareto index plot and the Gertensgarbe plot can be used to estimate the threshold defining the large losses. In the former case, the maximum likelihood estimator of the Pareto tail parameter is computed for increasing thresholds until it becomes approximately constant. The Gertensgarbe plot is based on the assumption that the optimal threshold can be found as a change point in the ordered series of claim costs and that the change point can be identified by means of a sequential version of the Mann-Kendall test as the intersection point between normalized progressive and retrograde rank statistics.

5.1.4 Measuring the Efficiency of the Bonus-Malus Scales

As explained in the preceding chapters, the basis of fair ratemaking in motor insurance is the fact that each policyholder is charged a premium that is proportional to the risk that he actually represents. The accident proneness of a policyholder being represented by the relative risk parameter Θ, we expect that a relative change in Θ will have the same impact on the premium paid to the insurance company. If this is the case then the system is said to be fully efficient.

Section 5.3 reviews two concepts of efficiency: Loimaranta efficiency and De Pril efficiency. Both intend to measure how the bonus-malus system responds to a change in the riskiness of the driver. Loimaranta efficiency is solely based on the stationary probabilities whereas De Pril efficiency is a transient concept and uses the time value of money (through discounting).

5.1.5 Bonus Hunger and Optimal Retention

Since the penalty induced by the bonus-malus system is independent of the claim amount, a crucial issue for the policyholder is therefore to decide whether it is profitable or not to report small claims (in order to avoid an increase in premium). Cheap claims are likely to be defrayed by the policyholders themselves, and not to be reported to the company. This phenomenon, known as the hunger for bonus after Philipson (1960), is studied in Section 5.4.

Section 5.4.1 is devoted to the censorship of claim amounts and claim frequencies arising from bonus-malus systems. Specifically, a statistical model is specified, that takes into account the fact that only 'expensive' claims are reported to the insurance company. We will consider that each policyholder has his own unknown retention limit, depending on the level occupied inside the bonus-malus scale as well as on observable characteristics (like age or gender, for instance). The policyholder reports the accident to the company only if its cost exceeds the retention limit. A regression model accounting for the fact that we observe the maximum between the accident cost and the retention limit is then fitted to the observed claim data. We then recover probability models for the accident costs (whereas formerly, we modelled claim costs). The claim frequencies can also be corrected in order to obtain accident frequencies (this is done in Section 5.4.2).

Section 5.4.3 examines the optimal claiming strategy, that should be followed by rational policyholders. A strategy for each policyholder can be defined by a vector $(rl_0, \ldots, rl_s)^T$ where rl_ℓ is the retention limit for the policyholder occupying level ℓ in the bonus-malus scale. This means that the cost of any accident of amount less than rl_ℓ is borne by the policyholder in level ℓ. The claims causing higher costs are reported to the insurer. The problem is to determine optimal values for the rl_ℓs. This can be done using the Lemaire algorithm. The optimal retention limits depend on the level occupied in the scale, on the annual expected claim frequency as well as on a discount rate. Note that the Lemaire algorithm gives the optimal retention limit obtained by means of dynamic programming. The resulting strategy should be adopted by rational policyholders, but may differ from the one empirically observed in insurance portfolios. The optimal retentions obtained from the Lemaire algorithm can also be seen as a measure of the toughness of the bonus-malus system. A system that induces large rl_ℓs is more severe than another one yielding moderate retention limits. As such, the rl_ℓs can also be used to measure the efficiency of the bonus-malus system.

The claim costs play an important role in this chapter. As explained above, modelling claim sizes is a difficult issue because of the strong heterogeneity and of the presence of large claims. Nevertheless, it seems reasonable to agree that large claims will always be reported to the company, so that only moderate claims are subject to bonus hunger.

In this chapter, we work within a given bonus-malus scale, whose levels are numbered from 0 to s, with fixed relativities $r_0, \ldots, r_s$ (the r_ℓs have been computed as explained in Chapter 4, or have been derived from marketing considerations, but are treated as given for the whole chapter).

5.1.6 Descriptive Statistics for Portfolio C

The numerical illustrations of this chapter are based on the observation of a Belgian motor third party liability insurance portfolio during the year 1997. This portfolio, henceforth referred to as Portfolio C, comprised 163 660 policies.

The following variables are available for portfolio C: As far as policyholders' characteristics are concerned, we know the Gender (male or female), the age (variable Ageph, four classes: 18–24, 25–30, 31–60 and > 60), the place of residence (variable City, rural or urban) and the Use of the car (private or professional). Concerning the insured vehicle, we know its age (variable Agev, four classes: 1–2, 3–5, 5–10 and > 10 years), the type of Fuel (petrol or gasoil) and its Power (three classes: < 66 kW, 66–110 kW and > 110kW). About the type of contract, we know whether the premium payment has been split up (variable Premium split, with payment once a year, or more than once a year) and the type of Coverage (motor third party liability only, or motor third party liability together with some more

Table 5.1 Descriptive statistics of the claim costs (only strictly positive values) for portfolio C.

Statistic	Value
# observations	18 176
Minimum	27.02
Maximum	1 989 567.9
Mean	1810.63
Standard deviation	17 577.83
25th percentile	145.02
Median	598.17
75th percentile	1464.75
90th percentile	3021.87
95th percentile	4268.06
99th percentile	19 893.68
Skewness	85.08

optional coverages). In addition to these covariates, we know the number of claims filed by each policyholder during 1997, the exposure-to-risk from which these claims originated, as well as the resulting total claim amount. The information recorded in the data base dates from the end of June 1998 (6 months after the end of the observation period). Hence, most of the 'small' claims are settled and their final cost is known. However, for the large claims, we work here with incurred losses (payments made plus reserve).

Descriptive statistics for claim costs are displayed in Table 5.1; we have at our disposal 18 176 observed individual claim costs, ranging from €27.02 to almost €2 000 000, with a mean of €1810.63. We see in Table 5.1 that 25 % of the recorded claim costs are below €145.02, that half of them are smaller than €598.17, and that 90 % of them are less than €3021.87. The interquartile range is €1319.73. The observed claim cost distribution is highly asymmetric, with a skewness coefficient of about 85.

5.2 Modelling Claim Severities

5.2.1 Claim Severities in Motor Third Party Liability Insurance

In nonlife business, the pure premium is the expected cost of all the claims that policyholders will file during the coverage period (under the assumption of the Law of Large Numbers). Let S_i, $i = 1, \ldots, n$, be the total claim amount relating to policy number i. The S_is are assumed to be independent and identically distributed with common mean μ. The Law of Large Numbers ensures that

$$\Pr\left[\overline{S}^{(n)} = \frac{1}{n}\sum_{i=1}^{n} S_i \to \mu \text{ as } n \to +\infty\right] = 1.$$

Under the conditions of the Law of Large Numbers, the pure premium is thus the expected claim amount.

The modelling of claim costs is much more difficult than claim frequencies. There are several reasons for this: In liability insurance, claims costs are often a mix of moderate and large claims. Usually, 'large claim' means exceeding some threshold, depending on the portfolio under study. This threshold can be selected using techniques from Extreme Value Theory, as described in CEBRIAN, DENUIT & LAMBERT (2003). Large liability claims need several years to be settled. Only estimates of the final cost appear in the file until the claim is closed. Moreover, the statistics available to fit a model for claim severities are much more limited than for claim frequencies, since only 10 % of the policies in the portfolio produced claims. Finally, the cost of an accident is for the most part beyond the control of a policyholder since the payments of the insurance company are determined by third-party characteristics. The degree of care exercised by a driver mostly influences the number of accidents, but in a much lesser way the cost of these accidents. The information contained in the available observed covariates is usually much less relevant for claim sizes than for claim counts.

In liability insurance, the settlement of larger claims often requires several years. Much of the data available for the recent accident years will therefore be incomplete, in the sense that the final claim cost will not be known. In this case, loss development factors can be used to obtain a final cost estimate. The average loss severity is then based on incurred loss data. In contrast to paid loss data (which are purely objective, representing the actual payments made by the company), incurred loss data include subjective reserve estimates.

The total claim amount generated by policyholder i covered for motor third party liability can be represented as

$$S_i = \sum_{k=1}^{N_i^{\text{small}}} C_{ik} + \sum_{k=1}^{N_i^{\text{large}}} L_{ik} \tag{5.1}$$

where

N_i^{small} is the number of standard (or small) claims filed by policyholder i
C_{ik} is the cost of the kth standard claim filed by policyholder i
N_i^{large} is the number of large claims filed by policyholder i
L_{ik} is the cost of the kth large claim filed by policyholder i.

All these random variables are assumed to be mutually independent. The random variables N_i^{small} and N_i^{large} are analysed as explained in Chapters 1–2. Here, we explain how to model the C_{ik}s and the L_{ik}s. The first question to be addressed is to separate standard claims and large claims.

5.2.2 Determining the Large Claims with Extreme Value Theory

Extreme Claim Amounts

Gamma, LogNormal and Inverse Gaussian distributions (as well as other parametric models) have often been used by actuaries to fit claim sizes. However, when the main interest is in the tail of loss severity distributions, it is essential to have a good model for the largest claims. Distributions providing a good overall fit can be particularly bad at fitting the tails. Extreme Value Theory and Generalized Pareto distributions focus on the tails, being supported by strong theoretical arguments.

We only give hereafter a short non technical description of the fundaments of Extreme Value Theory; for more details, we refer the reader to BEIRLANT *ET AL.* (2004). Considering a sequence of independent and identically distributed random variables (claim severities, say) $X_1, X_2, X_3, \ldots$, most classical results from probability and statistics that are relevant for insurance are based on sums $S_n = \sum_{i=1}^{n} X_i$. Let us mention the Law of Large Numbers and the Central Limit Theorem, for instance. Another interesting yet less standard statistic for the actuary is $M_n = \max\{X_1, \ldots, X_n\}$ the maximum of the n claims. Extreme Value Theory mainly addresses the following question: how does M_n behave in large samples (i.e. when n tends to infinity)? Of course, without further restriction, M_n obviously diverges to $+\infty$. Once M_n is appropriately centered and normalized, however, it may converge to some specific limit distribution (of three different types, according to the fatness of the tails of the X_is). In insurance applications, heavy tailed distributions are most often encountered. Such distributions have survival functions that decay like a power function (in contrast to the Gamma, Inverse Gaussian or LogNormal survival functions, for instance, which all decay exponentially to zero). A prominent example of a heavy tailed distribution is the Pareto distribution, widely used by actuaries.

Excess Over Threshold Approach and Generalized Pareto Distribution

The traditional approach to Extreme Value Theory is based on extreme value limit distributions. Here, a model for extreme losses is based on the possible parametric form of the limit distribution of maxima. A more flexible model is known as the 'Excesses Over Threshold' method. This approach appears as an alternative to maxima analysis for studying the extreme behaviour of some random variables. Basically, given a series $X_1, \ldots, X_n$ of independent and identically distributed random variables, the 'Excesses Over Threshold' method analyses the series $[X_i - u|X_i > u]$, $i = 1, \ldots, n$, of the exceedances of the variable over a high threshold u. Mathematical theory supports the Poisson distribution for the number of exceedances combined with independent excesses over the threshold.

Let F_u stand for the common cumulative distribution function of the $[X_i - u|X_i > u]$s; F_u thus represents the conditional distribution of the losses, given that they exceed the threshold u. The two-parameter Generalized Pareto distribution function $G_{\xi,\beta}(\cdot)$ provides a good approximation to the excess distribution F_u over large thresholds. This two-parameter family is defined as

$$G_{\xi,\beta}(x) = G_{\xi}\left(\frac{x}{\beta}\right), \quad \beta > 0,$$

where

$$G_{\xi}(x) = \begin{cases} 1 - (1 + \xi x)^{-1/\xi} \text{ if } \xi \neq 0 \\ 1 - \exp(-x) \text{ if } \xi = 0, \end{cases}$$

with $x \geq 0$ if $\xi \geq 0$ and $x \in [0, -1/\xi]$ if $\xi < 0$, is the the one-parameter Generalized Pareto distribution function. The parameter ξ is named the *Pareto index*. As particular cases of the Generalized Pareto distribution function $G_{\xi,\beta}$, we find some classical distributions, namely the Pareto distribution when $\xi > 0$, the type II Pareto distribution when $\xi < 0$ and the Exponential distribution when $\xi = 0$.

For some appropriate function $\beta(u)$ and some Pareto index ξ to be estimated from the data, the approximation

$$F_u(x) \approx G_{\xi,\beta(u)}(x), \quad x \geq 0 \tag{5.2}$$

holds for large u. The approximation (5.2) is justified by the following formula

$$\lim_{u \to +\infty} \sup_{x \geq 0} \left| F_u(x) - G_{\xi,\beta(u)}(x) \right| = 0 \tag{5.3}$$

which is true provided that F satisfies some rather general technical conditions. These conditions are verified by the heavy tailed distributions. In view of (5.2) the excesses $[X_i - u | X_i > u]$ can be treated as a random sample from the Generalized Pareto distribution provided the threshold u is large enough.

Choice of the Threshold
We have seen from (5.2) that if the heavy tailed character of the data is fulfilled, a high enough threshold is selected and enough data are available above that threshold, the use of Generalized Pareto distributions is justified to model large losses. The only practical problem in applying this result is how to determine what a 'high enough threshold' is; we deal with this problem in the present section.

Two factors have to be taken into account in the choice of an optimal threshold u:

- A value of u too large yields few exceedances and consequently imprecise estimates.
- A value of u too small implies that the generalized Pareto character does not hold for the moderate observations and it yields biased estimates. This bias can be important as moderate observations usually constitute the largest proportion of the sample.

Thus, our aim is to determine the minimum value of the threshold beyond which the Generalized Pareto distribution becomes a reasonable approximation to the tail of the distribution.

To identify the optimal threshold value, we apply here two methods:

Generalized Pareto Index Plot
In virtue of the stability property of the Generalized Pareto distribution, if X is Generalized Pareto distributed with distribution function $G_{\xi,\beta}$, the variable $[X - u | X > u]$ is Generalized Pareto distributed with distribution function $G_{\xi,\beta+\xi u}$, i.e. with the same index parameter ξ, for any $u > 0$. Consequently, in the plot of the index maximum likelihood estimators $\hat{\xi}$ resulting from using increasing thresholds, we will observe that estimation stabilizes when the smallest threshold for which the Generalized Pareto behaviour holds is reached.

Gertensgarbe Plot
This procedure proposed by GERTENSGARBE & WERNER (1989) is very powerful and provides an estimation of the optimal threshold. Briefly, the Gertensgarbe plot aims to select a proper threshold, based on the determination of the starting point of the extreme value region. More precisely, given the series of differences $\Delta_i = x_{(i)} - x_{(i-1)}$, $i = 2, 3, \ldots, n$ of a sorted sample, $x_{(1)} \leq x_{(2)} \leq \cdots \leq x_{(n)}$, the starting point of the extreme region will be detected as a change point of the series $\{\Delta_i, \ \ i = 2, 3, \ldots, n\}$. The key idea is that it may be reasonably

expected that the behaviour of the differences corresponding to the extreme observations will be different from the one corresponding to the non-extreme observations. This change of behaviour will appear as a change point of the series of differences.

To identify the change point in a series, a sequential version of the Mann-Kendall test is applied. In this test, the normalized series U_i is defined as

$$U_i = \frac{U_i^* - \frac{i(i-1)}{4}}{\sqrt{\frac{i(i-1)(2i+5)}{72}}},$$

where $U_i^* = \sum_{k=2}^{i} n_k$, and n_k is the number of values in $\Delta_2, \ldots, \Delta_k$ lesser than Δ_k. Another series, denoted by U_p, is calculated applying the same procedure to the series of the differences from the end to the start, $\Delta_n, \ldots, \Delta_2$, instead of from the start to the end. The intersection point between these two series, U_i and U_p, determines a probable change point that will be significant if it exceeds a high Normal percentile.

Since, usually, these techniques can only provide approximative information about the threshold, simultaneous application of them is highly recommended in order to get more reliable results.

Application to Claim Costs Recorded in Portfolio C

The Generalized Pareto index plot is shown in Figure 5.1. We see that the estimates of tail parameter ξ roughly stabilize after €85000. The Gertensgarbe plot gives a threshold of €104 397 (which corresponds to the 17th largest loss). The p-value of the Mann-Kendall

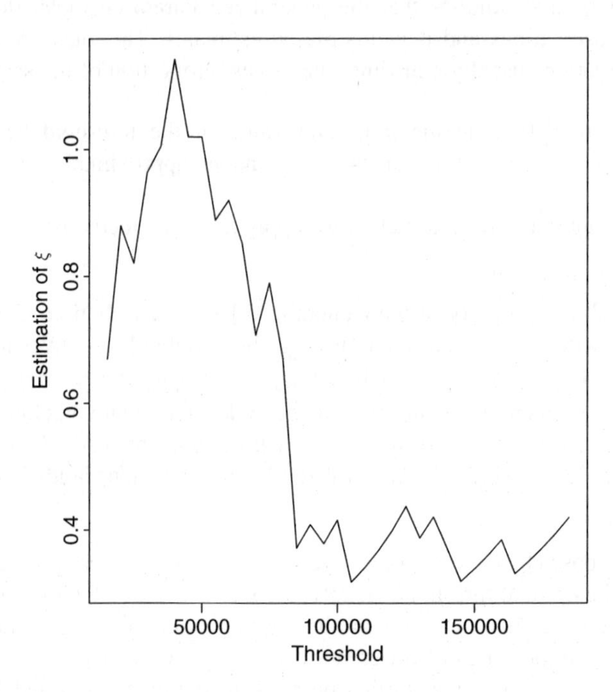

Figure 5.1 Generalized Pareto index plot for the claim costs in Portfolio C.

test is equal to 2.61 %, so that the result is significant at the 5 % level. Considering the two analyses, the threshold for being qualified as a large claim is set to €100 000.

5.2.3 Generalized Pareto Fit to the Costs of Large Claims

Maximum Likelihood

Now that the threshold defining the large losses has been determined as €100 000, we model the excesses over €100 000 with the help of a Generalized Pareto distribution. Descriptive statistics for the cost of large claims of Portfolio C are displayed in Table 5.2. The mean is equal to €364 606.2. The limited number of large losses (17 for Portfolio C) does not allow for incorporating exogeneous information in these amounts. Therefore, the same parameters ξ and β are used for all the large losses. These parameters are estimated by maximum likelihood.

Let us now fit the Generalized Pareto model to the excesses over €100 000. To this end, we use maximum likelihood theory, and we maximize the likelihood function given by

$$\mathcal{L}(\xi,\beta)=\prod_{i|x_i>100\,000}\left(\frac{1}{\beta}\left(1+\frac{\xi}{\beta}(x_i-100\,000)\right)^{-\frac{1}{\xi}-1}\right).$$

The log-likelihood to be maximized is

$$L(\xi,\beta)=-\ln\beta\#\{x_i|x_i>100\,000\}-\left(1+\frac{1}{\xi}\right)\sum_{i|x_i>100\,000}\ln\left(1+\frac{\xi}{\beta}(x_i-100\,000)\right)$$

where $\#\{x_i|x_i>100\,000\}=17$ in Portfolio C. This optimization problem requires numerical algorithms. There are different approaches to getting starting values for the parameters ξ and β. A natural approach consists of using moment conditions (that is, we equate sample mean and sample variance to their theoretical expressions involving ξ and β). The values

Table 5.2 Descriptive statistics of the cost of large claims (Portfolio C).

Statistic	Value
Length	17
Minimum	104 386.7
Maximum	1 989 567.9
Mean	364 606.2
Standard deviation	439 882.7
25th percentile	140 032.4
Median	252 231.7
75th percentile	407 477.6
90th percentile	499 727.4
95th percentile	797 847.5
99th percentile	1 751 223.8
Skewness	2.9

of β and ξ coming from the linear fit to the empirical mean excess function could also be used, as shown below.

Moment Method

The mean and the variance of the Generalized Pareto distribution are respectively given by $\beta/(1-\xi)$ provided $\xi < 1$, and by $\beta^2/[(1-\xi)^2(1-2\xi)]$ provided $\xi < 1/2$. Note that the moment of order r exists only if $\xi < 1/r$. Initial values for the maximum likelihood estimation obtained by the method of moments are then

$$\widehat{\xi}_0 = \frac{1}{2}\Big(1 - \frac{\bar{x}^2}{s^2}\Big) \text{ and } \widehat{\beta}_0 = \frac{1}{2}\bar{x}\Big(\frac{\bar{x}^2}{s^2} + 1\Big),$$

where $\bar{x}$ and s^2 are the sample mean and variance.

Linear Fit to the Empirical Mean Excess Function

Another convenient way to get starting values for the maximum likelihood estimates consists of using the mean excess function that is defined as

$$e(u) = \mathbb{E}[X - u|X > u] = \int_0^{+\infty} (1 - F_u(x))dx.$$

This function can be estimated from a random sample $\{x_1, x_2, \ldots, x_n\}$ by

$$\widehat{e}_n(u) = \frac{\sum_{i=1}^n x_i \mathbb{I}[x_i > u]}{\#\{x_i : x_i > u\}} - u = \frac{\sum_{i=1}^n (x_i - u)\mathbb{I}[x_i > u]}{\#\{x_i : x_i > u\}}$$

where $\mathbb{I}[A] = 1$ if the event A did occur and 0 otherwise. This means that $e(u)$ is estimated by the sum of exceedances over the threshold u divided by the number of data points exceeding the threshold u.

Usually, the mean excess function is evaluated on the observations of the sample. Denoting the sample observations arranged in ascending order as $x_{(1)} \leq x_{(2)} \leq \cdots \leq x_{[n]}$, we have in this case

$$\widehat{e}_n(x_{[k]}) = \frac{1}{n-k} \sum_{j=1}^{n-k} (x_{(k+j)} - x_{(k)}).$$

It is easily checked that when X has a Generalized Pareto distribution function $G_{\xi,\beta}$, the mean excess function is a linear function in u

$$e(u) = \frac{\beta}{1-\xi} + \frac{\xi}{1-\xi}u$$

provided $\beta + u\xi > 0$. Hence, the idea is to determine, on the basis of the graph of the empirical estimator of the excess function $\widehat{e}_n$, a region $[u, +\infty)$ where $\widehat{e}_n(t)$ becomes approximately linear for $t \geq u$. The intercept and slope of a straight line fit to $\widehat{e}_n$ determine the estimations of ξ and β.

Application to the Claim Costs Recorded in Portfolio C
The initial values obtained with the the moment method applied to large claims of Portfolio C are $\widehat{\xi}_0 = 0.319$ and $\widehat{\beta}_0 = 180\,176.7$. The maximum likelihood estimates of the Generalized Pareto parameters are $\widehat{\xi} = 0.4152$ and $\widehat{\beta} = 156\,737.4$. Different starting values have been used and the convergence always occurred.

Note that the limited sample size for the large losses does not allow us to draw reliable conclusions about major claims (in particular, large sample properties of the maximum likelihood estimators cannot be invoked with a sample size as small as 17). In practice, the insurance company must gather the large losses together in a data base to perform detailed analysis. The amounts of these large losses have to be corrected for different sources of inflation. The assistance of a reinsurance company is useful in this respect, especially for insurers with small to moderate portfolios.

5.2.4 Modelling the Number of Large Claims

The number of large claims N_i^{large} for policyholder i is modelled using the $\mathcal{P}oi(\lambda_i^{\text{large}})$ distribution. This is in line with the fact that large claims occur purely at random. Poisson regression is then used to incorporate the information available about policyholder i, summarized in a vector $\boldsymbol{x}_i^T = (x_{i1}, \ldots, x_{ip})$ consisting of explanatory variables (assumed here to be categorical and coded by means of binary variables), in the expected frequency of large claims through a linear predictor by means of an exponential link function.

The regression coefficients are estimated with Poisson regression. All the explanatory variables introduced in Section 5.1.6 have been excluded from the model. Only the intercept remained in the Poisson regression model. This is not surprising with such a small number of policies producing a large claim. We obtained $\widehat{\beta}_0 = -9.0555$, with a standard deviation equal to 0.2425. The resulting frequency of large claims is $\widehat{\lambda_i^{\text{large}}} = 0.0117\,\%$ for all policyholders.

Remark 5.1 (Logistic regression) In most cases, policyholders report 0 or just 1 large claim. Therefore, the number of large claims could also be modelled with a binary variable instead of a Poisson count. The probability that policyholder i reports (at least) a large claim can also be modelled with the help of logistic regression. Specifically, let us define the binary random variable J_i

$$J_i = \begin{cases} 0 \text{ if policyholder } i \text{ does not report any large claim during the observation period} \\ 1 \text{ if policyholder } i \text{ reports at least one large claim during the observation period.} \end{cases}$$

The aim is to model the probability $\Pr[J_i = 0] = q_i(\boldsymbol{x}_i)$ that policyholder i does not report any large claim during the coverage period.

Since $q_i(\boldsymbol{x}_i) \in [0, 1]$, we resort to some distribution function F to link $q_i(\boldsymbol{x}_i)$ to the linear predictor $\alpha_0 + \sum_{j=1}^{p} \alpha_j x_{ij}$, that is

$$q_i(\boldsymbol{x}_i) = F\left(\alpha_0 + \sum_{j=1}^{p} \alpha_j x_{ij}\right) \Leftrightarrow \alpha_0 + \sum_{j=1}^{p} \alpha_j x_{ij} = F^{-1}(q_i(\boldsymbol{x}_i)).$$

Theoretically, any distribution function F can be used; in practice, we often take for F the Normal or Logistic distribution function. For instance, the logistic regression model is specified as

$$\ln \frac{q_i(\boldsymbol{x}_i)}{1-q_i(\boldsymbol{x}_i)} = \alpha_0 + \sum_{j=1}^{p} \alpha_j x_{ij} \Leftrightarrow q_i = \frac{\exp\left(\alpha_0 + \sum_{j=1}^{p} \alpha_j x_{ij}\right)}{1+\exp\left(\alpha_0 + \sum_{j=1}^{p} \alpha_j x_{ij}\right)}.$$

The SAS®/STAT procedure GENMOD can be used to perform this analysis. As in the Poisson regression case, all the explanatory variables described in Section 5.1.6 are excluded from the model. The estimated probability that policyholder i does not report any large claim is 99.99013 %. Hence, the corresponding large claim frequency is $-\ln 0.9999013 = 0.00987\,\%$, which is not too far from the estimated λ_i^{large} obtained using Poisson regression.

5.2.5 Modelling the Costs of Moderate Claims

Different models can be used to describe the behaviour of the moderate claims (i.e. claims with an incurred cost less than €100 000) as a function of the observable characteristics of the policyholder; including Gamma, Inverse Gaussian and LogNormal distributions. We briefly review these three regression models next.

Gamma Distribution

Here we use a new parameterization of the Gamma probability density function (1.34). Specifically, we use the mean as parameter, together with a parameter related to the variation coefficient. The probability density function with the new parameters $\mu = \alpha/\beta$ and $\nu = \alpha$ is then given by

$$f(y|\mu, \nu) = \frac{1}{\Gamma(\nu)} \left(\frac{\nu y}{\mu}\right)^{\nu} \exp\left(-\frac{\nu y}{\mu}\right) \frac{1}{y}. \tag{5.4}$$

If Y has probability density function (5.4), then the first moments are given by

$$\mathbb{E}[Y] = \mu \text{ and } \mathbb{V}[Y] = \frac{\mu^2}{\nu}$$

so that the variance is proportional to the square of the mean. Gamma regression assumes a coefficient of variation constantly equal to $\nu^{-1/2}$. Thus it allows for heteroscedasticity (since the variance is proportional to the square of the mean, and is no more constant as in Gaussian regression models). Ideally, the Gamma regression model is best used with positive observations having a constant coefficient of variation. However, the model is robust to wide deviations from the latter assumption.

The parameter ν controls the shape of the probability density function. Specifically, (i) if $0 < \nu < 1$ then the probability density function admits a mode at the origin and decreases over $\mathbb{R}^+$; (ii) if $\nu = 1$ then the Gamma probability density function reduces to the Negative Exponential one; and (iii) if $\nu > 1$ then the probability density function vanishes at the origin and possesses a unique mode located at $y = \mu - \mu/\nu$. The skewness equals $\gamma_1 = 2\nu^{-1/2}$.

Let C_{ik} be the cost of the kth claim reported by policyholder i; we assume that the individual claim costs $C_{i1}, C_{i2}, \ldots$ are independent and identically distributed. Each C_{ik} conforms to the Gamma law with mean

$$\mu_i = \mathbb{E}[C_{ik}|\boldsymbol{x}_i] = \exp\left(\beta_0 + \sum_{j=1}^{p} \beta_j x_{ij}\right) \tag{5.5}$$

and variance $\mathbb{V}[C_{ik}|\boldsymbol{x}_i] = \mu_i^2/\nu$. Note that here we use an exponential link between the linear predictor $\beta_0 + \sum_{j=1}^{p} \beta_j x_{ij}$ and the expected value μ_i. For theoretical reasons, a reciprocal link function is sometimes preferable (but destroys the nice multiplicative structure of the resulting price list).

Let n_i be the number of claims reported by policyholder i, and let $c_{i1}, c_{i2}, \ldots, c_{in_i}$ be the corresponding claim costs. The likelihood associated with the observations is

$$\mathcal{L}(\boldsymbol{\beta}, \nu) = \prod_{i|n_i>0} \prod_{k=1}^{n_i} \left(\frac{1}{\Gamma(\nu)} \left(\frac{\nu c_{ik}}{\mu_i} \right)^{\nu} \exp\left(-\frac{\nu c_{ik}}{\mu_i} \right) \frac{1}{c_{ik}} \right).$$

The corresponding log-likelihood is given by

$$\begin{aligned} L(\boldsymbol{\beta}, \nu) &= \ln \mathcal{L}(\boldsymbol{\beta}, \nu) \\ &= \sum_{i|n_i>0} \left(-n_i \ln \Gamma(\nu) + \nu n_i \left(\ln \nu - \beta_0 - \sum_{j=1}^{p} \beta_j x_{ij} \right) + \nu \sum_{k=1}^{n_i} \ln c_{ik} - \frac{\nu}{\mu_i} \sum_{k=1}^{n_i} c_{ik} \right) \\ &\quad + \text{constant}. \end{aligned}$$

The likelihood equations are given by

$$\frac{\partial}{\partial \beta_j} L(\boldsymbol{\beta}, \nu) = 0 \Leftrightarrow \sum_{i|n_i>0} x_{ij} \left(n_i - \frac{c_{i\bullet}}{\mu_i} \right) = 0$$

for $j = 1, \ldots, p$, where $c_{i\bullet} = \sum_{k=1}^{n_i} c_{ik}$ is the total cost of the standard claims reported by policyholder i. The maximum likelihood estimators are obtained with the help of Newton-Raphson techniques. The estimation of ν can be performed by maximum likelihood as in Chapter 2, or it can be obtained from the Pearson- or deviance-based dispersion statistic.

Remark 5.2 Often, only the total claim amount $C_{i\bullet}$ is available, and not the individual C_{ik}s. In such a case, it is convenient to work with the mean claim amount $\overline{C}_i = C_{i\bullet}/n_i$, where n_i is the number of claims reported by policyholder i. Considering the Gamma likelihood equations, this is not restrictive since only the total claim amount is needed. Specifically, Formula (1.36) shows that the Gamma distributions are closed under convolution in some particular cases. In the new parameterization of the Gamma family used in the present chapter, the moment generating function of C_{ik} is

$$M(t) = \left(1 - \frac{t\mu_i}{\nu}\right)^{-\nu}$$

so that the moment generating function of $\overline{C}_i$ is

$$\prod_{j=1}^{n} M\left(\frac{t}{n_i}\right) = \left(1 - \frac{t\mu_i}{\nu_i'}\right)^{-\nu_i'}$$

where $\nu_i' = n_i\nu$. Hence, the arithmetic average of the C_{ik}s conforms to the $\mathcal{G}am(\mu_i, n_i\nu_i)$ distribution. This situation is accounted for in GENMOD by specifying an appropriate weight n_i.

Example 5.1 (Gamma Regression for the Moderate Claim Costs in Portfolio C) The Gamma regression performed on the claim costs recorded in Portfolio C leads to the results in Table 5.3 where the following variables have been eliminated: Fuel (p-value of 94.76 %), Gender (p-value of 88.89 %), Use (p-value of 27.48 %) and Power (p-value of 9.28 %). Moreover, for Agev, levels 6–10 and > 10 have been grouped together in a class > 5. For the variable Ageph, levels 31–60 and > 60 have been grouped in a class > 30. The resulting log-likelihood is equal to −147 629.10. Type 3 analysis is presented in the following table:

Source	DF	Chi-square	Pr>Chi-sq
Ageph	2	65.92	< .0001
Premium split	1	10.56	0.0012
City	1	16.37	< .0001
Agev	1	54.43	< .0001
Coverage	1	19.92	< .0001

We see that all the remaining variables are statistically relevant, and must be kept in the final model. The dispersion parameter is estimated at $\widehat{\nu} = 0.5465$.

Table 5.3 Results of the Gamma regression on the claim costs recorded in Portfolio C.

Variable	Level	Coeff β	Std error	Wald 95 %	conf limit	Chi-sq	Pr>Chi-sq
Intercept		7.0075	0.0230	6.9625	7.0525	93003.1	< .0001
Ageph	18–24	0.2210	0.0402	0.1423	0.2997	30.30	< .0001
Ageph	25–30	0.1682	0.0263	0.1167	0.2196	40.98	< .0001
Ageph	> 30	0	0	0	0	.	.
Premium split	No	0.0628	0.0193	0.0249	0.1007	10.53	0.0012
Premium split	Yes	0	0	0	0	.	.
City	Rural	0.0794	0.0197	0.0408	0.1179	16.27	< .0001
City	Urban	0	0	0	0	.	.
Agev	0–2	0.1862	0.0309	0.1256	0.2269	36.26	< .0001
Agev	3–5	−0.0513	0.0247	−0.0998	−0.0028	4.30	0.0380
Agev	> 5	0	0	0	0	.	.
Coverage	MTPL only	0.0968	0.0216	0.0545	0.1392	20.06	< .0001
Coverage	More	0	0	0	0	.	.

Remark 5.3 (Tweedie Generalized Linear Models) The Tweedie distributions are a three-parameter family. They allow for any power variance function and any power link. The Tweedie family includes the Gaussian, Poisson, Gamma and Inverse Gaussian families as special cases. Specifically, let $\mu_i = \mathbb{E}[Y_i]$ be the expectation of the ith response Y_i. We assume that

$$\mu_i^{q_1} = \beta_0 + \sum_{j=1}^{p} \beta_j x_{ij} \text{ and } \mathbb{V}[Y_i] = \phi \mu_i^{q_2}$$

where $\boldsymbol{x}_i$ is a vector of covariates and $\boldsymbol{\beta}$ is a vector of regression coefficients, for some ϕ, q_1 and q_2. A value of zero for q_1 is interpreted as $\ln(\mu_i) = \beta_0 + \sum_{j=1}^{p} \beta_j x_{ij}$. The variance power q_2 characterizes the distribution of the responses Y. The parameter q_2 is called the index parameter and determines the shape of the Tweedie distribution. For various values of q_2, we find the following particular cases: $q_2 = 0$ corresponds to the Normal distribution, $q_2 = 1$ corresponds to the Poisson distribution, $1 < q_2 < 2$ corresponds to the compound Poisson model with Gamma summands that is appropriate to model continuous data with exact zeros (see below), $q_2 = 2$ corresponds to the Gamma distribution and $q_2 > 2$ corresponds to stable distributions for positive continuous data.

As stated above, for $1 < q_2 < 2$, we get the compound Poisson model with Gamma summands, that is, the distribution of $Y = X_1 + \cdots + X_N$, where X_1 to X_N are independent Gamma distributed random variables and N is Poisson distributed. Note that the Y has a positive probability mass at zero, but otherwise has a continuous positive distribution. This distribution can be equivalently represented as a Poisson mixture of Gamma distributions. This distribution approaches Gamma as $q_2 \to 2$ and a scaled Poisson model as $q_2 \to 1$. This distribution is therefore genuinely intermediate between the Poisson and Gamma distributions.

Inverse Gaussian Distribution

The Inverse Gaussian distribution is the least frequently used generalized linear model. The resulting regression model is rarely discussed in textbooks. The Inverse Gaussian framework is most appropriate when modelling a nonnegative variable having a high initial peak, rapid drop, and a long right tail. We parallel here the treatment of the Gamma regression model.

Let us assume that the individual claim sizes have probability density function (1.39) with the new parameters μ and $\sigma^2 = \beta/\mu^2$. The Inverse Gaussian probability density function with these new parameters then becomes

$$f(y|\mu, \sigma^2) = \frac{1}{\sqrt{2\pi\sigma^2 y^3}} \exp\left(-\frac{(y-\mu)^2}{2(\sigma\mu)^2 y}\right).$$

Let us denote as $C_{i1}, C_{i2}, \ldots$ the costs of the claims reported by policyholder i. We assume that $C_{i1}, C_{i2}, \ldots$ are independent and identically distributed, with common probability density function $f(\cdot|\mu_i, \sigma^2)$. The expected claim amount μ_i is given by (5.5). The associated variance $\mathbb{V}[C_{ik}|\boldsymbol{x}_i] = \sigma^2\mu_i^3$ is proportional to the third power of the mean.

Let n_i be the number of claims reported by policyholder i. The likelihood is then given by

$$\mathcal{L}(\boldsymbol{\beta}, \sigma^2) = \prod_{i|n_i>0} \prod_{k=1}^{n_i} \left(\frac{1}{\sqrt{2\pi\sigma^2 c_{ik}^3}} \exp\left(-\frac{(c_{ik}-\mu_i)^2}{2(\sigma\mu_i)^2 c_{ik}}\right)\right).$$

The log-likelihood is

$$L(\boldsymbol{\beta}, \sigma^2) = \ln \mathcal{L}(\boldsymbol{\beta}, \sigma^2)$$
$$= \sum_{i|n_i>0} \left(-\frac{n_i}{2} \ln(2\pi\sigma^2) - \sum_{k=1}^{n_i} \frac{(c_{ik} - \mu_i)^2}{2(\sigma\mu_i)^2 c_{ik}} \right) + \text{constant},$$

so that the likelihood equations are given by

$$\frac{\partial}{\partial \beta_j} L(\boldsymbol{\beta}, \sigma^2) = 0 \Leftrightarrow \frac{\partial}{\partial \beta_j} \sum_{i|n_i>0} \sum_{k=1}^{n_i} \left(1 - \frac{c_{ik}}{\mu_i}\right)^2 \frac{1}{c_{ik}} = 0$$
$$\Leftrightarrow \sum_{i|n_i>0} \frac{x_{ij}}{\mu_i} \left(n_i - \frac{c_{i\bullet}}{\mu_i} \right) = 0.$$

The estimation of σ^2 can be performed by maximum likelihood as in Chapter 2, or it can be obtained from the Pearson- or deviance-based dispersion statistic.

Remark 5.4 As pointed out for the Gamma distribution in Remark 5.2, the actuary often only has at his disposal the total claim amount $C_{i\bullet}$, and not the individual C_{ik}s. This is not really a problem since the likelihood equations only involve $C_{i\bullet}$. Considering (1.40), we see that the moment generating function of the mean claim amount $\overline{C}_i$ in the new parameterization is given by

$$\exp\left(-\frac{n_i}{\sigma^2 \mu_i} \left(1 - \sqrt{1 - 2\sigma^2 \mu_i^2 \frac{t}{n_i}} \right) \right)$$

which corresponds to the Inverse Gaussian distribution with parameters μ_i and σ^2/n_i. As in the Gamma case, working with the average claim amounts is not restrictive for maximum likelihood estimation of the regression parameters, and this situation is accounted for in GENMOD by specifying an appropriate weight n_i.

Example 5.2 (Inverse Gaussian Regression for the Moderate Claim Costs in Portfolio C) The results of the Inverse Gaussian regression are given in Table 5.4 where the following variables have been eliminated: Fuel (p-value of 96.29 %), Gender (p-value of 84.58 %), Power (p-value of 78.32 %), Use (p-value of 56.40 %), Premium split (p-value of 23.04 %), City (p-value of 9.75 %) and Coverage (p-value of 5.37 %). Moreover, for Agev, levels 3–5, 6–10 and >10 have been grouped together in a class >2. For the variable Ageph, levels 31–60 and >60 have been grouped in a class >30. The resulting log-likelihood is equal to $-150\,214.60$. Type 3 analysis is presented in the following table:

Source	DF	Chi-square	Pr>Chi-sq
Ageph	2	10.56	0.0012
Agev	1	41.87	<.0001

The dispersion parameter is estimated at $\widehat{\sigma}^2 = 0.00897$.

Table 5.4 Results of the Inverse Gaussian regression on the claim costs recorded in Portfolio C.

Variable	Level	Coeff β	Std error	Wald 95 %	conf limit	Chi-sq	Pr>Chi-sq
Intercept		7.1169	0.0282	7.0616	7.1722	63634.3	<.0001
Ageph	18–24	0.2293	0.1104	0.0129	0.4457	4.31	0.0378
Ageph	25–30	0.1699	0.0697	0.0334	0.3065	5.95	0.0147
Ageph	> 30	0	0	0	0	.	.
Agev	0–2	0.1609	0.0756	0.0127	0.3091	4.53	0.0334
Agev	> 2	0	0	0	0	.	.

LogNormal Distribution

Before the generalized linear models gained popularity in the actuarial profession, claim sizes were often analysed using a Normal linear regression model after having been transformed to the log-scale. Although the results are usually quite similar for this method and Gamma regression, the latter approach is easier to interpret since it does not require any logarithmic transformation of the claim costs.

Assume that the moderate claim sizes for policyholder i are independent and LogNormally distributed, with parameters $\beta_0+\sum_{j=1}^{p}\beta_j x_{ij}$ and σ^2. Specifically, the C_{ik}s are independent, and identically distributed for fixed i, with $C_{ik}\sim \mathcal{L}Nor\ (\beta_0+\sum_{j=1}^{p}\beta_j x_{ij},\sigma^2)$. Let n_i be the number of claims reported by policyholder i, and let $c_{i1}, c_{i2}, \ldots, c_{in_i}$ be the corresponding claim costs. The likelihood associated with the observations is

$$\mathcal{L}(\boldsymbol{\beta})=\prod_{i|n_i>0}\prod_{k=1}^{n_i}\frac{1}{\sqrt{2\pi}c_{ik}\sigma}\exp\left(-\frac{1}{2\sigma^2}\left(\ln c_{ik}-\beta_0-\sum_{j=1}^{p}\beta_j x_{ij}\right)^2\right).$$

The maximum likelihood estimators are obtained with the help of Newton-Raphson techniques. The average cost of a standard claim for policyholder i is then obtained from the formula

$$\mathbb{E}[C_{ik}|\boldsymbol{x}_i]=\exp\left(\beta_0+\sum_{j=1}^{p}\beta_j x_{ij}+\frac{\sigma^2}{2}\right).$$

Note that, in contrast to the Gamma and Inverse Gaussian cases, we cannot easily deal with the situation where only the total amount of moderate claims is available. This is due to the fact that the LogNormal family of distributions is not closed under convolution. Therefore, we fit the model to the observations made on policyholders having filed a single standard claim.

Example 5.3 (LogNormal Regression for the Moderate Claim Costs in Portfolio C) The LogNormal regression cannot be performed with the help of SAS®/STAT procedure GENMOD (which does not support the LogNormal distribution). Often in practice, the data are first transformed on the log-scale, and a standard linear model is then fitted to the logarithms of the claim amounts. This ad-hoc procedure will be avoided here, and a maximum likelihood estimation procedure is performed on the original claim costs.

Table 5.5 Results of the LogNormal regression analysis for the claim costs recorded in Portfolio C.

Variable	Level	Coeff β	Std error	Wald 95 %	conf limit	t-value	Pr $>\|t\|$
Intercept		6.1223	0.0268	6.0702	6.1744	230.38	<.0001
Ageph	18–24	0.1746	0.0507	0.0753	0.2739	3.45	0.0006
Ageph	>60	0.1781	0.0312	0.1169	0.2392	5.71	<.0001
Ageph	25–60	0	0	0	0	.	.
City	Rural	0.0661	0.0237	0.0196	0.1125	2.79	0.0053
City	Urban	0	0	0	0	.	.
Agev	0–2	−0.1229	0.0398	−0.2009	−0.0448	−3.09	0.0020
Agev	3–5	−0.2116	0.0343	−0.2788	−0.1443	−6.16	<.0001
Agev	6–10	−0.1157	0.0301	−0.1747	−0.0568	−3.85	0.0001
Agev	>10	0	0	0	0	.	.

The procedure NLMIXED has been used to maximize the LogNormal likelihood. We obtained the results of Table 5.5 where the following variables have been eliminated: Gender (p-value of 77.46 %), Fuel (p-value of 68.55 %), Use (p-value of 28.23 %), Coverage (p-value of 18.66 %), Premium split (p-value of 10.86 %) and Power (p-value of 6.98 %). The levels 25–30 and 31–60 of the variable Ageph have been grouped together in a level 25–60. The resulting log-likelihood is equal to $-130\,832.3$ and the parameter σ was estimated at $\widehat{\sigma} = 1.4822$. Type 3 analysis for the final model is presented in the following table:

Source	DF	Chi-square	Pr>Chi-sq
Ageph	2	40.58	<.0001
City	1	7.70	0.0055
Agev	3	38.20	<.0001

All the three remaining variables are statistically significant, and must be kept in the model.

5.2.6 Resulting Price List for Portfolio C

Formula for the Pure Premium

Let $\boldsymbol{\beta}^{\text{freq}}$ be the vector of the regression coefficient for the claim frequencies, that is, the expected annual number of standard claims is $\exp(\beta_0^{\text{freq}} + \sum_{j=1}^{p} \beta_j^{\text{freq}} x_{ij})$. Similarly, let $\boldsymbol{\beta}^{\text{cost}}$ be the regression coefficient for the moderate claim sizes. We retain here the LogNormal modelling for the moderate claim sizes, so that the expected moderate claim amount is $\exp(\beta_0^{\text{cost}} + \sum_{j=1}^{p} \beta_j^{\text{cost}} x_{ij} + \sigma^2/2)$. Neglecting the large claims, the pure premium for policyholder i is given by

$$\mathbb{E}\left[\sum_{k=1}^{N_i^{\text{small}}} C_{ik}\right] = \mathbb{E}[N_i^{\text{small}}]\mathbb{E}[C_{i1}] = \exp\left(\beta_0^{\text{freq}} + \beta_0^{\text{cost}} + \sum_{j=1}^{p} (\beta_j^{\text{freq}} + \beta_j^{\text{cost}})x_{ij} + \frac{\sigma^2}{2}\right).$$

If all the components of $\boldsymbol{x}_i$ are binary, we then get a multiplicative price list. The total premium is then obtained by adding the expected cost of the large losses, i.e. $\mathbb{E}[N_i^{\text{large}}]\mathbb{E}[L_{i1}]$.

We still need to estimate $\boldsymbol{\beta}^{\text{freq}}$ to be able to compute the pure premium for the different categories of policyholders in Portfolio C.

Risk Classification for Claim Frequencies

Here we follow the method described in Chapter 2 for analysing the observed claim numbers. A Poisson regression is first performed on the claim frequencies of Portfolio C. This leads to the results of Table 5.6. Only the Use of the vehicle has been removed from the model (p-value of 34.48 %). The levels 0–2, 6–10 and > 10 of the variable Agev are grouped together. The log-likelihood of the final model is equal to −61 563.9 and the Type 3 analysis is presented in the following table:

Source	DF	Chi-square	Pr>Chi-sq
Ageph	3	892.66	< .0001
Gender	1	12.98	0.0003
Agev	1	13.09	0.0003
Fuel	1	153.84	< .0001
Premium split	1	366.51	< .0001
Coverage	1	63.70	< .0001
City	1	246.19	< .0001
Power	2	53.94	< .0001

Table 5.6 Results of the Poisson regression of the claim counts recorded in Portfolio C.

Variable	Level	Coeff β	Std error	Wald 95 % conf limit		Chi-sq	Pr>Chi-sq
Intercept		−1.4623	0.0690	−1.5976	−1.3270	448.69	< .0001
Ageph	18–24	0.6393	0.0299	0.5806	0.6980	456.17	< .0001
Ageph	25–30	0.3589	0.0198	0.3201	0.3977	329.06	< .0001
Ageph	> 60	−0.2234	0.0203	−0.2633	−0.1836	120.70	< .0001
Ageph	31–60	0	0	0	0	.	.
Gender	Female	0.0581	0.0161	0.0266	0.0896	13.08	0.0003
Gender	Male	0	0	0	0	.	.
Agev	3–5	−0.0629	0.0175	−0.0971	−0.0286	12.95	0.0003
Agev	Rest	0	0	0	0	.	.
Fuel	Petrol	−0.1910	0.0153	−0.2209	−0.1611	156.58	< .0001
Fuel	Gasoil	0	0	0	0	.	.
Premium split	No	−0.2748	0.0144	−0.3031	−0.2466	363.22	< .0001
Premium split	Yes	0	0	0	0	.	.
Coverage	MTPL only	0.1189	0.0150	0.0896	0.1482	63.25	< .0001
Coverage	More	0	0	0	0	.	.
City	Rural	−0.2265	0.0146	−0.2550	−0.1980	241.98	< .0001
City	Urban	0	0	0	0	.	.
Power	< 66 kW	−0.2969	0.0670	−0.4282	−0.1657	19.67	< .0001
Power	66–110 kW	−0.1929	0.0676	−0.3254	−0.0603	8.13	0.0044
Power	> 110 kW	0	0	0	0	.	.

As explained in Chapter 2, overdispersion can be detected by using the statistics T_1, T_2 or T_3 presented in Section 2.4.6. The values obtained with Portfolio C are $T_1 = 20.04$, $T_2 = 14.56$ and $T_3 = 12.89$. All the associated p-values are less than 10^{-4} leading to the rejection of the null hypothesis in favour of a mixed Poisson model.

In order to take the residual heterogeneity into account, we fitted a Negative Binomial regression model to the Portfolio C. The results are given in Table 5.7. All the variables and levels are still relevant. The log-likelihood is now equal to −61 393.3 and is better than with Poisson regression.

Resulting Price List

The resulting price list is obtained thanks to the Negative Binomial model for the frequency (presented in Table 5.7) and to the LogNormal model for the average cost of the standard claims (presented in Table 5.5).

Neglecting the large claims, the pure premium of the reference class is obtained by

$$\exp\left(\beta_0^{\text{freq}} + \beta_0^{\text{cost}} + \frac{\sigma^2}{2}\right) = €316.76. \tag{5.6}$$

This amount corresponds to the pure premium of a male policyholder living in an urban area, aged between 31 and 60, driving a car older than 10 years, using gasoil and with a power greater than 110 kW, paying his premium in several installments and having opted for a more extensive coverage than only the compulsory motor third party liability insurance.

To obtain the pure premium (neglecting the large claims) for a policyholder belonging to another risk class, the percentages of Table 5.8 must be applied to the base pure premium

Table 5.7 Results of the Negative Binomial regression of the claim counts recorded in Portfolio C.

Variable	Level	Coeff β	Std error	Wald 95 % conf limit		Chi-sq	Pr>Chi-sq
Intercept		−1.4626	0.0720	−1.6037	−1.3216	413.06	< .0001
Ageph	18–24	0.6436	0.0318	0.5812	0.7060	408.89	< .0001
Ageph	25–30	0.3604	0.0207	0.3198	0.4011	302.03	< .0001
Ageph	> 60	−0.2235	0.0209	−0.2645	−0.1825	114.04	< .0001
Ageph	31–60	0	0	0	0	.	.
Gender	Female	0.0603	0.0167	0.0276	0.0930	13.06	0.0003
Gender	Male	0	0	0	0	.	.
Agev	3–5	−0.0622	0.0181	−0.0977	−0.0267	11.78	0.0006
Agev	Rest	0	0	0	0	.	.
Fuel	Petrol	−0.1916	0.0159	−0.2227	−0.1605	145.55	< .0001
Fuel	Gasoil	0	0	0	0	.	.
Premium split	No	−0.2752	0.0150	−0.3045	−0.2459	338.40	< .0001
Premium split	Yes	0	0	0	0	.	.
Coverage	MTPL only	0.1192	0.0155	0.0888	0.1496	59.06	< .0001
Coverage	More	0	0	0	0	.	.
City	Rural	−0.2270	0.0151	−0.2566	−0.1974	226.08	< .0001
City	Urban	0	0	0	0	.	.
Power	< 66 kW	−0.2950	0.0698	−0.4318	−0.1582	17.87	< .0001
Power	66–110 kW	−0.1911	0.0705	−0.3293	−0.0529	7.34	0.0067
Power	> 110 kW	0	0	0	0	.	.

Table 5.8 Price list for the standard claims in Portfolio C.

Variable	Level	Influence of frequency	Influence of cost	Total influence
Ageph	18–24	190.33 %	119.08 %	226.64 %
Ageph	25–30	143.39 %	100.00 %	143.39 %
Ageph	> 60	79.97 %	119.49 %	95.56 %
Ageph	31–60	100.00 %	100.00 %	100.00 %
Gender	Female	106.22 %	100.00 %	106.22 %
Gender	Male	100.00 %	100.00 %	100.00 %
Agev	0–2	100.00 %	88.44 %	88.44 %
Agev	3–5	93.97 %	80.93 %	76.05 %
Agev	6–10	100.00 %	89.07 %	89.07 %
Agev	> 10	100.00 %	100.00 %	100.00 %
Fuel	Petrol	82.56 %	100.00 %	82.56 %
Fuel	Gasoil	100.00 %	100.00 %	100.00 %
Premium split	No	75.94 %	100.00 %	75.94 %
Premium split	Yes	100.00 %	100.00 %	100.00 %
Coverage	MTPL only	112.66 %	100.00 %	112.66 %
Coverage	More	100.00 %	100.00 %	100.00 %
City	Rural	79.69 %	106.83 %	85.14 %
City	Urban	100.00 %	100.00 %	100.00 %
Power	< 66 kW	74.45 %	100.00 %	74.45 %
Power	66–110 kW	82.60 %	100.00 %	82.60 %
Power	> 110 kW	100.00 %	100.00 %	100.00 %

(5.6) as a function of the characteristics of the policyholder. For example, the pure premium of a woman aged between 25 and 30, living in a rural region, driving a car older than 10 years, using petrol and with a power less than 66 kW, paying her premium once a year and being covered only for the MTPL will be equal to

$$\begin{aligned} €316.76 &\times 106.22\,\% \text{ (correction for being a female policyholder)} \\ &\times 143.39\,\% \text{ (correction for being aged between 25 and 30)} \\ &\times 85.14\,\% \text{ (correction for living in a rural area)} \\ &\times 82.56\,\% \text{ (correction for using petrol)} \\ &\times 74.45\,\% \text{ (correction for driving a low-power car)} \\ &\times 75.94\,\% \text{ (correction for paying the premium once a year)} \\ &\times 112.66\,\% \text{ (correction for buying MTPL coverage only)} \\ &= €216.01. \end{aligned}$$

Finally, to obtain the total pure premium of a policyholder, the expected cost of large claims (which is independent of the explanatory variables) must be added. The value is equal to

$$0.0117\,\% \times €368\,018.80 = €43.06.$$

The total pure premium of a policyholder belonging to the reference class is then

$$€316.76 + €43.06 = €359.82.$$

Note that the fitted mean equal to € 368 018.80 is close to the observed mean of the 17 large losses recorded in Table 5.2.

Recall that the analysis of large claims performed in this chapter is based on an estimation of their final cost six months after the end of the observation year 1997. We thus work with incurred losses (payments plus reserves). In practice, the company should maintain a data base recording the costs of large claims that have occurred in the past, corrected for the different sources of inflation (reinsurance companies can often provide valuable assistance to the ceding companies in this respect). The typical price for a reinsurance treaty covering motor third party liability insurance losses in excess of € 350 000 represents about 5 % of the total motor premium income of a Belgian insurance company, so that the expected cost of large claims computed in Portfolio C seems to be of the right order of magnitude.

Remark 5.5 If the sum of the individual pure premiums obtained above exceeds the observed total loss for the insurance portfolio during the reference period, or if we expect larger losses in the future (because, e.g., of different sources of inflation), we can then keep the same relative premium amounts applied to the anticipated future total claim cost. This allows us to incorporate in the individual premiums observed trends in the total claim amount.

5.3 Measures of Efficiency for Bonus-Malus Scales

The elasticity of a bonus-malus system measures its response to a change in the expected claim frequency or expected aggregate claim amount. We expect that the premium paid by the policyholders subject to bonus-malus scales is increasing in the expected claim frequency or total claim amount. The rate of increase is related to the concept of efficiency.

5.3.1 Loimaranta Efficiency

Definition

Let us denote as $\overline{r}(\vartheta)$ the average relativity once stationarity has been reached, for a policyholder with annual expected claim frequency ϑ, i.e.

$$\overline{r}(\vartheta) = \sum_{\ell=0}^{s} \pi_\ell(\vartheta) r_\ell.$$

The Loimaranta efficiency $\text{Eff}_{\text{Loi}}(\vartheta)$ is then defined as the elasticity of the relative premium induced by the bonus-malus system, that is,

$$\text{Eff}_{\text{Loi}}(\vartheta) = \frac{\dfrac{d\overline{r}(\vartheta)}{\overline{r}(\vartheta)}}{\dfrac{d\vartheta}{\vartheta}} = \frac{d\ln \overline{r}(\vartheta)}{d\ln \vartheta}.$$

Computation

The computation of $\text{Eff}_{\text{Loi}}(\vartheta)$ requires the determination of the derivative of $\overline{r}(\vartheta)$ with respect to the annual expected claim frequency ϑ. This derivative is given by

$$\frac{d\overline{r}(\vartheta)}{d\vartheta}=\sum_{\ell=0}^{s}\frac{d\pi_\ell(\vartheta)}{d\vartheta}r_\ell,$$

so that its computation requires the derivative of the stationary probabilities $\pi_\ell(\vartheta)$ with respect to the annual expected claim frequency ϑ. To get $d\pi_\ell(\vartheta)/d\vartheta$, it suffices to differentiate (4.8): we thus have to solve the linear system

$$\begin{cases}\dfrac{d\boldsymbol{\pi}^T(\vartheta)}{d\vartheta}=\dfrac{d\boldsymbol{\pi}^T(\vartheta)}{d\vartheta}\boldsymbol{P}(\vartheta)+\boldsymbol{\pi}^T(\vartheta)\dfrac{d\boldsymbol{P}(\vartheta)}{d\vartheta}\\ \sum_{\ell=0}^{s}\dfrac{d\pi_\ell(\vartheta)}{d\vartheta}=0\end{cases}$$

with respect to the $d\pi_\ell(\vartheta)/d\vartheta$s.

Global Efficiency

So far, we have defined the Loimaranta efficiency for a given value ϑ of the expected annual claim frequency. To get a value for the portfolio, we have to account for its composition with respect to rating factors as well as its residual heterogeneity. Hence, the Loimaranta efficiency for the portfolio is obtained by averaging over all the possible values for ϑ as

$$\text{Eff}_{\text{Loi}}=\mathbb{E}\Big[\text{Eff}_{\text{Loi}}\big(\Lambda\Theta\big)\Big].$$

Loimaranta Efficiency in Portfolio A

Table 5.9 displays the Loimaranta efficiencies for a good driver, with annual expected claim frequency 9.28 %, for an average driver with annual expected claim frequency 14.09 %, and for a bad driver with annual expected claim frequency 28.40 %. For the −1/top bonus-malus scale, the efficiency is larger for the average driver than for the good and bad ones. On the contrary, for the −1/+2 and −1/+3 bonus-malus scales, the efficiencies appear to increase from the good to the average driver, and from the average driver to the bad one. The global efficiencies listed in Table 5.9 are rather poor, ranging from 23.23 % for the −1/top bonus-malus scale to 28.39 % in the −1/+2 bonus-malus scale. This means that these bonus-malus systems weakly respond to a change in the underlying claim frequency.

Table 5.9 Loimaranta efficiency for three types of insured drivers (a good driver, with annual expected claim frequency 9.28 %, an average driver with annual expected claim frequency 14.09 %, and a bad driver with annual expected claim frequency 28.40 %) and global efficiency for Portfolio A, for the −1/top, −1/+2 and −1/+3 bonus-malus scales.

Frequency	Scale −1/top	Scale −1/+2	Scale −1/+3
0.0928	0.2865	0.2380	0.2987
0.1409	0.3144	0.3793	0.4008
0.2840	0.2901	0.6204	0.4733
Portfolio A	0.2323	0.2839	0.2775

Figure 5.2 displays the Loimaranta efficiency as a function of the annual expected claim frequencies for the three bonus-malus systems. We see that the efficiency first increases and then decreases, reaching its maximum value at about 15 % for the −1/top bonus-malus system, at about 30 % for the $-1/+2$ bonus-malus system, and at about 25 % for the $-1/+3$ bonus-malus system.

5.3.2 De Pril Efficiency

Definition

Loimaranta efficiency is an asymptotic concept that does not depend on the level presently occupied in the scale. De Pril efficiency is a transient concept that explicitly considers time value of money.

Let $v<1$ be the discount factor (present value on €1 paid in one year) and let us denote as $V_\ell^{(n)}(\vartheta)$ the average present value of all the premiums paid during the next n years by a policyholder with expected annual claim frequency ϑ and occupying level ℓ in the scale. Clearly, the $V_\ell^{(n)}(\vartheta)$s obey the following recurrence relation:

$$V_\ell^{(n)}(\vartheta)=b_\ell+v\sum_{k=0}^{+\infty}\Pr[N=k|\lambda\Theta=\vartheta]V_{T_k(\ell)}^{(n-1)}(\vartheta),\quad \ell=0,1,\ldots,s,$$

where b_ℓ is the premium paid by a policyholder with claim frequency ϑ occupying level ℓ in the scale (obtained by multiplying some base premium by the relativity r_ℓ).

Switching to an infinite time horizon, that is taking the limit

$$V_\ell(\vartheta)=\lim_{n\to\infty}V_\ell^{(n)}(\vartheta),$$

the $V_\ell(\vartheta)$s are the solution of

$$V_\ell(\vartheta)=b_\ell+v\sum_{k=0}^{+\infty}\Pr[N=k|\lambda\Theta=\vartheta]V_{T_k(\ell)}(\vartheta),\quad \ell=0,1,\ldots,s. \tag{5.7}$$

It can be shown (see, e.g., LEMAIRE (1995, page 87)) that the system (5.7) admits a unique solution.

In addition to the $V_\ell(\vartheta)$s, let us also consider the expected present value $C_\ell(\vartheta)$ of all the claim costs for a policyholder with annual expected claim frequency ϑ occupying level ℓ. The De Pril efficiency is then defined as

$$\mathrm{Eff}_{\mathrm{DeP}}(\ell,\vartheta)=\frac{d\ln V_\ell(\vartheta)}{d\ln C_\ell(\vartheta)}.$$

It measures how the bonus-malus system reacts to a change in the total expected claim amount.

Now, assume that $C_\ell(\vartheta)$ is equal to the product of the claim frequency ϑ times an expected claim cost that does not depend on ϑ, and that $V_\ell(\vartheta)$ is obtained by multiplying the same claim cost with an updated estimated claim frequency. De Pril efficiency then reduces to

$$\mathrm{Eff}_{\mathrm{DeP}}(\ell,\vartheta)=\frac{\dfrac{dV_\ell(\vartheta)}{V_\ell(\vartheta)}}{\dfrac{d\vartheta}{\vartheta}}=\frac{d\ln V_\ell(\vartheta)}{d\ln(\vartheta)},\quad \ell=0,1,\ldots,s.$$

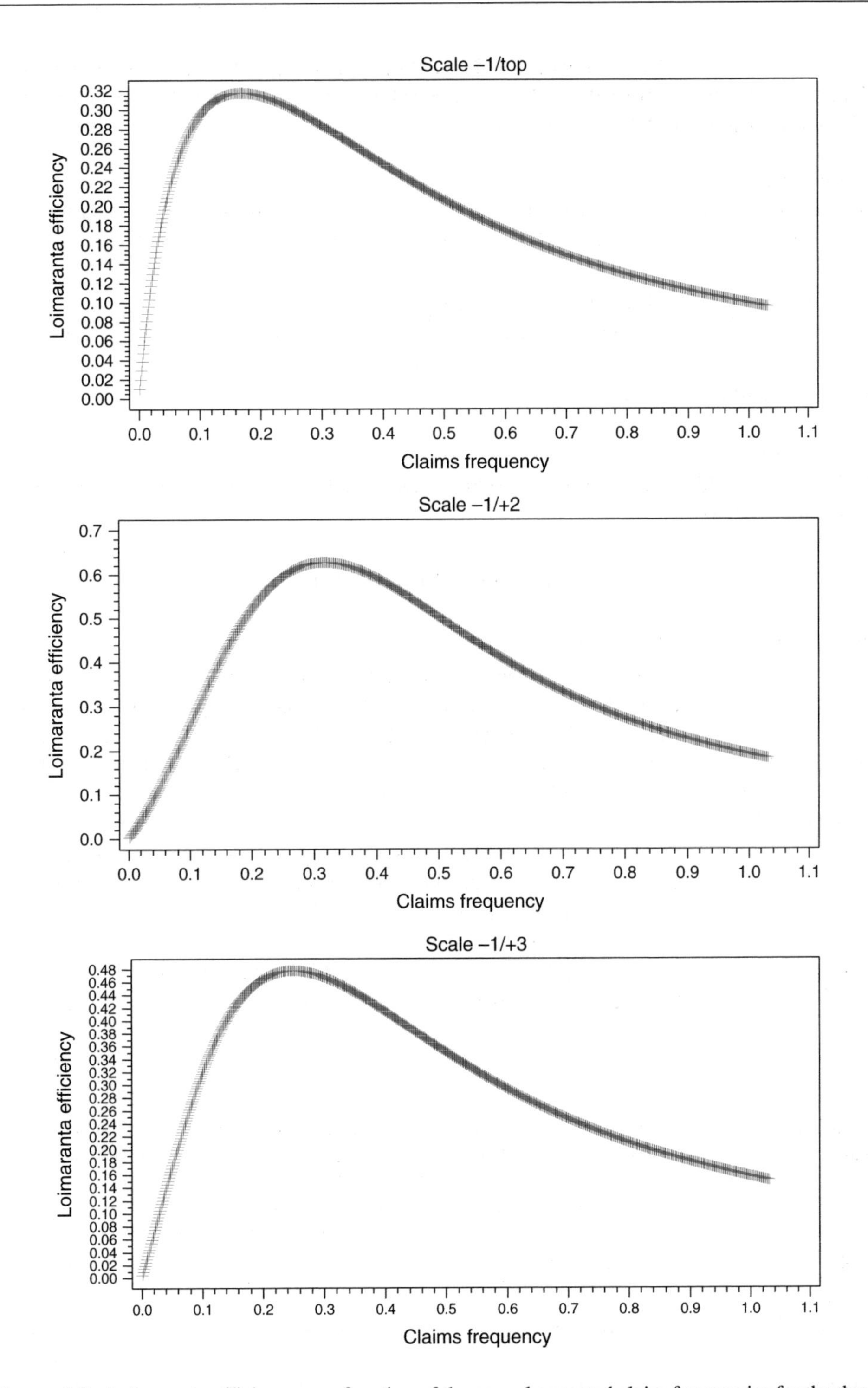

Figure 5.2 Loimaranta efficiency as a function of the annual expected claim frequencies for the three bonus-malus systems and Portfolio A.

De Pril efficiency Eff_{DeP} is thus defined analogously to Eff_{Loi}, substituting $V_\ell(\vartheta)$ for $\overline{r}(\vartheta)$. Note that $\text{Eff}_{\text{DeP}}(\ell, \vartheta)$ now depends on the starting class ℓ. The initial class can then be selected so as to maximize $\text{Eff}_{\text{DeP}}(\ell, \vartheta)$.

Computation

To compute $\text{Eff}_{\text{DeP}}(\ell, \vartheta)$, we need the derivatives $dV_\ell(\vartheta)/d\vartheta$ of the $V_\ell(\vartheta)$ satisfying (5.7). These derivatives can be obtained by solving the system

$$\frac{dV_\ell(\vartheta)}{d\vartheta} = v\sum_{k=0}^{\infty} \frac{\exp(-\vartheta)\vartheta^k}{k!} \left(\left(\frac{k}{\vartheta} - 1 \right) V_{T_k(\ell)}(\vartheta) + \frac{dV_{T_k(\ell)}(\vartheta)}{d\vartheta} \right), \quad \ell = 0, \ldots, s.$$

This system admits a unique solution.

Global Efficiency

At the portfolio level, the efficiency is then obtained by averaging over all the possible values for ϑ, that is,

$$\text{Eff}_{\text{DeP}}(\ell) = \mathbb{E}[\text{Eff}_{\text{DeP}}(\ell, \Lambda\Theta)].$$

De Pril Efficiency in Portfolio A

Table 5.10 displays the De Pril efficiencies associated with the highest level 5 for a good driver, with annual expected claim frequency 9.28 %; for an average driver with annual expected claim frequency 14.09 %; and for a bad driver with annual expected claim frequency 28.40 %. The discount factor is taken to be $v = 1/1.04$. De Pril efficiency behaves roughly as the Loimaranta efficiency displayed in Table 5.9.

Figure 5.3 displays the De Pril efficiency as a function of the annual expected claim frequencies for the three bonus-malus systems. We see that the efficiency first increases and then decreases, reaching its maximum value at about the same frequencies as the Loimaranta efficiency. The shape of both efficiencies is pretty much the same.

Let us now use the De Pril efficiency to select the optimal starting level. We have computed in Table 5.11 the values of $\text{Eff}_{\text{DeP}}(\ell, \vartheta)$ and $\text{Eff}_{\text{DeP}}(\ell)$ according to the initial level. We see that the optimal starting level is 0 for the three -1/top, $-1/+2$ and $-1/+3$ bonus-malus scales.

Table 5.10 De Pril efficiency for three types of insured drivers (a good driver, with annual expected claim frequency 9.28 %, an average driver with annual expected claim frequency 14.09 %, and a bad driver with annual expected claim frequency 28.40 %) and global efficiency for Portfolio A, for the -1/top, $-1/+2$ and $-1/+3$ bonus-malus scales (starting level: level 5).

Frequency	Scale -1/top	Scale $-1/+2$	Scale $-1/+3$
0.0928	0.2192	0.1871	0.2252
0.1409	0.2482	0.2880	0.3039
0.2840	0.2414	0.4633	0.3741
Portfolio A	0.1817	0.2186	0.2150

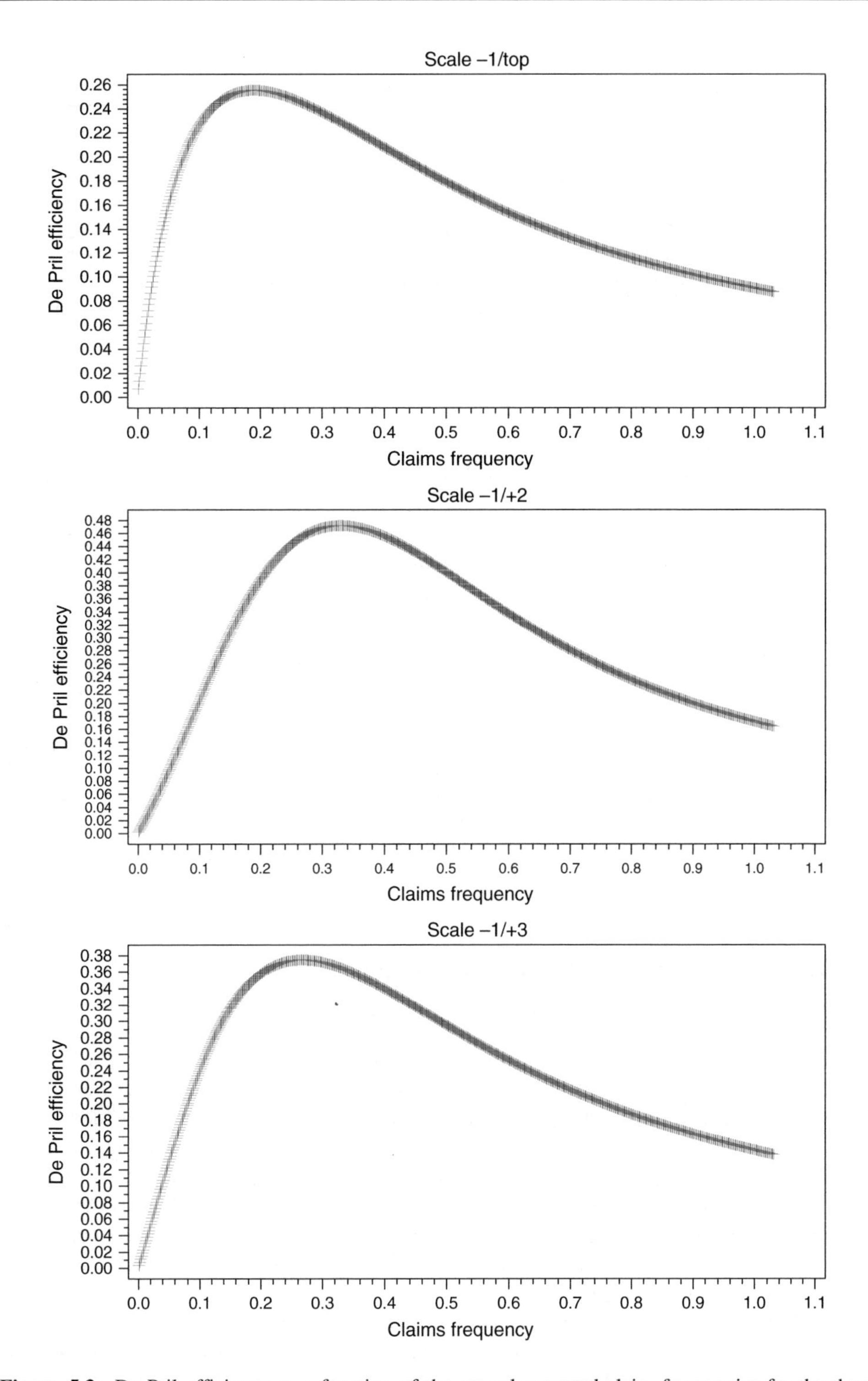

Figure 5.3 De Pril efficiency as a function of the annual expected claim frequencies for the three bonus-malus systems and Portfolio A.

Table 5.11 De Pril efficiency according to the starting level, for the -1/top, $-1/+2$ and $-1/+3$ bonus-malus scales.

Initial level		-1/top scale		
	$\text{Eff}_{\text{DeP}}(\ell, 0.0928)$	$\text{Eff}_{\text{DeP}}(\ell, 0.1409)$	$\text{Eff}_{\text{DeP}}(\ell, 0.2804)$	$\text{Eff}_{\text{DeP}}(\ell)$
0	0.2711379	0.3018164	0.287755	0.2230948
1	0.2644811	0.2952273	0.2826536	0.2179881
2	0.2558848	0.2863492	0.2746372	0.2109572
3	0.2454854	0.2755109	0.2648401	0.2025042
4	0.2332928	0.2628293	0.2537603	0.1927769
5	0.2191762	0.248209	0.2414355	0.1817001
Initial level		$-1/+2$ scale		
	$\text{Eff}_{\text{DeP}}(\ell, 0.0928)$	$\text{Eff}_{\text{DeP}}(\ell, 0.1409)$	$\text{Eff}_{\text{DeP}}(\ell, 0.2804)$	$\text{Eff}_{\text{DeP}}(\ell)$
0	0.2168178	0.3422846	0.5689437	0.262244
1	0.2159419	0.3399076	0.5617233	0.2597218
2	0.2156974	0.3369153	0.5488944	0.2558833
3	0.2120306	0.3274338	0.5247412	0.2474823
4	0.202593	0.311189	0.4963046	0.2351387
5	0.1870807	0.2878845	0.4633192	0.2186229
Initial level		$-1/+3$ scale		
	$\text{Eff}_{\text{DeP}}(\ell, 0.0928)$	$\text{Eff}_{\text{DeP}}(\ell, 0.1409)$	$\text{Eff}_{\text{DeP}}(\ell, 0.2804)$	$\text{Eff}_{\text{DeP}}(\ell)$
0	0.2733728	0.3689638	0.4526559	0.2605588
1	0.2691461	0.362999	0.4449703	0.2563432
2	0.2648276	0.3556598	0.4328504	0.2508412
3	0.2572639	0.3446637	0.417896	0.2429854
4	0.2429893	0.3261114	0.3971923	0.2302266
5	0.2252077	0.3038548	0.3740766	0.2149911

5.4 Bonus Hunger and Optimal Retention

5.4.1 Correcting the Estimations for Censoring

As explained in the introduction, the policyholders subject to a bonus-malus mechanism tend to self-defray minor accidents to avoid premium surcharges. This means that the number of accidents is a censored variable: the insurer only knows the number of claims filed by the insured drivers, and not the number of accidents they caused. We develop here a simple statistical model allowing for censorship in the observed claim costs (and thus also in the observed numbers of claims reported to the insurer). The claiming threshold is considered here as a random variable, specific to each policyholder and with a distribution depending on the level occupied in the bonus-malus scale at the beginning of the observation period as well as on observable characteristics.

Specifically, let us consider the LogNormal model for moderate claim sizes: the claim costs are then seen as independent and identically distributed realizations of LogNormal random variables in each risk class. Now, each policyholder in this class reports an accident to the insurer if its cost exceeds a random threshold, assumed to be LogNormally distributed with parameters specific to the level occupied in the scale. Note that we deal here with moderate claim sizes only. The reason is that large claims are not subject to bonus hunger and are systematically reported to the insurer. The frequency of large claims will have to be added to the corrected frequency of moderate claims to get the actual number of accidents caused each year.

Let ℓ_i be the level occupied by policyholder i in the bonus-malus scale at the beginning of the period, and let $RL_i(\ell_i) \sim \mathcal{L}Nor(\nu_i, \tau^2)$ be the random optimal retention, with a linear predictor of the form

$$\nu_i = \gamma_0 + \sum_{j=1}^{p} \gamma_j x_{ij} + f(\ell_i),$$

specific to policyholder i, where the function $f(\cdot)$ expresses the effect of occupying level ℓ_i in the bonus-malus scale. Considering the values obtained for the optimal retention in the literature (reaching a maximum somewhere in the middle of the scale, and decreasing when approaching uppermost and lowermost levels), we will use here a quadratic effect f of ℓ_i. Note that other approaches are possible (see the references in the closing section for more details).

This means that policyholder i will report all the accidents with a cost larger than $RL_i(\ell_i)$, and defray himself all those with a cost less than $RL_i(\ell_i)$. At the portfolio level, the $RL_i(\ell_i)$s are assumed to be independent. Now, let CA_{ik} be the cost of the kth accident caused by policyholder i. We assume that for each i the random variables $CA_{i1}, CA_{i2}, \ldots$ are independent and identically distributed, with $CA_{ik} \sim \mathcal{L}Nor(\mu_i, \sigma^2)$, where $\mu_i = \beta_0 + \sum_{j=1}^{p} \beta_j x_{ij}$. Moreover, the CA_{ik}s and $RL_i(\ell_i)$ are mutually independent. We consider here the explanatory variables selected in the LogNormal analysis of the censored claim costs, presented in Table 5.5.

Now, denoting as $c_{i1}, \ldots, c_{in_i}$ the costs of the n_i moderate claims filed by policyholder i, the likelihood is

$$\mathcal{L}(\boldsymbol{\beta}, \boldsymbol{\gamma}, \sigma^2, \tau) = \prod_{i|n_i>0} \prod_{k=1}^{n_i} f_i(c_{ik})$$

where $f_i(\cdot)$ denotes the probability density function of CA_{ik} given $CA_{ik} > RL_i(\ell_i)$. Each factor involved in the likelihood can be written is

$$f_i(c_{ik}) = \frac{\dfrac{1}{\sqrt{2\pi}\sigma c_{ik}} \exp\left(-\dfrac{(\ln c_{ik} - \mu_i)^2}{2\sigma^2}\right) \Phi\left(\dfrac{\ln c_{ik} - \nu_i}{\tau}\right)}{1 - \Phi\left(-\dfrac{\mu_i - \nu_i}{\sqrt{\sigma^2 + \tau^2}}\right)}.$$

The estimators of the parameters $\boldsymbol{\beta}$, $\boldsymbol{\gamma}$, σ^2, and τ are determined by maximizing the likelihood $\mathcal{L}(\boldsymbol{\beta}, \boldsymbol{\gamma}, \sigma^2, \tau)$.

This basic model could be refined in different respects. Firstly, the retention limit could depend on the number of claims previously filed by the policyholder during the same year. The retention for the second claim depends on the level to which the policyholder is transferred after the first claim has been reported. Here, we only use the observations related to the policyholders having filed a single standard claim during the observation period (so that we have the exact cost of this claim at our disposal). Also, the estimation of f could be performed in a nonparametric way, allowing for an effect $f(\ell_i)$ of being in level ℓ_i (and imposing some smoothness in the $f(\ell)$s if needed) before selecting an appropriate parametric specification.

Let us now apply this methodology to Portfolio C. The policyholders have been subject to the 23-level former compulsory Belgian bonus-malus scale. The estimation of the regression parameters in the model containing all the explanatory variables are displayed in Table 5.12. Here, we take $f(\ell) = (\ell - 13)^2$. The log-likelihood is $-130\,154.2565$. We see from this table that several covariates are not significant. Therefore, we adopt a backward selection procedure, and exclude the irrelevant covariates. This yields the results displayed in Table 5.13. The log-likelihood is now $-130\,155.4197$. The parameters σ and τ are estimated at $\widehat{\sigma} = 1.6821$ and $\widehat{\tau} = 1.0286$.

Compared to the LogNormal fit to the claim costs displayed in Table 5.5, we see that the intercept is now smaller, as expected. The age classes have been modified, and the young drivers seem to cause more expensive accidents. The effect of the covariate City remains approximately the same. The categories for the age of the vehicle have also been modified.

Table 5.12 Fit of the model for the accident costs subject to bonus hunger in Portfolio C, containing all the explanatory variables.

Variable	Level	Coeff β	Std error	Wald 95 %	conf limits	Chi-sq	Pr>Chi-sq
Intercept		5.8028	0.0453	5.7122	5.8934	16409.11	< 0.0001
Ageph	18–24	0.2624	0.0640	0.1344	0.3903	16.82	< 0.0001
Ageph	> 60	0.0717	0.0500	−0.0282	0.1716	2.06	0.1512
Ageph	25–60	0	0	0	0	.	.
City	Rural	0.0459	0.0275	−0.0091	0.1009	2.79	0.0948
City	Urban	0	0	0	0	.	.
Agev	0–2	0.0327	0.0652	−0.0977	0.1631	0.25	0.6158
Agev	3–5	−0.1721	0.0465	−0.2652	−0.0791	13.69	0.0002
Agev	6–10	−0.1445	0.0357	−0.2159	−0.0732	16.40	0.0001
Agev	> 10	0	0	0	0	.	.
Variable	**Level**	**Coeff γ**	**Std error**	**Wald 95 %**	**conf limits**	**Chi-sq**	**Pr>Chi-sq**
Intercept		3.5045	0.0904	3.3238	3.6852	1504.45	< 0.0001
Ageph	18–24	−0.3244	0.1637	−0.6519	0.0031	3.93	0.0476
Ageph	> 60	0.5419	0.1105	0.3209	0.7630	24.05	< 0.0001
Ageph	25–60	0	0	0	0	.	.
City	Rural	0.0796	0.0416	−0.0036	0.1628	3.66	0.0558
City	Urban	0	0	0	0	.	.
Agev	0–2	−0.7825	0.2621	−1.3067	−0.2582	8.91	0.0028
Agev	3–5	−0.3406	0.1107	−0.5620	−0.1193	9.47	0.0021
Agev	6–10	0.0190	0.0449	−0.0708	0.1087	0.18	0.6725
Agev	> 10	0	0	0	0	.	.
BM level		−0.0011	0.0006	−0.0022	0.0000	3.67	0.0555

Table 5.13 Fit of the final model for the accident costs subject to bonus hunger in Portfolio C.

Variable	Level	Coeff β	Std error	Wald 95 % conf limits		Chi-sq	Pr>Chi-sq
Intercept		5.8084	0.0402	5.7279	5.8889	20847.38	< 0.0001
Ageph	18–24	0.2522	0.0661	0.1201	0.3843	14.57	0.0001
Ageph	> 24	0	0	0	0	.	.
City	Rural	0.0692	0.0274	0.0145	0.1240	6.39	0.0115
City	Urban	0	0	0	0	.	.
Agev	3–5	−0.1866	0.0394	−0.2653	−0.1078	22.45	< 0.0001
Agev	6–10	−0.1582	0.0349	−0.2281	−0.0883	20.51	< 0.0001
Agev	0–2 & > 10	0	0	0	0	.	.
Variable	**Level**	**Coeff γ**	**Std error**	**Wald 95 % conf limits**		**Chi-sq**	**Pr>Chi-sq**
Intercept		3.5269	0.0870	3.3529	3.7010	1642.72	< 0.0001
Ageph	18–24	−0.3077	0.1430	−0.5937	−0.0217	4.63	0.0314
Ageph	> 60	0.6479	0.0852	0.4775	0.8183	57.81	< 0.0001
Ageph	25–60	0	0	0	0	.	.
Agev	0–2	−0.7126	0.1171	−0.9468	−0.4785	37.04	< 0.0001
Agev	3–5	−0.3170	0.0666	−0.4503	−0.1837	22.63	< 0.0001
Agev	> 5	0	0	0	0	.	.
BM level		−0.0011	0.0006	−0.0022	0.0000	3.70	0.0546

Concerning the retention levels, we see that young drivers are more likely to defray only relatively cheap accidents, whereas older drivers are ready to self-defray more expensive accidents. The more recent the vehicle, the less accidents are self-defrayed. This may be due to the fact that comprehensive coverage is often bought for new vehicles, so that claims are filed to both third party liability and comprehensive. The effect of the level occupied in the scale is as follows: policyholders occupying the middle of the scale are ready to defray more expensive accidents than policyholders at the top or at the bottom of the scale.

5.4.2 Number of Claims and Number of Accidents

Let M_i^{small} be the number of small accidents caused by policyholder i. The number of moderate claims filed by policyholder i is then given by

$$N_i^{\text{small}} = \sum_{k=1}^{M_i^{\text{small}}} \mathbb{I}[CA_{ik} > RL_i(\ell_i)].$$

By equating the expectations, we get

$$\begin{aligned}
\mathbb{E}[N_i^{\text{small}}] = \lambda_i &= \sum_{k=0}^{+\infty} \Pr[M_i^{\text{small}} = k] \sum_{j=1}^{k} \Pr[CA_{ij} > RL_i(\ell_i)] \\
&= \sum_{k=0}^{+\infty} \Pr[M_i^{\text{small}} = k] k \Pr[CA_{i1} > RL_i(\ell_i)] \\
&= \Pr[CA_{i1} > RL_i(\ell_i)] \mathbb{E}[M_i^{\text{small}}].
\end{aligned}$$

Hence, the expected number of moderate accidents is given by

$$\mathbb{E}[M_i^{\text{small}}] = \widetilde{\lambda}_i = \frac{\lambda_i}{\Pr[CA_{i1} > RL_i(\ell_i)]} = \frac{\lambda_i}{1 - \Phi\left(-\frac{\mu_i - \nu_i}{\sqrt{\sigma^2 + \tau^2}}\right)}.$$

The expected annual number of minor accidents $\widetilde{\lambda}_i$ caused by policyholder i in Portfolio C are displayed in Table 5.14.

The total number of accidents M_i caused by policyholder i is then equal to

$$M_i = M_i^{\text{small}} + N_i^{\text{large}}$$

since all the major accidents are reported to the company. The actual number of claims originating from the M_i accidents is

$$N_i = N_i^{\text{small}} + N_i^{\text{large}}.$$

Table 5.14 Expected annual claim frequency and corresponding expected annual accident frequency for the different risk classes in Portfolio C.

Risk class	Claim frequency	Accident frequency
18–24 + Rural + 0–2	0.3513403	0.3633521
18–24 + Rural + 3–5	0.3301527	0.3517856
18–24 + Rural + 6–10	0.3513403	0.3824574
18–24 + Rural+>10	0.3513403	0.3777624
18–24 + Urban + 0–2	0.4408723	0.4572087
18–24 + Urban + 3–5	0.4142855	0.4435002
18–24 + Urban + 6–10	0.4408723	0.4827648
18 – 24 + Urban+>10	0.4408723	0.4765034
>60 + Rural + 0–2	0.1476220	0.1659192
>60 + Rural + 3–5	0.1387197	0.1683992
>60 + Rural + 6–10	0.1476220	0.188461
>60 + Rural+>10	0.1476220	0.1831299
>60 + Urban + 0–2	0.1852406	0.209793
>60 + Urban + 3–5	0.1740696	0.2137084
>60 + Urban + 6–10	0.1852406	0.2396818
>60 + Urban+>10	0.1852406	0.2326217
25–60 + Rural + 0–2	0.1845933	0.1964037
25–60 + Rural + 3–5	0.1734614	0.193623
25–60 + Rural + 6–10	0.1845933	0.2129273
25–60 + Rural+>10	0.1845933	0.208954
25–60 + Urban + 0–2	0.2316332	0.2475869
25–60 + Urban + 3–5	0.2176646	0.2447353
25–60 + Urban + 6–10	0.2316332	0.2695796
25–60 + Urban+>10	0.2316332	0.264303

Moreover, the random variable N_i^{large} is independent from $(M_i^{\text{small}}, N_i^{\text{small}})$. Both M_i and N_i are mixed Poisson distributed. Specifically, given $\Theta_i = \theta$, $M_i \sim \mathcal{P}oi(\widetilde{\lambda}_i\theta + \lambda_i^{\text{large}})$ and $N_i \sim \mathcal{P}oi(\lambda_i\theta + \lambda_i^{\text{large}})$. Note that the variable that has been analysed in the preceding chapters is N_i, the number of accidents reported to the insurer, and not M_i.

5.4.3 Lemaire Algorithm for the Determination of Optimal Retention Limits

In the preceding section, we explained how to correct the cost of claims to obtain the accident costs. This also allowed us to switch from claim frequencies to accident frequencies. To this end, we estimated the retention limits that were used by the policyholders on the basis of the costs of the claims they filed to the insurance company. The aim of this section is somewhat different. Having the distribution of the accident costs and of the accident frequencies, we would like to determine the optimal claiming strategy (which may differ from the observed claiming strategy inferred in the previous section).

For each level of the scale, a critical claim size is determined: if the cost of the claim falls below this critical threshold then the rational policyholder should not report the accident to the company. Conversely if the cost exceeds this threshold, the rational policyholder should report the claim to the company. Note the close similarity with deductibles: under coherent behaviour, the bonus-malus scale is equivalent to a set of deductibles depending on the level occupied in the scale.

Cost of Non-Reported Accidents

Let $rl(\ell, \vartheta)$ be the optimal retention for a policyholder with expected annual accident frequency ϑ occupying level ℓ in the bonus-malus scale. Here, $rl(\ell, \vartheta)$ is not a random variable, but an unknown constant to be determined.

Assume that this policyholder has caused an accident with cost x at time t, $0 \leq t < 1$. If f is the probability density function of the accident cost, the probability $p_\ell(\vartheta)$ that the policyholder occupying level ℓ does not report this accident to the company is

$$p_\ell(\vartheta) = \int_{y=0}^{rl(\ell,\vartheta)} f(y)dy.$$

The probability that this policyholder reports exactly k accidents to the company during one year is denoted as $\overline{q}_\ell(k|\vartheta)$. This probability is given by the binomial expression

$$\overline{q}_\ell(k|\vartheta) = \sum_{h=k}^{\infty} \exp(-\vartheta)\frac{\vartheta^k}{k!}\binom{h}{k}\big(1 - p_\ell(\vartheta)\big)^k\big(p_\ell(\vartheta)\big)^{h-k}.$$

The average number of accidents reported to the company by a policyholder in level ℓ, denoted as $\overline{\vartheta}_\ell$, is given by

$$\overline{\vartheta}_\ell = \sum_{k=0}^{\infty} k\overline{q}_\ell(k|\vartheta).$$

The expected cost of a non-reported accident for a policyholder occupying level ℓ is

$$\mu_\ell(\vartheta) = \frac{1}{p_\ell(\vartheta)} \int_{y=0}^{rl(\ell,\vartheta)} yf(y)dy.$$

This policyholder will pay on average $\mu_\ell(\vartheta) \times (\vartheta - \overline{\vartheta_\ell})$ per period, because of the accidents not reported to the company. Let us assume that the accident occurrences are uniformly distributed over the year (so that on average they occur in the middle of the year). The average annual total cost borne by a policyholder in level ℓ is then

$$\overline{CT}(rl(\ell, \vartheta)) = b_\ell + v^{\frac{1}{2}} \mu_\ell(\vartheta)(\vartheta - \overline{\vartheta_\ell})$$

where b_ℓ is the premium paid at the beginning of the year, subject to the bonus-malus scale.

Let $V_\ell(\vartheta)$ be the present value of all the payments made by a policyholder with annual expected claim frequency ϑ occupying level ℓ. The $V_\ell(\vartheta)$s are obtained from

$$V_\ell(\vartheta) = \overline{CT}(rl(\ell, \vartheta)) + v \sum_{k=0}^{\infty} \overline{q}_\ell(k|\vartheta) V_{T_k(\ell)}(\vartheta), \ \ell = 0, 1, \ldots, s. \tag{5.8}$$

If the policyholder reports all the accidents to the company, the system (5.8) coincides with (5.7). The system (5.8) admits a unique solution. For a given set of optimal retentions, the $V_\ell(\vartheta)$s give the cost of the strategy, according to the level occupied in the scale.

Lemaire Algorithm

Let us consider a policyholder in level ℓ who just caused an accident with cost x at time t, $0 \leq t \leq 1$. There are two possibilities:

(1) Either he does not claim for the accident and the expected present cost is

$$v^{-t}\overline{CT}(rl(\ell, \vartheta)) + x + v^{1-t} \sum_{k=0}^{\infty} \overline{q}_\ell\big(k|\vartheta(1-t)\big) V_{T_{k+m}(\ell)}(\vartheta)$$

where m is the number of claims that the policyholder has already filed during the year.

(2) Or he reports the accident to the company and the expected present cost is

$$v^{-t}\overline{CT}(rl(\ell, \vartheta)) + v^{1-t} \sum_{k=0}^{\infty} \overline{q}_\ell\big(k|\vartheta(1-t)\big) V_{T_{k+m+1}(\ell)}(\vartheta).$$

The retention limit $rl(\ell, \vartheta)$ is the claim amount x for which the policyholder is indifferent between the two possibilities: the optimal retentions thus solve

$$rl(\ell, \vartheta) = v^{1-t} \sum_{k=0}^{\infty} \overline{q}_\ell\big(k|\vartheta(1-t)\big) \Big(V_{T_{k+m+1}(\ell)}(\vartheta) - V_{T_{k+m}(\ell)}(\vartheta) \Big), \tag{5.9}$$

for $\ell = 0, 1, \ldots, s$. Note that (5.9) does not provide an explicit expression for the optimal retention since $rl(\ell, \vartheta)$ also appears in the $\overline{q}_\ell(k|\vartheta(1-t))$s.

The optimal strategy is obtained using the following algorithm:

First Iteration

Part A Starting from $rl^{[0]}(\ell, \vartheta) = 0$ for $\ell = 0, \ldots, s$, the strategy consisting of reporting all the accidents to the insurer, (5.8) becomes

$$V_\ell(\vartheta) = b_\ell + v\sum_{k=0}^{\infty} \exp(-\vartheta)\frac{\vartheta^k}{k!} V_{T_k(\ell)}(\vartheta)$$

which gives the cost $\boldsymbol{V}^0(\vartheta)$ corresponding to the initial strategy.

Part B An improved strategy can then be obtained from (5.9) that reduces to

$$rl^{[1]}(\ell, \vartheta) = v^{1-t}\sum_{k=0}^{\infty} \exp(-(1-t)\vartheta)\frac{((1-t)\vartheta)^k}{k!}\Big(V_{T_{k+m+1}(\ell)}(\vartheta) - V_{T_{k+m}(\ell)}(\vartheta)\Big),$$

$\ell = 0, 1, \ldots, s$.

Second Iteration

Part A Inserting the $rl^{[1]}(\ell, \vartheta)$s in (5.8) gives the cost associated with this strategy. This cost will be smaller than the one associated with the initial strategy.

Part B Inserting the new cost in the system (5.9), we find an improved strategy $rl^{[2]}(\ell, \vartheta)$, $\ell = 0, \ldots, s$.

Subsequent Iterations

The successive insertion of updated retentions and costs in the systems (5.8)–(5.9) produces a sequence of strategies, with reduced costs.

In all the cases considered in Lemaire (1995), the sequence of the $rl_\ell^{[k]}$s converges to the optimal solution with miminum cost. The optimal retention limit is thus a function of the level ℓ occupied in the scale at the beginning of the insurance year, of the discount factor v, of the annual expected claim frequency ϑ, of the time t of occurrence of the accident, and of the number m of claims previously reported to the company from the beginning of the insurance period. The optimal strategy is an increasing function of t: the optimal retention increases as one approaches the end of the year (and the premium discount if no accidents are reported). The influence of t on the optimal retention limit is much weaker than the level ℓ, the discount factor v or ϑ. Putting $t = 0$ (and so $m = 0$) greatly simplifies the computation but leaves the retentions almost unchanged.

The optimal retentions coming from the Lemaire algorithm should not be considered as being the real threshold above which policyholders report the accident to the insurance company. Indeed, this algorithm postulates a high degree of rationality behind individual behaviours. It is enough to have a look at insurance statistics to see that some claims concern accidents bearing a cost much lower than the optimal retention, which contradicts the assumptions behind the Lemaire algorithm. The output of the Lemaire algorithm should be better understood as a measure of toughness for a particular bonus-malus system. Note that Walhin & Paris (2000,2001) softened this requirement by assuming that there was a proportion of the policyholders complying with Lemaire claiming rule, and the remainder reporting all the accidents, whatever their cost.

Note also that this approach requires knowledge of the uncensored distribution for claim costs and claim counts, which is usually not available in practice. The probability density function f corresponds to the cost of an accident (and not to the cost of a claim), and the distribution of the number of accidents is actually needed (not only the distribution of the number of claims that has been studied in the preceding chapters). Hence, the methodology described in Sections 5.4.1–5.4.2 has first to be applied to obtain the uncensored accident distribution.

Application of the Lemaire Algorithm to Portfolio C

The Lemaire algorithm can be applied with the distribution obtained for the cost of accidents (i.e. with the help of the LogNormal model with corrected regression coefficients) after having transformed the claim frequencies for Portfolio C into the accident frequencies.

Let us consider the $-1/+2$ bonus-malus scale, with relativities 62.4 % for level 0, 130.2 % for level 1, 142.9 % for level 2, 207.7 % for level 3, 241.4 % for level 4, and 309.1 % for level 5. We consider here a discount rate of 4 %.

Let us consider an individual aged between 25 and 60, living in a rural area, and driving a vehicle between 6 and 10 years old. His claim frequency is 18.46 %. His accident frequency is 21.29 %. The base pure premium is taken as the product of the claim frequency times the grand mean of all the claim sizes (large and moderate ones), that is, $0.1846 \times$ € 1810.63. The optimal retentions are as follows:

Level ℓ	Optimal retention
0	€574.75
1	€1050.39
2	€1341.16
3	€1760.77
4	€1260.69
5	€693.70

We see that this policyholder should defray accidents with a cost up to € 1760.77 if he occupied level 3.

Let us now consider an individual aged over 60, living in a urban area, and driving a vehicle between 3 and 5 years old. His claim frequency is 17.41 %. His accident frequency is 21.37 %. The base premium amounts to $0.1741 \times$ € 1810.63. The optimal retentions are as follows:

Level ℓ	Optimal retention
0	€539.89
1	€987.74
2	€1262.84
3	€1660.53
4	€1189.90
5	€655.34

The optimal retentions are now slightly smaller than before.

5.5 Further Reading and Bibliographic Notes

5.5.1 Modelling Claim Amounts in Related Coverages

In this chapter, only techniques for motor third party liability have been described. Besides the compulsory motor third party liability insurance, a number of related coverages are proposed to the drivers (like medical benefits, uninsured or underinsured motorist coverage, theft and collision and other than collision insurance). The problem caused by the late settlement of large claims generally disappears when optional coverages are considered. The annual claim amount S_i produced by policyholder i is then represented as

$$S_i = \sum_{k=1}^{N_i} C_{ik}$$

where N_i is the number of claims, and the C_{ik}s are the corresponding claim sizes. The S_is are assumed to be mutually independent, the C_{ik}s to be independent and identically distributed for fixed i, and independent of N_i.

The analysis of the N_is usually starts with a Poisson regression model. Then, the residual heterogeneity is taken into account by the inclusion of a random effect. The impact of the deductibles has to be carefully assessed for optional coverages. In case deductibles are specified in the insurance policies, the actuary has to keep in mind that the statistical summaries relate to conditional distributions (given that the claim costs exceed the corresponding deductibles).

According to the type of coverage, different models can be used for the C_{ik}s. The claim size is usually expressed as a percentage of the sum insured. For collision insurance, the C_{ik}s can be decomposed as

$$C_{ik} = \big(J_{ik} + (1 - J_{ik})P_{ik}\big)v_i$$

where v_i is the sum insured for policy i (the value of the vehicle fixed according to the rules contained in the policy); $J_{ik} = 1$ if the kth claim generates a total loss (that is, a loss that exhausts the sum insured, i.e. $C_{ik} = v_i$), and 0 otherwise; and $0 < P_{ik} < 1$ is the cost of the kth partial claim, expressed as a percentage of the sum insured. Beta regression can be used to incorporate observable covariates in the P_{ik}s, together with logistic regression for the J_{ik}s.

In the modelling of the claim sizes, the impact of the deductibles has to be carefully assessed. In particular, it is often important to take into account the different options with respect to deductibles. Signal theory tells us that the best drivers are likely to opt for the largest deductibles. The effect of the deductibles can be assessed empirically on the basis of summary statistics computed for each level of the deductible. Of course, this basic approach does not control for covariate effects. A more sophisticated comparison can be made on the basis of regression analyses. Also, explaining the choice of the deductible level on the basis of covariates may be interesting.

5.5.2 Tweedie Generalized Linear Model

The class of Tweedie exponential dispersion models generalizes, in a certain way, the class of extreme stable laws. At the same time, this class also includes such well-known continuous distributions as the Normal and Gamma, the purely discrete scaled Poisson distribution, as well as the class of mixed compound Poisson distribution with Gamma summands. The name Tweedie has been associated with this family by JORGENSEN (1987,1997) in honour of the pioneering works by TWEEDIE (1984).

In nonlife ratemaking, the Tweedie model is very convenient for risk classification. However, it does not allow the actuary to isolate the frequency part of the pure premium, and thus does not provide the actuary with the input for the design of bonus-malus scales. This is why in this book (which is mainly devoted to motor insurance pricing) we favoured the separate analysis of claim costs and claim sizes. For other insurance products, where only the total amount of claims is available for actuarial analysis, the Tweedie distribution is an excellent candidate for loss modelling.

In the actuarial literature, JORGENSEN & PAES DE SOUZA (1994) assumed Poisson arrival of claims and Gamma distributed costs for individual claims. These authors directly modelled the risk or expected cost of claims per insured unit using the Tweedie Generalized Linear Model. SMYTH & JORGENSEN (2002) observed that, when modelling the cost of insurance claims, it is generally necessary to model the dispersion of the costs as well as their mean. In order to model the dispersion, these authors used the framework of double generalized linear models. Modelling the dispersion increases the precision of the estimated tariffs. The use of double generalized linear models also allows the actuary to handle the case where only the total cost of claims and not the number of claims has been recorded.

5.5.3 Large Claims

The analysis of large losses performed in this chapter is based on CEBRIAN, DENUIT & LAMBERT (2003). Large losses are modelled using the Generalized Pareto distribution, and the main concern is to determine the threshold between small and large losses. An alternative has been developed by BUCH-KROMANN (2006) based on BUCH-LARSEN, NIELSEN, GUILLÉN & BOLANĆE (2005). This approach is based on a Champernowne distribution, corrected with a nonparametric estimator (that is obtained by transforming the data set with the estimated modified Champernowne distribution function and then estimating the density of the transformed data set using the classical kernel density estimator). Based on the analysis of a Danish data set, BUCH-KROMANN (2006) concluded that the Generalized Pareto approach performs better than the Champernowne one in terms of goodness-of-fit, whereas both methods are comparable in terms of predicting future claims.

Another approach is proposed by COORAY & ANANDA (2005) who combined a LogNormal probability density function together with a Pareto one. Specifically, these authors introduced a two-parameter smooth continuous composite LogNormal-Pareto model that is a two-parameter LogNormal density up to an unknown threshold value and a two-parameter Pareto density for the remainder. Continuity and differentiability are imposed at the unknown threshold to ensure that the resulting probability density function is smooth, reducing the number of parameters from four to two. The resulting two-parameter probability density function is similar in shape to the LogNormal density, yet its upper tail is thicker than the LogNormal density (and accomodates to the large losses observed in liability insurance). This approach clearly outperforms the one proposed in this chapter, in that all the parameters (including the threshold) are estimated in the same model. The approaches obtained with the methodology developed in this book can be used as starting values in the maximum likelihood maximization. Note however that COORAY & ANANDA (2005) did not consider the case where explanatory variables were available, so that their approach has to be extended to this more realistic situation.

5.5.4 Alternative Approaches to Risk Classification

There are numerous techniques applied to the modelling of insurance losses. Early references in actuarial science include TER BERG (1980a,b). Various mathematical and statistical models for estimation of automobile insurance pricing are reviewed in WEISBERG, TOMBERLIN & CHATTERJEE (1984). The methods are compared on their predictive ability based on two sets of automobile insurance data for two different states collected over two different periods. The issue of model complexity versus data availability is resolved through a comparison of the accuracy of prediction. The models reviewed range from the use of simple cell means to various multiplicative-additive schemes to the empirical Bayes approach. The empirical Bayes approach, with prediction based on both model-based and individual cell estimates, seems to yield the best forecast. See also JEE (1989).

WILLIAMS & HUANG (1996) applied KDD (for Knowledge Discovery in Databases) techniques for insurance risk assessment. DAENGDEJ, LUKOSE & MURISON (1999) considered CBR (for Case-Based Reasoning) techniques for claim predictions.

Classification techniques are also often used for risk classification. RETZLAFF-ROBERTS & PUELZ (1996) adopted an efficiency approach to the two-group linear programming method of discriminant analysis, using principles taken from data envelopment analysis, to predict group membership in an insurance underwriting scheme. YEO *ET AL.* (2001) applied clustering techniques before modelling insurance losses.

5.5.5 Efficiency

The efficiency (or elasticity) of a bonus-malus system was first studied by LOIMARANTA (1972) and DE PRIL (1978). Note however that in these papers, the unknown individual risk factors are not viewed as random variables. These authors work in the fixed effects model that is close in spirit to the limited fluctuations credibility theory. Their efficiency concepts are therefore entirely different from the notion of efficiency proposed in NORBERG (1976). Other efficiency measures have been proposed in the literature. For instance, HERAS *ET AL.* (2002) evaluated the asymptotic fairness of bonus-malus systems (i.e., their ability to assess the individual risk in the long run) assuming the simplest case when there is no hunger for bonus.

5.5.6 Optimal Retention Limits and Bonus Hunger

The problem of determining the optimal claim size has been the topic of several papers. DE LEVE & WEEDA (1968) considered a -1/top bonus-malus scale, so that the decision to file or not has to be made only if no claim has been made during the same period. LEMAIRE (1976,1977) studied the hunger for bonus and proposed a dynamic programming algorithm to determine the optimal claiming behaviour. DE PRIL (1979) considered that the claims were generated by a Poisson process and adapted a continuous-time approach. Specifically,

De Pril (1979) defined $L_n(\ell, k, t)$ as the amount that the actual accident must exceed in order to justify the filing of a claim, if the policyholder is at time t of period n in level ℓ and has already filed k claims. The optimal value of $L_n(\ell, k, t)$ is determined to minimize the discounted expectation of the total future costs (premiums and self-defrayed accidents) for the policyholder. After De Pril (1979), De Pril & Goovaerts (1983) determined bounds for the optimal critical claim size when only incomplete information about the claim amount distribution is available. They considered -1/top bonus-malus scales.

Dellaert *et al.* (1990) proved that under mild conditions the optimal decision rule is to claim for damages with amount above a certain limit. In some instances, policyholders are allowed to decide at the end of an insurance year which damages occurred during the year should be claimed; see, e.g., Martin-Lof (1973). This means that the policyholder has perfect information about the number of accidents and the corresponding damages at the moment he/she decides which damages to claim. This situation has been investigated by Dellaert *et al.* (1991). Let us also mention that Dellaert *et al.* (1993) considered damage insurance (where, in addition to the bonus hunger phenomenon, the optimal stopping rule to terminate the insurance has to be determined).

Holtan (2001) envisaged the loss of bonus after a claim as a rate of interest paid from the customer to the insurer, and studied the hunger for bonus from this viewpoint.

Optimal claiming rules have also been considered in Operational Research, using Markov decision processes. When a driver is involved in a motor accident decisions have to be made as to whether or not a claim should be made. Hastings (1976) considered this problem as a Markov decision process with the expected cost over a finite horizon (where the relevant costs are repair costs and premium costs) as objective function. See also Haehling von Lanzenauer (1974) and Hey (1985). Norman & Shearn (1980) proposed the following simple rule of thumb that is shown to work well in their case: irrespective of when the accident occurs, claim only if the amount of the claim exceeds the difference over the next 4 years between the total premiums payable if a claim is made and those payable if it is not, assuming that no further claims will be made. Chappell & Norman (1989) demonstrated that this simple rule was less efficient in the case of protected bonus. Kolderman & Volgenant (1985) examined the same problem under the assumption that only one claim is needed to change the insurance premium category, and any extra claims in the year have no effect on the current repair estimate for an accident.

Walhin & Paris (2000) derived the actual claim amount and frequency distributions within a bonus-malus system. As explained above, policyholders should defray the small claims to avoid the penalties induced by the bonus-malus system. Consequently, there are more accidents than the number of claims filed by the insurer: the insurance data are censored. The kind of censorship is nevertheless very particular, and much more complicated than the phenomena encountered in classical nonlife problems (where losses are censored because they exceed some policy limit, or fall below a given deductible). The procedure described in Section 5.4.1 is taken from Denuit, Maréchal, Pitrebois & Walhin (2007a), where alternative approaches to obtain uncensored accident distributions can be found.

Even if the bonus-hunger phenomenon has been extensively studied in connection with bonus-malus scales, the same idea applies to credibility systems. See, e.g., Norberg (1975) and Sundt (1988) for an illustration.

6

Multi-Event Systems

6.1 Introduction

The majority of bonus-malus systems in force throughout the world penalize the number of reported claims, without taking the cost of these claims into account. This can be considered as a shortcoming, since large claims should intuitively be more severely penalized. The first part of this chapter aims to develop credibility models that allow us to subdivide the claims into two categories, small and large losses (the extension to more than two categories is easy). Instead of determining a limiting amount to decide whether a loss should be qualified as large instead of small (such a criterion would lead to substantial practical problems, due to the time needed to evaluate the cost of the claim), we distinguish the accidents that caused property damage only from those that caused bodily injuries. Since the latter cost much more on average, this approach implicitly integrates the cost of the claim in *a posteriori* premium corrections. The Bayesian credibility approach turns out to lead to numerical integration. This is why we favour the linear credibility approach. Linear credibility formulas are developed to update the expected claim frequencies given past claim histories.

The second part of this chapter is devoted to bonus-malus scales with several types of events (assuming a Multinomial partitioning scheme). As mentioned above, all the classical bonus-malus systems are based on a single type of event: the occurrence of claims at fault, regardless of their severity or whether the policyholders are only partially liable for them. This over-simplification can be regarded as problematic for commercial purposes: it seems desirable to integrate the severity of the claims and to recognize the partial liability of the policyholder. For example, the bonus-malus system in force in France (studied in Chapter 9) entails a reduced penalty if the policyholder is only partially liable for the claim.

Prominent examples of *a posteriori* ratemaking mechanisms based on several types of events are provided by the experience rating systems in force in North America. These systems not only incorporate accidents at fault but also elements of the policyholders' driving record. For instance, the Massachusetts safe driver insurance plan encourages safe driving

Actuarial Modelling of Claim Counts: Risk Classification, Credibility and Bonus-Malus Systems M. Denuit, X. Maréchal, S. Pitrebois and J.-F. Walhin

by rewarding drivers who do not cause an accident, or incur a traffic law violation. It is based on several types of events (major and minor at-fault accidents and traffic violations). Specifically, each policyholder is assigned a level between 9 and 35, based on his driving record during the previous six years. A new driver begins at level 15 (relativity of 100 %). Occupying any level below 15 entails a premium discount, while above level 15 the driver pays a surcharge. For each incident-free year of driving, the policyholder goes down one level. The driver will move up a certain number of levels based on the type of incident: two levels for a minor traffic violation, three levels for a minor at-fault accident, four levels for a major at-fault accident and five levels for a major traffic violation. The Massachusetts system 'forgets' all incidents after six years.

This chapter addresses the actuarial modelling of such systems, with penalties depending on different types of events. As before, the modelling uses the concept of Markov Chains. We will see that under mild assumptions, the trajectory of each policyholder in the scale can be modelled with the aid of discrete-time Markov processes. The relativities associated with each level will then be computed using the maximum accuracy principle discussed in Chapter 4.

6.2 Multi-Event Credibility Models

6.2.1 Dichotomy

Let N_{it}^{mat} be the number of claims with material damage only, reported by policyholder i during period t. Similarly, let N_{it}^{bod} be the number of claims with bodily injuries, and let

$$N_{it}^{\text{tot}} = N_{it}^{\text{mat}} + N_{it}^{\text{bod}}$$

be the total number of claims. Policyholder i, $i = 1, \ldots, n$, is assumed to have been observed during T_i periods. In the previous chapters, N_{it}^{tot} was the variable of interest. Here, a dichotomy is operated, and we study N_{it}^{mat} and N_{it}^{bod} separately.

6.2.2 Multivariate Claim Count Model

Overdispersion and possible dependence between N_{it}^{mat} and N_{it}^{bod} are introduced via possibly correlated random effects Θ_i^{mat} and Θ_i^{bod} such that $\mathbb{E}[\Theta_i^{\text{mat}}] = \mathbb{E}[\Theta_i^{\text{bod}}] = 1$. Specifically, given $\Theta_i^{\text{mat}} = \theta_i^{\text{mat}}$, we assume that

$$N_{it}^{\text{mat}} \sim \mathcal{P}oi\big(\lambda_{it}^{\text{mat}} \theta_i^{\text{mat}}\big)$$

where $\lambda_{it}^{\text{mat}} = d_{it} \exp(\text{score}_{it}^{\text{mat}})$, and given $\Theta_i^{\text{bod}} = \theta_i^{\text{bod}}$, we assume that

$$N_{it}^{\text{bod}} \sim \mathcal{P}oi\big(\lambda_{it}^{\text{bod}} \theta_i^{\text{bod}}\big)$$

where $\lambda_{it}^{\text{bod}} = d_{it} \exp(\text{score}_{it}^{\text{bod}})$. Both scores are linear combinations of explanatory variables specific to policyholder i and year t (summarized in a vector $\boldsymbol{x}_{it}$). Specifically,

$$\text{score}_{it}^{\text{mat}} = \beta_0^{\text{mat}} + \sum_{j=1}^{p} \beta_j^{\text{mat}} x_{itj}$$

$$\text{score}_{it}^{\text{bod}} = \beta_0^{\text{bod}} + \sum_{j=1}^{p} \beta_j^{\text{bod}} x_{itj}.$$

We make the following assumptions about the dependence structure of the random variables:

(i) Given Θ_i^{bod}, the N_{it}^{bod}s, $t = 1, 2, \ldots, T_i$, are independent.
(ii) Given Θ_i^{mat}, the N_{it}^{mat}s, $t = 1, 2, \ldots, T_i$, are independent.
(iii) given $(\Theta_i^{\text{mat}}, \Theta_i^{\text{bod}})$, the sequences $\{N_{it}^{\text{bod}}, t = 1, 2, \ldots, T_i\}$ and $\{N_{it}^{\text{mat}}, t = 1, 2, \ldots, T_i\}$ are independent.

As explained in the previous chapter, overdispersion and serial correlation are induced by missing explanatory variables, whose effect is modelled with the help of the random effects Θ_i^{mat} and Θ_i^{bod}.

6.2.3 Bayesian Credibility Approach

The Bayesian approach requires numerical integration, which sometimes prevents the practical implementation of the resulting formulas (even if, nowadays, numerical integration has become straightforward with modern computers). It consists of deriving the conditional distribution of the random effects Θ_i^{mat} and Θ_i^{bod}, given the past claims history. This conditional distribution then drives *a posteriori* premium corrections.

Let us denote as

$$k_{i\bullet}^{\text{mat}} = \sum_{t=1}^{T_i} k_{it}^{\text{mat}}$$

the total number of claims with material damage only filed by policyholder i during the T_i coverage periods, and as

$$k_{i\bullet}^{\text{bod}} = \sum_{t=1}^{T_i} k_{it}^{\text{bod}}$$

the total number of claims with bodily injuries filed by this policyholder. The corresponding expected claim frequencies are

$$\lambda_{i\bullet}^{\text{mat}} = \sum_{t=1}^{T_i} \lambda_{it}^{\text{mat}} \text{ and } \lambda_{i\bullet}^{\text{bod}} = \sum_{t=1}^{T_i} \lambda_{it}^{\text{bod}}.$$

The joint distribution of the N_{it}^{mat}s and N_{it}^{bod}s is given by

$$\begin{aligned}
&\Pr[N_{it}^{\text{mat}} = k_{it}^{\text{mat}}, N_{it}^{\text{bod}} = k_{it}^{\text{bod}} \text{ for } t = 1, \ldots, T_i] \\
&= \int_0^{\infty} \int_0^{\infty} \exp(-\theta_i^{\text{mat}} \lambda_{i\bullet}^{\text{mat}} - \theta_i^{\text{bod}} \lambda_{i\bullet}^{\text{bod}}) (\theta_i^{\text{mat}})^{k_{i\bullet}^{\text{mat}}} (\theta_i^{\text{bod}})^{k_{i\bullet}^{\text{bod}}} \frac{\prod_{t=1}^{T_i} (\lambda_{it}^{\text{mat}})^{k_{it}^{\text{mat}}} (\lambda_{it}^{\text{bod}})^{k_{it}^{\text{bod}}}}{\prod_{t=1}^{T_i} (k_{it}^{\text{mat}}! k_{it}^{\text{bod}}!)} \\
&\quad \times f_{\Theta}(\theta_i^{\text{mat}}, \theta_i^{\text{bod}}) d\theta_i^{\text{mat}} d\theta_i^{\text{bod}}
\end{aligned}$$

where $f_{\Theta}(\cdot,\cdot)$ is the bivariate probability density function of the couple of random effects $(\Theta_i^{\text{mat}}, \Theta_i^{\text{bod}})$. The joint probability density function of Θ_i^{mat}, Θ_i^{bod}, N_{it}^{mat}, N_{it}^{bod}, $t = 1, \ldots, T_i$ is given by

$$\exp(-\theta_i^{\text{mat}}\lambda_{i\bullet}^{\text{mat}} - \theta_i^{\text{bod}}\lambda_{i\bullet}^{\text{bod}})(\theta_i^{\text{mat}})^{k_{i\bullet}^{\text{mat}}}(\theta_i^{\text{bod}})^{k_{i\bullet}^{\text{bod}}}\frac{\prod_{t=1}^{T_i}(\lambda_{it}^{\text{mat}})^{k_{it}^{\text{mat}}}(\lambda_{it}^{\text{bod}})^{k_{it}^{\text{bod}}}}{\prod_{t=1}^{T_i}(k_{it}^{\text{mat}}!k_{it}^{\text{bod}}!)}f_{\Theta}(\theta_i^{\text{mat}}, \theta_i^{\text{bod}})$$

so that the conditional probability density function of $(\Theta_i^{\text{mat}}, \Theta_i^{\text{bod}})$ given past claims history is

$$\frac{\exp(-\theta_i^{\text{mat}}\lambda_{i\bullet}^{\text{mat}} - \theta_i^{\text{bod}}\lambda_{i\bullet}^{\text{bod}})(\theta_i^{\text{mat}})^{k_{i\bullet}^{\text{mat}}}(\theta_i^{\text{bod}})^{k_{i\bullet}^{\text{bod}}}f_{\Theta}(\theta_i^{\text{mat}}, \theta_i^{\text{bod}})}{\int_0^{\infty}\int_0^{\infty}\exp(-\xi_1\lambda_{i\bullet}^{\text{mat}} - \xi_2\lambda_{i\bullet}^{\text{bod}})(\xi_1)^{k_{i\bullet}^{\text{mat}}}(\xi_2)^{k_{i\bullet}^{\text{bod}}}f_{\Theta}(\xi_1, \xi_2)d\xi_1 d\xi_2}. \tag{6.1}$$

The posterior distribution of $(\Theta_i^{\text{mat}}, \Theta_i^{\text{bod}})$ then allows for *a posteriori* corrections, as explained in Chapter 3.

6.2.4 Summary of Past Claims Histories

Denote as

$$N_{i\bullet}^{\text{bod}} = \sum_{t=1}^{T_i} N_{it}^{\text{bod}} \text{ and } N_{i\bullet}^{\text{mat}} = \sum_{t=1}^{T_i} N_{it}^{\text{mat}}$$

the total claim numbers of each category caused by policyholder i during the observation period. Since the random effects Θ_i^{mat} and Θ_i^{bod} do not vary with time, $N_{i\bullet}^{\text{mat}}$ and $N_{i\bullet}^{\text{bod}}$ are sufficient summaries of past claims histories (in the sense that the posterior distributions of Θ_i^{mat} and Θ_i^{bod} as well as the predictive distributions of N_{i,T_i+1}^{mat} and N_{i,T_i+1}^{bod} only depend on $N_{i\bullet}^{\text{mat}}$ and $N_{i\bullet}^{\text{bod}}$; see (6.1) where past claims histories enter through $k_{i\bullet}^{\text{mat}}$ and $k_{i\bullet}^{\text{bod}}$).

Clearly,

$$\lambda_{i\bullet}^{\text{mat}} = \mathbb{E}[N_{i\bullet}^{\text{mat}}] \text{ and } \lambda_{i\bullet}^{\text{bod}} = \mathbb{E}[N_{i\bullet}^{\text{bod}}].$$

It is then easy to see that given $\Theta_i^{\text{mat}} = \theta_i^{\text{mat}}$

$$N_{i\bullet}^{\text{mat}} \sim \mathcal{P}oi\big(\lambda_{i\bullet}^{\text{mat}}\theta_i^{\text{mat}}\big)$$

and that given $\Theta_i^{\text{bod}} = \theta_i^{\text{bod}}$

$$N_{i\bullet}^{\text{bod}} \sim \mathcal{P}oi\big(\lambda_{i\bullet}^{\text{bod}}\theta_i^{\text{bod}}\big),$$

invoking the conditional independence of the annual claim numbers of each category and the stability of the Poisson family under convolution. Therefore, $N_{i\bullet}^{\text{mat}}$ and $N_{i\bullet}^{\text{bod}}$ are both mixed Poisson distributed.

6.2.5 Variance-Covariance Structure of the Random Effects

For deriving linear credibility formulas, we only need the moment structure of the risk variables. Let us introduce the variance-covariance matrix of $(\Theta_i^{\text{bod}}, \Theta_i^{\text{mat}})$ that is denoted as

$$\boldsymbol{\Sigma}_{\Theta} = \begin{pmatrix} \sigma^2_{\text{bod}} & \sigma_{\text{bm}} \\ \sigma_{\text{bm}} & \sigma^2_{\text{mat}} \end{pmatrix}.$$

In words, σ^2_{bod} and σ^2_{mat} are the variances of Θ_i^{bod} and Θ_i^{mat}, respectively, and σ_{bm} is the covariance between Θ_i^{bod} and Θ_i^{mat}. Note that the following inequalities

$$\sigma^2_{\text{bod}} \geq 0, \quad \sigma^2_{\text{mat}} \geq 0 \text{ and } |\sigma_{\text{bm}}| \leq \sigma_{\text{bod}}\sigma_{\text{mat}}$$

must be fulfilled to ensure that $\boldsymbol{\Sigma}_{\Theta}$ is positive definite. The estimated variances and covariance have to fulfill the same constraints. If not, this rules out the linear credibility model.

6.2.6 Variance-Covariance Structure of the Annual Claim Numbers

Let us now compute the variance and covariance of $N_{i\bullet}^{\text{mat}}$ and $N_{i\bullet}^{\text{bod}}$. Since $N_{i\bullet}^{\text{mat}} \sim \mathcal{MPoi}(\lambda_{i\bullet}^{\text{mat}}, \Theta_i^{\text{mat}})$, we have

$$\mathbb{V}[N_{i\bullet}^{\text{mat}}] = \lambda_{i\bullet}^{\text{mat}} + \left(\lambda_{i\bullet}^{\text{mat}}\right)^2 \sigma^2_{\text{mat}}.$$

Similarly, from $N_{i\bullet}^{\text{bod}} \sim \mathcal{MPoi}(\lambda_{i\bullet}^{\text{bod}}, \Theta_i^{\text{bod}})$ we get

$$\mathbb{V}[N_{i\bullet}^{\text{bod}}] = \lambda_{i\bullet}^{\text{bod}} + \left(\lambda_{i\bullet}^{\text{bod}}\right)^2 \sigma^2_{\text{bod}}.$$

To have an idea about the dependence existing between the numbers of claims with material damage only and with bodily injuries, let us now compute the covariance between $N_{i\bullet}^{\text{mat}}$ and $N_{i\bullet}^{\text{bod}}$:

$$\begin{aligned}
\mathbb{C}[N_{i\bullet}^{\text{mat}}, N_{i\bullet}^{\text{bod}}] &= \mathbb{E}\Big[(N_{i\bullet}^{\text{mat}} - \lambda_{i\bullet}^{\text{mat}})(N_{i\bullet}^{\text{bod}} - \lambda_{i\bullet}^{\text{bod}})\Big] \\
&= \mathbb{E}\Big[\mathbb{E}\big[(N_{i\bullet}^{\text{mat}} - \lambda_{i\bullet}^{\text{mat}})(N_{i\bullet}^{\text{bod}} - \lambda_{i\bullet}^{\text{bod}})\big|\Theta_i^{\text{mat}}, \Theta_i^{\text{bod}}\big]\Big] \\
&= \mathbb{E}\Big[\mathbb{E}\big[N_{i\bullet}^{\text{mat}} - \lambda_{i\bullet}^{\text{mat}}\big|\Theta_i^{\text{mat}}\big]\mathbb{E}\big[N_{i\bullet}^{\text{bod}} - \lambda_{i\bullet}^{\text{bod}}\big|\Theta_i^{\text{bod}}\big]\Big] \\
&= \mathbb{E}\Big[\lambda_{i\bullet}^{\text{mat}}(\Theta_i^{\text{mat}} - 1)\lambda_{i\bullet}^{\text{bod}}(\Theta_i^{\text{bod}} - 1)\Big] \\
&= \lambda_{i\bullet}^{\text{mat}}\lambda_{i\bullet}^{\text{bod}}\sigma_{\text{bm}}.
\end{aligned}$$

As expected, the covariance σ_{bm} between Θ_i^{mat} and Θ_i^{bod} drives the covariance of $N_{i\bullet}^{\text{mat}}$ and $N_{i\bullet}^{\text{bod}}$.

6.2.7 Estimation of the Variances and Covariances

The formulas derived above for $\mathbb{V}[N_{i\bullet}^{\text{mat}}]$, $\mathbb{V}[N_{i\bullet}^{\text{bod}}]$ and $\mathbb{C}[N_{i\bullet}^{\text{mat}}, N_{i\bullet}^{\text{bod}}]$ suggest the following estimates for the parameters σ_{mat}^2, σ_{bod}^2 and σ_{bm}:

$$\widehat{\sigma_{\text{mat}}^2} = \frac{\sum_{i=1}^n \left(\left(k_{i\bullet}^{\text{mat}} - \widehat{\lambda_{i\bullet}^{\text{mat}}} \right)^2 - k_{i\bullet}^{\text{mat}} \right)}{\sum_{i=1}^n \left(\widehat{\lambda_{i\bullet}^{\text{mat}}} \right)^2}$$

$$\widehat{\sigma_{\text{bod}}^2} = \frac{\sum_{i=1}^n \left(\left(k_{i\bullet}^{\text{bod}} - \widehat{\lambda_{i\bullet}^{\text{bod}}} \right)^2 - k_{i\bullet}^{\text{bod}} \right)}{\sum_{i=1}^n \left(\widehat{\lambda_{i\bullet}^{\text{bod}}} \right)^2}$$

$$\widehat{\sigma_{\text{bm}}} = \frac{\sum_{i=1}^n \left(k_{i\bullet}^{\text{mat}} - \widehat{\lambda_{i\bullet}^{\text{mat}}} \right) \left(k_{i\bullet}^{\text{bod}} - \widehat{\lambda_{i\bullet}^{\text{bod}}} \right)}{\sum_{i=1}^n \widehat{\lambda_{i\bullet}^{\text{mat}}} \widehat{\lambda_{i\bullet}^{\text{bod}}}}$$

that are consistent in the random effects model.

The parameters σ_{mat}^2, σ_{bod}^2 and σ_{bm} have been estimated above on aggregate data, giving the estimators $\widehat{\sigma_{\text{mat}}^2}$, $\widehat{\sigma_{\text{bod}}^2}$ and $\widehat{\sigma_{\text{bm}}}$. Alternatively, these parameters could be estimated from individual data as follows:

$$\widetilde{\sigma_{\text{mat}}^2} = \frac{\sum_{i=1}^n \sum_{t=1}^{T_i} \left(\left(k_{it}^{\text{mat}} - \widehat{\lambda_{it}^{\text{mat}}} \right)^2 - k_{it}^{\text{mat}} \right)}{\sum_{i=1}^n \sum_{t=1}^{T_i} \left(\widehat{\lambda_{it}^{\text{mat}}} \right)^2}$$

$$\widetilde{\sigma_{\text{bod}}^2} = \frac{\sum_{i=1}^n \sum_{t=1}^{T_i} \left(\left(k_{it}^{\text{bod}} - \widehat{\lambda_{it}^{\text{bod}}} \right)^2 - k_{it}^{\text{bod}} \right)}{\sum_{i=1}^n \sum_{t=1}^{T_i} \left(\widehat{\lambda_{it}^{\text{bod}}} \right)^2}$$

$$\widetilde{\sigma_{\text{bm}}} = \frac{\sum_{i=1}^n \sum_{t=1}^{T_i} \left(k_{it}^{\text{mat}} - \widehat{\lambda_{it}^{\text{mat}}} \right) \left(k_{it}^{\text{bod}} - \widehat{\lambda_{it}^{\text{bod}}} \right)}{\sum_{i=1}^n \sum_{t=1}^{T_i} \widehat{\lambda_{it}^{\text{mat}}} \widehat{\lambda_{it}^{\text{bod}}}}.$$

The estimators $\widehat{\sigma_{\text{mat}}^2}$, $\widehat{\sigma_{\text{bod}}^2}$ and $\widehat{\sigma_{\text{bm}}}$ are preferred over $\widetilde{\sigma_{\text{mat}}^2}$, $\widetilde{\sigma_{\text{bod}}^2}$ and $\widetilde{\sigma_{\text{bm}}}$, respectively, since the variances of the former are smaller. As shown by PINQUET *ET AL.* (2001), the condition $0 < \widehat{\sigma_{\text{mat}}^2} < \widetilde{\sigma_{\text{mat}}^2}$ is necessary for the introduction of dynamic random effects.

6.2.8 Linear Credibility Premiums

Denote as

$$\lambda_{i,T_i+1}^{\text{mat}} = d_{i,T_i+1} \exp(\text{score}_{i,T_i+1}^{\text{mat}}) \text{ and as } \lambda_{i,T_i+1}^{\text{bod}} = d_{i,T_i+1} \exp(\text{score}_{i,T_i+1}^{\text{bod}})$$

the expected claim frequencies for policyholder i in period $T_i + 1$. The best linear predictor

$$c_{i0}^{\text{mat}} + \sum_{t=1}^{T_i} c_{it}^{\text{mat/mat}} N_{it}^{\text{mat}} + \sum_{t=1}^{T_i} c_{it}^{\text{bod/mat}} N_{it}^{\text{bod}}$$

of the true expected claim frequency $\lambda_{i,T_i+1}^{\text{mat}}\Theta_i^{\text{mat}}$ minimizes

$$Q^{\text{mat}} = \mathbb{E}\left[\left(\lambda_{i,T_i+1}^{\text{mat}}\Theta_i^{\text{mat}} - c_{i0}^{\text{mat}} - \sum_{t=1}^{T_i} c_{it}^{\text{mat/mat}} N_{it}^{\text{mat}} - \sum_{t=1}^{T_i} c_{it}^{\text{bod/mat}} N_{it}^{\text{bod}}\right)^2\right].$$

Similarly, the best linear predictor

$$c_{i0}^{\text{bod}} + \sum_{t=1}^{T_i} c_{it}^{\text{mat/bod}} N_{it}^{\text{mat}} + \sum_{t=1}^{T_i} c_{it}^{\text{bod/bod}} N_{it}^{\text{bod}}$$

of $\lambda_{i,T_i+1}^{\text{bod}}\Theta_i^{\text{bod}}$ minimizes

$$Q^{\text{bod}} = \mathbb{E}\left[\left(\lambda_{i,T_i+1}^{\text{bod}}\Theta_i^{\text{bod}} - c_{i0}^{\text{bod}} - \sum_{t=1}^{T_i} c_{it}^{\text{mat/bod}} N_{it}^{\text{mat}} - \sum_{t=1}^{T_i} c_{it}^{\text{bod/bod}} N_{it}^{\text{bod}}\right)^2\right].$$

The optima are obtained by setting to zero the derivatives of Q^{mat} with respect to c_{i0}^{mat} and to $c_{is}^{\text{mat/mat}}$, that is,

$$c_{i0}^{\text{mat}} = \lambda_{i,T_i+1}^{\text{mat}} - \sum_{t=1}^{T_i} c_{it}^{\text{mat/mat}} \lambda_{it}^{\text{mat}} - \sum_{t=1}^{T_i} c_{it}^{\text{bod/mat}} \lambda_{it}^{\text{bod}}$$

$$c_{is}^{\text{mat/mat}} = \lambda_{i,T_i+1}^{\text{mat}}\sigma_{\text{mat}}^2 - \sigma_{\text{mat}}^2 \sum_{t=1}^{T_i} c_{it}^{\text{mat/mat}} \lambda_{it}^{\text{mat}} - \sigma_{\text{bm}} \sum_{t=1}^{T_i} c_{it}^{\text{bod/mat}} \lambda_{it}^{\text{bod}}.$$

The last relation shows that $c_{is}^{\text{mat/mat}}$ does not depend on s. Similarly, one can check that $c_{is}^{\text{mat/bod}}$, $c_{is}^{\text{bod/mat}}$ and $c_{is}^{\text{bod/bod}}$ do not depend on s. This justifies the approach based on aggregate data $N_{i\bullet}^{\text{mat}}$ and $N_{i\bullet}^{\text{bod}}$ (that are an exhaustive summary of past claims histories in the credibility model with static random effects).

Denoting as $c_i^{\text{mat/mat}}$ ($c_i^{\text{bod/mat}}$, $c_i^{\text{mat/bod}}$ and $c_i^{\text{bod/bod}}$, respectively) the common values of the $c_{is}^{\text{mat/mat}}$ ($c_{is}^{\text{bod/mat}}$, $c_{is}^{\text{mat/bod}}$ and $c_{is}^{\text{bod/bod}}$, respectively), the best linear predictors are thus of the form

$$c_i^{\text{mat}} + c_i^{\text{mat/mat}} N_{i\bullet}^{\text{mat}} + c_i^{\text{bod/mat}} N_{i\bullet}^{\text{bod}}$$

for $\lambda_{i,T_i+1}^{\text{mat}}\Theta_i^{\text{mat}}$, and

$$c_i^{\text{bod}} + c_i^{\text{mat/bod}} N_{i\bullet}^{\text{mat}} + c_i^{\text{bod/bod}} N_{i\bullet}^{\text{bod}}$$

for $\lambda_{i,T_i+1}^{\text{bod}}\Theta_i^{\text{bod}}$. The meaning of the coefficients involved in these linear predictors is as follows:

$\boldsymbol{c_i^{\text{mat/mat}}}$ evaluates the information contained in past material claims on the occurrence of future material claims;

$\boldsymbol{c_i^{\text{bod/mat}}}$ evaluates the information contained in past claims with bodily injuries on the occurrence of future material claims;

$c_i^{\text{mat/bod}}$ evaluates the information contained in past material claims on the occurrence of future claims with bodily injuries;

$c_i^{\text{bod/bod}}$ evaluates the information contained in past claims with bodily injuries on the occurrence of future claims with bodily injuries.

The values of these coefficients are determined by minimizing simultaneously Q^{mat} and Q^{bod} that may be rewritten as

$$Q^{\text{mat}} = \mathbb{E}\Big[\big(\lambda_{i,T_i+1}^{\text{mat}}\Theta_i^{\text{mat}} - c_i^{\text{mat}} - c_i^{\text{mat/mat}}N_{i\bullet}^{\text{mat}} - c_i^{\text{bod/mat}}N_{i\bullet}^{\text{bod}}\big)^2\Big]$$

and as

$$Q^{\text{bod}} = \mathbb{E}\Big[\big(\lambda_{i,T_i+1}^{\text{bod}}\Theta_i^{\text{bod}} - c_i^{\text{bod}} - c_i^{\text{mat/bod}}N_{i\bullet}^{\text{mat}} - c_i^{\text{bod/bod}}N_{i\bullet}^{\text{bod}}\big)^2\Big].$$

Setting to zero the partial derivatives of Q^{mat} and Q^{bod} with respect to the six parameters gives:

$$c_i^{\text{mat}} = \lambda_{i,T_i+1}^{\text{mat}} - c_i^{\text{mat/mat}}\lambda_{i\bullet}^{\text{mat}} - c_i^{\text{bod/mat}}\lambda_{i\bullet}^{\text{bod}} \tag{6.2}$$

$$c_i^{\text{bod}} = \lambda_{i,T_i+1}^{\text{bod}} - c_i^{\text{mat/bod}}\lambda_{i\bullet}^{\text{mat}} - c_i^{\text{bod/bod}}\lambda_{i\bullet}^{\text{bod}} \tag{6.3}$$

$$\begin{aligned} 0 = {} & \lambda_{i,T_i+1}^{\text{mat}}\mathbb{E}[\Theta_i^{\text{mat}}N_{i\bullet}^{\text{mat}}] - c_i^{\text{mat}}\lambda_{i\bullet}^{\text{mat}} - c_i^{\text{mat/mat}}\mathbb{E}\big[(N_{i\bullet}^{\text{mat}})^2\big] \\ & - c_i^{\text{bod/mat}}\mathbb{E}[N_{i\bullet}^{\text{mat}}N_{i\bullet}^{\text{bod}}] \end{aligned} \tag{6.4}$$

$$\begin{aligned} 0 = {} & \lambda_{i,T_i+1}^{\text{mat}}\mathbb{E}[\Theta_i^{\text{mat}}N_{i\bullet}^{\text{bod}}] - c_i^{\text{mat}}\lambda_{i\bullet}^{\text{bod}} - c_i^{\text{mat/mat}}\mathbb{E}[N_{i\bullet}^{\text{mat}}N_{i\bullet}^{\text{bod}}] \\ & - c_i^{\text{bod/mat}}\mathbb{E}\big[(N_{i\bullet}^{\text{bod}})^2\big] \end{aligned} \tag{6.5}$$

$$\begin{aligned} 0 = {} & \lambda_{i,T_i+1}^{\text{bod}}\mathbb{E}[\Theta_i^{\text{bod}}N_{i\bullet}^{\text{mat}}] - c_i^{\text{bod}}\lambda_{i\bullet}^{\text{mat}} - c_i^{\text{mat/bod}}\mathbb{E}\big[(N_{i\bullet}^{\text{mat}})^2\big] \\ & - c_i^{\text{bod/bod}}\mathbb{E}[N_{i\bullet}^{\text{mat}}N_{i\bullet}^{\text{bod}}] \end{aligned} \tag{6.6}$$

$$\begin{aligned} 0 = {} & \lambda_{i,T_i+1}^{\text{bod}}\mathbb{E}[\Theta_i^{\text{bod}}N_{i\bullet}^{\text{bod}}] - c_i^{\text{bod}}\lambda_{i\bullet}^{\text{bod}} - c_i^{\text{mat/bod}}\mathbb{E}[N_{i\bullet}^{\text{mat}}N_{i\bullet}^{\text{bod}}] \\ & - c_i^{\text{bod/bod}}\mathbb{E}\big[(N_{i\bullet}^{\text{bod}})^2\big]. \end{aligned} \tag{6.7}$$

The expectancies involved in this system are given by

$$\mathbb{E}[\Theta_i^{\text{mat}}N_{i\bullet}^{\text{mat}}] = \lambda_{i\bullet}^{\text{mat}}\sigma_{\text{mat}}^2 + \lambda_{i\bullet}^{\text{mat}}$$

$$\mathbb{E}[\Theta_i^{\text{bod}}N_{i\bullet}^{\text{bod}}] = \lambda_{i\bullet}^{\text{bod}}\sigma_{\text{bod}}^2 + \lambda_{i\bullet}^{\text{bod}}$$

$$\mathbb{E}[\Theta_i^{\text{mat}}N_{i\bullet}^{\text{bod}}] = \lambda_{i\bullet}^{\text{bod}}\sigma_{\text{bm}} + \lambda_{i\bullet}^{\text{bod}}$$

$$\mathbb{E}[\Theta_i^{\text{bod}}N_{i\bullet}^{\text{mat}}] = \lambda_{i\bullet}^{\text{mat}}\sigma_{\text{bm}} + \lambda_{i\bullet}^{\text{mat}}$$

$$\mathbb{E}\big[(N_{i\bullet}^{\text{mat}})^2\big] = \lambda_{i\bullet}^{\text{mat}} + \big(\lambda_{i\bullet}^{\text{mat}}\big)^2(\sigma_{\text{mat}}^2 + 1)$$

$$\mathbb{E}\big[(N_{i\bullet}^{\text{bod}})^2\big] = \lambda_{i\bullet}^{\text{bod}} + \big(\lambda_{i\bullet}^{\text{bod}}\big)^2(\sigma_{\text{bod}}^2 + 1)$$

$$\mathbb{E}[N_{i\bullet}^{\text{mat}} N_{i\bullet}^{\text{bod}}] = \lambda_{i\bullet}^{\text{mat}} \lambda_{i\bullet}^{\text{bod}}(\sigma_{\text{bm}} + 1).$$

Inserting these expressions in (6.4)–(6.7), we get

$$\lambda_{i,T_i+1}^{\text{mat}} \sigma_{\text{mat}}^2 = c_i^{\text{mat/mat}}(1 + \lambda_{i\bullet}^{\text{mat}} \sigma_{\text{mat}}^2) + c_i^{\text{bod/mat}} \lambda_{i\bullet}^{\text{bod}} \sigma_{\text{bm}}$$

$$\lambda_{i,T_i+1}^{\text{mat}} \sigma_{\text{bm}} = c_i^{\text{mat/mat}} \lambda_{i\bullet}^{\text{mat}} \sigma_{\text{bm}} + c_i^{\text{bod/mat}}(1 + \lambda_{i\bullet}^{\text{bod}} \sigma_{\text{bod}}^2)$$

$$\lambda_{i,T_i+1}^{\text{bod}} \sigma_{\text{bm}} = c_i^{\text{mat/bod}}(1 + \lambda_{i\bullet}^{\text{mat}} \sigma_{\text{mat}}^2) + c_i^{\text{bod/bod}} \lambda_{i\bullet}^{\text{bod}} \sigma_{\text{bm}}$$

$$\lambda_{i,T_i+1}^{\text{bod}} \sigma_{\text{bod}}^2 = c_i^{\text{mat/bod}} \lambda_{i\bullet}^{\text{mat}} \sigma_{\text{bm}} + c_i^{\text{bod/bod}}(1 + \lambda_{i\bullet}^{\text{bod}} \sigma_{\text{bod}}^2)$$

and finally

$$c_i^{\text{bod/mat}} = \frac{\lambda_{i,T_i+1}^{\text{mat}} \sigma_{\text{bm}}}{(1 + \lambda_{i\bullet}^{\text{mat}} \sigma_{\text{mat}}^2)(1 + \lambda_{i\bullet}^{\text{bod}} \sigma_{\text{bod}}^2) - \lambda_{i\bullet}^{\text{mat}} \lambda_{i\bullet}^{\text{bod}} \sigma_{\text{bm}}^2}$$

$$c_i^{\text{mat/bod}} = \frac{\lambda_{i,T_i+1}^{\text{bod}} \sigma_{\text{bm}}}{(1 + \lambda_{i\bullet}^{\text{mat}} \sigma_{\text{mat}}^2)(1 + \lambda_{i\bullet}^{\text{bod}} \sigma_{\text{bod}}^2) - \lambda_{i\bullet}^{\text{mat}} \lambda_{i\bullet}^{\text{bod}} \sigma_{\text{bm}}^2}$$

$$c_i^{\text{mat/mat}} = \lambda_{i,T_i+1}^{\text{mat}} \frac{\sigma_{\text{mat}}^2 + \lambda_{i\bullet}^{\text{bod}}(\sigma_{\text{mat}}^2 \sigma_{\text{bod}}^2 - \sigma_{\text{bm}}^2)}{(1 + \lambda_{i\bullet}^{\text{mat}} \sigma_{\text{mat}}^2)(1 + \lambda_{i\bullet}^{\text{bod}} \sigma_{\text{bod}}^2) - \lambda_{i\bullet}^{\text{mat}} \lambda_{i\bullet}^{\text{bod}} \sigma_{\text{bm}}^2}$$

$$c_i^{\text{bod/bod}} = \lambda_{i,T_i+1}^{\text{bod}} \frac{\sigma_{\text{bod}}^2 + \lambda_{i\bullet}^{\text{mat}}(\sigma_{\text{mat}}^2 \sigma_{\text{bod}}^2 - \sigma_{\text{bm}}^2)}{(1 + \lambda_{i\bullet}^{\text{mat}} \sigma_{\text{mat}}^2)(1 + \lambda_{i\bullet}^{\text{bod}} \sigma_{\text{bod}}^2) - \lambda_{i\bullet}^{\text{mat}} \lambda_{i\bullet}^{\text{bod}} \sigma_{\text{bm}}^2}.$$

We see that $c_i^{\text{bod/mat}}$ and $c_i^{\text{mat/bod}}$ are increasing with σ_{bm}, and decreasing with σ_{mat}^2 and σ_{bod}^2. This is intuitively acceptable: the more the random effects are correlated, the more information is contained in the claims with material damage only about the claims with bodily injuries, and vice-versa. Inserting these solutions in the equations (6.2)–(6.3) gives

$$c_i^{\text{mat}} = \lambda_{i,T_i+1}^{\text{mat}} \frac{1 + \lambda_{i\bullet}^{\text{bod}}\big(\sigma_{\text{bod}}^2 - \sigma_{\text{bm}}\big)}{(1 + \lambda_{i\bullet}^{\text{mat}} \sigma_{\text{mat}}^2)(1 + \lambda_{i\bullet}^{\text{bod}} \sigma_{\text{bod}}^2) - \lambda_{i\bullet}^{\text{mat}} \lambda_{i\bullet}^{\text{bod}} \sigma_{\text{bm}}^2}$$

$$c_i^{\text{bod}} = \lambda_{i,T_i+1}^{\text{bod}} \frac{1 + \lambda_{i\bullet}^{\text{mat}}\big(\sigma_{\text{mat}}^2 - \sigma_{\text{bm}}\big)}{(1 + \lambda_{i\bullet}^{\text{mat}} \sigma_{\text{mat}}^2)(1 + \lambda_{i\bullet}^{\text{bod}} \sigma_{\text{bod}}^2) - \lambda_{i\bullet}^{\text{mat}} \lambda_{i\bullet}^{\text{bod}} \sigma_{\text{bm}}^2}.$$

To sum up, the best linear predictor for the expected frequency of claims with material damage only occurring in year $T_i + 1$ is

$$\lambda_{i,T_i+1}^{\text{mat}} \frac{1 + \lambda_{i\bullet}^{\text{bod}}\big(\sigma_{\text{bod}}^2 - \sigma_{\text{bm}}\big) + \sigma_{\text{bm}} N_{i\bullet}^{\text{bod}} + \big(\sigma_{\text{mat}}^2 + \lambda_{i\bullet}^{\text{bod}}(\sigma_{\text{mat}}^2 \sigma_{\text{bod}}^2 - \sigma_{\text{bm}}^2)\big) N_{i\bullet}^{\text{mat}}}{(1 + \lambda_{i\bullet}^{\text{mat}} \sigma_{\text{mat}}^2)(1 + \lambda_{i\bullet}^{\text{bod}} \sigma_{\text{bod}}^2) - \lambda_{i\bullet}^{\text{mat}} \lambda_{i\bullet}^{\text{bod}} \sigma_{\text{bm}}^2}$$

and the best linear predictor for the expected frequency of claims with bodily injuries in year T_i+1 is

$$\lambda_{i,T_i+1}^{\text{bod}} \frac{1+\lambda_{i\bullet}^{\text{mat}}\left(\sigma_{\text{mat}}^2-\sigma_{\text{bm}}\right)+\sigma_{\text{bm}}N_{i\bullet}^{\text{mat}}+\left(\sigma_{\text{bod}}^2+\lambda_{i\bullet}^{\text{mat}}(\sigma_{\text{mat}}^2\sigma_{\text{bod}}^2-\sigma_{\text{bm}}^2)\right)N_{i\bullet}^{\text{bod}}}{(1+\lambda_{i\bullet}^{\text{mat}}\sigma_{\text{mat}}^2)(1+\lambda_{i\bullet}^{\text{bod}}\sigma_{\text{bod}}^2)-\lambda_{i\bullet}^{\text{mat}}\lambda_{i\bullet}^{\text{bod}}\sigma_{\text{bm}}^2}.$$

6.2.9 Numerical Illustration for Portfolio A

A Priori Ratemaking

The observed annual frequency for claims with bodily injuries is 1.6 %. The observed annual frequency for claims with material damage only is 13.0 %.

In this portfolio, the two types of claims we consider are positively correlated. This can be seen from Table 6.1, where the conditional expectation of the number of claims of one type is computed given the number of claims of the other type. The more claims of one type reported, the higher this conditional expectation, resulting in positive dependence.

A Posteriori Corrections

Let us now update the claim frequencies with the help of the formulas obtained with linear credibility. To this end, we first estimate $\boldsymbol{\Sigma}_{\boldsymbol{\Theta}}$. This gives

$$\widehat{\sigma_{\text{mat}}^2}=0.8458$$
$$\widehat{\sigma_{\text{bod}}^2}=1.1188$$
$$\widehat{\sigma_{\text{bm}}}=0.6255.$$

Let us now consider two types of drivers:

- a good driver with $\lambda_{it}^{\text{mat}}=0.083$ and $\lambda_{it}^{\text{bod}}=0.010$, and
- a bad driver with $\lambda_{it}^{\text{mat}}=0.246$ and $\lambda_{it}^{\text{bod}}=0.030$.

Table 6.1 Observed annual frequency of claims with bodily injuries, given the number of claims with material damage only; and observed annual frequency of claims with material damage only, given the number of claims with bodily injuries; for portfolio A.

Conditional expectation of the number of claims with bodily injuries	Conditional expectation of the number of claims with material damage only
Given the number of claims with material damage only	Given the number of claims with bodily injuries
=0 is 1.5 %	=0 is 12.9 %
=1 is 2.7 %	=1 is 22.8 %
=2 is 3.6 %	=2 is 37.7 %

Table 6.2 Evolution of relativities and pure premiums (taking the average cost of a claim with material damage only (mat) as the monetary unit, and assuming that claims with bodily injuries (bod) are on average ten times more expensive) if no claim has been reported.

Time	Good driver			Bad driver		
	Relativity mat	Relativity bod	Pure premium	Relativity mat	Relativity bod	Pure premium
1	92.9 %	94.2 %	16.8 %	81.5 %	84.6 %	45.1 %
2	86.7 %	89.1 %	15.8 %	68.7 %	73.9 %	38.8 %
3	81.3 %	84.6 %	15.0 %	59.3 %	66.0 %	34.1 %
4	76.6 %	80.6 %	14.2 %	52.1 %	59.9 %	30.6 %
5	72.3 %	77.1 %	13.5 %	46.5 %	55.1 %	27.7 %
6	68.5 %	73.9 %	12.8 %	41.9 %	51.1 %	25.4 %
7	65.0 %	71.0 %	12.3 %	38.2 %	47.8 %	23.5 %
8	61.9 %	68.4 %	11.8 %	35.0 %	45.0 %	21.9 %
9	59.1 %	66.0 %	11.3 %	32.3 %	42.5 %	20.5 %
10	56.5 %	63.8 %	10.9 %	30.0 %	40.4 %	19.3 %

Table 6.2 displays the results for the case where no claim is reported for 10 years. The first column gives the coefficient to be applied on $\lambda_{it}^{\text{mat}}$, the second column gives the coefficient to be applied on $\lambda_{it}^{\text{bod}}$ and the third column gives the premium to be charged if the average cost of a material damage claim is 1 and the average cost of a bodily injury claim is 10 (the monetary unit is thus the average cost of a claim with material damage only, and claims with bodily injuries are assumed to be on average ten times more expensive than claims with material damage only). The first three columns are for the good driver and the next three are for the bad driver. We see that the correction coefficients are always smaller for the claims with material damage only than for the claims with bodily injuries. This is due to the fact that the former claims occur more frequently than the latter ones, so that not reporting any claim with material damage only entails more premium discount. As explained previously, the discounts are always larger for a bad driver than for a good one. However, the premiums always stay higher for the bad drivers.

Table 6.3 considers the case where the policyholder reported a single claim with material damage only, for 10 years. Finally, Table 6.4 considers the case where the policyholder reported a single claim with bodily injuries, for 10 years. Comparing these two tables, we see that the premium amount is larger if a claim with bodily injuries has been reported, compared to the case where a claim with material damage only has been reported. The cost of the claim is thus taken into account in the premium correction. The correction coefficients are always larger for the good driver than for the bad one, as explained previously. It is also interesting to note that reporting a claim of one type always increases the probability of reporting a claim of the other type. There is thus a double effect when updating the premium: the frequency of claims of the same type as the one that has been reported is increased, but the frequency of claims of the other type gets inflated, too.

Table 6.3 Evolution of relativities and pure premiums (taking the average cost of a claim with material damage only (mat) as the monetary unit, and assuming that claims with bodily injuries (bod) are on average ten times more expensive) if a single claim with material damage only has been reported during the first year.

Time	Good driver			Bad driver		
	Relativity mat	Relativity bod	Pure premium	Relativity mat	Relativity bod	Pure premium
1	171.6 %	152.0 %	29.0 %	150.7 %	134.9 %	77.0 %
2	160.3 %	142.8 %	27.2 %	127.3 %	115.7 %	65.6 %
3	150.4 %	134.7 %	25.6 %	110.1 %	101.6 %	57.2 %
4	141.7 %	127.6 %	24.1 %	97.0 %	90.7 %	50.8 %
5	133.9 %	121.2 %	22.9 %	86.7 %	82.2 %	45.7 %
6	126.9 %	115.5 %	21.7 %	78.3 %	75.2 %	41.6 %
7	120.6 %	110.3 %	20.7 %	71.4 %	69.4 %	38.1 %
8	114.9 %	105.7 %	19.8 %	65.6 %	64.6 %	35.3 %
9	109.7 %	101.4 %	18.9 %	60.7 %	60.4 %	32.8 %
10	105.0 %	97.5 %	18.2 %	56.5 %	56.8 %	30.7 %

Table 6.4 Evolution of relativities and pure premiums (taking the average cost of a claim with material damage only (mat) as the monetary unit, and assuming that claims with bodily injuries (bod) are on average ten times more expensive) if a single claim with bodily injuries has been reported during the first year.

Time	Good driver			Bad driver		
	Relativity mat	Relativity bod	Pure premium	Relativity mat	Relativity bod	Pure premium
1	150.7 %	201.9 %	32.1 %	131.7 %	185.5 %	87.3 %
2	140.5 %	193.1 %	30.4 %	110.4 %	166.8 %	76.6 %
3	131.5 %	185.4 %	28.9 %	94.8 %	152.9 %	68.6 %
4	123.5 %	178.5 %	27.6 %	82.9 %	142.0 %	62.4 %
5	116.5 %	172.3 %	26.4 %	73.6 %	133.2 %	57.5 %
6	101.1 %	166.7 %	25.3 %	66.0 %	125.8 %	53.5 %
7	104.4 %	161.7 %	24.3 %	59.8 %	119.6 %	50.1 %
8	99.2 %	157.0 %	23.5 %	54.6 %	114.3 %	47.3 %
9	94.5 %	152.8 %	22.7 %	50.2 %	109.6 %	44.8 %
10	90.2 %	148.9 %	21.9 %	46.4 %	105.5 %	42.6 %

6.3 Multi-Event Bonus-Malus Scales

6.3.1 Types of Claims

Here we adopt the same assumptions as in Chapter 4. Let us pick a policyholder at random from the portfolio and let us denote as N the number of claims reported during the year. Furthermore, let Λ be the (unknown) *a priori* expected claim frequency, with $\Pr[\Lambda = \lambda_k] = w_k$.

Denoting as Θ the (unknown) accident proneness of this policyholder, the conditional probability mass function of N is given by

$$\Pr[N=j|\Theta=\theta, \Lambda=\lambda_k]=\exp(-\lambda_k\theta)\frac{(-\lambda_k\theta)^j}{j!}, \quad j=0,1,2,\ldots$$

The risk profile of the portfolio is described by the distribution function F_Θ of Θ and we assume that $\mathbb{E}[\Theta]=1$. Since Θ represents the residual effect of unobserved characteristics, it seems reasonable to assume that Θ and Λ are mutually independent. Hence, the unconditional probability mass function of N is given by

$$\Pr[N=j]=\sum_k w_k\int_0^{+\infty}\Pr[N=j|\Theta=\theta, \Lambda=\lambda_k]dF_\Theta(\theta), \quad j=0,1,\ldots$$

We distinguish among m different types of claim reported by the policyholder. Each type of claim induces a specific penalty for the policyholder. For instance, one could think of

- claims with bodily injuries and claims with material damage only ($m=2$)
- claims with partial liability and claims with full liability ($m=2$)
- introducing claim severities (for instance, claims with amount less than €1000, between €1000 and €10 000, and claims above €10 000, so that $m=3$). In this case, we have to assume that claim severities and claim frequencies are mutually independent.

Here, we will assume that the claims are classified according to a multinomial scheme. Specifically, each time a claim is reported, it is classified in one of the m possible categories, with probabilities $q_1,\ldots,q_m$. Let us denote as N_i the number of claims of type i. Then, the random vector $(N_1,\ldots,N_m)$ is Multinomially distributed, with probability mass function

$$\Pr[N_1=k_1,\ldots,N_m=k_m]=\begin{cases}\frac{n!}{k_1!\cdots k_m!}q_1^{k_1}\cdots q_m^{k_m} \text{ if } k_1+\cdots+k_m=n\\ 0 \text{ otherwise,}\end{cases}$$

where n is the total number of claims.

Each of the m components separately has a Binomial distribution with parameters n and q_i, for the appropriate value of the subscript i, that is, $N_i\sim\mathcal{B}in(n,q_i)$. Because of the constraint that the sum of the components is n, that is, $N_1+\cdots+N_m=n$, they are negatively correlated.

The expected value is $\mathbb{E}[N_i]=nq_i$. The covariance matrix is as follows: Each diagonal entry is the variance of a Binomially distributed random variable, and is therefore $\mathbb{V}[N_i]=nq_i(1-q_i)$. The off-diagonal entries are the covariances. These are $\mathbb{C}[N_i,N_j]=-nq_iq_j$ for i, j distinct. This is a $m\times m$ nonnegative-definite matrix of rank $m-1$.

We will use the following result.

Property 6.1 *Let us assume that the total number of claims N is $\mathcal{P}oi(\lambda)$ distributed. Assume that the N claims may be classified into m categories, according to a multinomial partitioning scheme with probabilities $q_1,\ldots,q_m$. Let N_i represent the number of claims of* *type $i, i = 1, \ldots, m$. Then the random variables $N_1, \ldots, N_m$ are independent and Poisson distributed with respective parameters $\lambda q_1, \ldots, \lambda q_m$.*

Proof Since given $N = n$, $N_i \sim \mathcal{B}in(n, q_i)$, we can write

$$\begin{aligned}
\Pr[N_i = k] &= \sum_{n=k}^{\infty} \Pr[N_i = k | N = n] \Pr[N = n] \\
&= \sum_{n=k}^{\infty} \binom{n}{k} q_i^k (1 - q_i)^{n-k} \exp(-\lambda) \frac{\lambda^n}{n!} \\
&= \exp(-\lambda) \frac{(\lambda q_i)^k}{k!} \sum_{n=0}^{\infty} \frac{((1 - q_i)\lambda)^n}{n!} \\
&= \exp(-\lambda q_i) \frac{(\lambda q_i)^k}{k!}
\end{aligned}$$

which proves that $N_i \sim \mathcal{P}oi(\lambda q_i)$.

We now prove the independence property of the N_is. To this end, let us show that the joint probability mass function factors in the product of marginal probability mass functions:

$$\begin{aligned}
&\Pr[N_1 = n_1, \ldots, N_m = n_m] \\
&= \Pr[N_1 = n_1, \ldots, N_m = n_m | N = n_1 + \cdots + n_m] \exp(-\lambda) \frac{\lambda^{n_1 + \cdots + n_m}}{(n_1 + \cdots + n_m)!} \\
&= \frac{(n_1 + \cdots + n_m)!}{n_1! \cdots n_m!} q_1^{n_1} \cdots q_m^{n_m} \exp(-\lambda) \frac{\lambda^{n_1 + \cdots + n_m}}{(n_1 + \cdots + n_m)!} \\
&= \prod_{j=1}^{m} \exp(-\lambda q_j) \frac{(\lambda q_j)^{n_j}}{n_j!} \\
&= \prod_{j=1}^{m} \Pr[N_j = n_j].
\end{aligned}$$

which completes the proof. □

Let us now denote as $q_{k1}, q_{k2}, \ldots, q_{km}$ the probability that the claim is of type $1, 2, \ldots, m$, respectively, for a policyholder with $\Lambda = \lambda_k$. The identity $q_{k1} + q_{k2} + \cdots + q_{km} = 1$ obviously holds true. Now, let $N_1, N_2, \ldots, N_m$ be the number of claims of type $1, 2, \ldots, m$, respectively. Considering Property 6.1, given Θ and Λ, the random variables $N_1, N_2, \ldots, N_m$ are mutually independent, with respective conditional probability mass function

$$\Pr[N_l = j | \Theta = \theta, \Lambda = \Lambda_k] = \exp(-\lambda_k \theta q_{kl}) \frac{(-\lambda_k \theta q_{kl})^j}{j!}, \quad j = 0, 1, \ldots,$$

for $l = 1, \ldots, m$.

6.3.2 Markov Modelling for the Multi-Event Bonus-Malus Scale

The scale is assumed to have $s+1$ levels, numbered from 0 to s. A specified level is assigned to a new driver. Each claim free year is rewarded by a bonus point (i.e. the driver goes one level down). Each type of claim entails a specific penalty, expressed as a fixed number of levels per claim.

We assume that the scale possesses the following memoryless property: the knowledge of the present level and of the number of claims of each type filed during the present year suffices to determine the level to which the policy is transferred. This ensures that the bonus-malus system may be represented by a Markov chain (at least conditionally on the observable characteristics and random effects).

Let $p_{\ell_1\ell_2}(\vartheta; \boldsymbol{q})$ be the probability of moving from level ℓ_1 to level ℓ_2 for a policyholder with annual mean claim frequency ϑ and vector probability $\boldsymbol{q} = (q_1, \ldots, q_m)^T$; here q_j is the probability that the claim be of type j. Further, $\boldsymbol{P}(\vartheta; \boldsymbol{q})$ is the one-step transition matrix, i.e.

$$\boldsymbol{P}(\vartheta; \boldsymbol{q}) = \begin{pmatrix} p_{00}(\vartheta; \boldsymbol{q}) & \cdots & p_{0s}(\vartheta; \boldsymbol{q}) \\ \vdots & \ddots & \vdots \\ p_{s0}(\vartheta; \boldsymbol{q}) & \cdots & p_{ss}(\vartheta; \boldsymbol{q}) \end{pmatrix}.$$

Taking the nth power of $\boldsymbol{P}(\vartheta; \boldsymbol{q})$ yields the n-step transition matrix whose element $(\ell_1\ell_2)$, denoted as $p^{(n)}_{\ell_1\ell_2}(\vartheta; \boldsymbol{q})$, is the probability of moving from level ℓ_1 to level ℓ_2 in n transitions.

The transition matrix $\boldsymbol{P}(\vartheta; \boldsymbol{q})$ associated with such a bonus-malus system is assumed to be regular, i.e. there exists some integer $\xi_0 \geq 1$ such that all entries of $\big(\boldsymbol{P}(\vartheta; \boldsymbol{q})\big)^{\xi_0}$ are strictly positive. Consequently, the Markov chain describing the trajectory of a policyholder with expected claim frequency ϑ and vector probability $\boldsymbol{q}$ is ergodic and thus possesses a stationary distribution

$$\boldsymbol{\pi}(\vartheta; \boldsymbol{q}) = (\pi_0(\vartheta; \boldsymbol{q}), \pi_1(\vartheta; \boldsymbol{q}), \ldots, \pi_s(\vartheta; \boldsymbol{q}))^T.$$

Here, $\pi_\ell(\vartheta; \boldsymbol{q})$ is the stationary probability for a policyholder with mean frequency ϑ to be in level ℓ i.e.

$$\pi_{\ell_2}(\vartheta; \boldsymbol{q}) = \lim_{n\to+\infty} p^{(n)}_{\ell_1\ell_2}(\vartheta; \boldsymbol{q}).$$

The stationary probabilities are directly obtained from formula (4.9).

Let $L_{\vartheta;\boldsymbol{q}}$ be valued in $\{0, 1, \ldots, s\}$ and conform to the distribution $\boldsymbol{\pi}(\vartheta; \boldsymbol{q})$ i.e.

$$\Pr[L_{\vartheta;\boldsymbol{q}} = \ell] = \pi_\ell(\vartheta; \boldsymbol{q}), \quad \ell = 0, 1, \ldots, s.$$

The variable $L_{\vartheta;\boldsymbol{q}}$ thus represents the level occupied by a policyholder with annual expected claim frequency ϑ and probability vector $\boldsymbol{q}$ once the steady state has been reached.

Now, let L be the level occupied in the scale by a randomly selected policyholder once the steady state has been reached. The distribution of L can be written as

$$\Pr[L = \ell] = \sum_k w_k \int_0^{+\infty} \pi_\ell(\lambda_k\theta; \boldsymbol{q}_k) dF_\Theta(\theta), \quad \ell = 0, 1, \ldots, s. \tag{6.8}$$

6.3.3 Determination of the relativities

The relativity associated with level ℓ is denoted as r_ℓ, as before. The meaning is that an insured occupying that level pays an amount of premium equal to r_ℓ % of the reference premium determined on the basis of his observable characteristics.

As in Chapter 4, our aim is to minimize the expected squared difference between the 'true' relative premium Θ and the relative premium r_L applicable to this policyholder (after the stationary state has been reached), i.e. the goal is to minimize

$$\mathbb{E}\Big[(\Theta - r_L)^2\Big] = \sum_{\ell=0}^{s} \mathbb{E}\Big[(\Theta - r_\ell)^2 \Big| L = \ell\Big] \Pr[L=\ell]$$
$$= \sum_k w_k \int_0^{+\infty} \sum_{\ell=0}^{s} (\theta - r_\ell)^2 \pi_\ell(\lambda_k \theta; \boldsymbol{q}_k) dF_\Theta(\theta).$$

The solution is given by

$$r_\ell = \mathbb{E}[\Theta | L = \ell] = \frac{\sum_k w_k \int_0^{+\infty} \theta \pi_\ell(\lambda_k \theta; \boldsymbol{q}_k) dF_\Theta(\theta)}{\sum_k w_k \int_0^{+\infty} \pi_\ell(\lambda_k \theta; \boldsymbol{q}_k) dF_\Theta(\theta)}. \tag{6.9}$$

It is easily seen that $\mathbb{E}[r_L] = 1$, resulting in financial equilibrium once steady state is reached.

6.3.4 Numerical Illustrations

−1/+2/+3 Bonus-Malus Scale

Let us now consider the scale with six levels (numbered from 0 to 5) already used in the previous chapters. But now, instead of considering one type of claim, we penalize differently claims with bodily injuries and claims with material damage only. If no claims have been reported then the policyholder moves one level down. Claims with material damage only are penalized by two levels whereas claims with bodily injuries entail a penalty of three levels. If n_1 claims with bodily injuries and n_2 claims with material damage only are reported during the year then the policyholder moves $3n_1 + 2n_2$ levels up. This system is abbreviated as $-1/+2/+3$, in obvious notations.

The transition matrix for a policyholder with annual mean claim frequency ϑ and vector probability $\boldsymbol{q} = (q_1, q_2)^T$ is given by

$$\boldsymbol{P}(\vartheta; \boldsymbol{q}) = \begin{pmatrix} P_0 & 0 & P_1 & P_2 & P_3 & 1-\Sigma_1 \\ P_0 & 0 & 0 & P_1 & P_2 & 1-\Sigma_2 \\ 0 & P_0 & 0 & 0 & P_1 & 1-\Sigma_3 \\ 0 & 0 & P_0 & 0 & 0 & 1-\Sigma_4 \\ 0 & 0 & 0 & P_0 & 0 & 1-\Sigma_4 \\ 0 & 0 & 0 & 0 & P_0 & 1-\Sigma_4 \end{pmatrix}$$

where

$$P_0 = \exp(-\vartheta)$$
$$P_1 = \vartheta q_2 \exp(-\vartheta q_2) \exp(-\vartheta q_1)$$

$$P_2 = \exp(-\vartheta q_2)\vartheta q_1 \exp(-\vartheta q_1)$$

$$P_3 = \frac{(\vartheta q_2)^2}{2} \exp(-\vartheta q_2)\exp(-\vartheta q_1)$$

and Σ_i represents the sum of all the elements in row i. Specifically,

$$\Sigma_1 = P_0 + P_1 + P_2 + P_3$$

$$\Sigma_2 = P_0 + P_1 + P_2$$

$$\Sigma_3 = P_0 + P_1$$

$$\Sigma_4 = P_0.$$

Computation of the Relativities

Portfolio A Table 6.5 gives for each of the six levels of the bonus-malus scale the proportion of the portfolio in that level (column 2) and the relativity attached to that level (column 3) for the system $-1/+2/+3$. About 70 % of the portfolio is in level 0 and enjoys a discount of about 30 %. The rest of the portfolio is spread out among levels 1–5. The r_ℓs range from 68.0 % to 262.6 %.

In order to compare the results with those of traditional bonus-malus scales, we have also recalled the results given by the three other scales already considered in the previous chapters. For all of them, each claim-free year is rewarded by one level down in the scale. The first bonus-malus system penalizes each claim (with or without bodily injuries) by two levels up in the scale ($-1/+2$ bonus-malus scale), the second one by three levels up ($-1/+3$ bonus-malus scale) and the third one sends the policyholder to the maximal level after one claim (-1/top bonus-malus scale).

Clearly, the more the claims are penalized, the more the policyholders occupying the lowest levels are awarded discounts, and the less the policyholders in the upper part of the scale are penalized. The scale $-1/+2/+3$ is closer to the scale $-1/+2$. This is because the majority of the claims only induce material damage. Nevertheless, r_5 is reduced from 271.4 % to 262.6 % when the claims with bodily injuries are more severely penalized.

Table 6.5 Results for the bonus-malus systems $-1/+2/+3$, $-1/+2$, $-1/+3$ and -1/top for Portfolio A.

Level ℓ	$-1/+2/+3$		$-1/+2$		$-1/+3$		-1/top	
	π_ℓ	r_ℓ	π_ℓ	r_ℓ	π_ℓ	r_ℓ	π_ℓ	r_ℓ
5	4.7 %	262.6 %	4.4 %	271.4 %	7.3 %	230.8 %	12.8 %	181.2 %
4	4.7 %	215.8 %	4.7 %	218.5 %	5.9 %	200.9 %	9.7 %	159.9 %
3	4.9 %	182.1 %	4.4 %	192.5 %	9.0 %	145.2 %	7.7 %	143.9 %
2	8.6 %	138.0 %	8.7 %	138.8 %	7.3 %	133.1 %	6.2 %	131.3 %
1	7.1 %	127.8 %	7.1 %	128.6 %	6.0 %	123.0 %	5.2 %	120.9 %
0	70.0 %	68.0 %	70.6 %	68.5 %	64.5 %	64.2 %	58.5 %	61.2 %

Table 6.6 Results for the bonus-malus systems $-1/+2/+3$, $-1/+2$, $-1/+3$ and -1/top for Portfolio B.

Level ℓ	$-1/+2/+3$		$-1/+2$		$-1/+3$		-1/top	
	π_ℓ	r_ℓ	π_ℓ	r_ℓ	π_ℓ	r_ℓ	π_ℓ	r_ℓ
5	5.9 %	215.9 %	5.4 %	223.2 %	9.2 %	184.0 %	16.3 %	146.9 %
4	6.2 %	169.5 %	6.0 %	171.3 %	7.8 %	156.6 %	12.5 %	129.3 %
3	6.4 %	143.4 %	5.8 %	148.9 %	11.7 %	117.4 %	9.9 %	117.2 %
2	11.2 %	111.8 %	11.5 %	112.1 %	9.5 %	108.6 %	8.0 %	108.0 %
1	9.2 %	104.6 %	9.3 %	105.0 %	7.8 %	101.6 %	6.6 %	100.7 %
0	61.1 %	74.5 %	62.0 %	74.7 %	54.0 %	72.0 %	46.6 %	70.6 %

Portfolio B Table 6.6 gives for each of the six levels of the bonus-malus scale the proportion of the portfolio in that level (column 2) and the relativity attached to that level (column 3) for the system $-1/+2/+3$. About 60 % of the portfolio is in level 0 and enjoys a discount of about 25 %. The rest of the portfolio is spread out among levels 1–5. The r_ℓs range from 74.5 % to 215.9 %.

In order to compare the results with those of traditional bonus-malus scales, we have also recalled the results given by the three bonus-malus scales $-1/+2$, $-1/+3$ and -1/top already considered in the previous chapters. The scale $-1/+2/+3$ is close to the scale $-1/+2$ (again, this is because the majority of the claims only induce material damage). Nevertheless, r_5 is reduced from 223.2 % to 215.9 % when the claims with bodily injuries are more severely penalized.

6.4 Further Reading and Bibliographic Notes

The second part of this chapter is based on Pitrebois, Denuit & Walhin (2006a). Lemaire (1995, Chapter 13) applied a model proposed by Picard (1976) to Belgian data, distinguishing the accidents that caused property damage only, from those that caused bodily injuries. The credibility model proposed by Lemaire (1995) is based on a Poisson-Gamma mixture, and assumes that given the expected annual claim frequency of the policyholder, the frequency of claims with bodily injuries conforms to a Beta distribution. This approach can be extended to several categories of claims using a Dirichlet distribution (that is, using a suitable multivariate Beta distribution).

With the aid of multi-equation Poisson models with random effects, Pinquet (1998) designed an optimal credibility model for different types of claims. As an example, claims are separated into two groups according to fault with respect to a third party. See also Pinquet (1997) on allowing for the costs of the claims.

This chapter gives only basic methods to deal with different types of claims in *a posteriori* ratemaking. Advanced statistical and econometrics models could certainly improve the actuarial analysis. For instance, Wedel, Böckenholtb & Kamakurac (2003) developed a general class of factor-analytic models for the analysis of multivariate (truncated) count data. These models provide a parsimonious and easy-to-interpret representation of multivariate dependencies in counts that extend the general linear latent variable model.

7

Bonus-Malus Systems with Varying Deductibles

7.1 Introduction

In this chapter, we compare bonus-malus systems to deductibles. Specifically, we design a system in which the policyholders in the malus zone are allowed to choose at each renewal between a premium surcharge (induced by relativities associated with the bonus-malus scale) or a deductible in case of a claim during the forthcoming year. If the deductible is selected, this induces a strong incentive to careful driving. According to signal theory, drivers opting for the deductible are expected to be better drivers (on average) than those paying the premium surcharge induced by the upward mode in the bonus-malus scale.

Bonus-malus systems do not take claim amounts into account so that *a posteriori* corrections only rely on the number of claims. Holtan (1994) suggested the use of very high deductibles that may be borrowed by the policyholder from the insurance company. Although technically acceptable, this approach obviously causes considerable practical problems. While Holtan (1994) assumes a high deductible which is constant for all policyholders, and thus independent of the level they occupy in the bonus-malus scale at the claim occurrence time, the present chapter lets the deductible vary between the levels of the bonus-malus scales, and also considers a mixed case setup of both premium and deductible surcharge after a claim.

Specifically, the *a posteriori* premium correction induced by the bonus-malus scale is replaced with a deductible (in whole or in part). To each level of the bonus-malus scale is attached an amount of deductible, applied to the claims filed during the coverage period. Combining bonus-malus scales with varying deductibles presents a number of advantages:

(1) according to signal theory, policyholders choosing varying deductible should be good drivers;

(2) even if the policyholder leaves the company after a claim, he has to pay for the deductible (but in motor third party liability insurance, there may be difficulties linked to the collection of the deductible);
(3) in the mixed system, the r_ℓs and the severity of the deductibles may be tuned in an optimal way in order to attract the policyholders.

The numerical study conducted in this chapter shows, provided appropriate values for the parameters are selected, that the amounts of deductible are moderate in the mixed case (reduced relativities combined with deductibles per claim).

Although deductibles can be difficult to implement in motor third party liability insurance (companies indemnify the third parties directly and so cannot simply reduce the amount they pay), the system is nevertheless easy to implement for first party coverages (for which the payments are made directly to the policyholders), like material damage for instance. In the European Union, the premium for the material damage cover is either subject to the bonus-malus system applying to motor third party liability or to a specific bonus-malus system. In the latter case, using the techniques suggested in this chapter, one could replace the bonus-malus scale for material damage with deductibles determined by past claim history. This chapter will focus on optional coverages.

7.2 Distribution of the Annual Aggregate Claims

7.2.1 Modelling Claim Costs

As explained in the introduction, the strategy designed in this chapter is difficult to apply in motor third party liability insurance. To fix the ideas, we consider here first party coverages for material damages, so that the actuary is not faced with possible large losses. The total claim cost can thus be modelled using a compound mixed Poisson model. Specifically, let us denote as $C_1, C_2, \ldots, C_N$ the amounts of the N claims reported by the policyholder. The total claim amount for this policy is

$$S = \sum_{k=1}^{N} C_k, \tag{7.1}$$

with the convention that the empty sum equals 0. The severities $C_1, C_2, \ldots,$ are assumed to be independent and identically distributed, and independent of the claim frequency N. As in the preceding chapters, N is taken to be $\mathcal{MP}oi(\lambda, \Theta)$ distributed. This construction essentially states that the cost of an accident is for the most part beyond the control of a policyholder. The degree of care exercised by a driver mostly influences the number of accidents, but in a much lesser way the cost of these accidents. Nevertheless, this assumption seems acceptable in practice, at least as an approximation. Note that the severities $C_1, C_2, \ldots$ are also independent of Θ.

Considering Expression (7.1) for the total cost of claims S, the pure premium for a policyholder without claim history is $\lambda\mathbb{E}[C_1]$. This amount will be corrected by a specific relativity according to the level occupied in the bonus-malus scale, that is, according to the claims reported to the company in the past.

The distribution function of S is then given by

$$\Pr[S \leq x|\Lambda = \lambda, \Theta = \theta] = \sum_{n=0}^{\infty} \exp(-\lambda\theta)\frac{(\lambda\theta)^n}{n!}F^{\star(n)}(x), \quad x \geq 0, \tag{7.2}$$

where F is the common distribution function of the C_ks and

$$F^{\star(n)}(x) = \Pr[C_1 + \cdots + C_n \leq x]$$

is the n-fold convolution of F. Computing $F^{\star(n)}(x)$ amounts to performing an n-dimensional integration of the probability density function corresponding to F. Together with the sum over n in (7.2), this of course makes the formula to get $\Pr[S \leq x|\Lambda = \lambda, \Theta = \theta]$ very time consuming. Furthermore, we still have to integrate over the possible values of Λ and Θ to get $\Pr[S \leq x]$, that is,

$$\Pr[S \leq x] = \sum_k w_k \int_0^{+\infty} \Pr[S \leq x|\Lambda = \lambda_k, \Theta = \theta]dF_\Theta(\theta), \quad x \geq 0.$$

Fortunately, the Poisson distribution belongs to the Panjer's class of counting distributions for which there exists a recursive algorithm.

7.2.2 Discretization

The computation of the distribution function of S with the help of the Panjer algorithm requires the discretization of the individual claim amounts C_k. This amounts to replacing each C_k with a discrete claim cost C_k^Δ multiple of some fixed monetary unit Δ. Usually, actuaries use one of the following two methods to discretize the claim amounts:

Rounding up

In this case, a step Δ is selected and the discretized distribution function F^Δ is defined as

$$\begin{aligned} F^\Delta(x) &= F(k\Delta), \text{ for } k\Delta \leq x < (k+1)\Delta, \\ F^\Delta(k\Delta) &= F(k\Delta), \text{ for } k \in \mathbb{N}. \end{aligned} \tag{7.3}$$

We clearly have

$$F^\Delta(x) \leq F(x) \text{ for all } x \in \mathbb{R},$$

so that this approach is a prudent strategy (the discretized claim sizes C_k^Δ are larger than the original ones C_k in the $\preceq_{\mathrm{ST}}$-sense). However this approach is sometimes too conservative, in that it over-estimates the pure premium (since $\mathbb{E}[C_k^\Delta] > \mathbb{E}[C_k]$).

Mass Dispersion

It is also possible to discretize the C_ks keeping the expected claim cost unchanged. This is done by spreading the probability mass of each $(i\Delta, (i+1)\Delta]$ on the two extremities of the interval (instead of placing all the mass on $(i+1)\Delta$ as before). Define for $i = 0, 1, 2, \ldots,$

$$f_i^+ = \frac{1}{\Delta} \int_{x=i\Delta}^{(i+1)\Delta} \big((i+1)\Delta - x\big) dF(x).$$

Using integration by parts, f_i^+ can be cast into

$$\begin{aligned} f_i^+ &= \frac{1}{\Delta} \left(\Big[\big((i+1)\Delta - x\big) F(x) \Big]_{x=i\Delta}^{(i+1)\Delta} + \int_{x=i\Delta}^{(i+1)\Delta} F(x) dx \right) \\ &= \frac{1}{\Delta} \left(-\Delta F(i\Delta) + \int_{x=i\Delta}^{(i+1)\Delta} F(x) dx \right) \\ &= \frac{1}{\Delta} \int_{x=i\Delta}^{(i+1)\Delta} \big(F(x) - F(i\Delta)\big) dx. \end{aligned}$$

Similarly, for $i = 0, 1, 2, \ldots,$ define

$$\begin{aligned} f_{i+1}^- &= \frac{1}{\Delta} \int_{x=i\Delta}^{(i+1)\Delta} \big(x - i\Delta\big) dF(x) \\ &= \frac{1}{\Delta} \int_{x=i\Delta}^{(i+1)\Delta} \big(F((i+1)\Delta) - F(x)\big) dx. \end{aligned}$$

Let us now spread the probability mass $F((i+1)\Delta) - F(i\Delta)$ of the interval $(i\Delta, (i+1)\Delta]$ on its extremities $i\Delta$ (which receives f_i^+) and $(i+1)\Delta$ (which receives f_{i+1}^-). Then,

$$\begin{aligned} \Pr[C_1^\Delta = 0] &= f_0 = F(0) + f_0^+ \\ \Pr[C_1^\Delta = i\Delta] &= f_i = f_i^- + f_i^+ \text{ for } i \geq 1. \end{aligned} \tag{7.4}$$

When the discretization is performed according to (7.4), C_1^Δ and C_1 have the same mean. It is easy to see that for $k\Delta \leq x < (k+1)\Delta$,

$$F^\Delta(x) = \Pr[C_1^\Delta \leq x] = \sum_{i=0}^{k} f_i = \frac{1}{\Delta} \int_{y=k\Delta}^{(k+1)\Delta} F(y) dy.$$

Hence,

$$\begin{aligned} \mathbb{E}[C_1^\Delta] &= \int_0^{+\infty} \overline{F}^\Delta(x) dx \\ &= \sum_{k=0}^{+\infty} \int_{k\Delta}^{(k+1)\Delta} \overline{F}(x) dx \\ &= \int_0^{+\infty} \overline{F}(x) dx = \mathbb{E}[C_1]. \end{aligned}$$

7.2.3 *Panjer Algorithm*

Direct Convolution Approach

The Panjer algorithm then allows us to compute the distribution function of the discretized version of S, defined as

$$S^{\Delta} = \sum_{k=1}^{N} C_k^{\Delta}.$$

To realize the merit of the Panjer approach, let us first write the formulas in the direct approach. Let us denote as $f_i = \Pr[C_1^{\Delta} = i\Delta]$ the probability mass function of C_1^{Δ}. The probability mass function of $C_1^{\Delta} + \cdots + C_k^{\Delta}$ is

$$f_i^{\star(k)} = \Pr\left[\sum_{j=1}^{k} C_j^{\Delta} = i\Delta\right].$$

In the applications we have in mind, the distribution function for the claim costs satisfies $F(0) = 0$. Clearly, with Discretization Method (7.3), we then have $f_0 = 0$ so that $f_i^{\star(k)} = 0$ if $i \leq k-1$ (since $C_1^{\Delta} \geq \Delta$ with probability 1). The $f_i^{\star(k)}$s satisfy

$$f_i^{\star(k)} = \sum_{j=1}^{i-k+1} f_{i-j}^{\star(k-1)} f_j \text{ if } i \geq k. \tag{7.5}$$

Then, the probability mass function $g_i = \Pr[S^{\Delta} = i\Delta]$ of S^{Δ} satisfies

$$g_i = \sum_{k=0}^{i} \Pr[N = k] f_i^{\star(k)}, \quad i \in \mathbb{N}. \tag{7.6}$$

This direct computation of the probability mass function of S^{Δ} requires a lot of computation time. Things are even worse in the Discretization (7.4) for which $f_0 > 0$.

Panjer Family

The Panjer formula holds for a class of probability distributions referred to as the Katz family in statistical circles, and as the Panjer family in the actuarial literature. This family contains all the counting distributions such that the relation

$$p_k = \left(a + \frac{b}{k}\right) p_{k-1}, \quad k = 1, 2, \ldots, \tag{7.7}$$

is fulfilled for some a and b. The Panjer family contains three elements. Specifically, the probability distributions satisfying (7.7) are

(i) the Poisson distribution, obtained with $a = 0$ and $b > 0$;
(ii) the Negative Binomial distribution, obtained with $0 < a < 1$ and $a + b > 0$;
(iii) the Binomial distribution, obtained with $a < 0$ and $b = -a(m+1)$, for some positive integer m.

Panjer Algorithm with $f_0 = 0$

The Panjer algorithm is easily established with probability generating functions. Here, we use another approach with conditional expectations. The proof of the Panjer formula is based on the following technical lemma.

Lemma 7.1 *The relation*

$$f_j^{\star(n)} = \frac{n}{j}\sum_{i=1}^{j} i f_i f_{j-i}^{\star(n-1)},$$

is valid for $j \geq n$.

Proof The proof consists of noting that, on the one hand, for any $j \geq n$

$$\mathbb{E}\left[X_1 \left| \sum_{k=1}^{n} X_k = j \right.\right] = \frac{1}{n}\mathbb{E}\left[\sum_{k=1}^{n} X_k \left| \sum_{k=1}^{n} X_k = j \right.\right] = \frac{j}{n}.$$

and on the other hand, if $f_j^{\star(k)} > 0$,

$$\begin{aligned}
\mathbb{E}\left[X_1 \left| \sum_{k=1}^{n} X_k = j \right.\right] &= \sum_{i=1}^{j} i \Pr\left[X_1 = i \left| \sum_{k=1}^{n} X_k = j \right.\right] \\
&= \frac{\sum_{i=1}^{j} i \Pr[X_1 = i \text{ and } \sum_{k=2}^{n} X_k = j-i]}{\Pr[\sum_{k=1}^{n} X_k = j]} \\
&= \frac{\sum_{i=1}^{j} i f_i f_{j-i}^{\star(n-1)}}{f_j^{\star(n)}}.
\end{aligned}$$

Equating these two formulas yields the expected result. □

We are now ready to derive the Panjer algorithm if $f_0 = 0$ (as is the case with Discretization Method (7.3)).

Proposition 7.1 *If the probability distribution of N belongs to the Panjer family and if $f_0 = 0$, then the g_is are obtained recursively from*

$$g_j = \sum_{i=1}^{j}\left(a + i\frac{b}{j}\right) f_i g_{j-i} \tag{7.8}$$

starting with $g_0 = p_0$.

Proof Since $C_k \geq \Delta$ almost surely, we clearly have

$$\Pr[S^{\Delta} = 0] = \Pr[N = 0] = p_0.$$

For $j \geq 1$, Lemma 7.1 allows us to write

$$
\begin{aligned}
g_j &= \sum_{n=1}^{+\infty} \left(a + \frac{b}{n}\right) p_{n-1} f_j^{\star(n)} \\
&= \sum_{n=1}^{+\infty} a p_{n-1} \sum_{i=1}^{j} f_i f_{j-i}^{\star(n-1)} + \sum_{n=1}^{+\infty} \frac{b}{j} p_{n-1} \sum_{i=1}^{j} i f_i f_{j-i}^{\star(n-1)} \\
&= \sum_{i=1}^{j} \left(a + i\frac{b}{j}\right) f_i \sum_{n=1}^{+\infty} p_{n-1} f_{j-i}^{\star(n-1)} \\
&= \sum_{i=1}^{j} \left(a + i\frac{b}{j}\right) f_i g_{j-i},
\end{aligned}
$$

which ends the proof. □

Corollary 7.1 (Compound Poisson Case) *If $N \sim \mathcal{P}oi(\lambda)$ ($a = 0$ and $b = \lambda$), we get*

$$
g_i = \begin{cases} \exp(-\lambda) \text{ if } i = 0 \\ \frac{\lambda}{i} \sum_{j=1}^{i} j f_j g_{i-j} \text{ if } i \geq 1. \end{cases}
$$

Panjer Algorithm with $f_0 > 0$

Let us now consider that $f_0 > 0$ (resulting from Discretization Method (7.4)).

Proposition 7.2 *If the probability distribution of N belongs to the Panjer family and if $f_0 \neq 0$, then the g_is are obtained recursively from*

$$
g_j = \frac{1}{1 - a f_0} \sum_{i=1}^{j} \left(a + i\frac{b}{j}\right) f_i g_{j-i} \tag{7.9}
$$

starting from $g_0 = \varphi_N(f_0)$.

Proof Since $f_0^{\star(k)} = (f_0)^k$, $k \in \mathbb{N}$, we easily see that

$$
g_0 = \sum_{k=0}^{+\infty} p_k (f_0)^k = \varphi_N(f_0).
$$

Following the reasoning in the proof of Proposition 7.1, we get for $j > 0$

$$
\begin{aligned}
g_j &= \sum_{n=1}^{+\infty} a p_{n-1} \sum_{i=0}^{j} f_i f_{j-i}^{\star(n-1)} + \sum_{n=1}^{+\infty} \frac{b}{j} p_{n-1} \sum_{i=1}^{j} i f_i f_{j-i}^{\star(n-1)} \\
&= a f_0 \sum_{n=1}^{+\infty} p_{n-1} f_j^{\star(n-1)} + \sum_{i=1}^{j} \left(a + i\frac{b}{j}\right) f_i \sum_{n=1}^{+\infty} p_{n-1} f_{j-i}^{\star(n-1)} \\
&= a f_0 g_j + \sum_{i=1}^{j} \left(a + i\frac{b}{j}\right) f_i g_{j-i},
\end{aligned}
$$

which ends the proof. □

Computation of the Distribution Function of S
Clearly,

$$\Pr[S^{\Delta} > i\Delta] = 1 - \sum_{j=0}^{i} g_j, \quad i \in \mathbb{N}, \text{ where } g_j = \Pr[S^{\Delta} = j\Delta].$$

The recursive formula

$$\Pr[S^{\Delta} > i\Delta] = \begin{cases} 1, \text{ if } i = -1, \\ \Pr[S^{\Delta} > (i-1)\Delta] - g_i, \text{ if } \quad i \in \mathbb{N}, \end{cases}$$

gives the distribution of S.

7.3 Introducing a Deductible within *a Posteriori* Ratemaking

Often, the r_ℓs for high levels ℓ are so large that the system has to be softened before a possible commercial implementation, resulting in financial instability (since the company then faces a progressive decrease of the average premium level because of a clustering of the policyholders in the high-discount classes). To avoid this deficiency, the premium increase that the policyholder has to pay when he goes up in the scale could be (at least partly) replaced by a deductible that would be applied on claims filed by the policyholder during the following year. The company compensates for the reduced penalties in the malus zone with the deductibles paid by policyholders who report claims whilst in the malus zone. This can be commercially attractive since the policyholders are penalized only if they file claims in the future. The amount of these deductibles depends on the level attained by the policyholder and can be applied either annually or claim by claim.

7.3.1 Annual Deductible

Let us consider a policyholder occupying level ℓ in the scale. If this policyholder is subject to the *a posteriori* premium corrections induced by the bonus-malus scale, he will have to pay $r_\ell \lambda \mathbb{E}[C_1]$ to be covered by the company. If the policyholder is subject to the annual deductible instead, he will have to pay the pure premium $\lambda \mathbb{E}[C_1]$ as well as $\min\{S, d_\ell\}$. The indifference principle is now expressed for level ℓ by the equation

$$r_\ell \lambda \mathbb{E}[C_1] = \lambda \mathbb{E}[C_1] + \mathbb{E}[S|S < d_\ell] \Pr[S < d_\ell] + d_\ell \Pr[S > d_\ell], \tag{7.10}$$

where S is of the form (7.1) with the counting variable N distributed as a Negative Binomial with mean λr_ℓ (past claims history is used to reevaluate the expected annual claim frequency of the policyholder according to his position in the scale: the annual expected claim frequency equals λr_ℓ for a policyholder occupying level ℓ in the scale). Note that we work with zero interest rate, as is often the case in nonlife insurance problems, so that we do not use present values but nominal payments.

Equation (7.10) is to be solved for all levels ℓ such that $r_\ell > 100\,\%$ (that is, for all levels in the malus zone). So, the premium surcharge $(r_\ell - 1)\lambda\mathbb{E}[C_1]$ is replaced with an annual deductible d_ℓ. The relation (7.10) ensures that the substitution is actuarially fair.

In practice, equation (7.10) does not possess an explicit solution so that numerical techniques have to be used. Panjer's algorithm is employed to derive the distribution of S. The claim amounts are discretized according to Method (7.4) and the probability mass is concentrated on 500 points. Then, the equation can be solved using appropriate routines available in the IML package of SAS®.

7.3.2 *Per Claim Deductible*

Of course, the deductible could also be applied to each claim filed by the policyholder. The indifference principle invoked above will again be used to determine the amount of the deductible. Considering a policyholder in level ℓ, he will have to pay $\lambda r_\ell\mathbb{E}[C_1]$ if he is subject to the *a posteriori* corrections induced by the bonus-malus scale. If, on the contrary, a fixed deductible d_ℓ is applied per claim, he will have to pay $\lambda\mathbb{E}[C_1]$ as well as $\min\{C_k, d_\ell\}$ for each of the claims C_k reported to the company. Note that the expected number of claims is now $r_\ell\lambda$ because past claims history is used to update the claim frequency distribution. According to the indifference principle, the amount of deductible d_ℓ for a policyholder in level ℓ is the solution to the equation

$$\lambda r_\ell\mathbb{E}[C_1] = \lambda\mathbb{E}[C_1] + \lambda r_\ell\Big(\mathbb{E}[C_1|C_1 < d_\ell]\Pr[C_1 < d_\ell] + d_\ell\Pr[C_1 > d_\ell]\Big).$$

This equation can be simplified as

$$r_\ell\mathbb{E}[C_1] = \mathbb{E}[C_1] + r_\ell\Big(\mathbb{E}[C_1|C_1 < d_\ell]\Pr[C_1 < d_\ell] + d_\ell\Pr[C_1 > d_\ell]\Big). \tag{7.11}$$

Again, (7.11) has to be solved (numerically, in most cases) for all the levels ℓ for which $r_\ell > 100\,\%$.

7.3.3 *Mixed Case*

We could also mix both types of penalties. Specifically, the *a posteriori* corrections r_ℓ are softened and the policyholder is also subject to a deductible (either annual or per claim). This combination allows the actuary to get acceptable r_ℓs and to achieve financial stability thanks to the deductibles, as it will be seen from the numerical illustration proposed in the next section.

In the mixed case, the bonuses (i.e., the r_ℓs less than 1) are kept unchanged but the maluses (i.e., the r_ℓs larger than 1) are reduced by a fixed percentage: instead of r_ℓ, the policyholder will be subject to the penalty $r_\ell \times (1-\alpha)$ for some specified $0 < \alpha < 1$. To compensate for these reduced penalties, the policyholder is subject to deductibles d_ℓ varying according to the level ℓ occupied in the malus zone. The parameter α may be selected in order to achieve a good balance between premium increase and amounts of deductibles. The mixed system, combining relativities and deductibles, is expected to be the most relevant in practice. This will become clear from the numerical illustrations carried out in the next section.

Let us now give the equations providing the d_ℓs. In the case of an annual deductible, the indifference principle allows us to write

$$\begin{aligned}\lambda r_\ell \mathbb{E}[C_1] &= (1-\alpha)\lambda r_\ell \mathbb{E}[C_1] + \mathbb{E}[\min\{S, d_\ell\}] \\ &= (1-\alpha)\lambda r_\ell \mathbb{E}[C_1] + \mathbb{E}[S|S < d_\ell]\Pr[S < d_\ell] \\ &\quad + d_\ell \Pr[S > d_\ell]\end{aligned} \tag{7.12}$$

for the r_ℓs larger than 1 (i.e. in the malus zone). The left-hand side of this equation is the amount of premium paid by the policyholder when the standard bonus-malus system is in force, while the right-hand side is the average amount paid by the policyholder in the mixed bonus-malus system plus deductible system (that is, a reduced penalty plus the expected value of $\min\{S, d_\ell\}$). The solution cannot be obtained explicitly.

Let us now turn to the case where deductibles are applied per claim. The same reasoning yields

$$\lambda r_\ell \mathbb{E}[C_1] = (1-\alpha)\lambda r_\ell \mathbb{E}[C_1] + \lambda r_\ell \mathbb{E}[\min\{C_1, d_\ell\}]$$

so that the d_ℓs are the solution of

$$\alpha \mathbb{E}[C_1] = \mathbb{E}[\min\{C_1, d_\ell\}] = \mathbb{E}[C_1|C_1 < d_\ell]\Pr[C_1 < d_\ell] + d_\ell \Pr[C_1 > d_\ell]. \tag{7.13}$$

A noteworthy feature of this case is that the d_ℓs do not depend on ℓ. The same deductible applies to all the levels in the malus zone. Again, there is no explicit expression for the d_ℓs in general, and numerical techniques have to be used to solve this equation.

7.4 Numerical Illustrations

7.4.1 Claim Frequencies

In this section, the structure function is taken to be Gamma with unit mean, i.e. the probability density function of Θ is given by (1.35). We take the values of λ and a estimated from Portfolio A, that is $\lambda = 0.1474$ and $a = 0.889$.

7.4.2 Claim Severities

For the claim amounts, we work with two distributions with different tails (to investigate the influence of the claim sizes on the deductibles). Considering the inflated 1989 Taiwanese property damage loss distribution used by Lemaire & Zi (1994b), we take the C_ks to be LogNormally distributed with parameters $\mu = 9.2576$ and $\sigma^2 = 1.3569$. We also work with Negative Exponentially distributed C_ks with the same mean, equal to

$$\exp\left(9.2576 + \frac{1.3569}{2}\right) = 20\,662$$

for the sake of comparison. The pure premium amounts to

$$0.1474 \times 20\,662 = 3\,045.6$$

monetary units. The distribution of S is then determined by Panjer's algorithm. Claim amounts are discretized according to Method (7.4).

7.4.3 Annual Deductible

The relativities computed for the scale −1/top are displayed in the second column of Table 7.1 and for the scale $-1/+2$ in Table 7.2. In the pure bonus-malus case, it is thus clear that the r_ℓs associated with the upper levels are considerable (more than 300 % for level 5 in case of the scale $-1/+2$). Now, let us replace the r_ℓs in the malus zone (i.e. levels 1 to 5) with an annual deductible d_ℓ. In order to obtain the d_ℓs from equation (7.10), we first discretize the claim sizes using Method (7.4). Finally, (7.10) is solved numerically using routines available from the SAS®/IML package.

The third column of each table displays the new relativities. In this case, the maluses disappear ($r_\ell = 100\,\%$ for $\ell = 1, \ldots, 5$) and are compensated for by the deductibles listed in the two last columns. The fourth column shows the deductible to be applied if the loss amounts are Negative Exponentially distributed and the last column shows the deductible to be applied if the loss amounts are LogNormally distributed. Since the LogNormal distribution has a thicker tail than the Negative Exponential one, we expect larger amounts of deductible for the former. This is indeed the case, as can be seen from Tables 7.1 and 7.2.

The very high r_ℓs in the second column induce high amounts of deductible, even in the Negative Exponential case. Therefore, this solution seems to be difficult (if not impossible) to implement in practice.

Table 7.1 Results for an annual deductible varying according to the level occupied in the malus zone and for the scale −1/top.

Level ℓ	r_ℓ	r_ℓ with deductible	d_ℓ if $C_1 \sim \mathcal{E}xp$	d_ℓ if $C_1 \sim \mathcal{LN}or$
5	197.3 %	100 %	18 367	35 253
4	170.9 %	100 %	14 028	29 303
3	150.7 %	100 %	10 472	23 544
2	134.8 %	100 %	7470	17 934
1	122.0 %	100 %	4897	12 452
0	54.7 %	54.7 %	0	0

Table 7.2 Results for an annual deductible varying according to the level occupied in the malus zone and for the scale $-1/+2$.

Level ℓ	r_ℓ	r_ℓ with deductible	d_ℓ if $C_1 \sim \mathcal{E}xp$	d_ℓ if $C_1 \sim \mathcal{LN}or$
5	309.1 %	100 %	34 576	50 794
4	241.4 %	100 %	25 076	42 704
3	207.7 %	100 %	20 004	37 241
2	142.9 %	100 %	9027	20 933
1	130.2 %	100 %	6563	16 079
0	62.4 %	62.4 %	0	0

7.4.4 Per Claim Deductible

In this case, there is an analytical solution when the claims are Negative Exponentially distributed: the deductible d_ℓ involved in (7.11) is simply given by $\ln r_\ell$ times the expected claim cost. No explicit solution is available when claim amounts are LogNormally distributed, and numerical procedures have to be used in this case to find the deductibles d_ℓ.

Table 7.3 displays the results obtained for a deductible per claim for the scale -1/top and Table 7.4 for the scale $-1/+2$, when the premium paid by the policyholder is held constant whatever the claim history. As was the case for the annual deductible, the second column gives the relativities associated with each level of the scale in the case of a classical bonus-malus system. The third column gives the relative premium in the case of a scale with the deductible system. The fourth column shows the deductible to be applied if the loss amounts are Negative Exponentially distributed and the last column shows the deductible to be applied if the loss amounts are LogNormally distributed.

Again, the amounts of deductible displayed in the last two columns are very high compared to the annual premium, especially for the scale $-1/+2$. This results from the severe r_ℓs listed in column 2. In order to get acceptable amounts of deductible, keeping the financial stability of the system, in the next section we will combine softened penalties in the malus zone with moderate deductibles.

Table 7.3 Results for a deductible per claim varying according to the level occupied in the malus zone for the scale -1/top.

Level ℓ	r_ℓ	r_ℓ with deductible	d_ℓ if $C_1 \sim \mathcal{E}xp$	d_ℓ if $C_1 \sim \mathcal{LN}or$
5	197.3 %	100 %	14 041	16 254
4	170.9 %	100 %	11 073	12 191
3	150.7 %	100 %	8474	8941
2	134.8 %	100 %	6170	6288
1	122.0 %	100 %	4108	4080
0	54.7 %	54.7 %	0	0

Table 7.4 Results for a deductible per claim varying according to the level occupied in the malus zone for the scale $-1/+2$.

Level ℓ	r_ℓ	r_ℓ with deductible	d_ℓ if $C_1 \sim \mathcal{E}xp$	d_ℓ if $C_1 \sim \mathcal{LN}or$
5	309.1 %	100 %	23 317	31 592
4	241.4 %	100 %	18 209	22 633
3	207.7 %	100 %	15 101	17 803
2	142.9 %	100 %	7376	7651
1	130.2 %	100 %	5451	5502
0	62.4 %	62.4 %	0	0

7.4.5 Annual Deductible in the Mixed Case

Another solution would be to keep the system of maluses but to reduce the penalties r_ℓ applied to the policyholders in the malus zone, for example by choosing $\alpha = 20\,\%$. In exchange for these reduced r_ℓs, the policyholders are subject to an annual deductible on the claims they will eventually file. Equation (7.12) used to compute the annual deductible then becomes

$$\lambda r_\ell \mathbb{E}[C_1] = 80\,\% r_\ell \lambda \mathbb{E}[C_1] + \mathbb{E}[S|S < d_\ell]\Pr[S < d_\ell] + d_\ell \Pr[S > d_\ell] \tag{7.14}$$

to be solved for each level ℓ such that $r_\ell > 100\,\%$.

Tables 7.5 and 7.6 display the numerical results. The r_ℓs displayed in column 3 are equal to 80 % of those in column 2, except for the bonus level 0. Note that the reduced level 1 for the scale −1/top enters the bonus zone. The annual deductibles in the Negative Exponential case (column 4) are now reasonable (about twice the annual premium for most levels) but those in the LogNormal case remain considerable (up to five times the pure premium).

7.4.6 Per Claim Deductible in the Mixed Case

If we combine softened penalties for the bonus-malus system with deductibles per claim, Equation (7.13) becomes

$$20\,\% \mathbb{E}[C_1] = \mathbb{E}[C_1|C_1 < d_\ell]\Pr[C_1 < d_\ell] + d_\ell \Pr[C_1 > d_\ell]. \tag{7.15}$$

Table 7.5 Results for an annual deductible varying according to the level occupied in the malus zone, combined with reduced relativities r_ℓ, for the scale −1/top.

Level ℓ	r_ℓ	r_ℓ with deductible	d_ℓ if $C_1 \sim \mathcal{E}xp$	d_ℓ if $C_1 \sim \mathcal{LN}or$
5	197.3 %	157.8 %	6034	14 297
4	170.9 %	136.7 %	5845	14 127
3	150.7 %	120.6 %	5700	13 996
2	134.8 %	107.8 %	5584	13 894
1	122.0 %	97.6 %	5491	13 811
0	54.7 %	54.7 %	0	0

Table 7.6 Results for an annual deductible varying according to the level occupied in the malus zone, combined with reduced relativities r_ℓ, for the scale −1/+2.

Level ℓ	r_ℓ	r_ℓ with deductible	d_ℓ if $C_1 \sim \mathcal{E}xp$	d_ℓ if $C_1 \sim \mathcal{LN}or$
5	309.1 %	247.3 %	6832	15 017
4	241.4 %	193.1 %	6346	14 581
3	207.7 %	166.2 %	6108	14 364
2	142.9 %	114.3 %	5643	13 946
1	130.2 %	104.2 %	5551	13 864
0	62.4 %	62.4 %	0	0

Table 7.7 Results for a deductible per claim varying according to the level occupied in the malus zone, combined with reduced relativities r_ℓ, for the scale -1/top.

Level ℓ	r_ℓ	r_ℓ with deductible	d_ℓ if $C_1 \sim \mathcal{E}xp$	d_ℓ if $C_1 \sim \mathcal{LN}or$
5	197.3 %	157.8 %	4610	4604
4	170.9 %	136.7 %	4610	4604
3	150.7 %	120.6 %	4610	4604
2	134.8 %	107.8 %	4610	4604
1	122.0 %	97.6 %	4610	4604
0	54.7 %	54.7 %	0	0

Table 7.8 Results for a deductible per claim varying according to the level occupied in the malus zone, combined with reduced relativities r_ℓ, for the scale $-1/+2$.

Level ℓ	r_ℓ	r_ℓ with deductible	d_ℓ if $C_1 \sim \mathcal{E}xp$	d_ℓ if $C_1 \sim \mathcal{LN}or$
5	309.1 %	247.3 %	4610	4604
4	241.4 %	193.1 %	4610	4604
3	207.7 %	166.2 %	4610	4604
2	142.9 %	114.3 %	4610	4604
1	130.2 %	104.2 %	4610	4604
0	62.4 %	62.4 %	0	0

As already mentioned, since this equation does not depend on the level ℓ, the amount of deductible will be the same for each level of the scale.

Tables 7.7 and 7.8 display the numerical results. The third column gathers the relativities: those in the malus zone have been reduced by 20 % compared to column 2. The last two columns display the amounts of deductible. In this case, the amounts of deductible are reasonable and can be implemented in practice (about 150 % of the annual pure premium). Quite surprisingly, the LogNormal distribution now produces smaller deductibles than its Negative Exponential counterpart.

7.5 Further Reading and Bibliographic Notes

This chapter is based on Pitrebois, Denuit & Walhin (2005) for the most part.

The Panjer family of counting distributions is known as the Katz family in applied probability. This family attracted a lot of attention in the literature, due to the fact that it contains underdispersed (Binomial), equidispersed (Poisson), and overdispersed (Negative Binomial) distributions. As a result of Panjer's (1981) publication, a lot of other articles have appeared in the actuarial literature covering similar recursion relations. Multivariate versions of the Panjer algorithm will be used in Chapter 9.

Claim amounts have also been taken into account by Bonsdorff (2005), who studied bonus-malus systems where the transitions between bonus levels in the entire interval $[a, b]$

are determined by the number of claims of the previous year and the total amount of claims of the previous year.

VANDEBROEK (1993) analysed the efficiency of bonus-malus systems and partial coverages in preventing moral hazard problems, by means of stochastic dynamic programming. See also HOLTAN (1994) and LEMAIRE & ZI (1994a) for the trade-off between bonus-malus systems and deductibles.

8

Transient Maximum Accuracy Criterion

8.1 Introduction

8.1.1 From Stationary to Transient Distributions

All developments so far have been based on the stationary distribution of the Markov process describing the trajectory of the policyholder in the bonus-malus scale. As Borgan, Hoem & Norberg (1981) objected, an asymptotic criterion is moderately relevant for bonus-malus systems needing relatively long periods to reach their steady state, since policies are in force only during a limited number of insurance periods. These authors modified the criterion in order to take into account the rating error for new and young policies. As Norberg (1976), Borgan *et al.* (1981) measured the performances of a bonus-malus system by a weighted average of the expected squared rating errors for selected insurance periods.

In Chapter 4, the relativities associated with the levels of the bonus-malus scale were computed on the basis of an asymptotic criterion. The implicit assumption behind the results in the preceding chapters is thus that the Markov process reaches its steady state after a relatively short period, as is the case for the -1/top bonus-malus scale for instance. If a majority of the policies are far from the steady state, it seems desirable to modify the criterion so as to take into account the rating error for new policies and for policies of a moderate age as well.

8.1.2 A Practical Example: Creating a Special Scale for New Entrants

Before entering into the mathematical developments, let us describe a concrete situation where the asymptotic criterion used in the previous chapters is no longer relevant, and

---

Actuarial Modelling of Claim Counts: Risk Classification, Credibility and Bonus-Malus Systems M. Denuit, X. Maréchal, S. Pitrebois and J.-F. Walhin

where the transient regime must be considered. This concerns new entrants in a bonus-malus system (especially the young drivers). Often, age is included in the *a priori* ratemaking, raising the premium for young drivers (especially young males). The large premium surcharges imposed on young drivers pose social problems in many countries. As shown by BOUCHER & DENUIT (2006), the heterogeneity is huge inside classes of young drivers. Once individual factors have been accounted for (on the basis of a fixed effect model for panel data), young drivers even became less risky on average than mature and old ones in the empirical study conducted by BOUCHER & DENUIT (2006). In fact, the vast majoriy of claims reported by young drivers is concentrated on just a few policies.

In addition to the severe explicit penalties contained in the *a priori* tariff, young drivers enter the bonus-malus scale far above the average level they should occupy given their annual expected claim frequency. There is thus an implicit penalty for new drivers (added to the explicit penalty found in most commercial price lists), since the relativity corresponding to the access level of all bonus-malus systems is in every case substantially higher than the average stationary relativity. The implicit surcharge paid by newcomers can be evaluated by comparing the access level to the stationary level for the sub-population of the policyholders insured for a period of 20 years, say.

Young inexperienced drivers generally cause many more accidents than the other categories of the policyholders. At the same time, classes composed of young drivers are more heterogeneous than the other ones: the numerous claims are filed by a minority of insured drivers (causing several claims per year). There are basically two ways to take this phenomenon into account when designing bonus-malus systems:

- Either the more important residual heterogeneity is recognized (by a larger variance of the random effect) and particular transition rules (i.e. heavier penalties when a claim is filed) are imposed on young drivers during the first few years.
- Or young drivers are first placed in a special -1/top scale, and once the bottom level is attained they are sent to the regular bonus-malus scale (entering the scale at their average stationary level).

Let us follow the second approach. The bonus-malus scale is as follows: young drivers are first placed in the highest level of a -1/top scale (with six levels, say). Careful young drivers then reach level 0 in five years, and enter the regular scale at that time (the regular bonus-malus scale can be of the $-1/+2$ type, for instance).

In such a case, the levels of the initial -1/top scale form a transient class in the Markov chain describing the trajectory of the policyholders accross the bonus-malus scales. The policyholders will all leave the initial scale sooner or later and never come back to it, so that the associated stationary probabilities are all equal to 0. Applying the asymptotic criterion to compute the relativities of such a hybrid bonus-malus system does not account for the initial scale. The transient distribution will be influenced by the -1/top scale, and should therefore be used in this case to determine the relativities associated with the -1/top initial scale and with the regular $-1/+2$ scale.

8.1.3 Agenda

The transient regime is discussed in Section 8.2, where the convergence of bonus-malus systems is analysed. The modified criterion with a quadratic loss function is presented in Section 8.3. The exponential loss function is briefly discussed in Section 8.4.

In Section 8.5, we give the results obtained on the examples studied in the preceding chapters (former compulsory Belgian bonus-malus scale, −1/top scale and −1/+2 scale) when using the transient maximum accuracy criterion. All the results have been computed with the formulas taking the *a priori* ratemaking into account. We first examine the convergence to the steady state. Then we give the transient probability distributions and the transient relativities obtained when using a uniform initial distribution and a uniform distribution of the age of policy. We also compare the relativities computed using the transient maximum accuracy criterion to the relativities obtained with the help of the asymptotic maximum accuracy criterion. Finally, we give the evolution of the expected financial income.

Many EU insurers recently started to compete on the basis of bonus-malus systems. Because of marketing and competition for market shares, several insurers now offer the best level 'for life': provided the insured drivers reach level 0, they are allowed to stay in that level whatever the claims reported to the company. Note however that insurance companies remain free to cancel the policy after each claim. There is thus a super bonus level: the driver reaching level 0 of the scale is then allowed to 'claim for free'. These gifts to the best drivers are in contradiction to the actuarial and economic purposes of the *a posteriori* ratemaking systems. They are nevertheless very efficient from the marketing point of view, to keep the best drivers in the portfolio. There is thus an absorbing state in the Markov model describing the trajectory of the driver in the bonus-malus scale. Consequently, the stationary distribution is degenerated, placing a unit probability mass in level 0. Making level 0 absorbing for the Markov chain thus forbids the use of the stationary distribution. This particular case will be considered in Section 8.6.

All the numerical illustrations of this chapter are based on Portfolio A.

8.2 Transient Behaviour and Convergence of Bonus-Malus Scales

A method of computation of the convergence rate based on the eigenvalues of the transition matrix has been discussed in Chapter 4. Here, we examine the evolution of the total variation distance between the transient distribution $\{p_\ell^{(n)}(\vartheta), \ell = 0, 1, \ldots, s\}$ and the stationary distribution $\{\pi_\ell(\vartheta), \ell = 0, 1, \ldots, s\}$:

$$d_{TV}(\boldsymbol{p}^{(n)}(\vartheta), \boldsymbol{\pi}(\vartheta)) = \sum_{\ell=0}^{s} |p_\ell^{(n)}(\vartheta) - \pi_\ell(\vartheta)|, \quad n = 0, 1, 2, \ldots$$

for some given expected annual claim frequency ϑ. Considering a policyholder, picked at random in the portfolio, let us denote as L_n the level occupied by this policyholder in the bonus-malus scale after n years, and as L the level occupied once the stationary regime has been reached. The convergence can thus be assessed with

$$d_{TV}(\boldsymbol{p}^{(n)}, \boldsymbol{\pi}) = \sum_{\ell=0}^{s} \Big| \Pr[L_n = \ell] - \Pr[L = \ell] \Big|$$

$$= \sum_{\ell=0}^{s} \left| \sum_k w_k \int_0^{+\infty} \Big(p_\ell^{(n)}(\lambda_k \theta) - \pi_\ell(\lambda_k \theta) \Big) dF_\Theta(\theta) \right|, \quad (8.1)$$

for $n = 0, 1, 2, \ldots$ which is the sum on each level ℓ of the absolute difference between the probability for a policyholder to be in the level ℓ after n periods and the probability for this policyholder to be in level ℓ when the stationary state is reached.

We can assume that the convergence to the steady state is acceptable when we reach n_0 such that

$$d_{TV}(\boldsymbol{p}^{(n_0)}, \boldsymbol{\pi}) \leq \epsilon \quad (8.2)$$

for some fixed $\epsilon > 0$. The convergence could then be checked by computing and analysing the evolution of (8.1) with n.

Example 8.1 (Former compulsory Belgian bonus-malus scale) The convergence of the former compulsory Belgian bonus-malus scale is assessed using the evolution of

$$C_n = d_{TV}(\boldsymbol{p}^{(n)}, \boldsymbol{\pi})$$

displayed in Figure 8.1. We notice a fast convergence during the first 20 years (C_n decreasing from 0.9 to 0.2) and then a very slow convergence to reach $C_n < \epsilon = 0.05$ after about 38 years. This means that a policyholder must be in the portfolio for 38 years before approaching the stationary state.

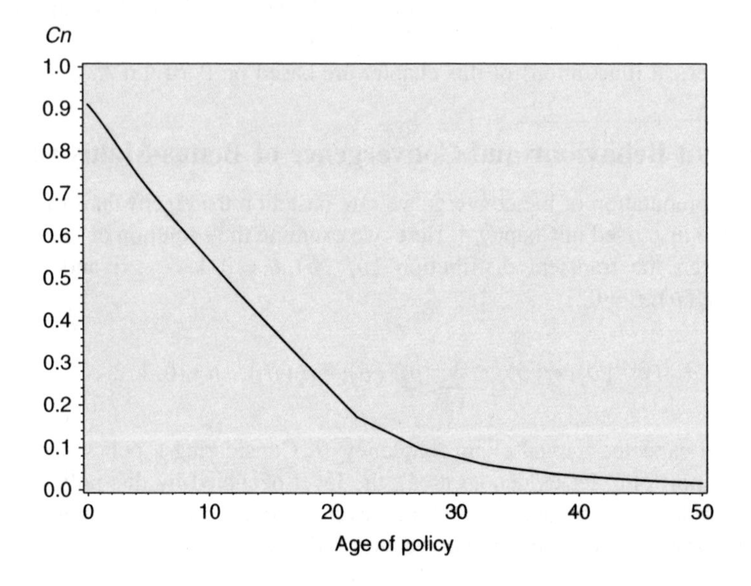

Figure 8.1 Convergence of the transient policyholders' distributions to the steady state distribution for the former compulsory Belgian bonus-malus scale.

8.3 Quadratic Loss Function

8.3.1 Transient Maximum Accuracy Criterion

In this chapter, a policyholder picked at random from the portfolio is now characterized by three random variables: Λ, Θ and A. As before, Λ is the expected claim frequency derived from the *a priori* ratemaking and Θ the residual effect due to the heterogeneity remaining inside each risk class. The integer-valued random variable A then represents the age of the policy in the portfolio and will enable us to take the transient behaviour of a bonus-malus system into account. The probability mass function of A is denoted as

$$\Pr[A = n] = a_n, \quad n = 1, 2, \ldots$$

In words, this means that a proportion a_n of the policies in the portfolio have been in force for n years.

We have already seen that the assumption of the independence between Λ and Θ is reasonable. As for Λ and A, it seems that they are likely to be correlated. Indeed, it seems probable that the age of the policyholder, which is included in the *a priori* ratemaking, is correlated with the age of the policy. However, in order to simplify the computation, we will further assume that Λ and A are independent. We also assume the independence between Θ and A.

Recall from Chapter 4 that the sequence of levels occupied by the policyholder in the bonus-malus scale is denoted as $\{L_0, L_1, L_2, \ldots\}$. Here, we denote as L_A the level occupied by a policyholder subject to the bonus-malus system for A years. Let us assume that the relativity applied to a policyholder picked at random from the portfolio is $r_{L_A}^{(A)}$. For a policyholder with age of policy n, this relativity becomes $r_{L_n}^{(n)}$. As a result, for each group of policyholders with age of policy n, the goal is then to minimize the expected squared rating error

$$\begin{aligned} Q_n &= \mathbb{E}\Big[(\Theta - r_{L_A}^{(A)})^2 \Big| A = n\Big] = \mathbb{E}\Big[(\Theta - r_{L_n}^{(n)})^2\Big] \\ &= \sum_k w_k \int_0^{+\infty} \sum_{\ell=0}^{s} (\theta - r_\ell^{(n)})^2 p_\ell^{(n)}(\lambda_k \theta) dF_\Theta(\theta). \end{aligned} \tag{8.3}$$

The solution is given by

$$\begin{aligned} r_\ell^{(n)} &= \mathbb{E}[\Theta | L_A = \ell, A = n] \\ &= \frac{\sum_k w_k \int_0^{+\infty} \theta p_\ell^{(n)}(\lambda_k \theta) dF_\Theta(\theta)}{\sum_k w_k \int_0^{+\infty} p_\ell^{(n)}(\lambda_k \theta) dF_\Theta(\theta)}. \end{aligned} \tag{8.4}$$

We see that (4.13) is a limiting form of equation (8.4) when n tends to infinity.

The asymptotic criterion Q seems reasonable when a majority of the risks are close to the steady state. In practice, however, real portfolios will often have a substantial fraction of comparatively young policies. Then it is desirable to obtain a solution for the relativities not only by means of Q but also by means of Q_n for various finite n. This is why we introduce a new criterion based on a weighted average of the form

$$\begin{aligned}\bar{Q} &= \mathbb{E}\Big[(\Theta - r_{L_A}^{(A)})^2\Big] = \sum_{n=1}^{\infty} a_n Q_n \\ &= \sum_{n=1}^{\infty} a_n \sum_k w_k \int_0^{+\infty} \sum_{\ell=0}^{s} (\theta - r_\ell^{(n)})^2 p_\ell^{(n)}(\lambda_k \theta) dF_\Theta(\theta).\end{aligned} \tag{8.5}$$

The solution of (8.5) is then given by

$$\begin{aligned}\bar{r}_\ell &= \mathbb{E}[\Theta | L_A = \ell] \\ &= \sum_{n=1}^{\infty} r_\ell^{(n)} \Pr[A = n | L_A = \ell] \\ &= \frac{\sum_{n=0}^{\infty} a_n \sum_k w_k \int_0^{+\infty} \theta p_\ell^{(n)}(\lambda_k \theta) dF_\Theta(\theta)}{\sum_{n=1}^{\infty} a_n \sum_k w_k \int_0^{+\infty} p_\ell^{(n)}(\lambda_k \theta) dF_\Theta(\theta)}.\end{aligned} \tag{8.6}$$

Remark 8.1 Let us mention that if the insurance company does not enforce any *a priori* ratemaking system, all the λ_ks are equal to $\overline{\lambda}$ and (8.6) reduces to the formula

$$\bar{r}_\ell = \frac{\sum_{n=1}^{\infty} a_n \int_0^{+\infty} \theta p_\ell^{(n)}(\overline{\lambda}\theta) dF_\Theta(\theta)}{\sum_{n=1}^{\infty} a_n \int_0^{+\infty} p_\ell^{(n)}(\overline{\lambda}\theta) dF_\Theta(\theta)}$$

that has been derived in BORGAN, HOEM & NORBERG (1981).

Example 8.2 (Former Compulsory Belgian Bonus-Malus Scale) We saw from Figure 8.1 that the steady state was approached at a slow rate for the 23-level Belgian bonus-malus scale. This is why we selected 50 years in the distribution of policy age. The transient relativities presented in Table 8.2 correspond to the transient distributions displayed in Table 8.1. They have all been computed with a uniform initial distribution ($\Pr[L^{(0)} = \ell] = 1/23$ for $\ell = 0, 1, \ldots, 22$).

We observe that the transient relativities slowly converge to the steady state relativities given in the last column of Table 8.2 but also that they generally underestimate these stationary relativities. This agrees with the fact that the transient probabilities of being in level 0 underestimate the steady state probability of being in this level. Therefore, the bonuses of levels 0, 1, 2, 3 and 4 must be greater and the maluses of levels 5 to 22 must be smaller to ensure financial balance. Moreover, we clearly see that, in the first steps of the transient behaviour, there are more levels granting a bonus in order to balance the lack (with respect to the stationary state) of policyholders in the lowest levels.

In addition to the uniform initial distribution, let us now consider a 'top' distribution (78 % of the policyholders are concentrated in level 22 at time 0 and each other level contains 1 % of the policyholders) and a 'bottom' distribution (78 % of the policyholders are placed in level 0 at time 0 and the others are spread uniformly over levels 1 to 22). Let us mention that all the policyholders could not be placed in level 0 or 22 because Equation (8.4) is not defined when $p_\ell^{(n)}$ equals 0. The results are displayed in Tables 8.3 and 8.4. We can see there

Table 8.1 Evolution of the uniform initial distribution for the former compulsory Belgian bonus-malus scale.

Level ℓ	$\Pr[L^{(0)}=\ell]$	$\Pr[L^{(10)}=\ell]$	$\Pr[L^{(20)}=\ell]$	$\Pr[L^{(30)}=\ell]$	$\Pr[L^{(40)}=\ell]$	$\Pr[L^{(50)}=\ell]$	$\Pr[L=\ell]$
22	4.3 %	4.6 %	5.1 %	5.3 %	5.3 %	5.3 %	5.4 %
21	4.3 %	3.4 %	3.6 %	3.7 %	3.8 %	3.8 %	3.8 %
20	4.3 %	2.8 %	2.8 %	2.9 %	2.9 %	2.9 %	2.9 %
19	4.3 %	2.4 %	2.3 %	2.3 %	2.3 %	2.3 %	2.3 %
18	4.3 %	2.3 %	2.1 %	2.0 %	1.9 %	1.9 %	1.9 %
17	4.3 %	2.5 %	1.9 %	1.8 %	1.7 %	1.7 %	1.6 %
16	4.3 %	2.6 %	1.8 %	1.6 %	1.5 %	1.5 %	1.5 %
15	4.3 %	2.6 %	1.8 %	1.5 %	1.4 %	1.4 %	1.3 %
14	4.3 %	2.7 %	1.7 %	1.5 %	1.4 %	1.3 %	1.3 %
13	4.3 %	2.8 %	1.8 %	1.5 %	1.3 %	1.3 %	1.2 %
12	4.3 %	4.1 %	2.0 %	1.5 %	1.4 %	1.3 %	1.2 %
11	4.3 %	4.0 %	2.0 %	1.5 %	1.4 %	1.3 %	1.2 %
10	4.3 %	4.0 %	2.0 %	1.6 %	1.4 %	1.3 %	1.2 %
9	4.3 %	4.0 %	2.2 %	1.7 %	1.5 %	1.4 %	1.4 %
8	4.3 %	4.0 %	2.3 %	1.8 %	1.7 %	1.6 %	1.6 %
7	4.3 %	4.0 %	2.8 %	2.0 %	1.8 %	1.8 %	1.7 %
6	4.3 %	3.9 %	2.8 %	2.1 %	1.9 %	1.8 %	1.8 %
5	4.3 %	3.7 %	2.8 %	2.1 %	1.9 %	1.8 %	1.8 %
4	4.3 %	4.7 %	4.5 %	4.2 %	4.2 %	4.2 %	4.3 %
3	4.3 %	4.2 %	4.0 %	3.8 %	3.8 %	3.8 %	3.8 %
2	4.3 %	3.8 %	4.4 %	3.5 %	3.4 %	3.4 %	3.4 %
1	4.3 %	3.4 %	4.0 %	3.2 %	3.1 %	3.1 %	3.1 %
0	4.3 %	23.6 %	39.3 %	46.9 %	49.0 %	49.7 %	50.3 %

Table 8.2 Evolution of the transient relativities for the uniform initial distribution for the former compulsory Belgian bonus-malus scale.

Level ℓ	$r_\ell^{(0)}$	$r_\ell^{(10)}$	$r_\ell^{(20)}$	$r_\ell^{(30)}$	$r_\ell^{(40)}$	$r_\ell^{(50)}$	r_ℓ
22	0.0 %	264.1 %	270.0 %	270.9 %	271.2 %	271.4 %	271.5 %
21	0.0 %	229.0 %	241.9 %	245.0 %	246.2 %	246.8 %	247.4 %
20	0.0 %	201.1 %	219.5 %	224.8 %	226.9 %	227.9 %	229.1 %
19	0.0 %	179.2 %	200.8 %	207.9 %	211.0 %	212.4 %	214.1 %
18	0.0 %	164.0 %	185.4 %	193.7 %	197.4 %	199.2 %	201.4 %
17	0.0 %	141.3 %	170.4 %	180.7 %	185.3 %	187.6 %	190.3 %
16	0.0 %	132.1 %	159.4 %	170.0 %	175.0 %	177.5 %	180.3 %
15	0.0 %	123.9 %	149.9 %	160.8 %	165.9 %	168.5 %	171.4 %
14	0.0 %	117.9 %	142.2 %	152.8 %	157.8 %	160.3 %	163.1 %
13	0.0 %	114.9 %	136.4 %	145.9 %	150.4 %	152.6 %	154.9 %
12	0.0 %	86.9 %	125.3 %	137.6 %	142.7 %	145.0 %	147.2 %
11	0.0 %	85.7 %	120.1 %	131.9 %	136.5 %	138.5 %	140.2 %

Table 8.2 (Continued).

Level ℓ	$r_\ell^{(0)}$	$r_\ell^{(10)}$	$r_\ell^{(20)}$	$r_\ell^{(30)}$	$r_\ell^{(40)}$	$r_\ell^{(50)}$	r_ℓ
10	0.0 %	83.1 %	115.1 %	126.6 %	131.0 %	132.7 %	134.0 %
9	0.0 %	82.1 %	110.7 %	120.6 %	123.8 %	124.9 %	125.5 %
8	0.0 %	82.6 %	107.3 %	115.0 %	117.0 %	117.4 %	117.3 %
7	0.0 %	81.5 %	94.7 %	107.8 %	110.9 %	111.5 %	111.4 %
6	0.0 %	79.1 %	91.3 %	103.8 %	106.5 %	107.0 %	106.7 %
5	0.0 %	75.5 %	87.6 %	100.0 %	102.6 %	103.0 %	102.8 %
4	0.0 %	77.4 %	80.9 %	83.3 %	83.1 %	82.9 %	82.6 %
3	0.0 %	74.1 %	78.2 %	80.7 %	80.6 %	80.4 %	80.2 %
2	0.0 %	70.4 %	64.8 %	76.7 %	77.9 %	77.9 %	77.8 %
1	0.0 %	66.4 %	61.8 %	74.2 %	75.4 %	75.5 %	75.5 %
0	0.0 %	51.5 %	43.8 %	43.5 %	44.2 %	44.6 %	45.1 %

Table 8.3 Evolution of the transient relativities for the top initial distribution for the former compulsory Belgian bonus-malus scale.

Level ℓ	$r_\ell^{(0)}$	$r_\ell^{(10)}$	$r_\ell^{(20)}$	$r_\ell^{(30)}$	$r_\ell^{(40)}$	$r_\ell^{(50)}$	r_ℓ
22	0.0 %	222.7 %	248.8 %	258.7 %	263.9 %	266.8 %	271.5 %
21	0.0 %	195.2 %	221.7 %	232.8 %	238.7 %	242.1 %	247.4 %
20	0.0 %	174.9 %	201.2 %	213.1 %	219.5 %	223.3 %	229.1 %
19	0.0 %	159.2 %	185.0 %	197.3 %	204.1 %	208.1 %	214.1 %
18	0.0 %	147.0 %	171.9 %	184.3 %	191.3 %	195.4 %	201.4 %
17	0.0 %	94.6 %	133.0 %	154.6 %	167.8 %	176.3 %	190.3 %
16	0.0 %	109.3 %	143.1 %	159.4 %	168.4 %	173.5 %	180.3 %
15	0.0 %	103.1 %	135.4 %	151.5 %	160.3 %	165.2 %	171.4 %
14	0.0 %	98.0 %	128.8 %	144.6 %	153.1 %	157.7 %	163.1 %
13	0.0 %	94.2 %	123.4 %	138.5 %	146.4 %	150.4 %	154.9 %
12	0.0 %	46.0 %	89.0 %	111.6 %	125.6 %	134.4 %	147.2 %
11	0.0 %	85.7 %	102.6 %	122.2 %	131.7 %	136.1 %	140.2 %
10	0.0 %	83.1 %	98.7 %	117.7 %	126.6 %	130.5 %	134.0 %
9	0.0 %	82.1 %	95.4 %	112.6 %	119.9 %	122.9 %	125.5 %
8	0.0 %	82.6 %	92.6 %	107.4 %	113.0 %	115.2 %	117.3 %
7	0.0 %	81.5 %	58.9 %	85.8 %	100.1 %	106.9 %	111.4 %
6	0.0 %	79.1 %	74.6 %	94.4 %	101.3 %	103.9 %	106.7 %
5	0.0 %	75.5 %	72.2 %	91.1 %	97.5 %	99.9 %	102.8 %
4	0.0 %	77.4 %	71.1 %	74.3 %	77.4 %	79.4 %	82.6 %
3	0.0 %	74.1 %	68.5 %	71.9 %	75.1 %	77.0 %	80.2 %
2	0.0 %	70.4 %	31.1 %	63.2 %	74.8 %	77.6 %	77.8 %
1	0.0 %	66.4 %	61.8 %	65.1 %	69.1 %	71.7 %	75.5 %
0	0.0 %	51.5 %	43.8 %	36.1 %	40.4 %	42.5 %	45.1 %

Table 8.4 Evolution of the transient relativities for the bottom initial distribution for the former compulsory Belgian bonus-malus scale.

Level ℓ	$r_\ell^{(0)}$	$r_\ell^{(10)}$	$r_\ell^{(20)}$	$r_\ell^{(30)}$	$r_\ell^{(40)}$	$r_\ell^{(50)}$	r_ℓ
22	0.0 %	304.9 %	290.2 %	281.8 %	277.4 %	275.0 %	271.5 %
21	0.0 %	268.1 %	262.9 %	256.6 %	252.9 %	250.8 %	247.4 %
20	0.0 %	242.9 %	240.2 %	236.1 %	233.4 %	231.8 %	229.1 %
19	0.0 %	207.5 %	224.0 %	221.5 %	218.9 %	217.2 %	214.1 %
18	0.0 %	190.1 %	207.1 %	206.6 %	205.0 %	203.8 %	201.4 %
17	0.0 %	177.6 %	192.9 %	193.3 %	192.6 %	191.9 %	190.3 %
16	0.0 %	181.3 %	184.1 %	182.7 %	181.9 %	181.4 %	180.3 %
15	0.0 %	186.6 %	178.8 %	174.5 %	172.9 %	172.2 %	171.4 %
14	0.0 %	140.8 %	160.7 %	163.3 %	163.7 %	163.6 %	163.1 %
13	0.0 %	148.2 %	156.6 %	156.2 %	155.7 %	155.4 %	154.9 %
12	0.0 %	134.1 %	151.3 %	149.9 %	148.6 %	148.0 %	147.2 %
11	0.0 %	142.6 %	149.3 %	145.5 %	143.0 %	141.7 %	140.2 %
10	0.0 %	148.6 %	147.4 %	141.9 %	138.4 %	136.4 %	134.0 %
9	0.0 %	115.3 %	129.3 %	128.5 %	127.2 %	126.4 %	125.5 %
8	0.0 %	122.3 %	126.3 %	123.1 %	120.6 %	119.2 %	117.3 %
7	0.0 %	125.5 %	120.3 %	118.5 %	115.8 %	114.0 %	111.4 %
6	0.0 %	126.8 %	118.1 %	115.2 %	112.0 %	109.9 %	106.7 %
5	0.0 %	127.0 %	115.9 %	112.2 %	108.7 %	106.4 %	102.8 %
4	0.0 %	99.4 %	92.0 %	88.4 %	86.1 %	84.7 %	82.6 %
3	0.0 %	98.1 %	90.1 %	86.3 %	83.9 %	82.4 %	80.2 %
2	0.0 %	96.6 %	84.8 %	83.8 %	81.6 %	80.1 %	77.8 %
1	0.0 %	95.0 %	82.7 %	81.6 %	79.4 %	77.9 %	75.5 %
0	0.0 %	56.6 %	49.8 %	47.5 %	46.5 %	46.0 %	45.1 %

that even if the ultimate distribution of the policyholders in the scale is the same whatever the initial distributions, the transient relativities are affected by these distributions. Starting with a uniform distribution of the policyholders in the scale (Table 8.2), the $r_\ell^{(n)}$s increase to the r_ℓs for levels 1 to 22, and they decrease to the limit for $\ell = 0$. The same phenomenon arises when starting with the top distribution (Table 8.3), but the difference between the $r_\ell^{(10)}$s and the asymptotic r_ℓs is now larger. On the contrary, if the bottom distribution is used (Table 8.4) then the $r_\ell^{(n)}$s decrease to their limit r_ℓ.

The relativities $\bar{r}_\ell$ computed using (8.6) are given in Table 8.5 according to the initial distribution of the policyholders in the scale and for a uniform distribution of age of policy over 50 years, that is, $a_n = 1/50$ for $n = 1, \ldots, 50$. For the sake of completeness, the steady state relativities are displayed in the last column. When a uniform initial distribution of the policyholders is used, the relativities based on the transient maximum accuracy criterion are smaller than the steady state relativities, except for level 0. This is the case for all levels if the top distribution is assumed. On the contrary, if the bottom distribution is used then the $\bar{r}_\ell$s are larger than the corresponding r_ℓs.

In order to figure out the impact of the maturity of the portfolio on the relativities, we considered two alternative age structures to the uniform age of policy distribution used so far. The three distributions (henceforth referred to as mature, young and old

Table 8.5 Relativities computed on the basis of the transient maximum accuracy criterion for the former compulsory Belgian bonus-malus scale, and for different initial distributions.

Level ℓ	Uniform distribution $\bar{r}_\ell$	Top distribution $\bar{r}_\ell$	Bottom distribution $\bar{r}_\ell$	Steady state distribution r_ℓ
22	266.1 %	241.1 %	283.5 %	271.5 %
21	234.4 %	185.4 %	255.2 %	247.4 %
20	208.6 %	164.0 %	232.6 %	229.1 %
19	187.4 %	147.5 %	213.7 %	214.1 %
18	170.1 %	134.5 %	198.3 %	201.4 %
17	156.1 %	124.2 %	185.6 %	190.3 %
16	144.8 %	115.7 %	175.0 %	180.3 %
15	135.6 %	108.8 %	166.0 %	171.4 %
14	128.1 %	102.9 %	158.9 %	163.1 %
13	121.9 %	98.0 %	153.7 %	154.9 %
12	116.6 %	93.7 %	148.3 %	147.2 %
11	111.8 %	89.8 %	142.6 %	140.2 %
10	107.5 %	86.4 %	137.1 %	134.0 %
9	103.8 %	83.5 %	133.7 %	125.5 %
8	100.1 %	80.8 %	128.7 %	117.3 %
7	96.3 %	78.1 %	122.8 %	111.4 %
6	92.7 %	75.3 %	117.2 %	106.7 %
5	89.2 %	72.7 %	112.0 %	102.8 %
4	81.8 %	68.9 %	99.7 %	82.6 %
3	78.0 %	65.8 %	94.1 %	80.2 %
2	74.5 %	63.0 %	89.3 %	77.8 %
1	71.3 %	60.2 %	85.0 %	75.5 %
0	45.8 %	40.1 %	52.5 %	45.1 %

portfolios, respectively) are summarized in Table 8.6. Table 8.7 displays the bonus-malus relativities obtained with these different age structures and a uniform initial distribution of the policyholders in the bonus-malus scale. We see that the older the portfolio, the closer the $\bar{r}_\ell$s to the corresponding r_ℓs.

8.3.2 Linear Scales

Sometimes, it is desirable to have the same relative penalty associated with each level. To this end, the actuary can linearize the $\bar{r}_\ell$s, as suggested by Gilde & Sundt (1989). The optimal linear relativity $\bar{r}_\ell^{lin} = \alpha + \beta\ell$, $\ell = 0, 1, \ldots, s$, in the transient case is thus the solution of the minimization of

$$\mathbb{E}\big[(\Theta - r_{L_A}^{(A)})^2\big] = \mathbb{E}\big[(\Theta - \alpha - \beta L_A)^2\big].$$

It is easy to check that the solution of this optimization problem is

$$\beta = \frac{\mathbb{C}[L_A, \Theta]}{\mathbb{V}[L_A]} \text{ and } \alpha = \mathbb{E}[\Theta] - \frac{\mathbb{C}[L_A, \Theta]}{\mathbb{V}[L_A]}\mathbb{E}[L_A].$$

Table 8.6 Values of the a_ns for three portfolios with different maturities.

Age of policy	Mature portfolio	Young portfolio	Old portfolio
1	2 %	10 %	0.5 %
2	2 %	10 %	0.5 %
3	2 %	5 %	0.5 %
4	2 %	5 %	0.5 %
5	2 %	5 %	0.5 %
6	2 %	5 %	0.5 %
7	2 %	5 %	1 %
8	2 %	2.5 %	1 %
9	2 %	2.5 %	1 %
10	2 %	2.5 %	1 %
11	2 %	2.5 %	1 %
12	2 %	2.5 %	1 %
13	2 %	2.5 %	1 %
14	2 %	2.5 %	1 %
15	2 %	2.5 %	1 %
16	2 %	2.5 %	1 %
17	2 %	2.5 %	1 %
18	2 %	1 %	1 %
19	2 %	1 %	1 %
20	2 %	1 %	1 %
21	2 %	1 %	1 %
22	2 %	1 %	1 %
23	2 %	1 %	1 %
24	2 %	1 %	1 %
25	2 %	1 %	1 %
26	2 %	1 %	1 %
27	2 %	1 %	1 %
28	2 %	1 %	1 %
29	2 %	1 %	1 %
30	2 %	1 %	1 %
31	2 %	1 %	1 %
32	2 %	1 %	1 %
33	2 %	1 %	1 %
34	2 %	1 %	2.5 %
35	2 %	1 %	2.5 %
36	2 %	1 %	2.5 %
37	2 %	1 %	2.5 %
38	2 %	1 %	2.5 %
39	2 %	1 %	2.5 %
40	2 %	1 %	2.5 %
41	2 %	1 %	2.5 %
42	2 %	1 %	2.5 %
43	2 %	1 %	2.5 %
44	2 %	1 %	5 %
45	2 %	0.5 %	5 %
46	2 %	0.5 %	5 %

Table 8.6 (Continued).

Age of policy	Mature portfolio	Young portfolio	Old portfolio
47	2 %	0.5 %	5 %
48	2 %	0.5 %	5 %
49	2 %	0.5 %	10 %
50	2 %	0.5 %	10 %

Table 8.7 Relativities computed on the basis of the transient maximum accuracy criterion for the former compulsory Belgian bonus-malus scale, and for different maturities of the portfolio.

Level ℓ	Mature portfolio $\bar{r}_\ell$	Young portfolio $\bar{r}_\ell$	Old portfolio $\bar{r}_\ell$	Steady state distribution r_ℓ
22	266.1 %	253.6 %	269.7 %	271.5 %
21	234.4 %	208.0 %	242.6 %	247.4 %
20	208.6 %	173.8 %	221.0 %	229.1 %
19	187.4 %	152.9 %	203.0 %	214.1 %
18	170.1 %	138.2 %	187.6 %	201.4 %
17	156.1 %	127.5 %	174.4 %	190.3 %
16	144.8 %	119.7 %	163.1 %	180.3 %
15	135.6 %	113.6 %	152.9 %	171.4 %
14	128.1 %	109.7 %	144.2 %	163.1 %
13	121.9 %	106.4 %	136.5 %	154.9 %
12	116.6 %	103.5 %	129.7 %	147.2 %
11	111.8 %	100.8 %	123.8 %	140.2 %
10	107.5 %	98.2 %	118.6 %	134.0 %
9	103.8 %	96.5 %	113.1 %	125.5 %
8	100.1 %	94.2 %	107.8 %	117.3 %
7	96.3 %	91.4 %	103.2 %	111.4 %
6	92.7 %	88.8 %	99.1 %	106.7 %
5	89.2 %	86.3 %	95.4 %	102.8 %
4	81.8 %	84.4 %	82.2 %	82.6 %
3	78.0 %	79.2 %	79.0 %	80.2 %
2	74.5 %	75.4 %	76.1 %	77.8 %
1	71.3 %	72.5 %	73.3 %	75.5 %
0	45.8 %	50.5 %	44.8 %	45.1 %

The linear premium scale is thus of the form

$$\bar{r}_\ell^{lin} = \mathbb{E}[\Theta] + \frac{\mathbb{C}[L_A, \Theta]}{\mathbb{V}[L_A]}(\ell - \mathbb{E}[L_A]),$$

where

$$\mathbb{C}[L_A, \Theta] = \sum_{n=1}^{+\infty} a_n \sum_{\ell=0}^{s} \ell \sum_k w_k \int_0^{+\infty} \theta p_\ell^{(n)}(\lambda_k \theta) dF_\Theta(\theta) - \mathbb{E}[L_A]\mathbb{E}[\Theta]$$

$$\mathbb{E}[L_A] = \sum_{n=1}^{\infty} a_n \sum_{\ell=0}^{s} \sum_k w_k \int_0^{+\infty} \ell p_\ell^{(n)}(\lambda_k \theta) dF_\Theta(\theta)$$

$$\mathbb{V}[L_A] = \sum_{n=1}^{\infty} a_n \sum_{\ell=0}^{s} \sum_k w_k \int_0^{+\infty} (\ell - \mathbb{E}[L_A])^2 p_\ell^{(n)}(\lambda_k \theta) dF_\Theta(\theta).$$

8.3.3 Financial Balance

We know from Chapter 4 that when the relativities are derived from the asymptotic criterion, the bonus-malus system is financially balanced when the steady state has been reached. The only way of keeping this financial balance during the transient behaviour of the bonus-malus scale is to change the relativities applied to the policyholders so that they are equal to the $r_\ell^{(n)}$s in each period n. For commercial reasons, it seems difficult to adopt such a strategy.

But, when the number of levels and the transistion rules of the bonus-malus system have been fixed, it is possible to check how the expected financial income evolves with respect to the financial balance for the different periods $n = 1, 2, \ldots$ until the steady state is reached.

If the relativities in force are the r_ℓs computed on the basis of the asymptotic maximum accuracy criterion, then the expected financial income for policies with age n is

$$\begin{aligned} I_n &= \sum_{\ell=0}^{s} r_\ell \Pr[L_n = \ell] \\ &= \sum_{\ell=0}^{s} \sum_k w_k \int_0^{+\infty} r_\ell \, p_\ell^{(n)}(\lambda_k \theta) dF_\Theta(\theta), \end{aligned} \quad (8.7)$$

where the $p_\ell^{(n)}(\cdot)$s are computed using (4.6).

Alternatively, if the relativities are the $\bar{r}_\ell$s computed on the basis of the transient maximum accuracy criterion, then the expected financial income for policies with age n is

$$\begin{aligned} \bar{I}_n &= \sum_{\ell=0}^{s} \bar{r}_\ell \Pr[L_n = \ell] \\ &= \sum_{\ell=0}^{s} \sum_k w_k \int_0^{+\infty} \bar{r}_\ell \, p_\ell^{(n)}(\lambda_k \theta) dF_\Theta(\theta). \end{aligned} \quad (8.8)$$

The evolution of the expected income, until steady state has been reached, is probably one of the most important parameters to take into account.

Example 8.3 (Former Belgian Compulsory Bonus-Malus Scale) Table 8.8 displays the evolution of the expected financial income I_n (computed with the steady state relativities) according to the initial distribution of the policyholders. It slowly converges to 100 %. The choice of the uniform initial distribution or the top initial distribution leads to an expected financial income greater than 100 %. Indeed, too many policyholders (with respect to the steady state situation) are in the malus levels, thus providing a greater income to the company. Conversely, with a bottom initial distribution, the expected financial income is smaller than 100 % as too many policyholders are in the bonus levels.

Table 8.8 Evolution of the expected financial income I_n based on r_ℓ (influence of the initial distribution).

Age of policy	Uniform distribution I_n	Top distribution I_n	Bottom distribution I_n
0	146.5 %	242.8 %	68.4 %
1	143.1 %	225.8 %	72.0 %
2	140.1 %	214.1 %	75.3 %
3	137.4 %	205.5 %	78.4 %
4	134.9 %	198.7 %	81.2 %
5	132.6 %	193.2 %	82.6 %
6	130.4 %	187.8 %	83.7 %
7	128.2 %	182.8 %	84.8 %
8	126.1 %	178.3 %	86.0 %
9	124.1 %	174.0 %	87.2 %
10	122.2 %	170.1 %	88.0 %
11	120.3 %	166.3 %	88.6 %
12	118.6 %	162.9 %	89.2 %
13	116.9 %	158.9 %	89.9 %
14	115.3 %	155.1 %	90.6 %
15	113.9 %	152.1 %	91.0 %
16	112.5 %	149.3 %	91.4 %
17	111.2 %	146.8 %	91.8 %
18	109.9 %	140.7 %	92.2 %
19	108.9 %	138.3 %	92.6 %
20	107.9 %	136.3 %	92.9 %
21	107.0 %	134.5 %	93.2 %
22	106.1 %	127.5 %	93.5 %
23	105.5 %	124.3 %	93.8 %
24	105.0 %	123.1 %	94.2 %
25	104.6 %	122.0 %	94.4 %
26	104.1 %	121.0 %	94.7 %
27	103.7 %	117.6 %	94.9 %
28	103.4 %	115.8 %	95.2 %
29	103.1 %	115.0 %	95.4 %
30	102.8 %	114.4 %	95.7 %
31	102.6 %	113.7 %	95.9 %
32	102.3 %	111.9 %	96.0 %
33	102.2 %	110.8 %	96.2 %
34	102.0 %	110.3 %	96.5 %
35	101.9 %	109.9 %	96.6 %
36	101.7 %	109.5 %	96.8 %
37	101.6 %	108.4 %	96.9 %
38	101.5 %	107.6 %	97.1 %
39	101.4 %	107.3 %	97.2 %
40	101.3 %	107.0 %	97.4 %
41	101.2 %	106.7 %	97.5 %
42	101.1 %	106.0 %	97.6 %
43	101.0 %	105.5 %	97.7 %
44	100.9 %	105.3 %	97.8 %
45	100.9 %	105.1 %	98.0 %

46	100.8 %	104.9 %	98.0 %
47	100.8 %	104.4 %	98.1 %
48	100.7 %	104.1 %	98.2 %
49	100.7 %	103.9 %	98.3 %
50	100.6 %	103.8 %	98.4 %

8.3.4 *Choice of an Initial Level*

As the number of years increases, the influence of the initial level diminishes for the individual policy, and it vanishes in the limit. Therefore, the choice of the initial level cannot be made part of the optimizing procedure based on the asymptotic criterion.

When the transient behaviour is taken into account, it is convenvient to select the optimal starting level by maximizing a measure of efficiency for the bonus-malus scale. In the transient case, the Q_n*-efficiency* is defined as

$$\begin{aligned} e_n &= \sum_{\ell=0}^{s} (r_\ell^{(n)})^2 \Pr[L_n = \ell] \\ &= \sum_{\ell=0}^{s} \sum_k w_k \int_0^{+\infty} (r_\ell^{(n)})^2 p_\ell^{(n)}(\lambda_k \theta) dF_\Theta(\theta). \end{aligned} \tag{8.9}$$

The Q_n-efficiency is equal to the variance $\mathbb{V}[r_{L_n}^{(n)}]$ up to a constant term, since

$$\mathbb{V}[r_{L_n}^{(n)}] = \mathbb{E}\Big[(r_{L_n}^{(n)})^2\Big] - \Big(\mathbb{E}\Big[r_{L_n}^{(n)}\Big]\Big)^2 = e_n - 1. \tag{8.10}$$

Looking for an optimum of e_n, it is then equivalent to use (8.9) or (8.10). The *Q-efficiency*, which is closely related to the variance $\mathbb{V}[r_L]$, is

$$\begin{aligned} e &= \sum_{\ell=0}^{s} (r_\ell)^2 \Pr[L = \ell] \\ &= \sum_{\ell=0}^{s} \sum_k w_k \int_0^{+\infty} (r_\ell)^2 \pi_\ell(\lambda_k \theta) dF_\Theta(\theta) = \mathbb{V}[r_L] + 1. \end{aligned}$$

Finally, the $\bar{Q}$*-efficiency*, which is equivalent to $\mathbb{V}[\bar{r}_{L_A}]$, is

$$\begin{aligned} \bar{e} &= \sum_{\ell=0}^{s} (\bar{r}_\ell)^2 \Pr[L_A = \ell] \\ &= \sum_{n=1}^{\infty} a_n \sum_{\ell=0}^{s} \sum_k w_k \int_0^{+\infty} (\bar{r}_\ell)^2 p_\ell^{(n)}(\lambda_k \theta) dF_\Theta(\theta) \\ &= \mathbb{V}[\bar{r}_{L_A}] + 1. \end{aligned}$$

To choose the initial level, it suffices to compute the relativities $\bar{r}_\ell$, with the help of (8.6), for each initial distribution $\boldsymbol{e}_k$, where $\boldsymbol{e}_k$ is the vector with 1 in the kth entry and 0 elsewhere. After having repeated the process for each $\boldsymbol{e}_k$, $k = 0, 1, \ldots, s$, we compute the $\bar{Q}$-efficiency of each solution. The optimal initial level is then the one that maximizes the $\bar{Q}$-efficiency.

8.4 Exponential Loss Function

Under an exponential loss function, the aim is to minimize the objective function

$$\mathcal{O}_n = \mathbb{E}\Big[\exp\big(-c(\Theta - r^{(n)}_{L_n})\big)\Big]$$

under the constraint $\mathbb{E}[r^{(n)}_{L_n}] = \mathbb{E}[\Theta]$. This yields

$$r^{(n)}_\ell = \mathbb{E}[\Theta] + \frac{1}{c}\Big(\mathbb{E}\Big[\ln \mathbb{E}[\exp(-c\Theta)|L_n]\Big] - \ln \mathbb{E}[\exp(-c\Theta)|L_n = \ell]\Big),$$

for $\ell = 0, 1, \ldots, s$.

Of course, there is no reason to focus on the particular nth period. Then, nonnegative weights $a_1, a_2, a_3, \ldots$ summing to 1 representing the age distribution of the policies in the portfolio are introduced and the aim of the actuary is to minimize

$$\overline{\mathcal{O}} = \sum_{n=1}^{+\infty} a_n \mathcal{O}_n.$$

Minimizing $\overline{\mathcal{O}}$ means minimizing the expected squared rating error for a randomly chosen policy.

8.5 Numerical Illustrations

8.5.1 Scale −1/Top

The transition rules of the −1/top scale are given in Table 4.1.

Initial Distribution

We have tested three different initial distributions $\Pr[L^{(0)} = \ell]$. In the first case (uniform distribution), the policyholders are uniformly spread in the scale (16.67 % of the policyholders in each of the six levels). In the second case (top distribution), 95 % of the policyholders are concentrated in the top of the scale (and the remaining 5 % are evenly spread over levels 0 to 4). In the last case (bottom distribution), 95 % of the policyholders start from the bottom of the scale (and the remaining 5 % are evenly spread over levels 1 to 5).

Convergence of the −1/Top Scale

Figure 8.2 represents the evolution of C_n with n for a uniform initial distribution. It gives an idea of the speed of convergence of the −1/top scale. We clearly see that $C_n = 0$ for $n \geq 5$, i.e. that the stationary state is reached after 5 years. This was known from Chapter 4.

Transient Relativities

Table 8.9 gives the evolution with n of the corresponding transient relativities for the three starting distributions. We see that the transient relativities do not depend on the initial

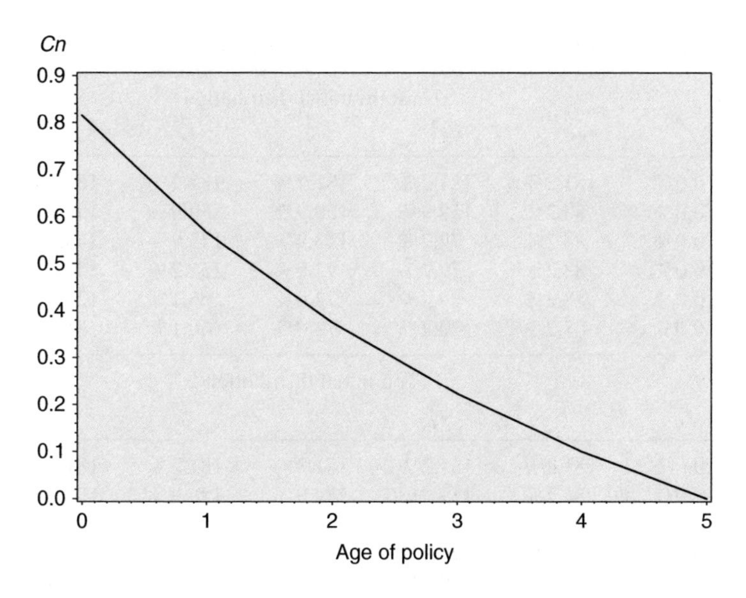

Figure 8.2 Convergence of the transient policyholders' distributions to the steady state distribution for the system −1/top.

distribution of the policyholders inside the scale. This comes from the fact that the transient distributions do not depend on the initial distribution for the −1/top scale.

This somewhat surprising situation can be explained as follows. The distribution of the policyholders in the −1/top scale after one year is given by

$$\begin{pmatrix} p_0^{(1)}(\vartheta) \\ p_1^{(1)}(\vartheta) \\ p_2^{(1)}(\vartheta) \\ p_3^{(1)}(\vartheta) \\ p_4^{(1)}(\vartheta) \\ p_5^{(1)}(\vartheta) \end{pmatrix} = \begin{pmatrix} p_0^{(0)} \\ p_1^{(0)} \\ p_2^{(0)} \\ p_3^{(0)} \\ p_4^{(0)} \\ p_5^{(0)} \end{pmatrix}^T \begin{pmatrix} \exp(-\vartheta) & 0 & 0 & 0 & 0 & 1-\exp(-\vartheta) \\ \exp(-\vartheta) & 0 & 0 & 0 & 0 & 1-\exp(-\vartheta) \\ 0 & \exp(-\vartheta) & 0 & 0 & 0 & 1-\exp(-\vartheta) \\ 0 & 0 & \exp(-\vartheta) & 0 & 0 & 1-\exp(-\vartheta) \\ 0 & 0 & 0 & \exp(-\vartheta) & 0 & 1-\exp(-\vartheta) \\ 0 & 0 & 0 & 0 & \exp(-\vartheta) & 1-\exp(-\vartheta) \end{pmatrix}$$

$$= \begin{pmatrix} (p_0^{(0)}+p_1^{(0)})\exp(-\vartheta) \\ p_2^{(0)}\exp(-\vartheta) \\ p_3^{(0)}\exp(-\vartheta) \\ p_4^{(0)}\exp(-\vartheta) \\ p_5^{(0)}\exp(-\vartheta) \\ 1-\exp(-\vartheta) \end{pmatrix}$$

Table 8.9 Transient relativities for the different initial distributions (−1/top scale).

Level ℓ	Uniform initial distribution						
	$r_\ell^{(0)}$	$r_\ell^{(1)}$	$r_\ell^{(2)}$	$r_\ell^{(3)}$	$r_\ell^{(4)}$	$r_\ell^{(5)}$	r_ℓ
5	0.0 %	181.2 %	181.2 %	181.2 %	181.2 %	181.2 %	181.2 %
4	0.0 %	88.2 %	159.9 %	159.9 %	159.9 %	159.9 %	159.9 %
3	0.0 %	88.2 %	79.2 %	143.9 %	143.9 %	143.9 %	143.9 %
2	0.0 %	88.2 %	79.2 %	72.0 %	131.2 %	131.2 %	131.2 %
1	0.0 %	88.2 %	79.2 %	72.0 %	66.1 %	120.9 %	120.9 %
0	0.0 %	88.2 %	79.2 %	72.0 %	66.1 %	61.2 %	61.2 %
Level ℓ	Top initial distribution						
	$r_\ell^{(0)}$	$r_\ell^{(1)}$	$r_\ell^{(2)}$	$r_\ell^{(3)}$	$r_\ell^{(4)}$	$r_\ell^{(5)}$	r_ℓ
5	0.0 %	181.2 %	181.2 %	181.2 %	181.2 %	181.2 %	181.2 %
4	0.0 %	88.2 %	159.9 %	159.9 %	159.9 %	159.9 %	159.9 %
3	0.0 %	88.2 %	79.2 %	143.9 %	143.9 %	143.9 %	143.9 %
2	0.0 %	88.2 %	79.2 %	72.0 %	131.2 %	131.2 %	131.2 %
1	0.0 %	88.2 %	79.2 %	72.0 %	66.1 %	120.9 %	120.9 %
0	0.0 %	88.2 %	79.2 %	72.0 %	66.1 %	61.2 %	61.2 %
Level ℓ	Bottom initial distribution						
	$r_\ell^{(0)}$	$r_\ell^{(1)}$	$r_\ell^{(2)}$	$r_\ell^{(3)}$	$r_\ell^{(4)}$	$r_\ell^{(5)}$	r_ℓ
5	0.0 %	181.2 %	181.2 %	181.2 %	181.2 %	181.2 %	181.2 %
4	0.0 %	88.2 %	159.9 %	159.9 %	159.9 %	159.9 %	159.9 %
3	0.0 %	88.2 %	79.2 %	143.9 %	143.9 %	143.9 %	143.9 %
2	0.0 %	88.2 %	79.2 %	72.0 %	131.2 %	131.2 %	131.2 %
1	0.0 %	88.2 %	79.2 %	72.0 %	66.1 %	120.9 %	120.9 %
0	0.0 %	88.2 %	79.2 %	72.0 %	66.1 %	61.2 %	61.2 %

The corresponding relativities are obtained by (8.4). They are given by

$$
\begin{pmatrix} r_0^{(1)} \\ r_1^{(1)} \\ r_2^{(1)} \\ r_3^{(1)} \\ r_4^{(1)} \\ r_5^{(1)} \end{pmatrix}
=
\begin{pmatrix}
\dfrac{\sum_k w_k \int_0^{+\infty} \theta (p_0^{(0)} + p_1^{(0)}) \exp(-\lambda_k \theta) dF_\Theta(\theta)}{\sum_k w_k \int_0^{+\infty} (p_0^{(0)} + p_1^{(0)}) \exp(-\lambda_k \theta) dF_\Theta(\theta)} \\
\dfrac{\sum_k w_k \int_0^{+\infty} \theta p_2^{(0)} \exp(-\lambda_k \theta) dF_\Theta(\theta)}{\sum_k w_k \int_0^{+\infty} p_2^{(0)} \exp(-\lambda_k \theta) dF_\Theta(\theta)} \\
\dfrac{\sum_k w_k \int_0^{+\infty} \theta p_3^{(0)} \exp(-\lambda_k \theta) dF_\Theta(\theta)}{\sum_k w_k \int_0^{+\infty} p_3^{(0)} \exp(-\lambda_k \theta) dF_\Theta(\theta)} \\
\dfrac{\sum_k w_k \int_0^{+\infty} \theta p_4^{(0)} \exp(-\lambda_k \theta) dF_\Theta(\theta)}{\sum_k w_k \int_0^{+\infty} p_4^{(0)} \exp(-\lambda_k \theta) dF_\Theta(\theta)} \\
\dfrac{\sum_k w_k \int_0^{+\infty} \theta p_5^{(0)} \exp(-\lambda_k \theta) dF_\Theta(\theta)}{\sum_k w_k \int_0^{+\infty} p_5^{(0)} \exp(-\lambda_k \theta) dF_\Theta(\theta)} \\
\dfrac{\sum_k w_k \int_0^{+\infty} \theta (1 - \exp(-\lambda_k \theta)) dF_\Theta(\theta)}{\sum_k w_k \int_0^{+\infty} (1 - \exp(-\lambda_k \theta)) dF_\Theta(\theta)}
\end{pmatrix}
$$

$$
= \begin{pmatrix}
\frac{\sum_k w_k \int_0^{+\infty} \theta \exp(-\lambda_k \theta) dF_\Theta(\theta)}{\sum_k w_k \int_0^{+\infty} \exp(-\lambda_k \theta) dF_\Theta(\theta)} \\
\frac{\sum_k w_k \int_0^{+\infty} \theta \exp(-\lambda_k \theta) dF_\Theta(\theta)}{\sum_k w_k \int_0^{+\infty} \exp(-\lambda_k \theta) dF_\Theta(\theta)} \\
\frac{\sum_k w_k \int_0^{+\infty} \theta \exp(-\lambda_k \theta) dF_\Theta(\theta)}{\sum_k w_k \int_0^{+\infty} \exp(-\lambda_k \theta) dF_\Theta(\theta)} \\
\frac{\sum_k w_k \int_0^{+\infty} \theta \exp(-\lambda_k \theta) dF_\Theta(\theta)}{\sum_k w_k \int_0^{+\infty} \exp(-\lambda_k \theta) dF_\Theta(\theta)} \\
\frac{\sum_k w_k \int_0^{+\infty} \theta \exp(-\lambda_k \theta) dF_\Theta(\theta)}{\sum_k w_k \int_0^{+\infty} \exp(-\lambda_k \theta) dF_\Theta(\theta)} \\
\frac{\sum_k w_k \int_0^{+\infty} \theta (1-\exp(-\lambda_k \theta)) dF_\Theta(\theta)}{\sum_k w_k \int_0^{+\infty} (1-\exp(-\lambda_k \theta)) dF_\Theta(\theta)}
\end{pmatrix}
$$

We see that the relativities at time 1 do not depend on the initial distribution $\{p_\ell^{(0)}, \ell = 0, 1, \ldots, 5\}$. Moreover, we observe that the values $r_0^{(1)}$ to $r_4^{(1)}$ are all equal. Only the value of $r_5^{(1)}$ is different from the others and is equal to the steady-state value r_5.

The distribution of the policyholders in the -1/top scale after two years is then given by

$$
\begin{pmatrix} p_0^{(2)}(\vartheta) \\ p_1^{(2)}(\vartheta) \\ p_2^{(2)}(\vartheta) \\ p_3^{(2)}(\vartheta) \\ p_4^{(2)}(\vartheta) \\ p_5^{(2)}(\vartheta) \end{pmatrix}
= \begin{pmatrix} p_0^{(0)} \\ p_1^{(0)} \\ p_2^{(0)} \\ p_3^{(0)} \\ p_4^{(0)} \\ p_5^{(0)} \end{pmatrix}^T
\begin{pmatrix}
\exp(-2\vartheta) & 0 & 0 & 0 & \exp(-\vartheta)(1-\exp(-\vartheta)) & 1-\exp(-\vartheta) \\
\exp(-2\vartheta) & 0 & 0 & 0 & \exp(-\vartheta)(1-\exp(-\vartheta)) & 1-\exp(-\vartheta) \\
\exp(-2\vartheta) & 0 & 0 & 0 & \exp(-\vartheta)(1-\exp(-\vartheta)) & 1-\exp(-\vartheta) \\
0 & \exp(-2\vartheta) & 0 & 0 & \exp(-\vartheta)(1-\exp(-\vartheta)) & 1-\exp(-\vartheta) \\
0 & 0 & \exp(-2\vartheta) & 0 & \exp(-\vartheta)(1-\exp(-\vartheta)) & 1-\exp(-\vartheta) \\
0 & 0 & 0 & \exp(-2\vartheta) & \exp(-\vartheta)(1-\exp(-\vartheta)) & 1-\exp(-\vartheta)
\end{pmatrix}
$$

$$
= \begin{pmatrix}
(p_0^{(0)} + p_1^{(0)} + p_2^{(0)}) \exp(-2\vartheta) \\
p_3^{(0)} \exp(-2\vartheta) \\
p_4^{(0)} \exp(-2\vartheta) \\
p_5^{(0)} \exp(-2\vartheta) \\
\exp(-\vartheta)(1-\exp(-\vartheta)) \\
1-\exp(-\vartheta)
\end{pmatrix}.
$$

The corresponding relativities are obtained by

$$\begin{pmatrix} r_0^{(2)} \\ r_1^{(2)} \\ r_2^{(2)} \\ r_3^{(2)} \\ r_4^{(2)} \\ r_5^{(2)} \end{pmatrix} = \begin{pmatrix} \frac{\sum_k w_k \int_0^{+\infty} \theta(p_0^{(0)} + p_1^{(0)} + p_2^{(0)}) \exp(-2\lambda_k\theta) dF_\Theta(\theta)}{\sum_k w_k \int_0^{+\infty} (p_0^{(0)} + p_1^{(0)} + p_2^{(0)}) \exp(-2\lambda_k\theta) dF_\Theta(\theta)} \\ \frac{\sum_k w_k \int_0^{+\infty} \theta p_3^{(0)} \exp(-2\lambda_k\theta) dF_\Theta(\theta)}{\sum_k w_k \int_0^{+\infty} p_3^{(0)} \exp(-2\lambda_k\theta) dF_\Theta(\theta)} \\ \frac{\sum_k w_k \int_0^{+\infty} \theta p_4^{(0)} \exp(-2\lambda_k\theta) dF_\Theta(\theta)}{\sum_k w_k \int_0^{+\infty} p_4^{(0)} \exp(-2\lambda_k\theta) dF_\Theta(\theta)} \\ \frac{\sum_k w_k \int_0^{+\infty} \theta p_5^{(0)} \exp(-2\lambda_k\theta) dF_\Theta(\theta)}{\sum_k w_k \int_0^{+\infty} p_5^{(0)} \exp(-2\lambda_k\theta) dF_\Theta(\theta)} \\ \frac{\sum_k w_k \int_0^{+\infty} \theta \exp(-\lambda_k\theta)(1-\exp(-\lambda_k\theta)) dF_\Theta(\theta)}{\sum_k w_k \int_0^{+\infty} \exp(-\lambda_k\theta)(1-\exp(-\lambda_k\theta)) dF_\Theta(\theta)} \\ \frac{\sum_k w_k \int_0^{+\infty} \theta(1-\exp(-\lambda_k\theta)) dF_\Theta(\theta)}{\sum_k w_k \int_0^{+\infty} (1-\exp(-\lambda_k\theta)) dF_\Theta(\theta)} \end{pmatrix}$$

$$= \begin{pmatrix} \frac{\sum_k w_k \int_0^{+\infty} \theta \exp(-2\lambda_k\theta) dF_\Theta(\theta)}{\sum_k w_k \int_0^{+\infty} \exp(-2\lambda_k\theta) dF_\Theta(\theta)} \\ \frac{\sum_k w_k \int_0^{+\infty} \theta \exp(-2\lambda_k\theta) dF_\Theta(\theta)}{\sum_k w_k \int_0^{+\infty} \exp(-2\lambda_k\theta) dF_\Theta(\theta)} \\ \frac{\sum_k w_k \int_0^{+\infty} \theta \exp(-2\lambda_k\theta) dF_\Theta(\theta)}{\sum_k w_k \int_0^{+\infty} \exp(-2\lambda_k\theta) dF_\Theta(\theta)} \\ \frac{\sum_k w_k \int_0^{+\infty} \theta \exp(-2\lambda_k\theta) dF_\Theta(\theta)}{\sum_k w_k \int_0^{+\infty} \exp(-2\lambda_k\theta) dF_\Theta(\theta)} \\ \frac{\sum_k w_k \int_0^{+\infty} \theta \exp(-\lambda_k\theta)(1-\exp(-\lambda_k\theta)) dF_\Theta(\theta)}{\sum_k w_k \int_0^{+\infty} \exp(-\lambda_k\theta)(1-\exp(-\lambda_k\theta)) dF_\Theta(\theta)} \\ \frac{\sum_k w_k \int_0^{+\infty} \theta(1-\exp(-\lambda_k\theta)) dF_\Theta(\theta)}{\sum_k w_k \int_0^{+\infty} (1-\exp(-\lambda_k\theta)) dF_\Theta(\theta)} \end{pmatrix}.$$

Once again, we can see that the relativities at time 2 do not depend on the initial distribution. Now, $r_4^{(2)}$ and $r_5^{(2)}$ are equal to the stationary relativities r_4 and r_5, respectively, and the values $r_0^{(2)}$ to $r_3^{(2)}$ are equal. Similar expressions can be computed for time 3, 4 and 5 to show that the transient relativities do not depend on the initial distribution.

The relativities $\bar{r}_\ell$ are displayed in Table 8.10 for the three initial distributions assuming a uniform distribution of age of policy $a_n = 1/5$, for $n = 1$ to 5 . The relativies computed from the bottom initial distribution are close to the steady state relativities given in the last column (except for level 0) whereas the relativities computed from a uniform or a top initial distribution are weaker than the stationary relativities for levels 1 to 5. So we see that the initial distribution can have a great influence on the resulting $\bar{r}_\ell$s.

Table 8.10 Relativies computed on the basis of the transient maximum accuracy criterion for the different initial distributions (−1/top scale).

Level ℓ	Uniform distribution $\bar{r}_\ell$	Top distribution $\bar{r}_\ell$	Bottom distribution $\bar{r}_\ell$	Steady state distribution r_ℓ
5	181.2 %	181.2 %	181.2 %	181.2 %
4	140.3 %	111.1 %	158.3 %	159.9 %
3	111.4 %	94.6 %	139.9 %	143.9 %
2	92.8 %	81.5 %	123.3 %	131.2 %
1	81.5 %	70.9 %	105.1 %	120.9 %
0	71.1 %	63.2 %	74.6 %	61.2 %

Influence of the Maturity of the Portfolio

Let us now examine the influence of the maturity of the portfolio. To this end, let us consider the three different distributions of the age of the policies ($\Pr[A = n] = a_n$) that are displayed in the following table:

Age of policy	Mature portfolio	Young portfolio	Old portfolio
1	20 %	30 %	10 %
2	20 %	25 %	15 %
3	20 %	20 %	20 %
4	20 %	15 %	25 %
5	20 %	10 %	30 %

The first one is a uniform distribution which represents a mature portfolio. The second distribution represents a relatively young portfolio or a portfolio which is growing, i.e. $\Pr[A = n]$ decreases with n. Finally, the third distribution represents an old portfolio or a portfolio which is declining, i.e. $\Pr[A = n]$ increases with n.

Asssuming a uniform initial distribution of the policyholders in the scale, we get the $\bar{r}_\ell$s displayed in Table 8.11.

Table 8.11 Relativies computed on the basis of the transient maximum accuracy criterion for different maturities of the portfolio (-1/top scale).

Level ℓ	Mature portfolio $\bar{r}_\ell$	Young portfolio $\bar{r}_\ell$	Old portfolio $\bar{r}_\ell$	Steady state distribution r_ℓ
5	181.2 %	181.2 %	181.2 %	181.2 %
4	140.3 %	131.8 %	149.6 %	159.9 %
3	111.4 %	103.0 %	121.4 %	143.9 %
2	92.8 %	88.3 %	98.4 %	131.2 %
1	81.5 %	81.1 %	81.9 %	120.9 %
0	71.1 %	74.4 %	68.3 %	61.2 %

Financial Balance

Finally, we compare the evolution of the expected financial income in two different cases. First, Table 8.12 presents the evolution of I_n (computed using (8.7)) when three different initial distributions are used. We see that, in each situation, the financial balance is reached after 5 years (the time needed to reach the steady state). We notice that the uniform and the top initial distribution ensure profit in the first years whereas the bottom initial distribution causes losses in the first years. Too many policyholders (with respect to the steady state situation) are in the malus levels of the scale in the first two cases whereas too many policyholders are in the bonus level in the last case.

Table 8.12 also gives the evolution of the expected financial income $\bar{I}_n$ computed using (8.8). The varying parameter is the distribution of the age of the policies. We see that this parameter has a little influence on the results. The most interesting point to notice is that the expected financial income does not converge to 100 % but goes down under 100 %. This is the result of the use of the $\bar{r}_\ell$s.

Choice of the Initial Level

On the basis of the concepts presented in Section 8.3.4, we now try to find the most efficient initial level for the −1/top scale. The procedure can be summarized as follows: for each initial distribution $\boldsymbol{p}^{(0)} = \boldsymbol{e}_k$ (where $\boldsymbol{e}_k$ is the vector with 0.95 in the kth entry and 0.01 elsewhere), we compute the relativities $\bar{r}_\ell$ with the help of (8.6), as well as the $\bar{Q}$-efficiency of each solution. The optimal initial level is then the one which maximises the $\bar{Q}$-efficiency.

Table 8.12 Evolution of the expected financial income I_n based on the r_ℓs and $\bar{I}_n$ based on the $\bar{r}_\ell$s.

n	Uniform distribution I_n	Top distribution I_n	Bottom distribution I_n
0	133.0 %	178.3 %	65.5 %
1	121.7 %	160.1 %	79.2 %
2	113.5 %	148.0 %	87.7 %
3	107.5 %	139.3 %	93.4 %
4	103.2 %	132.9 %	97.2 %
5	100.0 %	100.0 %	100.0 %
n	Mature portfolio $\bar{I}_n$	Young portfolio $\bar{I}_n$	Old portfolio $\bar{I}_n$
0	113.1 %	110.0 %	116.8 %
1	105.7 %	103.5 %	108.6 %
2	101.2 %	100.0 %	103.1 %
3	98.7 %	98.2 %	99.8 %
4	97.5 %	97.4 %	98.1 %
5	96.9 %	97.0 %	97.3 %

Table 8.13 Choice of the initial class for the −1/top scale.

Starting level	Level of the resulting scale						Efficiency $\bar{e}$
	5	4	3	2	1	0	
5	181.2 %	11.1 %	94.6 %	81.5 %	70.9 %	63.2 %	1.1231
4	181.2 %	158.3 %	100.2 %	86.7 %	75.7 %	64.5 %	1.155
3	181.2 %	158.3 %	139.9 %	93.6 %	81.8 %	67.1 %	1.1671
2	181.2 %	158.3 %	139.9 %	123.3 %	89.7 %	70.4 %	1.1692
1	181.2 %	158.3 %	139.9 %	123.3 %	105.1 %	74.6 %	1.1657
0	181.2 %	158.3 %	139.9 %	123.3 %	105.1 %	74.6 %	1.1657

The results are given in Table 8.13. We conclude that level $\ell = 2$ is the $\bar{Q}$-optimal initial level whereas level 5 is the usual starting level of this scale. The evolution of the expected financial income I_n when starting in level 2 is as follows:

$$
\begin{aligned}
I_0 &= 131.4\,\% \\
I_1 &= 128.2\,\% \\
I_2 &= 87.7\,\% \\
I_3 &= 93.4\,\% \\
I_4 &= 97.2\,\% \\
I_5 &= 100\,\%.
\end{aligned}
$$

8.5.2 −1/+2 Scale

The transition rules of the $-1/+2$ bonus-malus scale are given in Table 4.2.

Initial Distribution

As for the −1/top scale, we have tested three different initial distributions of the policyholders in the scale. In the first case (uniform distribution), the policyholders are uniformly spread in the scale (16.67 % of the policyholders in each of the six levels). In the second case (top distribution), 95 % of the policyholders are concentrated in the top of the scale (and the remaining 5 % are spread over levels 0 to 4). In the last case (bottom distribution), 95 % of the policyholders start from the bottom of the scale.

Convergence of the −1/+2 Scale

Figure 8.3 shows the convergence of the $-1/+2$ scale by plotting the evolution of C_n. The convergence is rather fast during the first five years and then slows down. The level $\epsilon = 0.05$ is reached after about eight years ($C_8 < \epsilon = 0.05$).

Transient Relativities

The three types of initial distribution (uniform, top and bottom) of the policyholders in the scale have been considered. Table 8.14 gives the transient relativities. The last column of this table indicates the steady state relativities.

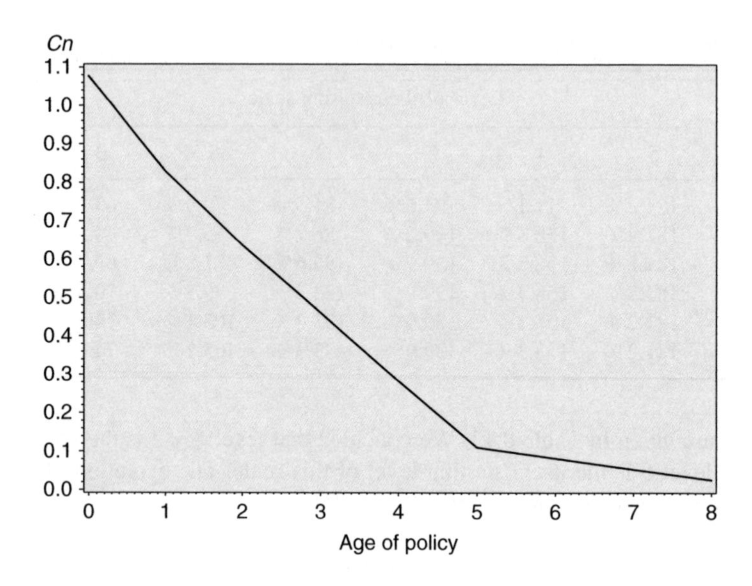

Figure 8.3 Convergence of the transient policyholders' distributions to the steady state distribution for the system $-1/+2$.

The corresponding relativies $\bar{r}_\ell$ are presented in Table 8.15 for a uniform distribution of age of policy. The last column gives the steady state relativities. We notice that the relativities are more severe when the initial distribution is closer to the bottom initial distribution. This seems reasonable since, with the bottom initial distribution, many policyholders are in the bonus level from the beginning. Therefore, the bonus must be weaker and the maluses stronger to ensure the financial balance. As for the -1/top bonus-malus scale, the relativities obtained starting from the bottom initial distribution are the closest to the stationary relativities.

Influence of the Maturity of the Portfolio
Let us consider the following three distributions of age of policy:

Age of policy	Mature portfolio	Young portfolio	Old portfolio
1	12.5 %	25 %	5 %
2	12.5 %	20 %	5 %
3	12.5 %	15 %	10 %
4	12.5 %	10 %	10 %
5	12.5 %	10 %	10 %
6	12.5 %	10 %	15 %
7	12.5 %	5 %	20 %
8	12.5 %	5 %	25 %

Table 8.14 Evolution of the transient relativities for the different initial distributions (−1/+2 scale).

Level ℓ	Uniform initial distribution						
	$r_\ell^{(0)}$	$r_\ell^{(1)}$	$r_\ell^{(2)}$	$r_\ell^{(5)}$	$r_\ell^{(7)}$	$r_\ell^{(8)}$	r_ℓ
5	0.0 %	188.6 %	204.8 %	250.3 %	263.1 %	265.2 %	271.4 %
4	0.0 %	99.4 %	167.3 %	203.1 %	208.8 %	214.2 %	218.5 %
3	0.0 %	97.2 %	93.5 %	158.6 %	182.1 %	183.8 %	192.5 %
2	0.0 %	97.2 %	97.1 %	132.3 %	137.9 %	137.8 %	138.8 %
1	0.0 %	88.2 %	86.3 %	122.2 %	121.0 %	127.6 %	128.6 %
0	0.0 %	88.2 %	79.2 %	67.4 %	67.8 %	68.0 %	68.5 %
Level ℓ	Top initial distribution						
	$r_\ell^{(0)}$	$r_\ell^{(1)}$	$r_\ell^{(2)}$	$r_\ell^{(5)}$	$r_\ell^{(7)}$	$r_\ell^{(8)}$	r_ℓ
5	0.0 %	181.5 %	182.0 %	227.8 %	254.6 %	254.7 %	271.4 %
4	0.0 %	88.3 %	160.2 %	198.7 %	188.8 %	212.6 %	218.5 %
3	0.0 %	97.2 %	79.4 %	129.2 %	177.2 %	166.1 %	192.5 %
2	0.0 %	97.2 %	97.1 %	131.3 %	135.8 %	135.8 %	138.8 %
1	0.0 %	88.2 %	86.3 %	121.0 %	108.2 %	125.5 %	128.6 %
0	0.0 %	88.2 %	79.2 %	61.6 %	61.6 %	66.0 %	68.5 %
Level ℓ	Bottom initial distribution						
	$r_\ell^{(0)}$	$r_\ell^{(1)}$	$r_\ell^{(2)}$	$r_\ell^{(5)}$	$r_\ell^{(7)}$	$r_\ell^{(8)}$	r_ℓ
5	0.0 %	242.0 %	281.7 %	276.4 %	273.5 %	273.4 %	271.4 %
4	0.0 %	185.0 %	215.4 %	217.8 %	218.0 %	218.3 %	218.5 %
3	0.0 %	97.2 %	155.4 %	188.7 %	191.0 %	191.5 %	192.5 %
2	0.0 %	162.7 %	144.8 %	140.8 %	141.5 %	139.4 %	138.8 %
1	0.0 %	88.2 %	144.7 %	134.9 %	128.1 %	130.7 %	128.6 %
0	0.0 %	88.2 %	79.2 %	72.4 %	70.8 %	69.9 %	68.5 %

Table 8.15 Relativities computed on the basis of the transient maximum accuracy criterion (influence of the initial distribution).

Level ℓ	Uniform distribution $\bar{r}_\ell$	Top distribution $\bar{r}_\ell$	Bottom distribution $\bar{r}_\ell$	Steady state distribution r_ℓ
5	232.6 %	207.1 %	274.9 %	271.4 %
4	164.3 %	123.1 %	214.0 %	218.5 %
3	131.8 %	107.4 %	184.7 %	192.5 %
2	114.5 %	95.2 %	144.7 %	138.8 %
1	99.4 %	83.9 %	131.7 %	128.6 %
0	71.0 %	63.4 %	75.9 %	68.5 %

Table 8.16 Relativities computed on the basis of the transient maximum accuracy criterion (influence of the maturity of the portfolio).

Level ℓ	Mature portfolio $\bar{r}_\ell$	Young portfolio $\bar{r}_\ell$	Old portfolio $\bar{r}_\ell$	Steady state distribution r_ℓ
5	232.6 %	218.6 %	245.9 %	271.4 %
4	164.3 %	144.0 %	184.6 %	218.5 %
3	131.8 %	116.4 %	152.1 %	192.5 %
2	114.5 %	106.6 %	122.7 %	138.8 %
1	99.4 %	93.8 %	107.0 %	128.6 %
0	71.0 %	73.8 %	69.3 %	68.5 %

The influence of the maturity of the portfolio is illustrated in Table 8.16. We notice that the relativities are more severe when the age of the portfolio is greater.

Choice of the Initial Level

We have computed the efficiency with different starting levels. Table 8.17 indicates that level 2 is the $\bar{Q}$-optimal level ($\bar{Q}$-efficiency of 1.2330). The evolution of the expected financial income when using an initial level $\ell = 2$ is as follows:

$$I_0 = 140.6\,\%$$
$$I_1 = 141.3\,\%$$
$$I_2 = 101.9\,\%$$
$$I_3 = 100.9\,\%$$
$$I_4 = 104.4\,\%$$
$$I_5 = 100.0\,\%$$
$$I_6 = 100.0\,\%$$
$$I_7 = 100.9\,\%$$
$$I_8 = 99.9\,\%.$$

Table 8.17 Choice of the initial class for the $-1/+2$ scale.

Starting level	Level of the resulting scale						Efficiency $\bar{e}$
	5	4	3	2	1	0	
5	207.1 %	123.1 %	107.4 %	95.2 %	83.9 %	63.4 %	1.1532
4	216.7 %	177.4 %	111.8 %	100.2 %	87.8 %	65.1 %	1.1942
3	231.3 %	185.1 %	158.2 %	105.9 %	94.1 %	67.5 %	1.2187
2	258.2 %	196.9 %	167.5 %	134.9 %	100.9 %	71.1 %	1.2330
1	267.5 %	210.2 %	177.7 %	140.4 %	127.6 %	74.9 %	1.2321
0	274.9 %	214.0 %	184.7 %	144.7 %	131.7 %	75.9 %	1.2251

8.6 Super Bonus Level

8.6.1 Mechanism

As explained above, the companies operating in Belgium continue to use the former compulsory bonus-malus scale, despite the deregulation of the *a posteriori* corrections. Slight modifications have nevertheless been brought to the system. For marketing purposes, and to keep the best drivers in the portfolio, many insurance companies operating in Belgium have allowed the insured drivers reaching level 0 to 'claim for free': whatever the number of accidents reported to the company, they stay in level 0. This makes the stationary distribution degenerate: all its probability mass is concentrated at 0. Therefore, computations must be based on the transient distribution.

The transition probabilities for the Belgian bonus-malus scale with a super bonus level 0 are the same as before, except for the line corresponding to level 0, which is replaced with a line of 0s and a single 1 on the diagonal.

8.6.2 Initial Distributions

We use the following initial distributions: the uniform distribution (1/23 of the portfolio in each level), the top distribution (78 % of the policyholders are concentrated in level 22 and the remaining 22 % are spread over levels 0 to 21) and the steady-state distribution displayed in Table 8.1. Starting with the stationary distribution gives an idea of the influence of the introduction of a super bonus level in an existing bonus-malus scale.

8.6.3 Transient Relativities

The results for the uniform initial distribution are given in Tables 8.18 (transient distributions) and 8.19 (corresponding relativities). We note that after 50 years, the majority of the policyholders (76.0 %) are in the super bonus level. Most of the other policyholders are in the upper levels. They correspond to the very bad drivers.

Concerning the transient relativities, we see that the super bonus level has a greater relativity than level 1 during the first 20 years. This side-effect of the introduction of a super bonus level is undesirable.

Table 8.20 presents the transient relativities for the top initial distribution. The relativities also show particular patterns as policyholders principally move by clusters in the scale. For example, the relativities of levels 0 to 2 are not ordered.

The transient relativities for the steady-state initial distribution are given in Table 8.21. Once again, the transient relativities are not always in ascending order.

Table 8.22 shows the $\bar{r}_\ell$s for the three initial distributions. The results for the steady-state initial distribution are not in ascending order. Indeed, the relativity associated with the super bonus level is larger than the relativities associated with levels 1 to 5.

Therefore, we see that the introduction of a super bonus level in an actual scale leads to some undesirable side-effects: concentration of the majority of the policyholders in the super bonus level after a few years and optimal relativities no longer ordered for the lowest levels (these side-effects could be overcome with the introduction of a linear scale).

Table 8.18 Evolution of the uniform initial distribution for the former Belgian compulsory bonus-malus scale with a super bonus level 0.

Level ℓ	$\Pr[L^{(0)} = \ell]$	$\Pr[L^{(10)} = \ell]$	$\Pr[L^{(20)} = \ell]$	$\Pr[L^{(30)} = \ell]$	$\Pr[L^{(40)} = \ell]$	$\Pr[L^{(50)} = \ell]$
22	4.35 %	4.43 %	4.65 %	4.62 %	4.54 %	4.46 %
21	4.35 %	3.28 %	3.34 %	3.27 %	3.18 %	3.09 %
20	4.35 %	2.68 %	2.59 %	2.48 %	2.38 %	2.30 %
19	4.35 %	2.39 %	2.15 %	2.00 %	1.89 %	1.80 %
18	4.35 %	2.25 %	1.88 %	1.69 %	1.57 %	1.47 %
17	4.35 %	2.45 %	1.75 %	1.50 %	1.35 %	1.24 %
16	4.35 %	2.48 %	1.64 %	1.35 %	1.19 %	1.07 %
15	4.35 %	2.51 %	1.57 %	1.24 %	1.06 %	0.95 %
14	4.35 %	2.59 %	1.55 %	1.18 %	0.98 %	0.85 %
13	4.35 %	2.65 %	1.54 %	1.13 %	0.91 %	0.78 %
12	4.35 %	3.92 %	1.68 %	1.13 %	0.88 %	0.73 %
11	4.35 %	3.85 %	1.68 %	1.10 %	0.83 %	0.68 %
10	4.35 %	3.76 %	1.68 %	1.07 %	0.79 %	0.63 %
9	4.35 %	3.72 %	1.69 %	1.05 %	0.76 %	0.59 %
8	4.35 %	3.65 %	1.69 %	1.02 %	0.72 %	0.55 %
7	4.35 %	3.55 %	2.07 %	1.09 %	0.72 %	0.53 %
6	4.35 %	3.44 %	2.05 %	1.06 %	0.68 %	0.49 %
5	4.35 %	3.30 %	2.01 %	1.03 %	0.64 %	0.46 %
4	4.35 %	3.08 %	1.86 %	0.93 %	0.57 %	0.40 %
3	4.35 %	2.87 %	1.72 %	0.84 %	0.50 %	0.34 %
2	4.35 %	2.67 %	2.49 %	0.93 %	0.49 %	0.32 %
1	4.35 %	2.49 %	2.33 %	0.85 %	0.44 %	0.28 %
0	4.35 %	31.98 %	54.39 %	67.44 %	72.93 %	76.00 %

Table 8.19 Evolution of the transient relativities for the uniform initial distribution for the former Belgian compulsory bonus-malus scale with a super bonus level 0.

Level ℓ	$r_\ell^{(0)}$	$r_\ell^{(10)}$	$r_\ell^{(20)}$	$r_\ell^{(30)}$	$r_\ell^{(40)}$	$r_\ell^{(50)}$
22	0.00 %	262.07 %	269.76 %	273.44 %	276.38 %	278.89 %
21	0.00 %	227.08 %	241.06 %	246.76 %	250.73 %	253.93 %
20	0.00 %	199.13 %	218.22 %	225.94 %	230.96 %	234.83 %
19	0.00 %	177.97 %	199.38 %	208.75 %	214.76 %	219.29 %
18	0.00 %	162.38 %	183.85 %	194.32 %	201.15 %	206.28 %
17	0.00 %	138.81 %	168.34 %	181.09 %	189.09 %	194.95 %
16	0.00 %	128.85 %	156.85 %	170.26 %	178.91 %	185.29 %
15	0.00 %	120.52 %	146.93 %	160.81 %	170.03 %	176.85 %
14	0.00 %	115.68 %	139.38 %	152.89 %	162.30 %	169.41 %
13	0.00 %	111.55 %	132.99 %	146.06 %	155.55 %	162.88 %

12	0.00 %	82.73 %	120.51 %	137.14 %	148.11 %	156.27 %
11	0.00 %	80.93 %	114.88 %	131.32 %	142.48 %	150.89 %
10	0.00 %	78.45 %	109.51 %	125.89 %	137.30 %	145.97 %
9	0.00 %	77.57 %	105.70 %	121.57 %	132.89 %	141.64 %
8	0.00 %	76.12 %	101.98 %	117.57 %	128.88 %	137.71 %
7	0.00 %	74.13 %	86.95 %	108.67 %	122.45 %	132.53 %
6	0.00 %	71.63 %	83.34 %	104.71 %	118.64 %	128.91 %
5	0.00 %	68.63 %	79.71 %	100.83 %	114.95 %	125.44 %
4	0.00 %	65.99 %	76.72 %	97.77 %	111.96 %	122.53 %
3	0.00 %	63.20 %	73.69 %	94.74 %	109.04 %	119.72 %
2	0.00 %	60.26 %	53.55 %	84.00 %	102.12 %	114.65 %
1	0.00 %	57.18 %	51.08 %	80.88 %	99.06 %	111.76 %
0	0.00 %	71.63 %	63.14 %	62.93 %	64.76 %	66.30 %

Table 8.20 Evolution of the transient relativities for the top initial distribution for the former Belgian compulsory bonus-malus scale with a super bonus level 0.

Level ℓ	$r_\ell^{(0)}$	$r_\ell^{(10)}$	$r_\ell^{(20)}$	$r_\ell^{(30)}$	$r_\ell^{(40)}$	$r_\ell^{(50)}$
22	0.00 %	222.21 %	248.39 %	258.92 %	264.90 %	269.02 %
21	0.00 %	194.77 %	221.25 %	232.80 %	239.52 %	244.20 %
20	0.00 %	174.51 %	200.74 %	212.98 %	220.29 %	225.44 %
19	0.00 %	158.98 %	184.55 %	197.19 %	204.95 %	210.49 %
18	0.00 %	146.72 %	171.43 %	184.26 %	192.36 %	198.21 %
17	0.00 %	94.41 %	132.50 %	154.14 %	168.27 %	178.47 %
16	0.00 %	108.65 %	142.42 %	159.20 %	169.47 %	176.75 %
15	0.00 %	102.37 %	134.64 %	151.30 %	161.70 %	169.14 %
14	0.00 %	97.45 %	128.08 %	144.50 %	154.98 %	162.56 %
13	0.00 %	93.22 %	122.35 %	138.56 %	149.11 %	156.81 %
12	0.00 %	45.84 %	88.39 %	111.05 %	126.23 %	137.79 %
11	0.00 %	80.93 %	100.89 %	121.77 %	134.74 %	143.91 %
10	0.00 %	78.45 %	96.79 %	117.41 %	130.47 %	139.76 %
9	0.00 %	77.57 %	93.31 %	113.57 %	126.67 %	136.05 %
8	0.00 %	76.12 %	90.05 %	110.05 %	123.21 %	132.70 %
7	0.00 %	74.13 %	58.11 %	85.28 %	102.42 %	114.95 %
6	0.00 %	71.63 %	70.05 %	95.00 %	111.25 %	122.56 %
5	0.00 %	68.63 %	67.53 %	92.07 %	108.38 %	119.80 %
4	0.00 %	65.99 %	65.07 %	89.52 %	105.88 %	117.39 %
3	0.00 %	63.20 %	62.70 %	87.07 %	103.48 %	115.08 %
2	0.00 %	60.26 %	30.03 %	63.63 %	84.04 %	98.31 %
1	0.00 %	57.18 %	51.08 %	71.54 %	91.75 %	105.60 %
0	0.00 %	71.63 %	63.14 %	45.42 %	50.95 %	54.87 %

Table 8.21 Evolution of the transient relativities for the steady-state initial distribution for the former Belgian compulsory bonus-malus scale with a super bonus level 0.

Level ℓ	$r_\ell^{(0)}$	$r_\ell^{(10)}$	$r_\ell^{(20)}$	$r_\ell^{(30)}$	$r_\ell^{(40)}$	$r_\ell^{(50)}$
22	0.00 %	257.93 %	269.51 %	274.15 %	277.44 %	280.11 %
21	0.00 %	220.81 %	239.92 %	246.91 %	251.34 %	254.73 %
20	0.00 %	192.52 %	216.80 %	225.92 %	231.39 %	235.43 %
19	0.00 %	174.66 %	197.88 %	208.13 %	214.60 %	219.40 %
18	0.00 %	160.52 %	182.99 %	193.97 %	201.07 %	206.36 %
17	0.00 %	124.72 %	162.58 %	178.29 %	187.56 %	194.09 %
16	0.00 %	117.95 %	151.98 %	167.93 %	177.67 %	184.58 %
15	0.00 %	111.34 %	142.53 %	158.79 %	169.04 %	176.34 %
14	0.00 %	115.43 %	139.02 %	152.24 %	161.58 %	168.76 %
13	0.00 %	111.81 %	132.87 %	145.80 %	155.21 %	162.52 %
12	0.00 %	66.21 %	111.64 %	132.00 %	144.79 %	154.00 %
11	0.00 %	68.49 %	107.89 %	127.22 %	139.89 %	149.15 %
10	0.00 %	68.99 %	103.57 %	122.36 %	135.14 %	144.58 %
9	0.00 %	80.04 %	105.88 %	121.30 %	132.35 %	140.95 %
8	0.00 %	79.91 %	102.09 %	117.34 %	128.54 %	137.29 %
7	0.00 %	78.24 %	76.19 %	101.95 %	117.93 %	129.33 %
6	0.00 %	75.25 %	75.20 %	99.48 %	115.11 %	126.43 %
5	0.00 %	71.18 %	73.30 %	96.51 %	112.02 %	123.41 %
4	0.00 %	78.33 %	76.60 %	97.28 %	111.37 %	121.84 %
3	0.00 %	73.15 %	73.69 %	94.24 %	108.54 %	119.21 %
2	0.00 %	67.74 %	42.42 %	76.29 %	96.72 %	110.74 %
1	0.00 %	62.28 %	42.83 %	75.07 %	94.93 %	108.74 %
0	0.00 %	92.41 %	87.94 %	84.93 %	84.85 %	85.14 %

Table 8.22 Relativies computed on the basis of the transient maximum accuracy criterion for the former Belgian compulsory bonus-malus scale with a super bonus level 0.

Level ℓ	Uniform distribution $\bar{r}_\ell$	Top distribution $\bar{r}_\ell$	Steady state distribution $\bar{r}_\ell$
22	267.5 %	241.0 %	265.2 %
21	234.6 %	184.8 %	227.8 %
20	207.9 %	163.3 %	199.0 %
19	185.9 %	146.8 %	176.8 %
18	168.0 %	133.8 %	159.5 %
17	163.5 %	123.4 %	146.1 %
16	141.8 %	114.9 %	135.6 %
15	132.3 %	107.8 %	127.2 %
14	124.4 %	101.8 %	120.4 %
13	117.7 %	96.6 %	114.6 %
12	111.9 %	92.1 %	109.7 %
11	106.9 %	88.0 %	105.5 %
10	102.4 %	84.4 %	101.7 %

9	98.1 %	81.0 %	97.9 %
8	94.0 %	77.8 %	94.4 %
7	90.1 %	74.8 %	91.2 %
6	86.5 %	71.8 %	88.1 %
5	83.2 %	69.1 %	85.2 %
4	78.3 %	65.6 %	79.8 %
3	74.0 %	62.4 %	75.9 %
2	70.3 %	59.5 %	72.3 %
1	67.1 %	56.6 %	69.2 %
0	65.5 %	51.4 %	87.6 %

8.7 Further Reading and Bibliographic Notes

There are relatively few papers dealing with the study of bonus-malus scales using transient distributions. The study of bonus-malus scales with transient distributions started with the seminal paper by BORGAN, HOEM & NORBERG (1981). GILDE & SUNDT (1989) studied linear scales in a transient regime. The examples mentioned in this chapter (special scale for new entrants, and absorbing level 0) are extensively treated in DENUIT, MARÉCHAL, PITREBOIS & WALHIN (2007b).

9

Actuarial Analysis of the French Bonus-Malus System

9.1 Introduction

As discussed in the preceding chapters, bonus-malus systems usually take the form of a scale comprising a number of levels. The policyholders move inside the scale according to the number of claims they report to the insurance company. To each level of the scale is attached a relativity (that is, a percentage, or relative premium). These relativities are applied to a base premium. Usually, bonus-malus systems may be modelled through a Markov chain, which makes mathematics easy for the actuary.

France is an exception. The French law imposes on the insurers operating in France a unique bonus-malus system that is not based on a scale. Instead the French bonus-malus system uses the concept of an increase-decrease coefficient (*coefficient de réduction-majoration* in French, henceforth abbreviated as CRM). More precisely, the French bonus-malus system implies a malus of 25 % per claim and a bonus of 5 % per claim-free year. So each policyholder is assigned a base premium and this base premium is adapted according to the number of claims reported to the insurer, multiplying the premium by 1.25 each time an accident at fault is reported to the company, and by 0.95 per claim-free year. In the case of shared responsibility, the increase is reduced by half (12.5 % instead of 25 %). Note that these increases are applied to the previous relativity: the first claim causes the premium to pass from 100 to 125, the second increases the premium to 156, the third to 195, and so on (all the numbers are rounded down). The penalties are thus convex in the number of claims reported by the driver, ensuring that the more claims are reported, the heavier they are penalized. The highest percentage is 350, and the lowest is 50 (attained after 13 consecutive claim-free years). According to the French special bonus rule, after two consecutive years without a claim, the driver goes back to the initial level 100 %. This special bonus rule is particularly generous.

In 1994, the European Union decreed that all its member countries had to drop their mandatory bonus-malus systems, claiming that such systems reduced competition between insurers and were in contradiction to the total rating freedom implemented by the Third Directive. However, the mandatory French system is still in force. Quite surprisingly, the European Court of Justice decided in 2004 that the mandatory bonus-malus systems of France and the Grand Duchy of Luxembourg were not in contradiction to the rating freedom imposed by the European legislation. These two countries were thus allowed to stick to their respective uniform bonus-malus mechanisms.

In this chapter, we show that the framework of credibility theory can be used to analyse the French bonus-malus system. Specifically, the greatest accuracy credibility approach presented in Chapter 3 is adapted to fit the CRM coefficients: the actuary resorts to a quadratic loss function but the shape of the credibility predictor is constrained ex ante to the form imposed by the French law. Let us mention that the approach developed in this chapter is not the only possible method to deal with CRM coefficients. It has been shown in KELLE (2000) that the French bonus-malus system corresponds to a scale comprising several hundreds of levels (530 levels, precisely), that can be analysed in the Markovian setting of Chapter 4. The large number of states needed is due to the malus reduction in the case of claims with shared responsibility, forcing the author to consider the pair (number of claims with whole responsibility, number of claims with partial liability) to make the computation. The form of the transition matrix is somewhat intricate and we believe that the alternative developed in this chapter offers an appropriate treatment of the CRMs.

Let us now detail the contents of this chapter. In Section 9.2, we model the CRMs and we compute the parameters involved in the French bonus-malus system. We also examine whether the bonus-malus system is financially balanced or not. Some numerical applications illustrate the methodological results. Section 9.3 discusses a special rule associated with the French bonus-malus system: claims for which the policyholder is only partially liable entail a reduced penalty. The impact of this reduction is evaluated, and numerical illustrations are discussed. The final Section 9.4 concludes with bibliographic notes.

9.2 French Bonus-Malus System

9.2.1 Modelling Claim Frequencies

We adopt here the framework of the preceding chapters. Let us pick at random a policyholder from the portfolio. We denote as N_t the number of claims reported by this policyholder in period t. We assume that N_t is Poisson distributed with parameter $\lambda\Theta$ where Θ is a random effect accounting for the heterogeneity present in the portfolio. By assumption, Θ is a positive random variable that represents the annual mean frequency in the portfolio (or in the risk class in the case of a segmented tariff). Given $\Theta = \theta$, the conditional probability mass function of N_t is $\mathcal{P}oi(\lambda\theta)$. We further assume that $\mathbb{E}[\Theta] = 1$, so that $\mathbb{E}[N] = \lambda$. The heterogeneity present in the portfolio is described by a structure function. Formally, the structure function is the probability density function f_Θ of Θ. Therefore, the unconditional probability mass function of N_t is $\mathcal{MP}oi(\lambda, \Theta)$. Furthermore, the random variables $N_1, N_2, N_3, \ldots$ are assumed to be independent and identically distributed given the risk proneness Θ of the policyholder. Since Θ is unknown to the insurer, this induces serial dependence among the N_ts.

9.2.2 *Probability Generating Functions of Random Vectors*

In this chapter we will use multivariate models for counting random vectors. Specifically, let us consider random vectors $\boldsymbol{M} = (M_1, \ldots, M_n)^T$ valued in $\mathbb{N}^n$. The multivariate probability mass function of $\boldsymbol{M}$ is

$$p_{\boldsymbol{M}}(k_1, \ldots, k_n) = \Pr[M_1 = k_1, \ldots, M_n = k_n].$$

Throughout the chapter we will extensively use the multivariate extension of the probability generating function introduced in Chapter 1, which is defined as

$$\begin{aligned}\varphi_{\boldsymbol{M}}(z) &= \mathbb{E}[z_1^{M_1} \cdots z_n^{M_n}]\\ &= \sum_{k_1=0}^{\infty} \cdots \sum_{k_n=0}^{\infty} z_1^{k_1} \cdots z_n^{k_n} p_{\boldsymbol{M}}(k_1, \ldots, k_n).\end{aligned}$$

Let us now point out several interesting properties of the multivariate probability generating functions. If any function that is known to be a multivariate probability generating function for a random vector $\boldsymbol{M}$ is expanded as a power series in z, then the coefficient of $z_1^{k_1} \cdots z_n^{k_n}$ must be $p_{\boldsymbol{M}}(k_1, \ldots, k_n)$. Furthermore,

- $z \mapsto \varphi_{\boldsymbol{M}}(z, z, \ldots, z)$ is the probability generating function of $M_1 + \cdots + M_n$;
- $z \mapsto \varphi_{\boldsymbol{M}}(z, 0, \ldots, 0)$ is the probability generating function of M_1;
- $\varphi_{\boldsymbol{M}}(z_1, \ldots, z_n) = \varphi_{M_1}(z_1) \cdots \varphi_{M_n}(z_n)$ when the random variables $M_1, \ldots, M_n$ are independent.

9.2.3 *CRM Coefficients*

We will assume that the CRM coefficients only depend on the observed number of reported claims and not on their severity. Therefore the base premium is simply λ multiplied by a constant (essentially the expected cost of a claim).

Let ϵ_t be the 'reduction' coefficient and δ_t be the 'majoration' coefficient applying to a policyholder who has been covered for t years. The CRM coefficient for years 1 to t then becomes

$$r_{\delta_t,\epsilon_t}(N_\bullet, I_\bullet, t) = (1+\delta_t)^{N_\bullet}(1-\epsilon_t)^{I_\bullet}$$

with

$$N_\bullet = \sum_{j=1}^{t} N_j \text{ and } I_\bullet = \sum_{j=1}^{t} I_j, \tag{9.1}$$

where I_j is defined as

$$I_j = \begin{cases} 1 \text{ if } N_j = 0 \\ 0 \text{ if } N_j \geq 1. \end{cases}$$

In words, $N_\bullet$ is the total number of claims reported by the policyholder during the period $(0, t)$ and $I_\bullet$ is the number of years without any claim reported to the company. Note that the CRM coefficients depend on t, so that the penalties and discounts may change for every $(0, t)$ period.

To obtain the parameters ϵ_t and δ_t, we minimize the expected squared difference between the 'true' relative premium Θ and the relative premium r_{δ_t,ϵ_t} applicable to the policyholder according to the French-type bonus-malus system. More specifically, for a policyholder observed during t years, and having filed $N_1, N_2, \ldots, N_t$ claims, we aim to determine ϵ_t and δ_t so as to minimize the objective function

$$\Psi_t(\delta, \epsilon) = \mathbb{E}[(\Theta - r_{\delta,\epsilon}(N_\bullet, I_\bullet, t))]^2$$

with respect to the arguments ϵ and δ. We therefore have to solve the first order conditions

$$\frac{\partial}{\partial \delta}\Psi_t(\delta, \epsilon) = 0 \quad \text{and} \quad \frac{\partial}{\partial \epsilon}\Psi_t(\delta, \epsilon) = 0$$

which rewrites as

$$\begin{cases} \mathbb{E}\left[N_\bullet(1+\delta)^{N_\bullet-1}(1-\epsilon)^{I_\bullet}(\Theta-(1+\delta)^{N_\bullet}(1-\epsilon)^{I_\bullet})\right]=0 \\ \mathbb{E}\left[I_\bullet(1+\delta)^{N_\bullet}(1-\epsilon)^{I_\bullet-1}(\Theta-(1+\delta)^{N_\bullet}(1-\epsilon)^{I_\bullet})\right]=0 \end{cases}$$

$$\Longleftrightarrow \begin{cases} \mathbb{E}\left[\Theta N_\bullet(1+\delta)^{N_\bullet-1}(1-\epsilon)^{I_\bullet}\right]=\mathbb{E}\left[N_\bullet(1+\delta)^{2N_\bullet-1}(1-\epsilon)^{2I_\bullet}\right] \\ \mathbb{E}\left[\Theta I_\bullet(1+\delta)^{N_\bullet}(1-\epsilon)^{I_\bullet-1}\right]=\mathbb{E}\left[I_\bullet(1+\delta)^{2N_\bullet}(1-\epsilon)^{2I_\bullet-1}\right]. \end{cases} \tag{9.2}$$

9.2.4 Computation of the CRMs at Time t

Let us define the conditional probability generating function of the random couple $(N_\bullet, I_\bullet)$ given $\Theta = \theta$ as

$$\varphi(\xi_1, \xi_2|\theta) = \mathbb{E}[\xi_1^{N_\bullet}\xi_2^{I_\bullet}|\Theta = \theta].$$

The conditional independence assumption of $N_1, N_2, \ldots, N_t$ allows us to write

$$\begin{aligned} \varphi(\xi_1, \xi_2|\theta) &= \prod_{j=1}^{t}\mathbb{E}\left[\xi_1^{N_j}\xi_2^{I_j}\Big|\Theta=\theta\right] \\ &= \left(e^{-\lambda\theta}(\xi_2-1)+e^{-\lambda\theta(1-\xi_1)}\right)^t. \end{aligned}$$

We can then rewrite the system (9.2) as

$$\begin{cases} 2\mathbb{E}[\Theta\varphi^{(1,0)}(1+\delta, 1-\epsilon|\Theta)] = \mathbb{E}[\varphi_2^{(1,0)}(1+\delta, 1-\epsilon|\Theta)] \\ 2\mathbb{E}[\Theta\varphi^{(0,1)}(1+\delta, 1-\epsilon|\Theta)] = \mathbb{E}[\varphi_2^{(0,1)}(1+\delta, 1-\epsilon|\Theta)] \end{cases}$$

where

$$\varphi^{(x,y)}(a, b|\theta) = \frac{\partial^x \partial^y}{\partial s^x \partial t^y} \varphi(s, t|\theta)\Big|_{s=a, t=b}$$
$$\varphi_2^{(x,y)}(a, b|\theta) = \frac{\partial^x \partial^y}{\partial s^x \partial t^y} \varphi(s^2, t^2|\theta)\Big|_{s=a, t=b}$$

for $x, y \in \{0, 1\}$.

We can rewrite the first order conditions as

$$\int_0^\infty \theta^2 \Big(e^{\lambda\theta\delta} - \epsilon e^{-\lambda\theta}\Big)^{t-1} e^{\lambda\theta\delta} f_\Theta(\theta) d\theta$$
$$= (1+\delta) \int_0^\infty \theta \Big(e^{\lambda\theta(2\delta+\delta^2)} + e^{-\lambda\theta}(\epsilon^2 - 2\epsilon)\Big)^{t-1} e^{\lambda\theta(2\delta+\delta^2)} f_\Theta(\theta) d\theta$$

and

$$\int_0^\infty \theta \Big(e^{\lambda\theta\delta} - \epsilon e^{-\lambda\theta}\Big)^{t-1} e^{-\lambda\theta} f_\Theta(\theta) d\theta$$
$$= (1-\epsilon) \int_0^\infty \Big(e^{\lambda\theta(2\delta+\delta^2)} + e^{-\lambda\theta}(\epsilon^2 - 2\epsilon)\Big)^{t-1} e^{-\lambda\theta} f_\Theta(\theta) d\theta.$$

These equations do not possess a closed form solution as is the case for the Markovian systems studied in Chapter 3. Nevertheless, they can be solved numerically using on the one hand a numerical integration algorithm or on the other hand either an algorithm allowing us to numerically solve a system of nonlinear equations or an optimisation algorithm, depending on what type of procedure is available. In the numerical illustrations proposed in this chapter, we have used an optimisation algorithm from the SAS/IML package, trying to minimize the sum of the squared differences between the left-hand side and the right-hand side of the two equations displayed above.

9.2.5 Global CRM

Note that we have obtained so far a numerical solution for each t: minimizing $\Psi_t(\delta, \epsilon)$ with respect to δ and ϵ gives the optimal solution (δ_t, ϵ_t) for the period $(0, t)$. However we want to obtain a unique set of CRM coefficients. These may be obtained in the transient setting developed in the preceding chapter. To this end, let us introduce the age structure of the portfolio. Specifically, we denote as A the number of years the driver is covered by the company, and as $N_1, N_2, \ldots, N_A$ the annual numbers of claims reported by this policyholder. Note that A is a random variable since we work with a policyholder picked at random from the portfolio. The idea is then to determine δ and ϵ so as to minimize $\mathbb{E}[\Psi_A(\delta, \epsilon)]$. The objective function then becomes

$$\Psi(\delta, \epsilon) = \mathbb{E}[\Psi_A(\delta, \epsilon)] = \sum_{t=1}^{\infty} a_t \Psi_t(\delta, \epsilon) \text{ where } a_t = \Pr[A = t],$$

to be minimized with respect to the parameters ϵ and δ.

Some algebra immediately leads to the following system of equations to solve:

$$\sum_{t=1}^{\infty} a_t t \int_0^{\infty} \theta^2 \Big(e^{\lambda\theta\delta} - \epsilon e^{-\lambda\theta}\Big)^{t-1} e^{\lambda\theta\delta} f_\Theta(\theta) d\theta$$

$$= (1+\delta) \sum_{t=1}^{\infty} a_t t \int_0^{\infty} \theta \Big(e^{\lambda\theta(2\delta+\delta^2)} + e^{-\lambda\theta}(\epsilon^2 - 2\epsilon)\Big)^{t-1} e^{\lambda\theta(2\delta+\delta^2)} f_\Theta(\theta) d\theta$$

and

$$\sum_{t=1}^{\infty} a_t t \int_0^{\infty} \theta \Big(e^{\lambda\theta\delta} - \epsilon e^{-\lambda\theta}\Big)^{t-1} e^{-\lambda\theta} f_\Theta(\theta) d\theta$$

$$= (1-\epsilon) \sum_{t=1}^{\infty} a_t t \int_0^{\infty} \Big(e^{\lambda\theta(2\delta+\delta^2)} + e^{-\lambda\theta}(\epsilon^2 - 2\epsilon)\Big)^{t-1} e^{-\lambda\theta} f_\Theta(\theta) d\theta.$$

Again, this system does not admit any closed-form solution, but can be solved numerically (using an appropriate SAS/IML optimization algorithm).

Remark 9.1 Note that here, we have made an averaging with respect to the age structure of the portfolio. In the case where the portfolio is partitioned into a series of risk classes, an average with respect to the composition of the portfolio (in terms of classification variables) could also be performed. If some explanatory variables are correlated with A, care must be taken in the second averaging.

9.2.6 Multivariate Panjer and De Pril Recursive Formulas

Notations

In Sections 9.2.7 and 9.3.3, we will need the bivariate and trivariate extensions of the Panjer algorithm that was described in Section 7.2. The present section is devoted to the presentation of this method as well as a particular case for the sum of independent and identically distributed random vectors, known as multivariate De Pril's recursive formula.

Assume independent and identically distributed realizations of possibly dependent losses $\boldsymbol{X}_i = (X_{i1}, \ldots, X_{ik})^T$ affected by a common event, denoted by the counting variable N. For example N may count the number of hurricanes hitting the United States and the X_js may represent the cost of the hurricane in state numbered j, $j = 1, \ldots, k$. It is natural to try to obtain the distribution of the aggregate claim:

$$\boldsymbol{S} = (S_1, \ldots, S_k)^T = \left(\sum_{i=1}^{N} X_{i1}, \ldots, \sum_{i=1}^{N} X_{ik}\right)^T. \tag{9.3}$$

Even when the components of $\boldsymbol{X}$ are independent, the components of $\boldsymbol{S}$ will have some positive dependence due to the common counter N.

When N belongs to the Panjer family of counting random variables, let us show that a multivariate version of Panjer's recursive formula emerges. To this end, we will use the following notations:

- the probability mass function of $\boldsymbol{S}$ is denoted as

$$g(\boldsymbol{s}) = \Pr[S_1 = s_1, \ldots, S_k = s_k];$$

- the probability mass function of $\boldsymbol{X}$ is denoted as

$$f(\boldsymbol{x}) = \Pr[X_1 = x_1, \ldots, X_k = x_k];$$

- the difference between vectors has to be understood componentwise, that is,

$$\boldsymbol{s} - \boldsymbol{x} = (s_1 - x_1, \ldots, s_k - x_k)^T;$$

- and, finally,

$$\sum_{\boldsymbol{x} \neq \boldsymbol{0}}^{\boldsymbol{s}} f(\boldsymbol{x}) = \left(\sum_{x_1=0}^{s_1} \cdots \sum_{x_k=0}^{s_k} f(x_1, \ldots, x_k) \right) - f(0, \ldots, 0).$$

Multivariate Panjer Algorithm
We are now ready to state and prove the following result.

Property 9.1 *Let* $\boldsymbol{S}$ *be as in* (9.3) *with* $\boldsymbol{X}_i = (X_{i1}, \ldots, X_{ik})^T$, $i = 1, 2, \ldots$, *independent and identically distributed, arithmetic and independent of* N. *Furthermore, we assume that* N *belongs to Panjer's class, i.e. its probability mass function satisfies* (7.7). *Then, if* φ_N *denotes the probability generating function of* N, *we have*

$$g(\boldsymbol{0}) = \varphi_N(f(\boldsymbol{0})) \tag{9.4}$$

$$g(\boldsymbol{s}) = \frac{1}{1 - af(\boldsymbol{0})} \sum_{\boldsymbol{x} \neq \boldsymbol{0}}^{\boldsymbol{s}} \left(a + b\frac{x_i}{s_i} \right) g(\boldsymbol{s} - \boldsymbol{x}) f(\boldsymbol{x}), \quad s_i \geq 1, \quad i = 1, \ldots, k. \tag{9.5}$$

Proof Let $\varphi_{\boldsymbol{X}}(\cdot)$ and $\varphi_{\boldsymbol{S}}(\cdot)$ be the probability generating functions of $\boldsymbol{X}$ and $\boldsymbol{S}$, respectively. From

$$g(\boldsymbol{s}) = \sum_{n=0}^{\infty} \Pr[N = n] f^{\star(n)}(\boldsymbol{s}),$$

we get

$$\varphi_{\boldsymbol{S}}(\boldsymbol{u}) = \sum_{n=0}^{\infty} \Pr[N = n] \big(\varphi_{\boldsymbol{X}}(\boldsymbol{u})\big)^n$$

from which (9.4) follows immediately. By hypothesis, one has

$$n \Pr[N = n] = a(n-1) \Pr[N = n-1] + (a+b) \Pr[N = n-1], \quad n \geq 1.$$

Multiplying on both sides of the equality by $\varphi_X^{n-1}(\boldsymbol{u})u_i\partial/\partial u_i\varphi_X(\boldsymbol{u})$ and summing over $n=1$ to $+\infty$, we get

$$u_i\frac{\partial}{\partial u_i}\varphi_S(\boldsymbol{u}) = a\varphi_X(\boldsymbol{u})u_i\frac{\partial}{\partial u_i}\varphi_S(\boldsymbol{u}) + (a+b)u_i\frac{\partial}{\partial u_i}\varphi_X(\boldsymbol{u})\varphi_S(\boldsymbol{u}).$$

Comparing identical powers of $\boldsymbol{u}$ on both sides of the equality gives:

$$s_i g(\boldsymbol{s}) = a\sum_{\boldsymbol{x}=\boldsymbol{0}}^{\boldsymbol{s}} f(\boldsymbol{x})(s_i - x_i)g(\boldsymbol{s}-\boldsymbol{x}) + (a+b)\sum_{\boldsymbol{x}=\boldsymbol{0}}^{\boldsymbol{s}} f(\boldsymbol{x})x_i g(\boldsymbol{s}-\boldsymbol{x})$$

$$\Leftrightarrow s_i g(\boldsymbol{s}) = as_i f(\boldsymbol{0})g(\boldsymbol{s}) + \sum_{\boldsymbol{x}\neq\boldsymbol{0}}^{\boldsymbol{s}} f(\boldsymbol{x})g(\boldsymbol{s}-\boldsymbol{x})\big(as_i + bx_i\big)$$

$$\Leftrightarrow g(\boldsymbol{s}) = \frac{1}{1-af(\boldsymbol{0})}\sum_{\boldsymbol{x}\neq\boldsymbol{0}}^{\boldsymbol{s}} f(\boldsymbol{x})g(\boldsymbol{s}-\boldsymbol{x})\left(a + b\frac{x_i}{s_i}\right),$$

which ends the proof. □

The Panjer formula in dimension 1 immediately follows from Property 9.1 by putting $k=1$.

Multivariate De Pril Algorithm

Another interesting problem is to compute the multivariate convolution of a random vector. This is actually the result we will need in the following sections. Let $\boldsymbol{X}_i = (X_{i1}, \ldots, X_{ik})^T$, $i = 1, 2, \ldots, n$, be independent and identically distributed realizations of the random vector $\boldsymbol{X} = (X_1, \ldots, X_k)^T$. We want to obtain the distribution of the random vector

$$\boldsymbol{S} = (S_1, \ldots, S_k)^T = \left(\sum_{i=1}^{n} X_{i1}, \ldots, \sum_{i=1}^{n} X_{ik}\right)^T. \tag{9.6}$$

The probabiliy mass function of S is the n-fold convolution $f^{\star(n)}(\boldsymbol{x})$. The multivariate version of De Pril's formula provides a recursive formula to derive the distribution of $\boldsymbol{S}$.

Property 9.2 *Let $\boldsymbol{X}_i = (X_{i1}, \ldots, X_{ik})^T$ be independent and identically distributed realizations of the random vector $\boldsymbol{X}$ defined on the nonnegative integers and with probability function such that $f(\boldsymbol{0}) > 0$. Then, the following recursion holds :*

$$f^{\star(n)}(\boldsymbol{0}) = f^n(\boldsymbol{0})$$

$$f^{\star(n)}(\boldsymbol{s}) = \frac{1}{f(\boldsymbol{0})}\sum_{\boldsymbol{x}\neq\boldsymbol{0}}^{\boldsymbol{s}}\left(\frac{n+1}{s_i}x_i - 1\right)f^{\star(n)}(\boldsymbol{s}-\boldsymbol{x})f(\boldsymbol{x}), \quad s_i \geq 1, \quad i = 1, \ldots, k.$$

Proof Let us introduce an auxilliary random vector $\boldsymbol{W}$ with probability mass function $h(\boldsymbol{0}) = 0$ and

$$h(\boldsymbol{x}) = \frac{f(\boldsymbol{x})}{1-f(\boldsymbol{0})}, \quad \boldsymbol{x} > \boldsymbol{0},$$

and the auxilliary random variable $N \sim \mathcal{B}in(n, 1 - f(\mathbf{0}))$.

The probability generating function of $(W_{11} + \cdots + W_{1N}, \ldots, W_{k1} + \cdots + W_{kN})^T$ is given by

$$\Big(1 - (1 - f(\mathbf{0}))(1 - \varphi_{\boldsymbol{W}}(\boldsymbol{u}))\Big)^n = (\varphi_{\boldsymbol{X}}(\boldsymbol{u}))^n,$$

from which we conclude that $\boldsymbol{S} = (W_{11} + \cdots + W_{1N}, \ldots, W_{k1} + \cdots + W_{kN})^T$.

Applying Property 9.1 with

$$a = \frac{f(\mathbf{0}) - 1}{f(\mathbf{0})} \text{ and } b = \frac{1 - f(\mathbf{0})}{f(\mathbf{0})}(n+1),$$

we get the desired result. □

9.2.7 *Analysis of the Financial Equilibrium of the French Bonus-Malus System*

An interesting property of the relativities associated with Markovian bonus-malus systems and obtained through Norberg's least-squares criterion is that they make the bonus-malus system financially balanced, i.e. the premium income of the insurer does not increase nor decrease over time (on average). In this section, we would like to check whether or not the French-type bonus-malus system enjoys this property.

More precisely, once ϵ_t and δ_t have been obtained, we would like to verify whether $\mathbb{E}[r_{\epsilon_t,\delta_t}(N_\bullet, I_\bullet, t)]$ is equal to 1, where $N_\bullet$ and $I_\bullet$ are as defined in (9.1). The computation of $\mathbb{E}[r_{\epsilon_t,\delta_t}(N_\bullet, I_\bullet, t)]$ requires knowledge of the joint distribution of the random couple $(N_\bullet, I_\bullet)$.

Let us denote as

$$f(x, y|\theta) = \Pr[N_1 = x, I_1 = y|\Theta = \theta]$$

the joint discrete mass function of the random couple (N_1, I_1), conditional on $\Theta = \theta$, and as

$$f^{\star(t)}(x, y|\theta) = \Pr[N_\bullet = x, I_\bullet = y|\Theta = \theta]$$

the joint discrete mass function of the random couple $(N_\bullet, I_\bullet)$ defined in (9.1), conditional on $\Theta = \theta$. We then have the following result.

Property 9.3 *For fixed θ, the following recursive formulas*

$$g^{\star(t)}(x, y|\theta) = f^{\star(t)}(x, t - y|\theta) \text{ for } 0 \leq y \leq t \text{ and } x > 0$$

$$g^{\star(t)}(0, 0|\theta) = e^{-\lambda\theta t}$$

$$f(x, 0|\theta) = e^{-\lambda\theta}\frac{(\lambda\theta)^x}{x!} \text{ for } x > 0$$

$$f(0, 1|\theta) = e^{-\lambda\theta}$$

$$g^{\star(t)}(x, y|\theta) = e^{\lambda\theta}\sum_{u=1}^{x}\left(\frac{t+1}{y} - 1\right)g^{\star(t)}(x - u, y - 1|\theta)g(u, 1|\theta) \text{ for } y \geq 1 \text{ and } x \geq 1,$$

hold true, with the convention that the functions take the value 0 *where they have not been defined.*

Proof It is trivial that for $t = 1$ we have

$$f(x, 0|\theta) = e^{-\lambda\theta}\frac{(\lambda\theta)^x}{x!}, \quad x > 0$$
$$f(0, 1|\theta) = e^{-\lambda\theta}.$$

As $f^{\star(t)}$ is the t-fold convolution of a lattice random vector, we are in a position to apply the bivariate extension of De Pril's algorithm given in Property 9.2. As that algorithm needs a mass at the origin, we define an auxilliary probability mass function

$$g^{\star(t)}(x, y|\theta) = f^{\star(t)}(x, t-y|\theta).$$

We find

$$g^{\star(t)}(0, 0|\theta) = e^{-\lambda\theta t}$$
$$g^{\star(t)}(x, y|\theta) = e^{\lambda\theta}\left(\sum_{u=0}^{x}\left(\frac{t+1}{x}u - 1\right)g^{\star(t)}(x-u, y-1|\theta)g(u, 1|\theta)\right.$$
$$\left. + \sum_{u=1}^{x}\left(\frac{t+1}{x}u - 1\right)g^{\star(t)}(x-u, y|\theta)g(u, 0|\theta)\right), \quad x \geq 1$$
$$g^{\star(t)}(x, y|\theta) = e^{\lambda\theta}\left(\sum_{u=0}^{x}\left(\frac{t+1}{y} - 1\right)g^{\star(t)}(x-u, y-1|\theta)g(u, 1|\theta)\right.$$
$$\left. + \sum_{u=1}^{x}(-1)g^{\star(t)}(x-u, y|\theta)g(u, 0|\theta)\right), \quad y \geq 1.$$

Because $g(0, 1|\theta) = 0$ and $g(x, 0|\theta) = 0$ for $x > 0$, we obtain

$$g^{\star(t)}(0, 0|\theta) = e^{-\lambda\theta t}$$
$$g^{\star(t)}(x, y|\theta) = e^{\lambda\theta}\sum_{u=1}^{x}\left(\frac{t+1}{x}u - 1\right)g^{\star(t)}(x-u, y-1|\theta)g(u, 1|\theta), \quad x \geq 1$$
$$g^{\star(t)}(x, y|\theta) = e^{\lambda\theta}\sum_{u=1}^{x}\left(\frac{t+1}{y} - 1\right)g^{\star(t)}(x-u, y-1|\theta)g(u, 1|\theta), \quad y \geq 1.$$

Because $g^{\star(t)}(x, 0|\theta) = 0$ for $x > 0$, only the second recursive formula has to be used. This formula is numerically stable because $(t+1)/y - 1 > 0$. □

In order to obtain the unconditional probability mass function

$$f^{\star(t)}(x, y) = \Pr[N_\bullet = x, I_\bullet = y]$$

of $N_\bullet$ and $I_\bullet$, it suffices to integrate the conditional mass function $f^{\star(t)}(x, y|\theta)$ with respect to the structure function f_Θ, that is,

$$f^{\star(t)}(x, y) = \int_0^\infty f^{\star(t)}(x, y|\theta) f_\Theta(\theta) d\theta, \quad x > 0, \quad 0 \leq y \leq t.$$

These quantities can then be used to evaluate $\mathbb{E}[r_{\epsilon_t,\delta_t}(N_\bullet, I_\bullet, t)]$.

9.2.8 *Numerical Illustration*

We will assume that Θ is Gamma distributed with probability density function given by (1.35). The parameters a and λ are estimated on the basis of Portfolio A, that is, $\widehat{\lambda} = 0.1474$ and $\widehat{a} = 0.889$.

Table 9.1 displays, for different values of t, the coefficients δ_t and ϵ_t obtained by solving the system given in Section 9.2.4. We observe a dramatic decrease of the values of ϵ_t and δ_t over time. The last column of the table allows us to verify the financial equilibrium of the system. The total premium income first decreases to 97.61 % and then increases to 107.29 % after 30 years. The discount per claim-free year decreases from 14.23 % to about 1 %. Similarly, the penalty induced by each reported claim decreases from 61.45 % to 6.09 %. The *a posteriori* corrections are therefore considerably softened with time.

The decrease of ϵ_t and δ_t with time t that is apparent from Table 9.1 can be explained as follows: The aim is that r_{ϵ_t,δ_t} be as close as possible to the unknown risk parameter Θ. Since Θ does not depend on t whereas $N_\bullet$ and $I_\bullet$ are almost surely nondecreasing with t, the optimal parameters ϵ_t and δ_t must decrease to compensate for the increase in $N_\bullet$ and $I_\bullet$. This is why averaging over time is needed.

Table 9.2 gives the CRM coefficient $r_{\epsilon_t,\delta_t}(x, y) = (1+\delta_t)^x(1-\epsilon_t)^y$ for different periods of length t and for different values of the total number of claims x. The index $t.y$ means that we have y claimsfree years during the period $(0, t)$. For the sake of comparison, Table 9.3 gives the CRM coefficients obtained from classical Bayesian credibility. In this case, the *a priori* annual expected claim frequency is multiplied by $(a+x)/(a+t)$ as discussed in Chapter 3. We observe some large discrepancies between the values listed in Tables 9.2 and 9.3.

We see from Tables 9.2 and 9.3 that the discounts awarded to the policyholders who did not report any claim (column $x = 0$ in Tables 9.2 and 9.3) are larger with Bayesian

Table 9.1 Parameters δ_t and ϵ_t and financial equilibrium for different values of t.

t	δ_t	ϵ_t	Financial equilibrium
1	0.6145	0.1423	0.9761
2	0.4595	0.0955	0.9985
3	0.3690	0.0727	1.0001
4	0.3092	0.0589	1.0099
10	0.1585	0.0279	1.0431
20	0.0880	0.0149	1.0638
30	0.0609	0.0102	1.0729

Table 9.2 CRM coefficients $r_{\epsilon_t,\delta_t}(x,y)$ for different values of t, x and y.

$t.y$	x						
	0	1	2	3	4	5	6
1	86 %	161 %	261 %	421 %	680 %	1097 %	1771 %
2.≥1	82 %	132 %	193 %	281 %	410 %	599 %	874 %
2.0			213 %	311 %	454 %	662 %	967 %
3.≥2	80 %	118 %	161 %	221 %	302 %	414 %	566 %
3.1			174 %	238 %	326 %	446 %	610 %
3.0				257 %	351 %	481 %	658 %

Table 9.3 Premium update coefficients derived from Bayesian credibility in the Poisson-Gamma model.

t	x						
	0	1	2	3	4	5	6
1	86 %	182 %	279 %	375 %	472 %	568 %	665 %
2	75 %	160 %	244 %	329 %	413 %	498 %	582 %
3	67 %	142 %	217 %	292 %	367 %	442 %	518 %

credibility than with CRM coefficients. From the approximate financial stability evidenced in Table 9.1, the penalties induced by CRM coefficients must therefore be softer compared to Bayesian credibility. For instance, policyholders who reported a single claim (column $x = 1$ in Tables 9.2 and 9.3) have a premium surcharge ranging from 118 to 161 % with CRM coefficients, and ranging from 142 to 182 % with Bayesian credibility. However, policyholders reporting many claims are more heavily penalized with CRM coefficients than with Bayesian credibility. This comes from the convex behaviour of the CRM coefficients, whereas Bayesian credibility corrections are linear in the past number of claims. In the Poisson-Gamma (or Negative Binomial) case, the penalties corresponding to CRM coefficients are thus convex functions of the number of claims reported in the past, whereas corrections induced by credibility mechanisms are linear in this number. Compared with credibility, the bonus-malus system grants less discounts, penalizes policyholders reporting a single claim to a lesser extent but induces more severe premium corrections for those reporting at least two claims.

To obtain unique values for the CRM coefficients, we have to decide about an age structure of the policies comprised in the portfolio. Here, we take the following hypothetical distribution of the portfolio:

$$a_1 = 10\,\%$$

$$a_5 = 20\,\%$$

$$a_{12} = 30\,\%$$

$$a_{20} = 20\,\%$$

$$a_{25} = 10\,\%$$
$$a_{30} = 10\,\%$$
$$a_t = 0 \text{ for all other } t.$$

The minimization of

$$\mathbb{E}[\Psi_A(\epsilon, \delta)] = \sum_{t=1}^{\infty} a_t \Psi_t(\epsilon, \delta)$$

with respect to ϵ and δ then gives the optimal solutions

$$\delta = 0.0710 \text{ and } \epsilon = 0.0133.$$

The financial equilibrium is achieved, as the total premium income tends to 104.29 % of the initial one. When working with a weighted average of the $\Psi_t(\epsilon, \delta)$s, the values associated with large t play the prominent role, resulting in values for δ and ϵ similar to those obtained for $t > 20$ in Table 9.1.

With optimal CRM coefficients, the discount for claim-free policyholders is rather modest (1.33 % per claim-free year), but the penalty in case of a claim is also moderate (7.1 %). The large differences compared with the official values of today's bonus-malus system in France (5 % of discount per claim-free year, and 25 % increase per claim) can be explained by the fact that all the penalties are suppressed after two claim-free years according to the terms of the French law, which is particularly generous.

We have also tested two different sets of a_ts, to study the influence of the age structure of the portfolio on the optimal CRM coefficients. With the age structure of an 'old' portfolio, that is,

$$a_1 = 10\,\%$$
$$a_5 = 10\,\%$$
$$a_{12} = 10\,\%$$
$$a_{20} = 20\,\%$$
$$a_{25} = 20\,\%$$
$$a_{30} = 30\,\%$$
$$a_t = 0 \text{ for all other } t,$$

the minimization of $\mathbb{E}[\Psi_A(\epsilon, \delta)]$ gives

$$\delta = 0.0658 \text{ and } \epsilon = 0.0115.$$

With the age structure of a 'young' portfolio, that is,

$$a_1 = 30\,\%$$
$$a_5 = 20\,\%$$

$$
\begin{aligned}
a_{12} &= 20\,\% \\
a_{20} &= 10\,\% \\
a_{25} &= 10\,\% \\
a_{30} &= 10\,\% \\
a_t &= 0 \text{ for all other } t,
\end{aligned}
$$

the minimization of $\mathbb{E}[\Psi_A(\epsilon,\delta)]$ gives

$$
\delta = 0.0694 \text{ and } \epsilon = 0.0132.
$$

The influence of the age structure on the optimal CRM coefficients is thus rather moderate.

9.3 Partial Liability

9.3.1 Reduced Penalty and Modelling Claim Frequencies

The French bonus-malus system possesses many particular rules. This section is devoted to the study of one of them. Specifically, according to the terms of the French law, if the policyholder is partially liable for the claim then the premium is multiplied by 1.125 instead of 1.25. To take such a rule into account, we have to model the random couple (N_{1t}, N_{2t}) where N_{1t} counts the number of full liability claims filed during year t and N_{2t} counts the number of partial liability claims filed during the same year. Clearly, $N_{1t}+N_{2t}$ is the total number of claims N_t used in the preceding section.

Let q be the probability that the policyholder is only partially liable for the claim he files. Further, let us assume a Bernoulli scheme for the claim types. This ensures that, conditionally on Θ, N_{1t} and N_{2t} are independent and both conform to the Poisson distribution (see Property 6.1). Specifically, we have now

$$
\Pr[N_{1t}=k|\Theta=\theta] = e^{-\lambda\theta(1-q)}\frac{(\lambda\theta(1-q))^k}{k!}, \quad k=0,1,2,\ldots,
$$

$$
\Pr[N_{2t}=k|\Theta=\theta] = e^{-\lambda\theta q}\frac{(\lambda\theta q)^k}{k!}, \quad k=0,1,2,\ldots
$$

The random variables N_{1t} and N_{2t} are obviously dependent if the risk proneness Θ is unknown. The joint probability mass for the random couple (N_{1t}, N_{2t}) is given by

$$
\Pr[N_{1t}=k_1, N_{2t}=k_2] = \int_0^{+\infty} \Pr[N_{1t}=k_1|\Theta=\theta]\Pr[N_{2t}=k_2|\Theta=\theta] f_\Theta(\theta)d\theta.
$$

This is a mixed bivariate Poisson model.

9.3.2 Computations of the CRMs at Time t

Let us consider a policyholder covered for t years. In addition to the parameter δ_t giving the magnitude of the penalty in case of a full-liability claim, we introduce the new parameter γ_t

giving the reduced penalty in case of a partial liability claim. Now the CRM coefficient for the time period $(0, t)$ is

$$r_{\delta_t,\gamma_t,\epsilon_t}(N_{1\bullet}, N_{2\bullet}, I_{12\bullet}, t) = (1+\delta_t)^{N_{1\bullet}}(1+\gamma_t)^{N_{2\bullet}}(1-\epsilon_t)^{I_{12\bullet}}$$

with

$$N_{1\bullet} = \sum_{j=1}^{t} N_{1j}$$

$$N_{2\bullet} = \sum_{j=1}^{t} N_{2j}$$

$$I_{12\bullet} = \sum_{j=1}^{t} I_j \text{ with } I_j = \begin{cases} 1 \text{ if } N_{1j} = N_{2j} = 0 \\ 0 \text{ otherwise.} \end{cases}$$

We will assume that $\delta_t = \alpha\gamma_t$ with α fixed by the actuary. The value of α describes the way a claim with full liability is penalized, compared to a claim with partial liability. Then the CRM coefficient becomes

$$r_{\gamma_t,\epsilon_t}(N_{1\bullet}, N_{2\bullet}, I_{12\bullet}, t) = (1+\alpha\gamma_t)^{N_{1\bullet}}(1+\gamma_t)^{N_{2\bullet}}(1-\epsilon_t)^{I_{12\bullet}}.$$

In order to obtain ϵ_t and γ_t we have now to minimize the objective function

$$\Psi_t(\gamma, \epsilon) = \mathbb{E}\left[\left(\Theta - r_{\gamma,\epsilon}(N_{1\bullet}, N_{2\bullet}, I_{12\bullet}, t)\right)^2\right]$$

with respect to the parameters γ and ϵ. The first order conditions are

$$\begin{cases} \mathbb{E}\left[\Theta(1-\epsilon)^{I_{12\bullet}}\left(\alpha N_{1\bullet}(1+\alpha\gamma)^{N_{1\bullet}-1}(1+\gamma)^{N_{2\bullet}} + N_{2\bullet}(1+\gamma)^{N_{2\bullet}-1}(1+\alpha\gamma)^{N_{1\bullet}}\right)\right] \\ = \mathbb{E}\left[(1-\epsilon)^{2I_{12\bullet}}\left(\alpha N_{1\bullet}(1+\alpha\gamma)^{2N_{1\bullet}-1}(1+\gamma)^{2N_{2\bullet}} + N_{2\bullet}(1+\gamma)^{2N_{2\bullet}-1}(1+\alpha\gamma)^{2N_{1\bullet}}\right)\right] \\ \mathbb{E}\left[\Theta(1+\alpha\gamma)^{N_{1\bullet}}(1+\gamma)^{N_{2\bullet}} I_{12\bullet}(1-\epsilon)^{I_{12\bullet}-1}\right] \\ = \mathbb{E}\left[(1+\alpha\gamma)^{2N_{1\bullet}}(1+\gamma)^{2N_{2\bullet}} I_{12\bullet}(1-\epsilon)^{2I_{12\bullet}-1}\right]. \end{cases}$$

Let us define

$$\varphi(\xi_1, \xi_2, \xi_3|\theta) = \mathbb{E}\left[\xi_1^{N_{1\bullet}}\xi_2^{N_{2\bullet}}\xi_3^{I_{12\bullet}}\middle|\Theta = \theta\right]$$

to be the conditional probability generating function of the random vector $(N_{1\bullet}, N_{2\bullet}, I_{12\bullet})$ given $\Theta = \theta$. We clearly have that

$$\varphi(\xi_1, \xi_2, \xi_3|\theta) = \left(e^{-\lambda\theta}(\xi_3 - 1) + e^{\lambda\theta\left((1-q)\xi_1 + q\xi_2 - 1\right)}\right)^t.$$

It can be verified that the first order conditions are as follows:

$$\begin{cases} 2\mathbb{E}[\Theta\alpha\varphi^{(1,0,0)}(1+\alpha\gamma, 1+\gamma, 1-\epsilon|\Theta)+\Theta\varphi^{(0,1,0)}(1+\alpha\gamma, 1+\gamma, 1-\epsilon|\Theta)] \\ = \mathbb{E}[\alpha\varphi_2^{(1,0,0)}(1+\alpha\gamma, 1+\gamma, 1-\epsilon|\Theta)+\varphi_2^{(0,1,0)}(1+\alpha\gamma, 1+\gamma, 1-\epsilon|\Theta)] \\ 2\mathbb{E}[\Theta\varphi^{(0,0,1)}(1+\alpha\gamma, 1+\gamma, 1-\epsilon|\Theta)] = \mathbb{E}[\varphi_2^{(0,0,1)}(1+\alpha\gamma, 1+\gamma, 1-\epsilon|\Theta)] \end{cases}$$

where

$$\varphi^{(x,y,z)}(a, b, c|\theta) = \frac{\partial^x \partial^y \partial^z}{\partial s^x \partial t^y \partial u^z}\varphi(s, t, u|\theta)\Big|_{s=a,t=b,u=c}$$

$$\varphi_2^{(x,y,z)}(a, b, c|\theta) = \frac{\partial^x \partial^y \partial^z}{\partial s^x \partial t^y \partial u^z}\varphi(s^2, t^2, u^2|\theta)\Big|_{s=a,t=b,u=c}$$

for $x, y, z \in \{0, 1\}$. Again, numerical procedures are needed to find the solution of this optimization problem.

9.3.3 Financial Equilibrium

Analyzing the financial equilibrium of the system now amounts to checking whether $\mathbb{E}[r_{\gamma_t,\epsilon_t}(N_{1\bullet}, N_{2\bullet}, I_{12\bullet}, t)]$ is equal to 1 with the optimal values γ_t and ϵ_t. To this end, we need the joint distribution of the random vector $(N_{1\bullet}, N_{2\bullet}, I_{12\bullet})$. The joint probability mass function of this vector is given in the following result that extends Property 9.3 in the present setting.

Property 9.4 *For fixed θ, the following recursive formulas*

$$g^{\star(t)}(x, y, z|\theta) = f^{\star(t)}(x, y, t-z|\theta), \text{ for } 0 \leq z \leq t, \quad x, y \geq 0 \text{ and } x+y > 0$$

$$g^{\star(t)}(0, 0, 0|\theta) = e^{-\lambda\theta t}$$

$$f(x, y, 0|\theta) = e^{-\lambda\theta}\frac{(\lambda\theta)^{x+y}(1-q)^x q^y}{x!y!} \text{ for } x, y \geq 0 \text{ and } x+y > 0,$$

$$f(0, 0, 1|\theta) = e^{-\lambda\theta}$$

$$g^{\star(t)}(x, y, z|\theta) = e^{\lambda\theta}\sum_{u=0}^{x}\sum_{v=0}^{y}\left(\frac{t+1}{z}-1\right)g^{\star(t)}(x-u, y-v, z-1|\theta)g(u, v, 1|\theta),$$

$$\text{for } 1 \leq z \leq t, \quad x, y \geq 0 \text{ and } x+y > z-1$$

hold true with the convention that the defined functions take the value 0 where they have not been defined.

Proof It is trivial that for $t = 1$ we have

$$f(x, 0, z|\theta) = e^{-\lambda\theta(1+\gamma)}\frac{(\lambda\theta)^x(\lambda\theta\gamma)^z}{x!z!}, \quad x, z > 0$$

$$f(0, 1, 0|\theta) = e^{-\lambda\theta(1+\gamma)}.$$

As $f^{\star(t)}$ is the t-fold convolution of a lattice random vector, we will apply the trivariate extension of De Pril's algorithm described in Property 9.2. As this algorithm needs a mass at the origin, we define an auxilliary density function

$$g^{\star(t)}(x, y, z|\theta) = f^{\star(t)}(x, t-y, z|\theta).$$

Using similar arguments as before we obtain the following recursion

$$g^{\star(t)}(x, y, z|\theta) = e^{\lambda\theta(1+\gamma)} \sum_{u=0}^{x} \sum_{w=0}^{z} \left(\frac{t+1}{y} - 1\right) g^{\star(t)}(x-u, y-1, z-w|\theta) g(u, 1, w|\theta). \quad \square$$

9.3.4 *Numerical Illustrations*

To illustrate this special case, we use the same parameters as in Section 9.2.8. We assume that 20 % of the claims concern partial liability, that is $q = 0.2$. We numerically solve the system of two equations for different values of α. Table 9.4 gives the results. As before, γ_t and ϵ_t decrease with t and an averaging is needed to get a unique set of parameters.

The total income of the company is not much influenced by the value of α, and is quite close to the values listed in Table 9.1.

To obtain unique values for the CRM coefficients, we choose the first age distribution of Section 9.2.8. The minimization of

$$\mathbb{E}[\Psi_A(\gamma, \epsilon)] = \sum_{t=1}^{\infty} a_t \Psi_t(\gamma, \epsilon)$$

then gives the values displayed in Table 9.5. The same comments apply to this case. Specifically, the $\Psi_t(\gamma, \epsilon)$s with large values of t play the prominent role, giving optimal CRM coefficients close to the values obtained with $t > 20$ in Table 9.1.

The values of the optimal CRM coefficients displayed in Table 9.5 are again much smaller than those implemented by the French law. As before, this is due to the special bonus rule of the French system (after two consecutive years without claim, the driver goes back to the initial level of 100 %).

Table 9.4 Parameters γ_t and ϵ_t and financial equilibrium for different values of t and α.

t	$\alpha = 1.5$			$\alpha = 2.0$			$\alpha = 2.5$		
	γ_t	ϵ_t	Financial equilibrium	γ_t	ϵ_t	Financial equilibrium	γ_t	ϵ_t	Financial equilibrium
1	0.4336	0.1423	0.9747	0.3316	0.1423	0.9729	0.2674	0.1423	0.9714
2	0.3253	0.0954	0.9870	0.2498	0.0953	0.9850	0.2022	0.0953	0.9832
3	0.2616	0.0727	0.9984	0.2014	0.0726	0.9965	0.1633	0.0726	0.9946
4	0.2194	0.0589	1.0083	0.1692	0.0589	1.0062	0.1374	0.0588	1.0047
10	0.1128	0.0279	1.0419	0.0873	0.0279	1.0402	0.0711	0.0279	1.0386
20	0.0627	0.0149	1.0634	0.0486	0.0149	1.0624	0.0397	0.0149	1.0617
30	0.0435	0.0102	1.0725	0.0337	0.0102	1.0707	0.0276	0.0102	1.0712

Table 9.5 Parameters γ and ϵ and financial equilibrium for different values of α.

$\alpha = 1.5$			$\alpha = 2.0$			$\alpha = 2.5$		
γ	ϵ	Financial equilibrium	γ	ϵ	Financial equilibrium	γ	ϵ	Financial equilibrium
0.0507	0.0133	1.0419	0.0393	0.0133	1.0403	0.0321	0.0133	1.0393

9.4 Further Reading and Bibliographic Notes

This chapter is based on PITREBOIS, DENUIT & WALHIN (2006b). Despite its apparent difference with bonus-malus scales, the French bonus-malus system can be treated as a scale with many levels. KELLE (2000) followed this route, and used a Markov chain with 530 states to analyse the French system. The large amount of states needed is due to the malus reduction in the case of claims with shared responsibility, forcing the author to consider the pair (number of claims with whole responsibility, number of claims with partial liability) to make the computation.

In this chapter, we did not consider all the characteristics of the bonus-malus system in force in France. We have disregarded the special bonus rule (which suppresses all the penalties after two claim-free years). The French law imposes other specific rules on insurance companies. For instance, the French bonus-malus system is such that drivers never pay more than 350 % of the base premium nor less than 50 % of the base premium. Therefore the minimization process has to be carried with an adapted CRM coefficient of the form

$$r^*_{\epsilon,\delta} = \max(0.5, \min(3.5, r_{\epsilon,\delta})).$$

Several simplifying assumptions can be considered to ease the numerical computations. For instance, we could work with binary annual claim numbers: either the policyholder does not report any claim or he reports a single claim to the company. Such an assumption replacing N_t by $\min\{N_t, 1\}$ leads to smaller discounts and higher penalties, which is a prudent strategy for the insurer.

Even if the vast majority of bonus-malus systems appear as scales in which policyholders move according to their claims history, there are some exceptions (such as the system in force in France). We refer the reader to NEUHAUS (1988) for another example, where the malus after a claim is expressed by a fixed monetary amount (instead of a relativity). This interesting mechanism restores some fairness in case of differentiated *a priori* price lists.

The multivariate version of Panjer's recursive formula has been derived by SUNDT (1999) and AMBAGASPITYA (1999). SUNDT (1999) provided a proof based on conditional expectations whereas AMBAGASPITYA (1999) used a proof based on generating functions. SUNDT (1999) also showed that the following recursive formula can be used :

$$f_S(\boldsymbol{s}) = \frac{1}{1 - af_X(\boldsymbol{0})} \sum_{\boldsymbol{x} \neq \boldsymbol{0}}^{\boldsymbol{s}} \left(a + b\frac{s_1 + \cdots + s_k}{x_1 + \cdots + x_k} \right) f_X(\boldsymbol{x}) f_S(\boldsymbol{s} - \boldsymbol{x}), \quad \boldsymbol{x} > \boldsymbol{0}.$$

The multivariate version of De Pril's recursive formula for the convolution of independent and identically distributed random vectors has been derived by SUNDT (1999) as a particular case of the multivariate Panjer algorithm, and by WALHIN (2001) as a particular case of the multivariate version of DHAENE & VANDEBROEK'S (1995) recursive formula for the multivariate individual risk model. It can also be deduced from the multivariate extension of De Pril's methodology as shown in DICKSON & WATERS (1999).

Bibliography

ALBRECHT, P. (1983a). Parametric multiple regression risk models: Connections with tariffication, especially in motor insurance. *Insurance: Mathematics and Economics* **2**, 113–117.

ALBRECHT, P. (1983b). Parametric multiple regression risk models: Theory and statistical analysis. *Insurance: Mathematics and Economics* **2**, 49–66.

ALBRECHT, P. (1983c). Parametric multiple regression risk models: Some connections with IBNR. *Insurance: Mathematics and Economics* **2**, 69–73.

ALBRECHT, P. (1984). Laplace transforms, Mellin transforms and mixed Poisson processes. *Scandinavian Actuarial Journal*, 58–64.

ALBRECHT, P. (1985). An evolutionary credibility model for claim numbers. *ASTIN Bulletin* **15**, 1–17.

AMBAGASPITYA, R.S. (1999). On the distributions of two classes of correlated aggregate claims. *Insurance: Mathematics and Economics* **24**, 255–263.

ANDRADE E SILVA, J.M., & CENTENO, M. (2005). A note on bonus scales. *Journal of Risk and Insurance* **72**, 601–607.

ANGERS, J.-F., DESJARDINS, D., DIONNE, G., & GUERTIN, F. (2006). Vehicle and fleet random effects in a model of insurance rating for fleets of vehicles. *ASTIN Bulletin* **36**, 25–77.

ANTONIO, K., & BEIRLANT, J. (2007). Actuarial statistics with generalized linear mixed models. *Insurance: Mathematics and Economics* **40**, 58–76.

ARNOLD, B.C., CASTILLO, E., & SARABIA, J.M. (1999). *Conditional Specification of Statistical Models*. Springer, New York.

BEIRLANT, J., DERVEAUX, V., DE MEYER, A.M., GOOVAERTS, M.J., LABIES, E., & MAENHOUDT, B. (1991). Statistical risk evaluation applied to (Belgian) car insurance. *Insurance: Mathematics and Economics* **10**, 289–302.

BEIRLANT, J., GOEGEBEUR, Y., SEGERS, J., & TEUGELS, J. (2004). *Statistics of Extremes: Theory and Applications*. Wiley Series in Probability and Statistics. John Wiley & Sons, Ltd.

BERMÚDEZ, L., DENUIT, M., & DHAENE, J. (2000). Exponential bonus-malus systems integrating a priori risk classification. *Journal of Actuarial Practice* **9**, 84–112.

BESSON, J.L., & PARTRAT, C. (1990). Loi de Poisson inverse gaussienne et systèmes de bonus-malus. *Proceedings of the Astin Colloquium, Montreux* **81**, 418–419.

BOLANCÉ, C., GUILLÉN, M., & PINQUET, J. (2003). Time-varying credibility for frequency risk models: Estimation and tests for autoregressive specification on the random effect. *Insurance: Mathematics and Economics* **33**, 273–282.

BONSDORFF, H. (1992). On the convergence rate of bonus-malus systems. *ASTIN Bulletin* **22**, 217–223.

Actuarial Modelling of Claim Counts: Risk Classification, Credibility and Bonus-Malus Systems M. Denuit, X. Maréchal, S. Pitrebois and J.-F. Walhin

BONSDORFF, H. (2005). On asymptotic properties of bonus-malus systems based on the number and the size of the claims. *Scandinavian Actuarial Journal*, 309–320.

BORGAN, O., HOEM, J.M., & NORBERG, R. (1981). A nonasymptotic criterion for the evaluation of automobile bonus systems. *Scandinavian Actuarial Journal*, 165–178.

BOSKOV, M., & VERRALL, R.J. (1994). Premium rating by geographical area using spatial models. *ASTIN Bulletin* **24**, 131–143.

BOUCHER, J.-PH., & DENUIT, M. (2006). Fixed versus random effects in Poisson regression models for claim counts: a case study with motor insurance. *ASTIN Bulletin* **36**, 285–301.

BOUCHER, J.-PH., DENUIT, M., & GUILLÉN, M. (2006). Risk classification for claim counts: Zero-inflated mixed Poisson and hurdle models. Working Paper 06-06, Institut des Sciences Actuarielles, Université Catholique de Louvain, Louvain-la-Neuve, Belgium.

BROCKMAN, M.J., & WRIGHT, T.S. (1992). Statistical motor rating: Making effective use of your data. *Journal of the Institute of Actuaries* **119**, 457–543.

BROUHNS, N., DENUIT, M., MASUY, B., & VERRALL, R.J. (2002). Ratemaking by geographical area in the Boskov and Verrall model: a case study using Belgian car insurance data. *actu-L* **2**, 3–28.

BROUHNS, N., GUILLÉN, M., DENUIT, M., & PINQUET, J. (2003). Bonus-malus scales in segmented tariffs with stochastic migration between segments. *Journal of Risk and Insurance* **70**, 577–599.

BUCH-KROMANN, T. (2006). Estimation of large insurance losses: A case study. *Journal of Actuarial Practice* **13**, 191–211.

BUCH-LARSEN, T., NIELSEN, J.P., GUILLÉN, M., & BOLANCÉ, C. (2005). Kernel density estimation for heavy-tailed distributions using the Champernowne transformation. *Statistics* **39**, 503–518.

BÜHLMANN, H. (1967). Experience rating and credibility. *ASTIN Bulletin* **4**, 199–207.

BÜHLMANN, H. (1970). *Mathematical Methods in Risk Theory*. Springer, New York.

BÜHLMANN, P., & BÜHLMANN, H. (1999). Selection of credibility regression models. *ASTIN Bulletin* **29**, 245–270.

BÜHLMANN, H., & GISLER, A. (2005). *A Course in Credibility Theory and its Applications*. Springer, Berlin.

BURNHAM, K.P., & ANDERSON, D.R. (2002). *Model Selection and Multi-Model Inference: A Practical Information-Theoretic Approach*. Springer, New York.

BUTLER, P. (1993). Cost-based pricing of individual automobile risk transfer: Carmile exposure unit analysis. *Journal of Actuarial Practice* **1**, 51–84 (with discussion).

CAMERON, A.C., & TRIVEDI, P.K. (1998). *Regression Analysis of Count Data*. Cambridge University Press.

CARRIÈRE, J. (1993a). Nonparametric tests for mixed Poisson distributions. *Insurance: Mathematics and Economics* **12**, 3–8.

CARRIÈRE, J. (1993b). A semiparametric estimator of a risk distribution. *Insurance: Mathematics and Economics* **13**, 75–81.

CARTER, M., & VAN BRUNT, B. (2000). *The Lebesgue-Stieltjes Integral. A Practical Introduction*. Springer, New York.

CEBRIAN, A., DENUIT, M., & LAMBERT, PH. (2003). Generalized Pareto fit to the society of Actuaries' large claims database. *North American Actuarial Journal* **7**, 18–36.

CENTENO, M., & ANDRADE E SILVA, J.M.A. (2001). Bonus systems in an open portfolio. *Insurance: Mathematics and Economics* **28**, 341–350.

CENTENO, M., & ANDRADE E SILVA, J.M.A. (2002). Optimal bonus scales under path-dependent bonus rules. *Scandinavian Actuarial Journal*, 615–627.

CHAPPELL, D., & NORMAN, J.M. (1989). Optimal, near-optimal and rule-of-thumb claiming rules for a protected bonus vehicle insurance policy. *European Journal of Operations Research* **41**, 151–156.

CONSUL, P.C. (1990). A model for distributions of injuries in auto-accidents. *Bulletin of the SWISS Association of Actuaries*, 161–168.

COORAY, K., & ANANDA, M.M.A. (2005). Modeling actuarial data with a composite lognormal-Pareto model. Scandinavian Actuarial Journal, 321–334.

COUTTS, S. (1984). Motor premium rating. *Insurance: Mathematics and Economics* **3**, 73–96.

CUMMINS, D.J., DIONNE, G., MCDONNALD, J.B., & PRITCHETT, M.B. (1990). Application of the GB2 family of distributions in modeling insurance loss processes. *Insurance: Mathematics and Economics* **9**, 257–272.

DAENGDEJ, J., LUKOSE, D., & MURISON, R. (1999). Using statistical models and case-based reasoning in claims prediction: Experience from a real-world problem. *Knowledge-based Systems* **12**, 239–245.

DALGAARD, P. (2002). *Introductory Statistics with R*. Springer, New York.

DANNENBURG, D.R., KAAS, R., & GOOVAERTS, M.J. (1996). *Practical Actuarial Credibility Models*. Institute of Actuarial Science and Econometrics, University of Amsterdam, Amsterdam, The Netherlands.

DEAN, C.B., LAWLESS, J.F., & WILLMOT, G.E. (1989). A mixed Poisson-inverse-Gaussian regression model. *The Canadian Journal of Statistics* **17**, 171–182.

DE BOOR, C. (1978). *A Practical Guide to Splines*. Springer, New York.

DE LEVE, G., & WEEDA, P.J. (1968). Driving with Markov programming. *ASTIN Bulletin* **5**, 62–86

DELLAERT, N.P., FRENK, J.B.G., KOUWENHOVEN, A., & VAN DER LAAN, B.S. (1990). Optimal claim behaviour for third party liability insurances, or to claim or not to claim: that is the question. *Insurance: Mathematics and Economics* **9**, 59–76.

DELLAERT, N.P., FRENK, J.B.G., & VAN RIJSOORT, L.P. (1993). Optimal claim behaviour for vehicle damage insurance. *Insurance: Mathematics and Economics* **12**, 225–244.

DELLAERT, N.P., FRENK, J.B.G., & VOSKOL, E. (1991). Optimal claim behaviour for third party liability insurances with perfect information. *Insurance: Mathematics and Economics* **10**, 145–151.

DENUIT, M. (1997). A new distribution of Poisson-type for the number of claims. *ASTIN Bulletin* **27**, 229–242.

DENUIT, M. (2002). S-convex extrema, Taylor-type expansions and stochastic approximations. *Scandinavian Actuarial Journal*, 45–67.

DENUIT, M., DE VYLDER, F.E., & LEFÈVRE, CL. (1999). Extremal generators and extremal distributions for the continuous s-convex stochastic orderings. *Insurance: Mathematics and Economics* **24**, 201–217.

DENUIT, M., & DHAENE, J. (2001). Bonus-Malus scales using exponential loss functions. *German Actuarial Bulletin* **25**, 13–27.

DENUIT, M., DHAENE, J., GOOVAERTS, M.J., & KAAS, R. (2005). *Actuarial Theory for Dependent Risks: Measures, Orders and Models*. John Wiley & Sons, Inc., New York.

DENUIT, M., & LAMBERT, PH. (2001). Smoothed NPML estimation of the risk distribution underlying Bonus-Malus systems. *Proceedings of the Casualty Actuarial Society* **88**, 142–174.

DENUIT, M., & LANG, S. (2004). Nonlife ratemaking with Bayesian GAM's. *Insurance: Mathematics and Economics* **35**, 627–647.

DENUIT, M., MARÉCHAL, X., PITREBOIS, S., & WALHIN, J.-F. (2007a). Claiming behaviour in motor insurance with bonus-malus systems. Working Paper, Institut des Sciences Actuarielles, UCL, Louvain-la-Neuve, Belgium.

DENUIT, M., MARÉCHAL, X., PITREBOIS, S., & WALHIN, J.-F. (2007b). Actuarial analysis of some special rules in bonus-malus systems. Working Paper, Institut des Sciences Actuarielles, UCL, Louvain-la-Neuve, Belgium.

DE PRIL, N. (1978). The efficiency of a Bonus-Malus system. *ASTIN Bulletin* **10**, 59–72.

DE PRIL, N. (1979). Optimal claim decisions for a Bonus-Malus system: A continuous approach. *ASTIN Bulletin* **10**, 215–222.

DE PRIL, N., & GOOVAERTS, M.J. (1983). Bounds for the optimal critical claim size of a bonus system. *Insurance: Mathematics and Economics* **2**, 27–32.

DER, G., & EVERITT, B. (2002). *A Handbook of Statistical Analyses using SAS®*. Chapmann & Hall/CRC, Boca Raton.

DESJARDINS, D., DIONNE, G., & PINQUET, J. (2001). Experience rating schemes for fleets of vehicles. *ASTIN Bulletin* **31**, 81–105.

DE VYLDER, F.E. (1985). Non-linear regression in credibility theory. *Insurance: Mathematics and Economics* **4**, 163–172.

DE VYLDER, F.E. (1996). *Advanced Risk Theory. A Self-Contained Introduction*. Editions de l'Université Libre de Bruxelles - Swiss Association of Actuaries, Bruxelles.

DE WIT, G.W., & VAN EEGHEN, J. (1984). Rate making and society's sense of fairness. *ASTIN Bulletin* **14**, 151–163.

DHAENE, J., & VANDEBROEK, M. (1995). Recursions for the individual model. *Insurance: Mathematics and Economics* **16**, 31–38.

Dharmadhikari, S.W., & Joag-Dev, K. (1988). *Unimodality, Convexity and Applications*. Academic Press, New York.

Dickson, D.C.M., & Waters, H.R. (1999). Multi-period aggregate loss distributions for a life portfolio. *ASTIN Bulletin* **29**, 295–309.

Dimakos, X.K., & Rattalma, A.F. (2002). Bayesian premium rating with latent structure. *Scandinavian Actuarial Journal*, 162–184.

Dionne, G., Artis, M., & Guillén, M. (1996). Count data models for a credit scoring system. *Journal of Empirical Finance* **3**, 303–325.

Dionne, G., & Vanasse, C. (1989). A generalization of actuarial automobile insurance rating models: the Negative Binomial distribution with a regression component. *ASTIN Bulletin* **19**, 199–212.

Dionne, G., & Vanasse, C. (1992). Automobile insurance ratemaking in the presence of asymmetrical information. *Journal of Applied Econometrics* **7**, 149–165.

Dixon, M., Kelsey, R., & Verrall, R. (2000). Postcode insurance rating: spatial modelling and performance evaluation. Paper presented at the 4th IME Congress, Barcelona.

Dufresne, F. (1988). Distribution stationnaire d'un système bonus-malus et probabilité de ruine. *ASTIN Bulletin* **18**, 31–46.

Dufresne, F. (1995). The efficiency of the Swiss Bonus-Malus system. *Bulletin of the Swiss Association of Actuaries*, 29–41.

Elvers, E. (1991). A note on the Generalized Poisson distribution. *ASTIN Bulletin* **21**, 167.

Fahrmeir, L., Lang, S., & Spies, F. (2003). Generalized geoadditive models for insurance claims data. *German Actuarial Bulletin* **26**, 7–23.

Fahrmeir, L., & Tutz, G. (2001). *Multivariate Statistical Modelling Based on Generalized Linear Models*. Springer, New York.

Faraway, J.J. (2006). *Extending the Linear Model with R*. Chapman & Hall/CRC, Boca Raton.

Feller, W. (1971). *An Introduction to Probability Theory and its Applications (Vol. II)*. John Wiley & Sons, Inc., New York.

Ferreira, J. (1977). Identifying equitable insurance premiums for risk classes: an alternative to the classical approach. Lecture presented at the 23rd international meeting of the Institute of Management Sciences, Athens, Greece.

Franklin, C.H. (2005). Maximum likelihood estimation. In: *Encyclopedia of Social Measurement*, Vol. 2, pp. 653–664. John Wiley & Sons, Inc., New York.

Frees, E.W. (2003). Multivariate credibility for aggregate loss models. *North American Actuarial Journal* **7**, 13–27.

Frees, E.W., & Wang, P. (2005). Credibility using copulas. *North American Actuarial Journal* **9**, 31–48.

Frees, E.W., & Wang, P. (2006). Copula credibility for aggregate loss models. *Insurance: Mathematics and Economics* **38**, 360–373.

Frees, E.W., Young, V.R., & Luo, Y. (1999). A longitudinal data analysis interpretation of credibility models. *Insurance: Mathematics and Economics* **24**, 229–247.

French, E., & Jones, J.B. (2004). On the distribution and dynamics of health care costs. *Journal of Applied Econometrics* **19**, 705–722.

Gerber, H.U., & Jones, D. (1975). Credibility formulas of the updating type. *Transactions of the Society of Actuaries* **27**, 31–52.

Gertensgarbe, F.W., & Werner, P.C. (1989). A method for the statistical definition of extreme-value regions and their application to meteorological time series. *Zeitschrift fur Meteorologie* **39**, 224–226.

Gilde, V., & Sundt, B. (1989). On bonus systems with credibility scales. *Scandinavian Actuarial Journal*, 13–22.

Golden, R.M. (2003). Discrepancy risk model selection test theory for comparing possibly misspecified or nonnested models. *Psychometrika* **68**, 229–249.

Goovaerts, M.J., & Hoogstad, W. (1987). *Credibility Theory*. Survey in Actuarial Studies, Nationale Nederlanden N.V., Rotterdam.

Gossiaux, A.-M., & Lemaire, J. (1981). Méthodes d'ajustement de distributions de sinistres. *Bulletin of the Swiss Association of Actuaries*, 87–95.

Grandell, J. (1997). *Mixed Poisson Processes*. Chapman & Hall, New York.

GREENWOOD, M., & YULE, G.U. (1920). An inquiry into the nature of frequency distributions representative of multiple happenings with particular reference to the occurrence of multiple attacks of disease or of repeated accidents. *Journal of the Royal Statistical Society, Series A* **83**, 255–279.

GRENANDER, U. (1957a). On the heterogeneity in non-life insurance. *Scandinavian Actuarial Journal*, 153–179.

GRENANDER, U. (1957b). Some remarks on bonus systems in automobile insurance. *Scandinavian Actuarial Journal*, 180–197.

HACHEMEISTER, C.A. (1975). Credibility for regression models with application to trend. In: P.M. Kahn, Editor, *Credibility: Theory and Applications*, Academic Press, New York, pp. 129–163.

HAEHLING VON LANZENAUER, C. (1974). Optimal claim decisions by policyholders in automobile insurance with merit rating structures. *Operations Research* **22**, 979–990.

HASTINGS, N.A.J. (1976). Optimal claiming on vehicle insurance. *Operational Research Quarterly* **27**, 908–913.

HENRIET, D., & ROCHET, J.C. (1986). La logique des systèmes bonus-malus en assurance automobile: une approche théorique. *Annales d'Économie et de Statistique* **1**, 133–152.

HERAS, A., GIL, J.A., GARCIA-PINEDA, P., & VILAR, J.C. (2004). An application of linear programming to bonus-malus system design. *ASTIN Bulletin* **34**, 435–456.

HERAS, A., VILAR, J.L., & GIL, J.A. (2002). Asymptotic fairness of bonus-malus systems and optimal scales of premiums. *The Geneva Papers on Risk and Insurance - Theory* **27**, 61–82.

HERZOG, T.N. (1994). *Introduction to Credibility Theory*. ACTEX Publications.

HEY, J.D. (1985). No claims bonus. *The Geneva Papers on Risk and Insurance* **10**, 209–228.

HINDE, J. (1982). Compound Poisson regression models. In: R. Gilchrist, Editor, *GLIM 82: Proceedings of the International Conference on Generalised Linear Models*, Springer, New York.

HOLTAN, J. (1994). Bonus made easy. *ASTIN Bulletin* **24**, 61–74.

HOLTAN, J. (2001). Optimal loss financing under bonus-malus contracts. *ASTIN Bulletin* **31**, 161–173.

HSIAO, C. (2003). *Analysis of Panel Data*. Cambridge University Press, Cambridge.

HUANG, X., SONG, L., & LIANG, Y. (2003). Semiparametric credibility ratemaking using a piecewise linear prior. *Insurance: Mathematics and Economics* **33**, 585–593.

HÜRLIMANN, W. (1990) On maximum likelihood estimation for count data models. *Insurance: Mathematics and Economics* **9**, 39–49.

ISLAM, M.N., & CONSUL, P.C. (1992). A probabilistic model for automobile claims. *Bulletin of the Swiss Association of Actuaries* 85–93.

JEE, B. (1989). A comparative analysis of alternative pure premium models in the automobile risk classification system. *Journal of Risk and Insurance* **56**, 434–459.

JEWELL, W.S. (1975). Model variations in credibility theory. In: P.M. Kahn, Editor, *Credibility*, Academic Press, New York, pp. 199–244.

JOHNSON, N.L., KOTZ, S., & KEMP, A.W. (1992). *Univariate Discrete Distributions*, John Wiley & Sons, Inc., New York.

JORGENSEN, B. (1987). Exponential dispersion models. *Journal of the Royal Statistical Society, Series B* **49**, 127–162.

JORGENSEN, B. (1997). *The Theory of Regression Models*. Chapman & Hall, London.

JORGENSEN, B., & PAES DE SOUZA, M.C. (1994). Fitting Tweedie's compound Poisson model to insurance claims data. *Scandinavian Actuarial Journal*, 69–93.

KARLIS, D. (2005). EM algorithm for mixed Poisson and other discrete distributions. *ASTIN Bulletin* **35**, 3–24.

KELLE, M. (2000). Modélisation du système de bonus malus français. *Bulletin Français d'Actuariat* **4**, 45–64.

KENDALL, M., & STUART, A. (1977). *Advanced Theory of Statistics, Vol. I*. Griffin, London.

KESTEMONT, R.-M., & PARIS, J. (1985). Sur l'ajustement du nombre de sinistres. *Bulletin of the Swiss Association of Actuaries*, 157–163.

KLUGMAN, S. (1992). *Bayesian Statistics in Actuarial Science*. Kluwer, Boston.

KLUGMAN, S., PANJER, H., & WILLMOT, G. (2004). *Loss Models: From Data to Decisions*. John Wiley & Sons, Inc., New York.

KOLDERMAN, J., & VOLGENANT, A. (1985). Optimal claiming in an automobile insurance system with bonus-malus structure. *Journal of the Operational Research Society* **36**, 239–247.

Lambert, D. (1992). Zero-inflated Poisson regression, with an application to defects in manufacturing. *Technometrics* **34**, 1–14.

Lee, Y., & Nelder, J.A. (1996). Hierarchical generalized linear models. *Journal of the Royal Statistical Society, Series B* **58**, 619–678.

Lefèvre, Cl., & Picard, Ph. (1996). On the first crossing of a Poisson process in a lower boundary. In: C.C. Heyde, Yu V. Prohorov, R. Pyke and S.T. Rachev Editors, *Athens Conference on Applied Probability and Time Series, Vol. I, Applied Probability*. Lecture Notes in Statistics, Springer, Berlin, **114**, pp. 159–175.

Lemaire, J. (1976). Driver versus company: Optimal behaviour of the policyholder. *Scandinavian Actuarial Journal*, 209–219.

Lemaire, J. (1977). La soif du bonus. *ASTIN Bulletin* **9**, 181–190.

Lemaire, J. (1979). How to define a Bonus-Malus system with an exponential utility function. *ASTIN Bulletin* **10**, 274–282.

Lemaire, J. (1995). *Bonus-Malus Systems in Automobile Insurance*. Kluwer Academic Publisher, Boston.

Lemaire, J., & Vandermeulen, E. (1983). Une propriété du principe de l'espérance mathématique. *Bulletin Trimestriel de l'Institut des Actuaires Français*, 5–14.

Lemaire, J., & Zi, H. (1994a). High deductibles instead of Bonus-Malus. Can it work? *ASTIN Bulletin* **24**, 75–88.

Lemaire, J., & Zi, H. (1994b). A comparative analysis of 30 Bonus-Malus systems. *ASTIN Bulletin* **24**, 287–309.

Liang, K.Y., & Zeger, S.L. (1986). Longitudinal data analysis using generalized linear models. *Biometrika* **73**, 13–22.

Lindsay, B. (1989a). On the determinants of moment matrices. *Annals of Statistics* **17**, 711–721.

Lindsay, B. (1989b). Moment matrices: applications in mixtures. *Annals of Statistics* **17**, 722–740.

Lindsay, B. (1995). *Mixture Models: Theory, Geometry and Applications*. Institute of Mathematical Statistics and the American Statistical Association.

Lo, C.H., Fung, W.K., & Zhu, Z.Y. (2006). Generalized estimating equations for variance and covariance parameters in regression credibility models. *Insurance: Mathematics and Economics* **39**, 99–113.

Loimaranta, K. (1972). Some asymptotic properties of bonus systems. *ASTIN Bulletin* **6**, 223–245.

Lord, D. (2006). Modeling motor vehicle crashes using Poisson-gamma models: Examining the effects of low sample mean values and small sample size on the estimation of the fixed dispersion parameter. *Accident Analysis and Prevention* **38**, 751–766.

Luo, Y., Young, V.R., & Frees, E.W. (2004). Credibility ratemaking using collateral information. *Scandinavian Actuarial Journal*, 448–461.

Makov, U.E. (2002). Principal applications of Bayesian methods in actuarial science: A perspective. *North American Actuarial Journal* **5**, 53–73 (with discussion).

Makov, U.E., Smith, A.F.M., & Liu, Y.H. (1996). Bayesian methods in actuarial science. *The Statistician* **45**, 503–517.

Martin-Lof, A. (1973). A method for finding the optimal decision rule for a policy holder of an insurance with a bonus system. *Scandinavian Actuarial Journal*, 23–29.

Mc Cullagh, P., & Nelder, J.A. (1989). *Generalized Linear Models*. Chapman & Hall, New York.

Morillo, I., & Bermudez, L. (2003). Bonus-malus systems using an exponential loss function with an Inverse Gaussian distribution. *Insurance: Mathematics and Economics* **33**, 49–57.

Mowbray, A.H. (1914). How extensive a payroll exposure is necessary to give a dependable pure premium. *Proceedings of the Casualty Actuarial Society* **1**, 24–30.

Nelder, J.A., & Verral, R.J. (1997). Credibility theory and generalized linear models. *ASTIN Bulletin* **27**, 71–82.

Nelder, J.A., & Wedderburn, R.W.M. (1972). Generalized linear models. *Journal of the Royal Statistical Society, Series A* **135**, 370–384.

Neuhaus, W. (1988). A bonus-malus system in automobile insurance. *Insurance: Mathematics and Economics* **7**, 103–112.

Norberg, R. (1975). Credibility premium plans which make allowance for bonus hunger. *Scandinavian Actuarial Journal*, 73–86.

NORBERG, R. (1976). A credibility theory for automobile bonus system. *Scandinavian Actuarial Journal*, 92–107.

NORBERG, R. (1980). Empirical Bayes credibility. *Scandinavian Actuarial Journal*, 177–194.

NORBERG, R. (1986). Hierarchical credibility: Analysis of a random effect linear model with nested classification. *Scandinavian Actuarial Journal*, 204–222.

NORBERG, R. (2004). Credibility Theory. In: J. Teugels and B. Sundt Editors, *Encyclopedia of Actuarial Science*, John Wiley & Sons, Ltd, pp. 398–406.

NORMAN, J.M., & SHEARN, D.C.S. (1980). Optimal claiming on vehicle insurance revisited. *Journal of the Operational Research Society* **31**, 181–186.

PANJER, H.H. (1981). Recursive evaluation of a family of compound distributions. *ASTIN Bulletin* **12**, 22–26.

PANJER, H.H. (1987). Models of claim frequency. In: I.B. Mac Neill and G.J. Umphrey Editors, *Advances in the Statistical Sciences, Vol. VI, Actuarial Sciences*, The University of Western Ontario Series in Philosophy of Science, Reidel, Dordrecht **39**, pp. 115–122.

PHILIPSON, C. (1960). The Swedish system of bonus. *ASTIN Bulletin* **1**, 134–141.

PICARD, P. (1976). Généralisation de l'étude sur la survenance des sinistres en assurance automobile. *Bulletin Trimestriel de l'Institut des Actuaires Français*, 204–267.

PINQUET, J. (1997). Allowance for cost of claims in Bonus-Malus systems. *ASTIN Bulletin* **27**, 33–57.

PINQUET, J. (1998). Designing optimal bonus-malus systems from different types of claims, *ASTIN Bulletin*, 205–220.

PINQUET, J. (2000). Experience rating through heterogeneous models. In: G. Dionne, Editor, *Handbook of Insurance*, Kluwer Academic Publishers.

PINQUET, J., GUILLÉN, M., & BOLANCÉ, C. (2001). Allowance for the age of claims in bonus-malus systems. *ASTIN Bulletin* **31**, 337–348.

PITREBOIS, S., DENUIT, M., & WALHIN, J.-F. (2003a). Fitting the Belgian Bonus-Malus system. *Belgian Actuarial Bulletin* **3**, 58–62.

PITREBOIS, S., DENUIT, M., & WALHIN, J.-F. (2003b). Setting a bonus-malus scale in the presence of other rating factors: Taylor's work revisited. *ASTIN Bulletin* **33**, 419–436.

PITREBOIS, S., DENUIT, M., & WALHIN, J.-F. (2004). Bonus-malus scales in segmented tariffs: Gilde & Sundt's work revisited. *Australian Actuarial Journal* **10**, 107–125.

PITREBOIS, S., DENUIT, M., & WALHIN, J.-F. (2005). Bonus-malus systems with varying deductibles. *ASTIN Bulletin* **35**, 261–274.

PITREBOIS, S., DENUIT, M., & WALHIN, J.-F. (2006a). Multi-event bonus-malus scales. *Journal of Risk and Insurance* **73**, 517–528.

PITREBOIS, S., DENUIT, M., & WALHIN, J.-F. (2006b). An actuarial analysis of the French bonus-malus system. *Scandinavian Actuarial Journal*, 247–264.

PITREBOIS, S., WALHIN, J.-F., & DENUIT, M. (2006c). How to transfer policyholders from one bonus-malus scale to the other? *German Actuarial Bulletin* **27**, 607–618.

POHLMEIER, W., & ULRICH, V. (1995). An econometric model of the two-part decision making process in the demand for health care. *Journal of Human Resources* **30**, 339–361.

PURCARU, O., & DENUIT, M. (2003). Dependence in dynamic claim frequency credibility models. *ASTIN Bulletin* **33**, 23–40.

QIAN, W. (2000). An application of nonparametric regression estimation in credibility theory. *Insurance: Mathematics and Economics* **27**, 169–176.

QUIGLEY, J., BEDFORD, T., & WALLS, L. (2006). Estimating rate of occurrence of rare events with empirical bayes: A railway application. *Reliability Engineering and System Safety*, in press.

RAFTERY, A.E. (1995). Bayesian model selection in social research. In: P. Marsden, Editor, *Sociological Methodology*, The American Sociological Association, Washington, pp. 111–163.

REJESUS, R.M., COBLE, K.H., KNIGHT, T.O., & JIN, Y. (2006). Developing experience-based premium rate discounts in crop insurance. *American Journal of Agricultural Economics* **88**, 409–419.

RETZLAFF-ROBERTS, D., & PUELZ, R. (1996). Classification in automobile insurance using a DEA and discriminant analysis hybrid. *Journal of Productivity Analysis* **7**, 417–427.

ROBERTS, A.W., & VARBERG, D.E. (1973). *Convex Functions*. Academic Press, New York.

ROLSKI, T., SCHMIDLI, H., SCHMIDT, V., & TEUGELS, J. (1999). *Stochastic Processes for Insurance and Finance*. John Wiley & Sons, Inc., New York.

RUOHONEN, M. (1987). On a model for the claim number process. *ASTIN Bulletin* **18**, 57–68.

SANTOS SILVA, J.M.C., & WINDMEIJER, F. (2001). Two-part multiple spell models for health care demand. *Journal of Econometrics* **104**, 67–89.

SARABIA, J.M., GOMEZ-DENIZ, E., & VAZQUEZ-POLO, F.J. (2004). On the use of conditional specification models in claim count distributions: An application to bonus-malus systems. *ASTIN Bulletin* **34**, 85–98.

SHAKED, M. (1980). On mixtures from exponential families. *Journal of the Royal Statistical Society, Series B* **42**, 192–198.

SHARIF, A.H., & PANJER, H.H. (1993). A probabilistic model for automobile claims: a comment on the article by M.N. Islam and P.C. Consul. *Bulletin of the Swiss Association of Actuaries*, 279–282.

SHENGWANG, M., WEI, Y., & WHITMORE, G.A. (1999). Accounting for individual over-dispersion in a Bonus-Malus automobile insurance system. *ASTIN Bulletin* **29**, 327–337.

SIMAR, L. (1976). Maximum likelihood estimation of a compound Poisson process. *The Annals of Statistics* **4**, 1200–1209.

SMYTH, G.J., & JORGENSEN, B. (2002). Fitting Tweedie's compound Poisson model to insurance claims data: Dispersion modelling. *ASTIN Bulletin* **32**, 143–157.

SUBRAMANIAN, K. (1998). Bonus-malus systems in a competitive environment. *North American Actuarial Journal* **2**, 38–44.

SUNDT, B. (1981). Recursive credibility estimation. *Scandinavian Actuarial Journal*, 3–21.

SUNDT, B. (1983). Parameter estimation in some credibility models. *Scandinavian Actuarial Journal*, 239–255.

SUNDT, B. (1988). Credibility estimators with geometric weights. *Insurance: Mathematics and Economics* **7**, 113–122.

SUNDT, B. (1999). On multivariate Panjer recursions. *ASTIN Bulletin* **29**, 29–45.

TAYLOR, G.C. (1989). Use of spline functions for premium rating by geographic area. *ASTIN bulletin* **19**, 91–122.

TAYLOR, G.C. (1996). Geographic premium rating by Whittaker spatial smoothing. *ASTIN Bulletin* **31**, 147–160.

TAYLOR, G. (1997). Setting a bonus-malus scale in the presence of other rating factors. *ASTIN Bulletin* **27**, 319–327.

TER BERG, P. (1980a). Two pragmatic approaches to loglinear claim cost analysis. *ASTIN Bulletin* **11**, 77–90

TER BERG, P. (1980b). On the loglinear Poisson and gamma model. *ASTIN Bulletin* **11**, 35–40.

TER BERG, P. (1996). A loglinear lagrangian Poisson model. *ASTIN Bulletin* **26**, 123–129.

TITTERINGTON, D.M., SMITH, A.F.M., & MAKOV, U.E. (1985). *Statistical Analysis of Finite Mixture Distributions*. John Wiley & Sons Ltd, Chichester.

TREMBLAY, L. (1992). Using the Poisson-inverse gaussian distribution in bonus-malus systems. *ASTIN Bulletin* **22**, 97–106.

TUCKER, H.G. (1963). An estimate of the compounding distribution of a compound Poisson distribution. *Theory of Probability and Its Applications* **8**, 195–200.

TWEEDIE, M.C.K. (1984). An index which distinguishes between some important exponential families. In: J.K. Ghosh and J. Roy, Editors, *Statistics: Applications and New Directions, Proceedings of the Indian Statistical Institute Golden Jubilee International Conference*, pp. 579–604.

VANDEBROEK, M. (1993). Bonus-malus system or partial coverage to oppose moral hazard problems? *Insurance: Mathematics and Economics* **13**, 1–5.

VISWANATHAN, K.S., & LEMAIRE, J. (2005). Bonus-malus systems in a deregulated environment: Forecasting market shares using diffusion models. *ASTIN Bulletin* **35**, 299–319.

VUONG, Q.H. (1989). Likelihood ratio tests for model selection and non-nested hypotheses. *Econometrica* **57**, 307–333.

WALHIN, J.-F. (2001). Some comments on the individual risk model and multivariate extension. *German Actuarial Bulletin* **25**, 257–270.

WALHIN, J.-F., & PARIS, J. (1999). Using mixed Poisson distributions in connection with Bonus-Malus systems. *ASTIN Bulletin* **29**, 81–99.

WALHIN, J.-F., & PARIS, J. (2000). The true claim amount and frequency distributions within a bonus-malus system. *ASTIN Bulletin* **30**, 391–403.

WALHIN, J.-F., & PARIS, J. (2001). The practical replacement of a bonus-malus system. *ASTIN Bulletin* **31**, 317–335.

WEDEL, M., BÖCKENHOLTB, U., & KAMAKURAC, W.A. (2003). Factor models for multivariate count data. *Journal of Multivariate Analysis* **87**, 356–369.

WEISBERG, H.I., TOMBERLIN, T.J., & CHATTERJEE, S. (1984). Predicting insurance losses under cross-classification: A comparison of alternative approaches. *Journal of Business & Economic Statistics* **2**, 170–178.

WHITNEY, W. (1918). The theory of experience rating. *Proceedings of the Casualty Actuarial Society* **4**, 274–292.

WILLIAMS, G.J., & HUANG, Z. (1996). KDD for insurance risk assessment: A case study. CSIRO Mathematical and Information Sciences.

WILLMOT, G.E. (1987). The Poisson-inverse gaussian distribution as an alternative to the negative binomial. *Scandinavian Actuarial Journal*, 113–127.

WINKELMANN, R. (2003). *Econometric Analysis of Count Data*. Springer, New York.

YEO, A.C., SMITH, K.A., WILLIS, R.J., & BROOKS, M. (2001). Clustering technique for risk classification and prediction of claim costs in the automobile insurance industry. *International Journal of Intelligent Systems in Accounting, Finance and Management* **10**, 39–50.

YEO, K.L., & VALDEZ, E.A. (2006). Claim dependence with common effects in credibility models. *Insurance: Mathematics and Economics* **38**, 609–629.

YIP, K.C.H., & YAU, K.K.W. (2005). On modeling claim frequency data in general insurance with extra zeros. *Insurance: Mathematics and Economics* **36**, 153–163.

YOUNG, V.R. (1997). Credibility using semiparametric models. *ASTIN Bulletin* **27**, 273–285.

YOUNG, V.R. (1998a). Credibility using a loss function from spline theory: parametric models with a one-dimensional sufficient statistic. *North American Actuarial Journal* **2**, 101–117 (with discussion).

YOUNG, V.R. (1998b). Robust Bayesian credibility using semiparametric models. *ASTIN Bulletin* **28**, 187–203.

YOUNG, V.R. (2000). Credibility using semiparametric models and a loss function with a constancy penalty. *Insurance: Mathematics and Economics* **26**, 151–156.

YOUNG, V.R., & DE VYLDER, F.E. (2000). Credibility in favor of unlucky insureds. *North American Actuarial Journal* **4**, 107–113.